Dissipative Lattice Dynamical Systems

INTERDISCIPLINARY MATHEMATICAL SCIENCES*

Published

Vol. 22 *Dissipative Lattice Dynamical Systems*
by Xiaoying Han & Peter Kloeden

Vol. 21 *An Introduction to Nonautonomous Dynamical Systems and their Attractors*
by Peter Kloeden & Meihua Yang

Vol. 20 *Stochastic PDEs and Modelling of Multiscale Complex System*
eds. Xiaopeng Chen, Yan Lv & Wei Wang

Vol. 19 *Kernel-based Approximation Methods using MATLAB*
by Gregory Fasshauer & Michael McCourt

Vol. 18 *Global Attractors of Non-Autonomous Dynamical and Control Systems (Second Edition)*
by David N Cheban

Vol. 17 *Festschrift Masatoshi Fukushima: In Honor of Masatoshi Fukushima's Sanju*
eds. Zhen-Qing Chen, Niels Jacob, Masayoshi Takeda & Toshihiro Uemura

Vol. 16 *Hilbert–Huang Transform and Its Applications (Second Edition)*
eds. Norden E Huang & Samuel S P Shen

Vol. 15 *Recent Advances in Applied Nonlinear Dynamics with Numerical Analysis: Fractional Dynamics, Network Dynamics, Classical Dynamics and Fractal Dynamics with Their Numerical Simulations*
eds. Changpin Li, Yujiang Wu & Ruisong Ye

Vol. 14 *Recent Developments in Computational Finance: Foundations, Algorithms and Applications*
eds. Thomas Gerstner & Peter Kloeden

Vol. 13 *Stochastic Analysis and Applications to Finance: Essays in Honour of Jia-an Yan*
eds. Tusheng Zhang & Xunyu Zhou

Vol. 12 *New Trends in Stochastic Analysis and Related Topics: A Volume in Honour of Professor K D Elworthy*
eds. Huaizhong Zhao & Aubrey Truman

*For the complete list of titles in this series, please go to http://www.worldscientific.com/series/ims

Interdisciplinary Mathematical Sciences – Vol. 22

Dissipative Lattice Dynamical Systems

Xiaoying Han
Auburn University, USA

Peter Kloeden
Universität Tübingen, Germany

World Scientific

NEW JERSEY • LONDON • SINGAPORE • BEIJING • SHANGHAI • HONG KONG • TAIPEI • CHENNAI • TOKYO

Published by

World Scientific Publishing Co. Pte. Ltd.
5 Toh Tuck Link, Singapore 596224
USA office: 27 Warren Street, Suite 401-402, Hackensack, NJ 07601
UK office: 57 Shelton Street, Covent Garden, London WC2H 9HE

British Library Cataloguing-in-Publication Data
A catalogue record for this book is available from the British Library.

Interdisciplinary Mathematical Sciences — Vol. 22
DISSIPATIVE LATTICE DYNAMICAL SYSTEMS

ISBN 978-981-126-775-8 (hardcover)
ISBN 978-981-126-776-5 (ebook for institutions)
ISBN 978-981-126-777-2 (ebook for individuals)

For any available supplementary material, please visit
https://www.worldscientific.com/worldscibooks/10.1142/13182#t=suppl

Typeset by Stallion Press
Email: enquiries@stallionpress.com

Printed in Singapore

Dedicated to my parents (XH)

Dedicated to the memory of Karin Wahl-Kloeden (PEK)

Preface

Lattice dynamical systems (LDS) are basically infinite dimensional systems of ordinary differential equations, either autonomous or nonautonomous, and are formulated as ordinary differential equations on Hilbert spaces of bi-infinite sequences. There have been many generalisations to include delayed, random and stochastic terms as well as multi-valued terms. LDS arise in a wide range of applications with intrinsic discrete structures such as chemical reaction, pattern recognition, image processing, living cell systems, cellular neural networks, etc. Sometimes they are derived as spatial discretisations of models based on partial differential equations, but they need not arise in this way.

There is an extensive literature on lattice dynamical systems. During the 1990s there was a strong emphasis on travelling waves in such systems and in recent decades on attractors. This book focuses on dissipative lattice dynamical systems and their attractors of various forms such as autonomous, nonautonomous and random. The existence of such attractors is established by showing that the corresponding dynamical system has an appropriate kind of absorbing set and is asymptotically compact in some way.

Asymptotic compactness is usually established by showing that the system satisfies an asymptotic tails property inside the absorbing set, which essentially leads to a total boundedness property. This approach is based on a seminal paper of Bates, Lu and Wang [Bates *et al.* (2001)], which has since been used and extended many times in a broad variety of situations. In each case the technical details are different, but the basic idea is similar.

There is now also a very large literature on dissipative lattice dynamical systems, especially on attractors of all kinds in such systems. We cannot hope to do justice to all of these papers here. Instead we have focused on key papers of representative types of lattice systems and various types of attractors. Our selection is biased by our own interests, in particular to those dealing with biological applications. Nevertheless, we believe that this book will provide the reader with a solid introduction to field, its main results and the methods that are used to obtain them.

At the end of each chapter we have included a section with some problems. These are not meant to be exercises for students, although some could serve that purpose. Their main goal is to draw the reader's attention to important issues for clarification and extension of the material and proofs in the book. Some are fairly straightforward, but others are serious research problems, in some cases very difficult ones.

Auburn,
Tübingen
June 2022

Xiaoying Han
Peter Kloeden

Contents

PART 1

Background

Chapter 1

Lattice dynamical systems: a preview

1.1 Introduction

Lattice dynamical systems (LDS), as considered in this book, are essentially infinite dimensional systems of ordinary differential equations (ODEs). In particular, they can be formulated as ordinary differential equations on a Hilbert or Banach space of bi-infinite sequences. The infinite dimensionality of this state space takes their investigation beyond the usual qualitative theory of ODEs, but its special nature often means that such an investigation is not as technically complicated as for the corresponding partial differential equation (PDE) from which an LDS may have been derived. This allows a greater focus on the dynamical behaviour of such systems. Not all lattice dynamical systems originate by discretising an underlying PDE. Some may arise by discretising integral equations, others are intrinsically discrete.

1.2 Examples of lattice dynamical systems

Lattice dynamical systems may arise from discretisation of continuum models or as infinite dimensional counterparts of finite ODE models.

1.2.1 *PDE based models*

A classical lattice dynamical system is based on a reaction-diffusion equation

$$\frac{\partial u}{\partial t} = \nu \frac{\partial^2 u}{\partial x^2} - \lambda u + f(u) + g(x), \tag{1.1}$$

where λ and ν are positive constants, on a one-dimensional domain $\mathbb{R}$. It is obtained by using a central difference quotient to discretise the Laplacian. Setting the stepsize scaled to equal 1 leads to the infinite dimensional system of ordinary differential equations

$$\frac{\mathrm{d}u_i}{\mathrm{d}t} = \nu\left(u_{i-1} - 2u_i + u_{i+1}\right) - \lambda u_i + f(u_i) + g_i, \quad i \in \mathbb{Z}, \tag{1.2}$$

where $u_i(t)$, g_i and $f(u_i(t))$ correspond to $u(x_i,t)$, $g(x_i)$ and $f(u(x_i,t))$ for each $i \in \mathbb{Z}$. When the function f in (1.1) depends also on x, then the corresponding term in (1.2) becomes $f_i(u_i) = f(x_i, u(x_i,t))$.

Similarly, the spatial discretisation of a wave-like equation

$$\frac{\partial^2 u}{\partial t^2} = \nu \frac{\partial^2 u}{\partial x^2} - \lambda u + f\left(x, u, \frac{\partial u}{\partial t}\right) + g(x),$$

leads to an LDS consisting of an infinite dimensional system of second order ordinary differential equations such as

$$\frac{\mathrm{d}^2 u_i}{\mathrm{d}t^2} = \nu\left(u_{i-1} - 2u_i + u_{i+1}\right) - \lambda u_i + f_i\left(u_i, \frac{\mathrm{d}u_i}{\mathrm{d}t}\right) + g_i, \quad i \in \mathbb{Z}.$$

This can be reformulated as an infinite dimensional system of a pair of first order ordinary differential equations

$$\frac{\mathrm{d}u_i}{\mathrm{d}t} = v_i$$
$$\frac{\mathrm{d}v_i}{\mathrm{d}t} = \nu\left(u_{i-1} - 2u_i + u_{i+1}\right) - \lambda u_i + f_i\left(u_i, v_i\right) + g_i.$$

The appearance of switching effects and recovery delays in systems of excitable cells leads to reaction-diffusion systems which are technically very difficult to analyse [Kloeden and Lorenz (2017)]. This motivated [Han and Kloeden (2016)] to study the following lattice system with a reaction term which is switched off when a certain threshold is exceeded and restored after a suitable recovery time:

$$\frac{\mathrm{d}u_i}{\mathrm{d}t} = \nu(u_{i-1} - 2u_i + u_{i+1}) + f_i(t, u_i)\mathrm{H}[\varsigma_i - \max_{-\theta \le s \le 0} u_i(t+s)], \; i \in \mathbb{Z}, \tag{1.3}$$

$$u_i(t) = \phi_i(t - t_0), \quad \forall\, t \in [t_0 - \theta, t_0], \quad i \in \mathbb{Z}, \quad t_0 \in \mathbb{R}.$$

Here $\nu = 1/\varkappa > 0$ is the coupling coefficient where $\varkappa$ is the intercellular resistance, while $\varsigma_i \in \mathbb{R}$ is the threshold triggering the switch-off at the i-th site and $u_i(t + \cdot) \in C([-\theta, 0], \mathbb{R})$ is the segment of u_i on time interval $[t - \theta, t]$ where θ is a positive constant.

In addition, H is the Heaviside operator

$$\mathrm{H}(x) = \begin{cases} 1, \; x \ge 0, \\ 0, \; x < 0, \end{cases} \qquad x \in \mathbb{R}. \tag{1.4}$$

To facilitate the mathematical analysis, the Heaviside function is often replaced by a set-valued mapping χ defined on $\mathbb{R}$ by

$$\chi(s) = \begin{cases} \{0\}, & s < 0, \\ [0,1], & s = 0, \\ \{1\}, & s > 0, \end{cases} \qquad s \in \mathbb{R}. \tag{1.5}$$

Then the lattice system (1.3) can be reformulated as the lattice differential inclusion

$$\frac{\mathrm{d}}{\mathrm{d}t}u_i(t) \in \nu(u_{i-1} - 2u_i + u_{i+1}) + f_i(t, u_i)\chi\big(\varsigma_i - \max_{-\theta \le s \le 0} u_i(t+s)\big).$$

1.2.2 *Neural field models*

Lattice dynamical systems need not originate by discretising an underlying PDE as above, but could arise from an integro-differential equation such as the Amari neural field model [Amari (1977)] (see also Chapter 3 of [Coombes *et al.* (2014)] by Amari):

$$\frac{\partial u(t,x)}{\partial t} = -u(t,x) + \int_{\Omega} \mathcal{K}(x-y)\mathrm{H}\left(u(t,y) - \varsigma\right) \mathrm{d}y, \qquad x \in \Omega \subset \mathbb{R},$$

where $\varsigma > 0$ is a given threshold and H is the Heaviside function defined as in (1.4).

Such continuum neural models may lose their validity in capturing detailed dynamics at discrete sites when the discrete structures of neural systems become dominant, so a lattice model may be more appropriate. The following lattice version of the Amari model was introduced in [Han and Kloeden (2019a)],

$$\frac{\mathrm{d}}{\mathrm{d}t}u_{\mathrm{i}}(t) = f_{\mathrm{i}}(u_{\mathrm{i}}(t)) + \sum_{\mathrm{j} \in \mathbb{Z}^d} \kappa_{\mathrm{i},\mathrm{j}}\mathrm{H}(u_{\mathrm{j}}(t) - \varsigma) + g_{\mathrm{i}}(t), \quad \mathrm{i} \in \mathbb{Z}^d. \tag{1.6}$$

When the Heaviside function is replaced by the set-valued mapping χ defined in (1.5), the lattice system (1.6) can be reformulated as the lattice differential inclusion

$$\frac{\mathrm{d}}{\mathrm{d}t}u_{\mathrm{i}}(t) \in f_{\mathrm{i}}(u_{\mathrm{i}}(t)) + \sum_{\mathrm{j} \in \mathbb{Z}^d} \kappa_{\mathrm{i},\mathrm{j}}\chi(u_{\mathrm{j}}(t) - \varsigma) + g_{\mathrm{i}}(t), \quad \mathrm{i} \in \mathbb{Z}^d.$$

The Heaviside function can also be approximated by a simplifying sigmoidal function such as

$$\sigma_\varepsilon(x) = \frac{1}{1 + e^{-x/\varepsilon}}, \qquad x \in \mathbb{R}, \quad 0 < \varepsilon < 1.$$

This avoids the need to introduce a differential inclusion as above. This sigmoidal function is globally Lipschitz with the Lipschitz constant $L_\sigma = \frac{1}{\varepsilon}$ and does not lead to an inclusion equation. For example, Wang, Kloeden & Yang [Wang *et al.* (2020a)] considered the autonomous neural field lattice system with delays

$$\frac{\mathrm{d}}{\mathrm{d}t}u_{\mathrm{i}}(t) = f_{\mathrm{i}}(u_{\mathrm{i}}(t)) + \sum_{\mathrm{j} \in \mathbb{Z}^d} \kappa_{\mathrm{i},\mathrm{j}}\sigma_\varepsilon(u_{\mathrm{j}}(t - \theta_{\mathrm{j}}) - \varsigma) + g_{\mathrm{i}}, \quad \mathrm{i} \in \mathbb{Z}^d.$$

Delays are often included in neural field models to account for the finite transmission time of signals between neurons.

1.2.3 *Intrinsically discrete models*

Some LDSs arise naturally without involving discretisation. Instead, they may be developed as the infinite dimensional counterparts to a finite dimensional ODE system. For example, based on the Hopfield neural network [Hopfield (1984)] modeled by an n-dimensional system of ODEs

$$\mu_i \frac{\mathrm{d}u_i(t)}{\mathrm{d}t} = -\frac{u_i(t)}{\varkappa_i} + \sum_{j=1}^{n} \lambda_{i,j} f_j(u_j(t)) + g_i, \qquad i = 1, ..., n,$$

where u_i is the mean soma potential of neuron i, μ_i and $\varkappa_i$ are the input capacitance of the cell membrane and transmembrane resistance, respectively. Han, Usman & Kloeden [Han *et al.* (2019)] considered the random Hopfield neural lattice model:

$$\mu_i \frac{\mathrm{d}u_i(t)}{\mathrm{d}t} = -\frac{u_i(t)}{\varkappa_i} + \sum_{j=i-n}^{i+n} \lambda_{i,j} f_j(u_j(t)) + g_i(\vartheta_t(\omega)), \qquad i \in \mathbb{Z},$$

where $\vartheta_t(\omega)$ is a sample path of a noise process.

1.3 Sequence spaces

An LDS can be formulated as an ordinary differential equation on an appropriate space of infinite sequences.

Let ℓ^2 be the Hilbert space of real-valued square summable bi-infinite sequences $\boldsymbol{u} = (u_i)_{i\in\mathbb{Z}}$ with norm and inner product

$$\|\boldsymbol{u}\| := \left(\sum_{i\in\mathbb{Z}} u_i^2\right)^{1/2}, \quad \langle \boldsymbol{u}, \boldsymbol{v}\rangle := \sum_{i\in\mathbb{Z}} u_i v_i \text{ for } \boldsymbol{u} = (u_i)_{i\in\mathbb{Z}},\ \boldsymbol{v} = (v_i)_{i\in\mathbb{Z}} \in \ell^2.$$

For $p \geq 1$, ℓ^p denotes the Banach space of real-valued p-summable bi-infinite sequences $\boldsymbol{u} = (u_i)_{i\in\mathbb{Z}}$ with norm

$$\|\boldsymbol{u}\|_p := \left(\sum_{i\in\mathbb{Z}} |u_i|^p\right)^{1/p}, \quad \text{for } \boldsymbol{u} = (u_i)_{i\in\mathbb{Z}} \in \ell^p.$$

Its dual space is ℓ^q, where $\frac{1}{p} + \frac{1}{q} = 1$, with the dual coupling

$$[\![\boldsymbol{u}, \boldsymbol{v}]\!] := \sum_{i\in\mathbb{Z}} u_i v_i \quad \text{for } \boldsymbol{u} = (u_i)_{i\in\mathbb{Z}} \in \ell^p, \boldsymbol{v} = (v_i)_{i\in\mathbb{Z}} \in \ell^q.$$

Similarly, ℓ^∞ is the Banach space of real-valued bounded bi-infinite sequences with norm $\|\boldsymbol{u}\|_\infty := \sup_{i\in\mathbb{Z}} |u_i|$.

One can show that $\ell^2 \subset \ell^p \subset \ell^\infty$ for $p \geq 2$. (Note that these inclusions are in the opposite direction to the Lebesgue integral spaces $\mathcal{L}^p$).

Weighted norm sequence spaces

Since $u_i \to 0$ as $i \to \pm\infty$ for $\boldsymbol{u} = (u_i)_{i\in\mathbb{Z}} \in \ell^2$, the Hilbert space ℓ^2 does not include traveling wave solutions or solutions with non-zero constant components. Similarly, in neural models the values at distant neurons need not vanish.

Weighted sequence spaces are used to handle such dynamical behaviour. For greater applicability these will be defined for weighted space of bi-infinite sequences with vectorial integer indices $\mathrm{i} = (i_1, \cdots, i_d) \in \mathbb{Z}^d$ and any $p \geq 1$. In particular, given a positive sequence of weights $(\rho_\mathrm{i})_{\mathrm{i}\in\mathbb{Z}^d}$, ℓ^p_ρ denotes the Banach space

$$\ell^p_\rho := \left\{\boldsymbol{u} = (u_\mathrm{i})_{\mathrm{i}\in\mathbb{Z}^d} : \sum_{\mathrm{i}\in\mathbb{Z}^d} \rho_\mathrm{i}|u_\mathrm{i}|^p < \infty, \quad u_\mathrm{i} \in \mathbb{R}\right\}$$

with the norm

$$\|\boldsymbol{u}\|_{p,\rho} := \left(\sum_{\mathrm{i}\in\mathbb{Z}^d} \rho_\mathrm{i}|u_\mathrm{i}|^p\right)^{1/p} \quad \text{for } \boldsymbol{u} = (u_\mathrm{i})_{\mathrm{i}\in\mathbb{Z}^d} \in \ell^p_\rho.$$

For the special case with $p = 2$, $\|\boldsymbol{u}\|_{2,\rho}$ is written as $\|\boldsymbol{u}\|_\rho$ in short.

The weights ρ_i are often assumed to satisfy the following assumption.

Assumption 1.1. $\rho_\mathrm{i} > 0$ for all $\mathrm{i} \in \mathbb{Z}^d$ and $\rho_\Sigma := \sum_{\mathrm{i}\in\mathbb{Z}^d} \rho_\mathrm{i} < \infty$.

Lemma 1.1. *Let Assumption 1.1 hold. Then* $\ell^2 \subset \ell^2_\rho$ *and* $\|\boldsymbol{u}\|_\rho \leq \sqrt{\rho_\Sigma}\|\boldsymbol{u}\|$ *for* $\boldsymbol{u} \in \ell^2$.

Proof. Let $\boldsymbol{u} \in \ell^2$. By Assumption 1.1, $0 < \rho_\mathrm{i} < \rho_\Sigma$ for each $\mathrm{i} \in \mathbb{Z}^d$. Hence

$$\|\boldsymbol{u}\|^2_\rho = \sum_{\mathrm{i}\in\mathbb{Z}^d} \rho_\mathrm{i} u_\mathrm{i}^2 \leq \sum_{\mathrm{i}\in\mathbb{Z}^d} \rho_\Sigma u_\mathrm{i}^2 = \rho_\Sigma \sum_{\mathrm{i}\in\mathbb{Z}^d} u_\mathrm{i}^2 = \rho_\Sigma \|\boldsymbol{u}\|^2. \qquad \square$$

Lemma 1.2 ([Han *et al.* (2011)]). *Let Assumption 1.1 hold. Then* ℓ^p_ρ *is separable. In particular,* ℓ^2_ρ *is a separable Hilbert space.*

Proof. Separability holds because $\bigcup_{N\geq 1} \ell_N$ is a countable dense subset of ℓ^p_ρ, where

$$\ell_N = \{\boldsymbol{u} = (u_\mathrm{i})_{\mathrm{i}\in\mathbb{Z}^d} \ : \ u_\mathrm{i} \in \mathbb{Q} \text{ for } \mathrm{i} \in \mathbb{Z}^d \text{ and } u_\mathrm{i} = 0 \quad \text{for} \quad |\mathrm{i}| > N\}.$$

First, it is clear that $\bigcup_{N\geq 1} \ell_N$ is a countable subset of ℓ^p_ρ. Then, given any element $\boldsymbol{u} = (u_\mathrm{i})_{\mathrm{i}\in\mathbb{Z}^d} \in \ell^p_\rho$ and any $\varepsilon > 0$, there exists a positive integer $I(\varepsilon) \in \mathbb{N}$ such that

$$\sum_{|\mathrm{i}|>I(\varepsilon)} \rho_\mathrm{i}|u_\mathrm{i}|^p < \varepsilon^p/2.$$

Choose $\bar{\boldsymbol{u}} = (\bar{u}_\mathrm{i})_{\mathrm{i}\in\mathbb{Z}^d}$ such that $\bar{u}_\mathrm{i} \in \mathbb{Q}$ for $|\mathrm{i}| \leq I(\varepsilon)$ and $\bar{u}_i = 0$ for $|\mathrm{i}| > I(\varepsilon)$ with

$$\sum_{|\mathrm{i}|\leq I(\varepsilon)} \rho_\mathrm{i}|u_\mathrm{i} - \bar{u}_\mathrm{i}|^p < \varepsilon^p/2.$$

Then $\bar{\boldsymbol{u}} \in \bigcup_{N\geq 1} \ell_N$ and

$$\|\boldsymbol{u} - \bar{\boldsymbol{u}}\|_{p,\rho} < \varepsilon.$$

This implies that $\bigcup_{N\geq 1} \ell_N$ is dense in ℓ^p_ρ and hence ℓ^p_ρ is separable. □

The following additional assumption on the weights with indices $\mathrm{i} \in \mathbb{Z}^d$ will often also be used.

Assumption 1.2. There exist positive constants γ_0 and γ_1 such that

$$\rho_{\mathrm{i}\pm 1_j} \leq \gamma_0 \rho_{\mathrm{i}}, \quad \left|\rho_{\mathrm{i}\pm 1_j} - \rho_{\mathrm{i}}\right| \leq \gamma_1 \rho_{\mathrm{i}} \quad \text{for all } \mathrm{i} = (i_1, \ldots, i_d) \in \mathbb{Z}^d, \quad j = 1, \cdots, d,$$

where 1_j represents the vector in $\mathbb{Z}^d$ with the jth element equals 1, and all other elements equal 0. For example when $d = 1$, [Wang (2006)] considered the weights $\rho_i = (1 + i^2)^{-c}$ with $c > \frac{1}{2}$ for $i \in \mathbb{Z}$.

1.4 An illustrative lattice reaction-diffusion model

The paper of Bates, Lu & Wang [Bates *et al.* (2001)], has had a seminal influence on the investigation of attractors in lattice dynamical systems. The main ideas will be briefly outlined here in simplified form.

The authors assumed that the nonlinear function $f : \mathbb{R} \to \mathbb{R}$ in the LDS (1.2) is continuously differentiable, hence locally Lipschitz, with $f(0) = 0$ and satisfies the dissipativity condition

$$xf(x) \leq 0, \quad x \in \mathbb{R}. \tag{1.7}$$

In addition, it was assumed that $\boldsymbol{g} = (g_i)_{i\in\mathbb{Z}} \in \ell^2$.

To simplify the exposition we will assume here that f is globally Lipschitz with Lipschitz constant L_f. Then the function F defined component wise by $F_i(\boldsymbol{u}) := f(u_i)$ for $i \in \mathbb{Z}$ is globally Lipschitz with

$$\|F(\boldsymbol{u}) - F(\boldsymbol{v})\|^2 = \sum_{i\in\mathbb{Z}} |f(u_i) - f(v_i)|^2 \leq L_f^2 \sum_{i\in\mathbb{Z}} |u_i - v_i|^2$$

and takes values in ℓ^2 since

$$\|F(\boldsymbol{u})\|^2 = \sum_{i\in\mathbb{Z}} |f(u_i) - f(0)|^2 \leq L_f^2 \sum_{i\in\mathbb{Z}} |u_i|^2 = L_f^2 \|\boldsymbol{u}\|^2.$$

Moreover, $\langle F(\boldsymbol{u}), \boldsymbol{u}\rangle \leq 0$ due to (1.7).

Define the operator $\Lambda : \ell^2 \to \ell^2$ by

$$(\Lambda \boldsymbol{u})_i = u_{i-1} - 2u_i + u_{i+1}, \quad i \in \mathbb{Z}$$

and the operators D^+, $\mathrm{D}^- : \ell^2 \to \ell^2$ by

$$(\mathrm{D}^+\boldsymbol{u})_i = u_{i+1} - u_i, \quad (\mathrm{D}^-\boldsymbol{u})_i = u_{i-1} - u_i, \quad i \in \mathbb{Z}.$$

It is straightforward to check that

$$-\Lambda = \mathrm{D}^+\mathrm{D}^- = \mathrm{D}^-\mathrm{D}^+ \quad \text{and} \quad \langle \mathrm{D}^-\boldsymbol{u}, \boldsymbol{v}\rangle = \langle \boldsymbol{u}, \mathrm{D}^+\boldsymbol{v}\rangle \quad \forall\, \boldsymbol{u}, \boldsymbol{v} \in \ell^2,$$

and hence $\langle \Lambda\boldsymbol{u}, \boldsymbol{u}\rangle = -\|\mathrm{D}^+\boldsymbol{u}\|^2 \leq 0$ for any $\boldsymbol{u} \in \ell^2$.

In ℓ^2 this means Λ is negative definite since $\|D^+\boldsymbol{u}\| = 0$ implies that all components u_i are identical and hence $\boldsymbol{u}$ is zero in ℓ^2. Moreover, Λ is a bounded linear operator and generates a uniformly continuous semi-group. Λ is often called the discrete Laplace operator.

The lattice system (1.1) can be written as an ODE

$$\frac{d\boldsymbol{u}(t)}{dt} = \nu\Lambda\boldsymbol{u} - \lambda\boldsymbol{u} + F(\boldsymbol{u}) + \boldsymbol{g} \tag{1.8}$$

on ℓ^2, where $\boldsymbol{g} = (g_i)_{i\in\mathbb{Z}}$, $F : \ell^2 \to \ell^2$ is given component wise by $F_i(\boldsymbol{u}) := f(u_i)$ for some continuously differentiable globally Lipschitz function $f : \mathbb{R} \to \mathbb{R}$ with $f(0) = 0$. It follows that the function on the RHS of the infinite dimensional ODE (1.8) maps ℓ^2 into itself and is globally Lipschitz on ℓ^2.

Existence and uniqueness theorems for ODEs on Banach spaces (see e.g., [Deimling (1977)]) ensure the global existence and uniqueness of a solution $\boldsymbol{u}(t) = \boldsymbol{u}(t;\boldsymbol{u}_o)$ in ℓ^2 given initial datum $\boldsymbol{u}(0) = \boldsymbol{u}_o$. Moreover, $\boldsymbol{u}(t;\boldsymbol{u}_o)$ generates a semi-group $\{\varphi(t)\}_{t\geq 0}$, i.e., an autonomous semi-dynamical system, on ℓ^2.

Existence of an absorbing set

It is easy to show that the semi-group $\{\varphi(t)\}_{t\geq 0}$ has a positive invariant absorbing set. In fact, taking the inner product in ℓ^2 of (1.8) with $\boldsymbol{u} = \boldsymbol{u}(t;\boldsymbol{u}_o)$ gives

$$\frac{d}{dt}\|\boldsymbol{u}\|^2 + 2\nu\|D^+\boldsymbol{u}\|^2 + 2\lambda\|\boldsymbol{u}\|^2 = 2\langle F(\boldsymbol{u}),\boldsymbol{u}\rangle + 2\langle \boldsymbol{g},\boldsymbol{u}\rangle \leq -\lambda\|\boldsymbol{u}\|^2 + \frac{1}{\lambda}\|\boldsymbol{g}\|^2,$$

and hence

$$\frac{d}{dt}\|\boldsymbol{u}\|^2 \leq -\lambda\|\boldsymbol{u}\|^2 + \frac{1}{\lambda}\|\boldsymbol{g}\|^2.$$

The Gronwall inequality then gives

$$\|\boldsymbol{u}(t)\|^2 \leq \|\boldsymbol{u}_o\|^2 e^{-\lambda t} + \frac{1}{\lambda}\|\boldsymbol{g}\|^2,$$

and hence the closed and bounded subset of ℓ^2

$$Q := \left\{\boldsymbol{u} \in \ell^2 : \|\boldsymbol{u}\|^2 \leq 1 + \frac{1}{\lambda}\|\boldsymbol{g}\|^2\right\}$$

is a positively invariant absorbing set for the semi-group $\{\varphi(t)\}_{t\geq 0}$ on ℓ^2.

When the function f is assumed to be locally rather than globally Lipschitz the above inequality shows that the solutions cannot blow up and hence can be extended without restriction into the future.

Asymptotic tails and asymptotic compactness

A very significant contribution of the paper [Bates *et al.* (2001)] was to show that the semi-group generated by the LDS (1.8) is asymptotically compact, from which it follows that it has a global attractor $\mathcal{A}$ in ℓ^2. Their method of proof has since been adapted and used repeatedly in a large number of other papers including almost all of those discussed in this chapter.

The first step of the proof is to derive an *asymptotic tails estimate* for the solutions $\boldsymbol{u}(t;\boldsymbol{u}_o)$ of the LDS in the absorbing set Q.

Lemma 1.3. *For every $\varepsilon > 0$ there exist $T(\varepsilon) > 0$ and $I(\varepsilon) \in \mathbb{N}$ such that*

$$\sum_{|i|>I(\varepsilon)} |u_i(t;\boldsymbol{u}_o)|^2 \leq \varepsilon$$

for all $\boldsymbol{u}_o \in Q$ and $t \geq T(\varepsilon)$.

The proof requires a smooth cut-off function $\xi : \mathbb{R}^+ \to [0,1]$ with $\xi(s) = 0$ for $0 \leq s \leq 1$, $\xi(s) \in [0,1]$ for $1 \leq s \leq 2$ and $\xi(s) = 1$ for $s \geq 1$. For a large positive fixed integer k (to be determined in the proof) the proof uses

$$v_i(t) = \xi_k(|i|)u_i(t) \quad \text{with} \quad \xi_k(|i|) = \xi\left(\frac{|i|}{k}\right), \qquad i \in \mathbb{Z}.$$

Multiplying equation (1.8) by $v_i(t) = \xi_k(|i|)u_i(t)$ and summing over $i \in \mathbb{Z}$ gives

$$\begin{aligned}\frac{1}{2}\frac{\mathrm{d}}{\mathrm{d}t}\sum_{i\in\mathbb{Z}} \xi_k(|i|)|u_i(t)|^2 + \nu\langle \mathrm{D}^+\boldsymbol{u}, \mathrm{D}^+\boldsymbol{v}\rangle + \lambda\sum_{i\in\mathbb{Z}} \xi_k(|i|)u_i^2(t)\\ = \sum_{i\in\mathbb{Z}} \xi_k(|i|)u_i(t)f(u_i(t))u_i(t) + \sum_{i\in\mathbb{Z}} \xi_k(|i|)g_i.\end{aligned}$$

After some skillful estimates this leads to

$$\frac{\mathrm{d}}{\mathrm{d}t}\sum_{i\in\mathbb{Z}} \xi_k(|i|)|u_i(t)|^2 + \lambda\sum_{i\in\mathbb{Z}} \xi_k(|i|)u_i^2(t) \leq \frac{C}{k} + \frac{1}{\lambda}\sum_{|i|\geq k} g_i^2 \leq \frac{1}{2}\varepsilon$$

for $k \geq I(\varepsilon)$ since $\boldsymbol{g} = (g_i)_{i\in\mathbb{Z}} \in \ell^2$. Finally, by the Gronwall inequality,

$$\sum_{|i|\geq 2k} |u_i(t)|^2 \leq \sum_{i\in\mathbb{Z}} \xi_k(|i|)|u_i(t)|^2 \leq \varepsilon$$

for $t \geq T(\varepsilon)$ (to handle the initial condition) and $k \geq I(\varepsilon)$.

To obtain asymptotic compactness a sequence $\boldsymbol{u}(t_n;\boldsymbol{u}_o^{(n)})$ with $\boldsymbol{u}_o^{(n)} \in Q$ and $t_n \to \infty$ is considered. Since Q is closed and bounded subset of the Hilbert space ℓ^2 it is weakly compact. This gives a weakly convergent subsequence with a limit in Q. The asymptotic tail estimate is then used to separate a finite number of terms from the small tail to show that the weak limit is in fact a strong limit. The existence of a global attractor then follows by standard results in dynamical systems theory.

1.5 Outline of this book

This monograph consists of 21 chapters divided into 6 parts: Background, Laplacian LDS, A selection of lattice models, Stochastic and random LDS, Hopfield lattice models and LDS in biology. The main emphasis is on establishing the existence of attractors in such systems.

The Background part consists of two chapters, including this introductory chapter and another chapter on dynamical systems which provide background material on various kinds of dynamical systems and their attractors.

The Laplacian LDS part contains 3 chapters. In Chapter 3 we investigate the existence of global attractors in the autonomous case of the basic Laplacian lattice model of [Bates *et al.* (2001)] in some detail, in particular the asymptotic tails and asymptotic compactness arguments. Chapter 4 concerns the approximation of such attractors, first by finite dimensional versions of the lattice model and secondly by Euler numerical approximations. In Chapter 5 a non-autonomous Laplacian lattice model and its pullback attractor are considered on weighted sequence spaces.

Part III collects a selection of different lattice models to provide the reader with an overview of broad range of different kinds of lattice models as well as to provide a technical background for later applications that involve these types of models. There are 5 chapters. Lattice models based on delay differential equations are considered in Chapter 6 and on set-valued differential equations in Chapter 7, while Chapter 8 deals with lattice models based on second order differential equations. In Chapter 9 discrete time lattice models, i.e., described by difference equations rather than differential equations, are investigated, which are based on models motivated by spatial ecology. The resulting systems involve compact rather than asymptotically compact operators, as elsewhere in the book. The final Chapter 10 briefly presents and states without proofs results from the literature on the finite dimension of attractors, exponential attractors and travelling waves. The aim is to provide the reader with a quick overview of some important topics which are tangential to our main interests and the methods used in the book.

Stochastic and random LDS are the focus of Part IV. Chapter 11 introduces random dynamical systems and random ordinary differential equations which generate them. Random lattice models are then considered in detail in Chapters 12 and 13. These are generated by stochastic differential equations with additive or linear multiplicative noise which can be transformed to random ordinary differential equations by using Ornstein-Uhlenbeck processes. Finally, in Chapter 14 an LDS driven by fractional Brownian motion is considered.

Part V on Hopfield lattice models has 3 chapters. Chapter 15 and Chapter 16 consider deterministic Hopfield models on unweighted and weighted sequence spaces, respectively. Approximations of attractors are investigated depending on the number of connections of each neuron going to infinity. Chapter 17 examines the effects of noise on lattice Hopfield models.

Finally in Part VI we consider LDS in biology, which is of personal interest for us, so the models there are biased to those we ourselves have investigated. Chapter 18 considers the Fitzhugh-Nagumo lattice model in weighted sequence spaces. Then in Chapter19 we look at Amari lattice models, where the Heaviside function is formulated as a set-valued mapping or approximated by a sigmoidal function, while Chapter 20 deals with a neural lattice model with nonlinear state dependent noise coefficients. Finally, Chapter 21 focuses on lattice systems with switching effects and delayed recovery.

1.6 Endnotes

The proof of Lemma 1.2 was taken from [Han *et al.* (2011)], where sequence spaces with weighted norms are considered, see also Chapter 5 and elsewhere in this book.

Further details of the asymptotic tails argument of [Bates *et al.* (2001)] sketched above will be given in later chapters. An alternative compactness argument via the quasi-stability concept [Chueshov (2015)] was used by [Czaja (2022)] in the sequence space ℓ^2. See [Diestel *et al.* (1993)] for weak compactness in the space $\mathcal{L}^2$, and [Kisielewicz (1992)] for weak compactness in spaces $\mathcal{C}$.

Other applications of lattice models are given in [Afraimovich and Nekorkin (1994); Amigo *et al.* (2010); Bates and Chmaj (2003); Chow and Mallet-Paret (1995); Han and Kloeden (2019b); Kapral (1991); McBride *et al.* (2010)] and referenced therein.

1.7 Problems

Problem 1.1. Prove that ℓ^2 is dense in ℓ^p for $p > 2$ or give a counter example otherwise.

Problem 1.2. Determine a lattice version of the scalar porous media operator $\frac{\partial}{\partial x}\left(u\frac{\partial u}{\partial x}\right)$. Does the corresponding lattice system (1.2) with the discretised porous media operator instead of the discretised Laplacian operator have a global attractor?

Problem 1.3. Prove the existence and uniqueness of a global solution to the ODE (1.8), given f is only locally Lipschitz with appropriate growth conditions.

Chapter 2

Dynamical systems

Background material on autonomous and non-autonomous dynamical systems is summarised in this Chapter for the reader's convenience. More details and proofs can be found in the literature mentioned in the Endnotes.

Throughout this chapter, let $(\mathfrak{X}, \mathfrak{d}_{\mathfrak{X}})$ be a complete metric space, and let $\mathcal{P}_{cc}(\mathfrak{X})$ denote the collection of all *non-empty compact subsets* of $\mathfrak{X}$. The distance between two points x, $y \in \mathfrak{X}$ is given by

$$\mathfrak{d}_{\mathfrak{X}}(x, y) = \mathfrak{d}_{\mathfrak{X}}(y, x) \qquad \text{(symmetric)}.$$

We define the distance between a point $x \in \mathfrak{X}$ and a non-empty compact subset B in $\mathfrak{X}$ by

$$\operatorname{dist}_{\mathfrak{X}}(x, B) := \inf_{b \in B} \mathfrak{d}_{\mathfrak{X}}(x, b).$$

Remark 2.1. The mapping $b \mapsto \mathfrak{d}_{\mathfrak{X}}(x, b)$ is continuous for x fixed, in fact

$$|\mathfrak{d}_{\mathfrak{X}}(x, b) - \mathfrak{d}_{\mathfrak{X}}(x, c)| \leq \mathfrak{d}_{\mathfrak{X}}(b, c),$$

and the subset B is non-empty and compact, so the inf can be replaced by min here, i.e., it is actually attained.

Then we define the distance of a compact subset A from a compact subset B by

$$\operatorname{dist}_{\mathfrak{X}}(A, B) := \sup_{a \in A} \operatorname{dist}_{\mathfrak{X}}(a, B) = \sup_{a \in A} \inf_{b \in B} \mathfrak{d}_{\mathfrak{X}}(a, b),$$

which is sometimes written as $\mathcal{H}^*_{\mathfrak{X}}(A, B)$ and called the *Hausdorff separation* or *semi-distance* of A from B.

Remark 2.2. The function $a \mapsto \operatorname{dist}_{\mathfrak{X}}(a, B)$ is continuous for fixed B and the set A is compact, so the sup here can be replaced by max.

The Hausdorff separation, $\operatorname{dist}_{\mathfrak{X}}(A, B)$ satisfies the *triangle inequality*

$$\operatorname{dist}_{\mathfrak{X}}(A, B) \leq \operatorname{dist}_{\mathfrak{X}}(A, C) + \operatorname{dist}_{\mathfrak{X}}(C, B).$$

However, $\mathrm{dist}_{\mathfrak{X}}(A, B)$ is *not* a metric, since it can be equal to zero without the sets being equal, i.e., $\mathrm{dist}_{\mathfrak{X}}(A, B) = 0$ if $A \subset B$.

Define

$$\mathcal{H}_{\mathfrak{X}}(A, B) := \max\{\mathrm{dist}_{\mathfrak{X}}(A, B), \mathrm{dist}_{\mathfrak{X}}(B, A)\}.$$

This is a metric on $\mathcal{P}_{\mathrm{cc}}(\mathfrak{X})$ called the *Hausdorff metric.*

Theorem 2.1. *$(\mathcal{P}_{\mathrm{cc}}(\mathfrak{X}), \mathcal{H}_{\mathfrak{X}})$ is a complete metric space.*

2.1 Abstract dynamical systems

In this section we introduce the concepts of autonomous and non-autonomous dynamical systems, respectively. In particular, definitions of single-valued and set-valued autonomous dynamical systems are given in Sect. 2.1.1, the process formulation of single-valued and set-valued non-autonomous dynamical systems are given in Sect. 2.1.2, and the skew product formulation of single-valued and set-valued non-autonomous dynamical systems are given in Sect. 2.1.3.

2.1.1 *Autonomous dynamical systems*

Definition 2.1. An autonomous dynamical system on a metric space $(\mathfrak{X}, \mathfrak{d}_{\mathfrak{X}})$ is given by mapping $\varphi : \mathbb{R} \times \mathfrak{X} \to \mathfrak{X}$, which satisfies the properties:

(i) *initial condition*: $\varphi(0, x_0) = x_0$ for all $x_0 \in \mathfrak{X}$,
(ii) *group under composition*:

$$\varphi(s + t, x_0) = \varphi(s, \varphi(t, x_0)) \quad \text{for all } s, t \in \mathbb{R},\ x_0 \in \mathfrak{X},$$

(iii) *continuity*: the mapping $(t, x) \mapsto \varphi(t, x)$ is continuous at all points $(t_0, x_0) \in \mathbb{R} \times \mathfrak{X}$.

Throughout this book, define $\mathbb{R}^+ := \{t \in \mathbb{R} : t \geq 0\}$.

Definition 2.2. An autonomous semi-dynamical dynamical system on a metric space $(\mathfrak{X}, \mathfrak{d}_{\mathfrak{X}})$ is given by mapping $\varphi : \mathbb{R}^+ \times \mathfrak{X} \to \mathfrak{X}$, which satisfies the properties:

(i) *initial condition*: $\varphi(0, x_0) = x_0$ for all $x_0 \in \mathfrak{X}$,
(ii) *semi-group under composition*:

$$\varphi(s + t, x_0) = \varphi(s, \varphi(t, x_0)) \quad \text{for all } s, t \in \mathbb{R}^+,\ x_0 \in \mathfrak{X},$$

(iii) *continuity*: the mapping $(t, x) \mapsto \varphi(t, x)$ is continuous at all points $(t_0, x_0) \in \mathbb{R}^+ \times \mathfrak{X}$.

Next we provide the definition of set-valued autonomous dynamical systems. There is a large literature for autonomous set-valued dynamical systems, which are often called set-valued semi-groups or *general dynamical systems*, see e.g., [Szegö and Treccani (1969)]. Such systems are often generated by differential inclusions or differential equations without uniqueness [Aubin and Cellina (1984); Smirnov (2002)].

Definition 2.3. A set-valued autonomous dynamical system on a metric space $(\mathfrak{X}, \mathfrak{d}_{\mathfrak{X}})$ is defined in terms of an attainability set mapping $(t,x) \mapsto \Phi(t,x)$ on $\mathbb{R}^+ \times \mathfrak{X}$ satisfying

(i) *compactness*: $\Phi(t,x_0)$ is a non-empty compact subset of $\mathfrak{X}$ for all $(t,x_0) \in \mathbb{R}^+ \times \mathfrak{X}$,
(ii) *initial condition*: $\Phi(0,x_0) = \{x_0\}$ for all $x_0 \in \mathfrak{X}$,
(iii) *semi-group*: $\Phi(s+t,x_0) = \Phi(s,\Phi(t,x_0))$ for all $t,\ s \in \mathbb{R}^+$ and all $x_0 \in \mathfrak{X}$,
(iv) *upper semi-continuity in initial conditions*: $(t,x) \mapsto \Phi(t,x)$ is upper semi-continuous in $(t,x) \in \mathbb{R}^+ \times \mathfrak{X}$ with respect to the Hausdorff semi-distance $\mathrm{dist}_{\mathfrak{X}}$, i.e.,

$$\mathrm{dist}_{\mathfrak{X}}\left(\Phi(t,x), \Phi(t_0,x_0)\right) \to 0 \quad \text{as} \quad (t,x) \to (t_0,x_0) \;\text{ in }\; \mathbb{R}^+ \times \mathfrak{X},$$

(v) $t \mapsto \Phi(t,x_0)$ is continuous in $t \in \mathbb{R}^+$ with respect to the Hausdorff metric $\mathcal{H}_{\mathfrak{X}}$ uniformly in x_0 in compact subsets $B \in \mathcal{P}_{cc}(\mathfrak{X})$, i.e.,

$$\sup_{x_0 \in B} \mathcal{H}_{\mathfrak{X}}\left(\Phi(t,x_0), \Phi(t_0,x_0)\right) \to 0 \quad \text{as} \quad t \to t_0 \;\text{ in }\; \mathbb{R}^+.$$

2.1.2 *Two-parameter non-autonomous dynamical systems*

Two abstract formulations of non-autonomous dynamical systems will be considered in this book, presented in this section and Sect. 2.1.3, respectively. The first is a more direct generalisation of the definition of an abstract autonomous semi-dynamical system and is based on the properties of the solution mappings of non-autonomous differential equations. It is called a *process* or *two-parameter semi-group*. First define

$$\mathbb{R}^2_{\geq} := \{(t,t_0) \in \mathbb{R} \times \mathbb{R} \,:\, t \geq t_0\}.$$

Definition 2.4. (Process) A *process* on a metric space $(\mathfrak{X}, \mathfrak{d}_{\mathfrak{X}})$ is a mapping $\psi : \mathbb{R}^2_{\geq} \times \mathfrak{X} \to \mathfrak{X}$ with the following properties:

(i) *initial condition:* $\psi(t_0,t_0,x_0) = x_0$ for all $x_0 \in \mathfrak{X}$ and $t_0 \in \mathbb{R}$.
(ii) *two-parameter semi-group property*:

$$\psi(t_2,t_0,x_0) = \psi(t_2,t_1,\psi(t_1,t_0,x_0))$$

for all (t_1,t_0), $(t_2,t_1) \in \mathbb{R}^2_{\geq}$ and $x_0 \in \mathfrak{X}$.
(iii) *continuity:* the mapping $(t,t_0,x_0) \mapsto \psi(t,t_0,x_0)$ is continuous.

Remark 2.3. We can consider a process ψ as a two-parameter family of mappings $\psi_{t,t_0}(\cdot)$ on $\mathfrak{X}$ that forms a two-*parameter semi-group* under composition, i.e.,

$$\psi_{t_2,t_0}(x) = \psi_{t_2,t_1} \circ \psi_{t_1,t_0}(x), \quad \forall\ t_0 \leq t_1 \leq t_2 \text{ in } \mathbb{R}.$$

Remark 2.4. For an autonomous system, a process reduces to

$$\psi(t, t_0, x_0) = \varphi(t - t_0, x_0),$$

since the solutions depend only on the elapsed time $t - t_0$, i.e., just one parameter instead of independently on the actual time t and the initial time t_0, i.e., two parameters.

Definition 2.5. (Set-valued process) A *set-valued process* on metric space $(\mathfrak{X}, \mathfrak{d}_{\mathfrak{X}})$ is given by a mapping $\mathbb{R}^2_{\geq} \times \mathfrak{X} \ni (t, t_0, x) \mapsto \Psi(t, t_0, x_0) \in \mathcal{P}_{cc}(\mathfrak{X})$ such that

(i) $\Psi(t_0, t_0, x_0) = \{x_0\}$ for all $x_0 \in \mathfrak{X}$ and all $t_0 \in \mathbb{R}$,
(ii) $\Psi(t_2, t_0, x_0) = \Psi(t_2, t_1, \Psi(t_1, t_0, x_0))$ for all $t_0 \leq t_1 \leq t_2$ in $\mathbb{R}$ and all $x_0 \in \mathfrak{X}$,
(iii) $(t, t_0, x_0) \mapsto \Psi(t, t_0, x_0)$ is upper semi-continuous in $(t, t_0, x_0) \in \mathbb{R}^2_{\geq} \times \mathfrak{X}$ with respect to the Hausdorff semi-distance $\text{dist}_{\mathfrak{X}}$, i.e.,

$$\text{dist}_{\mathfrak{X}} \left(\Psi(s, s_0, y_0), \Psi(t, t_0, x_0)\right) \to 0$$

as $(s, s_0, y_0) \to (t, t_0, x_0)$ in $\mathbb{R}^2_{\geq} \times \mathfrak{X}$,
(iv) $t \mapsto \Psi(t, t_0, x_0)$ is continuous in $t \in \mathbb{R}$ with respect to the Hausdorff metric uniformly in (t_0, x_0) in compact subsets of $\mathbb{R} \times \mathfrak{X}$, i.e.,

$$\sup_{(t_0, x_0) \in B} \mathcal{H}_{\mathfrak{X}} \left(\Psi(s, t_0, x_0), \Psi(t, t_0, x_0)\right) \to 0 \quad \text{as} \quad s \to t \ \text{ in } \ \mathbb{R}$$

for each $B \in \mathcal{P}_{cc}(\mathbb{R} \times \mathfrak{X})$.

2.1.3 *Skew product flows*

A skew product flow consists of an *autonomous dynamical system* (full group) on a base space $\mathfrak{P}$, which is the source of the non-autonomity in a *cocycle mapping* acting on a state space $\mathfrak{X}$. The autonomous dynamical system is often called the *driving system.* Throughout this section, suppose that $(\mathfrak{P}, \mathfrak{d}_{\mathfrak{P}})$ is a complete metric space and consider the time set $\mathbb{R}$.

Definition 2.6. An autonomous dynamical system $\vartheta = (\vartheta_t)_{t \in \mathbb{R}}$ acting on the base space $(\mathfrak{P}, \mathfrak{d}_{\mathfrak{P}})$ is a *driving dynamical system* if

(i) $\vartheta_0(p) = p$ all $p \in \mathfrak{P}$,
(ii) $\vartheta_{s+t}(p) = \vartheta_s \circ \vartheta_t(p)$ for all $p \in \mathfrak{P}$ and $s,\, t \in \mathbb{R}$,
(iii) $(t, p) \mapsto \vartheta_t(p)$ is continuous for all $p \in \mathfrak{P}$ and $s,\, t \in \mathbb{R}$.

Definition 2.7. A skew product flow (ϑ, π) on $\mathfrak{P} \times \mathfrak{X}$ consists of a driving dynamical system $\vartheta = \{\vartheta_t\}_{t \in \mathbb{R}}$ acting on the base space $(\mathfrak{P}, \mathfrak{d}_{\mathfrak{P}})$ and a cocycle mapping π : $\mathbb{R}^+ \times \mathfrak{P} \times \mathfrak{X} \to \mathfrak{X}$ acting on the state space $(\mathfrak{X}, \mathfrak{d}_{\mathfrak{X}})$ satisfying

(i) *initial condition:* $\pi(0, p, x) = x$ for all $p \in \mathfrak{P}$ and $x \in \mathfrak{X}$,
(ii) *cocycle property:* for all $s,\, t \in \mathbb{R}^+$, $p \in \mathfrak{P}$ and $x \in \mathfrak{X}$,

$$\pi(s + t, p, x) = \pi(s, \vartheta_t(p), \pi(t, p, x)),$$

(iii) *continuity:* $(t, p, x) \mapsto \pi(t, p, x)$ is continuous.

Remark 2.5. The base system ϑ serves as a driving system which makes the cocycle mapping non-autonomous. Skew product flows often have very nice properties, in particular, when the base space $\mathfrak{P}$ is compact. This occurs when the driving system is, for example, periodic or almost periodic. It provides more detailed information about the dynamical behaviour of the system. George Sell, a pioneering researcher in the area, described the effect of a compact base space as being equivalent to *compactifying time*, see e.g., [Sell (1971)].

Remark 2.6. The skew product flow can also be used to define a random dynamical system, in which the driving system ϑ is an ergodic dynamical system $(\Omega, \mathcal{F}, \mathbb{P}, \{\vartheta_t\}_{t\in\mathbb{R}})$, i.e., the base space $(\Omega, \mathcal{F}, \mathbb{P})$ is a probability space and $(t, \omega) \mapsto \vartheta_t(\omega)$ is a measurable flow which is ergodic under $\mathbb{P}$, and the cocycle mapping $\pi : (t, \omega, x) \mapsto \pi(t, \omega, x)$ is measurable. More details on random dynamical systems will be given in Chapter 11.

Definition 2.8. A set-valued skew product flow (ϑ, Π) on $\mathfrak{P}\times\mathfrak{X}$ consists of a driving dynamical system ϑ and a cocycle attainability set mapping $\Pi : \mathbb{R}^+ \times \mathfrak{P} \times \mathfrak{X} \to \mathcal{P}_{cc}(\mathfrak{X})$ satisfying

(i) *compactness*: $\Pi(t, p, x)$ is a non-empty compact subset of $\mathfrak{X}$ for all $t \geq 0$, $p \in \mathfrak{P}$ and $x \in \mathfrak{X}$,

(ii) *initial condition:* $\Pi(0, p, x) = \{x\}$ for all $p \in \mathfrak{P}$ and $x \in \mathfrak{X}$,

(iii) *cocycle property:* for all s, $t \geq 0$, $p \in \mathfrak{P}$ and $x \in \mathfrak{X}$,

$$\Pi(s + t, p, x) = \Pi(s, \vartheta_t(p), \Pi(t, p, x)),$$

(iv) *continuity in time*: $\lim_{t\to s} \mathcal{H}_{\mathfrak{X}}(\Pi(t, p, x), \Pi(s, p, x)) = 0$ for all $t, s \geq 0$, $p \in \mathfrak{P}$ and $x \in \mathfrak{X}$,

(v) *upper semi-continuity in parameter and initial conditions*

$$\lim_{p\to p_0, x\to x_0} \text{dist}_{\mathfrak{X}} \left(\Pi(t, p, x), \Pi(t, p_0, x_0)\right) = 0$$

uniformly in $t \in [T_0, T_1]$ for any $0 \leq T_0 < T_1 < \infty$ for all $(p_0, x_0) \in \mathfrak{P} \times \mathfrak{X}$.

2.2 Invariant sets and attractors of dynamical systems

We are interested in the long term, i.e., asymptotic, behaviour of an underlying dynamical system. The *invariant sets* of a dynamical system provide us with a lot of useful information about the dynamical behaviour of the system, in particular its asymptotic behaviour. In this section we first provide the definitions of invariant sets for autonomous and non-autonomous dynamical systems, and then introduce concepts and the theory of attractors for autonomous dynamical systems, processes, and skew product flows, respectively.

Definition 2.9. Let $\varphi : \mathbb{R} \times \mathfrak{X} \to \mathfrak{X}$ be an autonomous dynamical system on $\mathfrak{X}$. A non-empty subset D of $\mathfrak{X}$ is said to be *invariant (positively invariant, negatively* *invariant (resp.)) under* φ if

$$\varphi(t,D) = (\subset, \supset (resp.))D \quad \text{for all } t \in \mathbb{R}, \quad \text{where} \quad \varphi(t,D) := \bigcup_{x\in D} \{\varphi(t,x)\}.$$

Definition 2.10. Let $\psi : \mathbb{R}^2_{\geq} \times \mathfrak{X} \to \mathfrak{X}$ be a process on $\mathfrak{X}$. A family of non-empty subsets $\mathcal{D} = (D_t)_{t\in\mathbb{R}}$ of $\mathfrak{X}$ is said to be *invariant (positively invariant, negatively invariant (resp.)) under* ψ if

$$\psi(t,t_0,D_{t_0}) = (\subset, \supset (resp.))D_t \quad \text{for all } (t,t_0) \in \mathbb{R}^2_{\geq}.$$

Definition 2.11. Let (ϑ, π) be a skew product flow on $\mathfrak{P} \times \mathfrak{X}$. A family $\mathcal{D} = (D_p)_{p\in\mathfrak{P}}$ of non-empty subsets D_p of $\mathfrak{X}$ is said to be *invariant (positively invariant, negatively invariant (resp.)) under* π if

$$\pi(t,p,D_p) = (\subset, \supset (resp.))D_{\vartheta_t(p)} \qquad \text{for all } p \in \mathfrak{P} \text{ and } t \in \mathbb{R}^+.$$

There are two types of invariance concepts for set-valued dynamical systems, one depending on the full sets, and the other involving only certain trajectories, referred to as *strong* and *weak* invariance, respectively. In this book we only consider the strong invariance. In fact, replacing the dynamical system φ, process ψ and skew product flow π by set-valued dynamical system Φ, set-valued process Ψ and set-valued skew product flow Π, respectively, in the above definitions, give the corresponding definitions of strongly invariance, strongly positive invariance and strongly negative invariance under set-valued dynamical systems, processes, and skew product flows, respectively.

2.2.1 *Attractors of autonomous semi-dynamical systems*

Definition 2.12. An *entire path* of a semi-dynamical system $\{\varphi(t)\}_{t\geq 0}$ on a complete metric state space $(\mathfrak{X}, \mathfrak{d}_{\mathfrak{X}})$ is a mapping $\mathfrak{e} : \mathbb{R} \to \mathfrak{X}$ with the property that

$$\mathfrak{e}(t) = \varphi(t-s, \mathfrak{e}(s)) \quad \forall\ (t,s) \in \mathbb{R}^2_{\geq}.$$

Note that $t-s \in \mathbb{R}^+$, the time set on which the semi-dynamical system φ is defined. However, the entire solution $\mathfrak{e}$ is defined for all $t \in \mathbb{R}$, not just in $\mathbb{R}^+$.

Lemma 2.1. *Let K be a compact invariant set w.r.t. a semi-dynamical system $\{\varphi(t)\}_{t\geq 0}$. Then for every $x \in K$ there exists an entire solution $\mathfrak{e}_x : \mathbb{R} \to K$ with $\mathfrak{e}_x(0) = x$.*

The ω-limit sets of a semi-dynamical system characterise its asymptotic behaviour as $t \to \infty$.

Definition 2.13. (Omega-limit sets) The *ω-limit set* of a bounded set $B \subset \mathfrak{X}$ is defined by

$$\omega(B) = \{x \in \mathfrak{X} : \exists\, t_k \to \infty, y_k \in B \text{ with } \varphi(t_k, y_k) \to x\}.$$

The ω-limit sets have the following properties.

Theorem 2.2. *For a non-empty bounded subset B of $\mathfrak{X}$,*

$$\omega(B) = \bigcap_{t \geq 0} \overline{\bigcup_{s \geq t} \varphi(s, B)} .$$

An attractor is an invariant set of special interest since it contains all the long term dynamics of a dissipative dynamical system, i.e., it is where every thing ends up. In particular, it contains the omega-limit set $\omega(B)$ of every non-empty bounded subset B of the state space $\mathfrak{X}$. In addition, an attractor is the omega-limit set of a neighbourhood of itself, i.e., it attracts a neighbourhood of itself. This additional stability property distinguishes an attractor from omega-limit sets in general.

Definition 2.14. A global attractor of a semi-dynamical system $\{\varphi(t)\}_{t \geq 0}$ is a non-empty compact invariant set $\mathcal{A}$ of $\mathfrak{X}$ which attracts all non-empty bounded subsets B of $\mathfrak{X}$, i.e.,

$$\operatorname{dist}_{\mathfrak{X}} (\varphi(t, B), \mathcal{A}) \to 0 \quad \text{as } t \to \infty.$$

An attractor may have a very complicated geometrical shape, e.g., the fractal dimensional set in the Lorenz ODE system. It is often easier to determine a closed and bounded absorbing set with a simpler geometrical shape such as a ball, in particular in infinite dimensional spaces, where closed and bounded subsets are much more common and easily determined than compact subsets.

Definition 2.15. A non-empty subset Q of $\mathfrak{X}$ is called an *absorbing set* of φ if for every non-empty bounded subset B of $\mathfrak{X}$ there exists a $T_B \geq 0$ such that

$$\varphi(t, B) \subset Q \quad \forall\, t \geq T_B.$$

All of the future dynamics is in Q, which need not be invariant, but often it is *positively invariant*, i.e., $\varphi(t, Q) \subset Q$ for all $t \in \mathbb{R}^+$. Some additional compactness property of the semi-group φ in Q such as its asymptotic compactness is then needed to ensure the non-emptiness of the attractor.

Definition 2.16. A semi-dynamical system $\{\varphi(t)\}_{t \geq 0}$ on a complete metric space $(\mathfrak{X}, \mathfrak{d}_{\mathfrak{X}})$ is said to be *asymptotically compact* if, for every sequence $\{t_k\}_{k \in \mathbb{N}}$ in $\mathbb{R}^+$ with $t_k \to \infty$ as $k \to \infty$ and every bounded sequence $\{x_k\}_{k \in \mathbb{N}}$ in $\mathfrak{X}$, the sequence $\{\varphi(t_k, x_k)\}_{k \in \mathbb{N}}$ has a convergent subsequence.

Theorem 2.3. *Let $\{\varphi(t)\}_{t \geq 0}$ be an autonomous semi-dynamical system on a complete metric space $(\mathfrak{X}, \mathfrak{d}_{\mathfrak{X}})$ which is asymptotically compact and has a closed and bounded absorbing set $Q \subset \mathfrak{X}$. Then φ has an attractor $\mathcal{A}$, which is contained in Q and is given by*

$$\mathcal{A} = \bigcap_{t \geq 0} \overline{\bigcup_{s \geq t} \varphi(s, Q)}.$$

Moreover, if Q is positively invariant then

$$\mathcal{A} = \bigcap_{t \geq 0} \varphi(t, Q).$$

In particular, $\mathcal{A} = \omega(Q)$.

An attractor, when it exists, is characterised by the bounded entire paths of the systems.

Corollary 2.1. *Let $\mathcal{A}$ be an attractor of a semi-dynamical system $\{\varphi(t)\}_{t\geq 0}$. Then for every $a \in \mathcal{A}$ there exists an entire solution $\mathfrak{e}_a : \mathbb{R} \to \mathcal{A}$ with $\mathfrak{e}_a(0) = a$.*

Definition 2.17. A compact subset $\mathcal{A}$ is said to be a (strong) global attractor for a set-valued dynamical system Φ, if it satisfies

(i) $\Phi(t, \mathcal{A}) = \mathcal{A}$ for all t.
(ii) $\mathcal{A}$ attracts every bounded subset of $\mathfrak{X}$.

2.2.2 *Attractors of processes*

Now consider a two-parameter semi-group or process $\{\psi(t, t_0)\}_{(t,t_0)\in\mathbb{R}^2_{\geq}}$ on the metric state space $\mathfrak{X}$ and time set $\mathbb{R}$.

An entire path of a two-parameter semi-group or process ψ is defined analogously to an entire solution of an autonomous semi-dynamical system.

Definition 2.18. An *entire solution* of a process $\{\psi(t, t_0)\}_{(t,t_0)\in\mathbb{R}^2_{\geq}}$ on a complete metric space $(\mathfrak{X}, \mathfrak{d}_{\mathfrak{X}})$ is a mapping $\mathfrak{e} : \mathbb{R} \to \mathfrak{X}$ with the property that $\mathfrak{e}(t) = \psi(t, t_0, \mathfrak{e}(t_0))$ for all $(t, t_0) \in \mathbb{R}^2_{\geq}$.

Steady state solutions are entire solutions, but there may be other interesting *bounded* entire solutions.

Lemma 2.2. *Let $\mathcal{D} = (D_t)_{t\in\mathbb{R}}$ be a ψ-invariant family of subsets of $\mathfrak{X}$. Then for any $t_0 \in \mathbb{R}$ and any $x_0 \in D_{t_0}$, there exists an entire solution $\mathfrak{e}_{x_0,t_0} : \mathbb{R} \to \mathfrak{X}$ of ψ such that $\mathfrak{e}_{x_0,t_0}(t_0) = x_0$ and*

$$\mathfrak{e}_{x_0,t_0}(t) \in D_t \quad \textit{for all } t \in \mathbb{R}.$$

An attractor for a process ψ should thus be a family $\mathscr{A} = (A_t)_{t\in\mathbb{R}}$ of non-empty compact subsets A_t of $\mathfrak{X}$, which is ψ-invariant, i.e.,

$$\psi(t, t_0, A_{t_0}) = A_t \quad \text{for all } (t, t_0) \in \mathbb{R}^2_{\geq}.$$

There is, however, a problem with convergence. There are two possibilities, one with pullback convergence and one with forward convergence.

Definition 2.19. A family $\mathscr{A} = (A_t)_{t\in\mathbb{R}}$ of non-empty compact subsets of $\mathfrak{X}$, which is ψ-invariant, is called a

- *pullback attractor* if it pullback attracts all bounded subsets B of $\mathfrak{X}$, i.e.,
$$\lim_{t_0 \to -\infty} \text{dist}_{\mathfrak{X}} \left(\psi(t, t_0, B), A_t\right) = 0, \qquad (\text{fixed } t)$$
- *forward attractor* if it forward attracts all bounded subsets B of $\mathfrak{X}$, i.e.,
$$\lim_{t \to \infty} \text{dist}_{\mathfrak{X}} \left(\psi(t, t_0, B), A_t\right) = 0, \qquad (\text{fixed } t_0).$$

We say that a pullback attractor $\mathcal{A} = (A_t)_{t \in \mathbb{R}}$ is *uniformly bounded* if $\bigcup_{t \in \mathbb{R}} A_t$ is bounded or, equivalently, if there is a common bounded subset B of $\mathfrak{X}$ such that $A_t \subseteq B$ for all $t \in \mathbb{R}$. We have the following characterisation of a uniformly bounded pullback attractor.

Proposition 2.1. *A uniformly bounded pullback attractor $\mathcal{A} = (A_t)_{t \in \mathbb{R}}$ of a process ψ is uniquely determined by the bounded entire solutions of the process, i.e.,*
$$a_0 \in A_{t_0} \quad \Longleftrightarrow \quad \exists \text{ a bounded entire solution } \mathfrak{e}(\cdot) \text{ with } \mathfrak{e}(t_0) = a_0.$$

To handle non-uniformities in the dynamics, which are typical in non-autonomous behaviour, we consider a pullback absorbing family of sets instead of a single set and assume that it absorbs a family of non-empty bounded subsets, which do not grow too quickly. The following definition is needed to ensure that the component sets in the non-autonomous family do not grow too quickly.

Definition 2.20. A family $\mathcal{B} = (B_t)_{t \in \mathbb{R}}$ of non-empty bounded subsets B_t of $\mathfrak{X}$ is said to have *subexponential growth* if
$$\limsup_{|t| \to \infty} \|B_t\| e^{c|t|} = 0 \qquad \forall c > 0, \quad \text{where} \quad \|B_t\| = \sup_{b \in B_t} \|b\|.$$
In this case it is called a *tempered* family.

Definition 2.21. A family $\mathcal{Q} = (Q_t)_{t \in \mathbb{R}}$ of non-empty subsets of $\mathfrak{X}$ is called a *pullback absorbing family* for a process ψ on $\mathfrak{X}$ if for each $t \in \mathbb{R}$ and every tempered family $\mathcal{B} = (B_t)_{t \in \mathbb{R}}$ of non-empty bounded subsets of $\mathfrak{X}$ there exists a $T_{t,\mathcal{B}} \in \mathbb{R}^+$ such that
$$\psi\left(t, t_0, B_{t_0}\right) \subseteq Q_t \qquad \text{for all } t_0 \leq t - T_{t,\mathcal{B}}.$$

Definition 2.22. A process ψ on a Banach space $\mathfrak{X}$ is said to be *pullback asymptotically compact* if, for each $t \in \mathbb{R}$, each sequence $\{t_k\}_{k \in \mathbb{N}}$ in $\mathbb{R}$ with $t_k \leq t$ and $t_k \to -\infty$ as $k \to \infty$, and each bounded sequence $\{x_k\}_{k \in \mathbb{N}}$ in $\mathfrak{X}$, the sequence $\{\psi(t, t_k, x_k)\}_{k \in \mathbb{N}}$ has a convergent subsequence.

Theorem 2.4. *(Existence of a pullback attractor) Suppose that a process ψ on a complete metric space $(\mathfrak{X}, \mathfrak{d}_{\mathfrak{X}})$ is pullback asymptotically compact and has a ψ-positive invariant pullback absorbing family $\mathcal{Q} = (Q_t)_{t \in \mathbb{R}}$ of compact sets. Then ψ has a global pullback attractor $\mathcal{A} = (A_t)_{t \in \mathbb{R}}$ with component subsets determined by*
$$A_t = \bigcap_{t_0 \leq t} \psi\left(t, t_0, Q_{t_0}\right) \qquad \textit{for each } \; t \in \mathbb{R}.$$
Moreover, if $\mathcal{A}$ is uniformly bounded then it is unique.

Remark 2.7. Theorem 2.4 characterises and gives the existence of a pullback attractor. Notice that the actual construction assumes nothing about the dynamics outside the absorbing sets, i.e., in particular that it is pullback absorbing. Thus forward attractors can be constructed by a similar pullback argument within a forward absorbing set, but this provides only a necessary condition for the family of sets obtained so to be a forward attractor. Moreover, when they exist, forward attractors need not be unique.

To define pullback attractors for set-valued processes, denote by $\mathcal{P}(\mathfrak{X})$ the collection of all families of non-empty subsets of $\mathfrak{X}$ and let $\mathcal{D} = \{D_t : D_t \subset \mathfrak{X}, D_t \neq \emptyset\}_{t\in\mathbb{R}}$. For any $\mathcal{D}, \tilde{\mathcal{D}} \in 2^{\mathfrak{X}}$, the notation $\tilde{\mathcal{D}} \subset \mathcal{D}$ means $\tilde{D}_t \subset D_t$ for every $t \in \mathbb{R}$.

Definition 2.23. A subset $\mathfrak{D}$ of $\mathcal{P}(\mathfrak{X})$ is *inclusion closed* if for $\mathcal{D} \in \mathfrak{D}$ and $\tilde{\mathcal{D}} \in \mathcal{P}(\mathfrak{X})$, then $\tilde{\mathcal{D}} \subset \mathcal{D}$ implies that $\tilde{\mathcal{D}} \in \mathcal{P}(\mathfrak{X})$.

Such a collection $\mathfrak{D}$ defined in Def. 2.23 is called a *universe.*

Definition 2.24. Let $\{\Psi(t,t_0)\}_{(t,t_0)\in\mathbb{R}^2_{\geq}}$ be a set-valued process on $\mathfrak{X}$. A family of non-empty bounded sets $\mathfrak{Q} := (Q_t)_{t\in\mathbb{R}}$ is said to be $\mathfrak{D}$-pullback absorbing for the set-valued process Ψ, if for any $\mathcal{D} = (D_t)_{t\in\mathbb{R}} \in \mathfrak{D}$ and each $t \in \mathbb{R}$, there exists some time $T_{\mathcal{D}}(t) > 0$ such that

$$\Psi(t, t-\tau, D_{t-\tau}) \subset Q_t, \quad \text{for all } \tau \geq T_{\mathcal{D}}.$$

A family of non-empty bounded sets $\mathfrak{Q} := (Q_t)_{t\in\mathbb{R}}$ is said to be $\mathfrak{D}$-pullback attracting for the set-valued process Ψ, if every $\mathcal{D} = (D_t)_{t\in\mathbb{R}} \in \mathfrak{D}$ satisfies

$$\lim_{\tau\to\infty} \operatorname{dist}_{\mathfrak{X}}(\Psi(t, t-\tau, D_{t-\tau}), Q_t) = 0.$$

Definition 2.25. Let $\{\Psi(t,t_0)\}_{(t,t_0)\in\mathbb{R}^2_{\geq}}$ be a set-valued process on $\mathfrak{X}$ and let $\mathfrak{D}$ be a universe. A family $\mathfrak{A} = (A_t)_{t\in\mathbb{R}}$ is said to be a global $\mathfrak{D}$-pullback attractor for Ψ if

(i) $A_t \subset \mathfrak{X}$ is compact for any $t \in \mathbb{R}$;
(ii) $\mathfrak{A}$ is invariant;
(iii) $\mathfrak{A}$ is $\mathfrak{D}$-pullback attracting.

The existence of a pullback attractor usually relies on some asymptotically compactness. In this work we will use the following definition.

Definition 2.26. A set-valued process $\{\Psi(t,t_0)\}_{(t,t_0)\in\mathbb{R}^2_{\geq}}$ is said to be $\mathfrak{D}$-pullback *asymptotically upper semi-compact* in $\mathfrak{X}$ if for any fixed time $t \in \mathbb{R}$, any sequence $y_n \in \Psi(t, t-\tau_n, x_n)$ has a convergent subsequence in $\mathfrak{X}$ whenever $\tau_n \to \infty$ as $n \to \infty$ and $x_n \in D_{t-\tau_n}$ with $\mathcal{D} = (D_t)_{t\in\mathbb{R}} \in \mathfrak{D}$.

The following proposition from [Caraballo and Kloeden (2009)] gives the existence of pullback attractors.

Proposition 2.2. *Let* $\{\Psi(t,t_0)\}_{(t,t_0)\in\mathbb{R}^2_{\geq}}$ *be a set-valued process on* $\mathfrak{X}$ *and let* $\mathfrak{D}$ *be a universe. Assume that*

(i) $\Psi(t, t_0, x)$ *is upper semi-continuous in* x *for any* $(t, t_0) \in \mathbb{R}^2_{\geq}$,
(ii) $\Psi(t, t_0)$ *is* $\mathfrak{D}$*-pullback asymptotically upper semi-compact in* $\mathfrak{X}$,
(iii) $\Psi(t, t_0)$ *has a* $\mathfrak{D}$*-pullback absorbing set* $\mathfrak{Q} = (Q_t)_{t\in\mathbb{R}} \in \mathfrak{D}$.

Then the set-valued process $\{\Psi(t, t_0)\}_{(t,t_0)\in\mathbb{R}^2_{\geq}}$ *has a unique* $\mathfrak{D}$*-pullback attractor* $\mathfrak{A} = (A_t)_{t\in\mathbb{R}}$ *with its components given by*

$$A_t = \bigcap_{s\geq 0} \overline{\bigcup_{\tau\geq s} \Psi(t, t-\tau, Q_{t-\tau})}.$$

When investigating set-valued processes it is often convenient to consider their single-valued trajectories.

Definition 2.27. A *trajectory* of a set-valued process $\{\Psi(t, t_0)\}_{(t,t_0)\in\mathbb{R}^2_{\geq}}$ is a single-valued function $\psi : [t_0, t_1] \cap \mathbb{R} \to \mathfrak{X}$ for some $(t_1, t_0) \in \mathbb{R}^2_{\geq}$ such that

$$\psi(t) \in \Psi(t, s, \psi(s)) \quad \text{for all} \quad t_0 \leq s \leq t \leq t_1 \quad \text{in} \quad \mathbb{R}.$$

A trajectory is called an *entire trajectory* if it is a trajectory on the whole time set $\mathbb{R}$.

In the discrete time case, trajectories are simply parts of sequences. Note that in the continuous time case trajectories are not assumed to be continuous but this follows from the next theorem, which is a generalisation of a theorem by Barbashin.

Theorem 2.5. (Barbashin's Theorem) *Let* $\{\Psi(t, t_0)\}_{(t,t_0)\in\mathbb{R}^2_{\geq}}$ *be a set-valued process on a complete metric space* $(\mathfrak{X}, \mathfrak{d}_{\mathfrak{X}})$. *Then*

(1) *there exists a trajectory from* x_0 *to* $x_1 \in \Psi(t_1, t_0, x_0)$ *for each* $(t_1, t_0) \in \mathbb{R}^2_{\geq}$ *and* $x_0 \in \mathfrak{X}$;
(2) *trajectories of a set-valued processes are continuous functions;*
(3) *the set* $\mathcal{J}(t_1, t_0, K)$ *of all trajectories joining* x_0 *to an arbitrary* $x_1 \in \Psi(t_1, t_0, x_0)$ *with* $x_0 \in K$ *is compact in* $C([t_0, t_1]; \mathfrak{X})$ *for all* $(t_1, t_0) \in \mathbb{R}^2_{\geq}$ *and any non-empty compact subset* K *of* $\mathfrak{X}$.

Definition 2.28. A family $\mathcal{D} = (D_t)_{t\in\mathbb{R}}$ of non-empty sets of $\mathfrak{X}$ is said to be *invariant* for a set-valued process Ψ if $\Psi(t, t_0, D_{t_0}) = D_t$ for all $(t, t_0) \in \mathbb{R}^2_{\geq}$; *positively invariant* if $\Psi(t, t_0, D_{t_0}) \subset D_t$ for all $(t, t_0) \in \mathbb{R}^2_{\geq}$; and *strongly negatively invariant* if $D_t \subset \Psi(t, t_0, D_{t_0})$ for all $(t, t_0) \in \mathbb{R}^2_{\geq}$.

Theorem 2.6. *Let* $\{\Psi(t, t_0)\}_{(t,t_0)\in\mathbb{R}^2_{\geq}}$ *be a set-valued process on a complete metric space* $(\mathfrak{X}, \mathfrak{d}_{\mathfrak{X}})$ *and let* $\mathcal{K} = (K_t)_{t\in\mathbb{R}}$ *be a family of non-empty compact subsets of* $\mathfrak{X}$, *which is* Ψ*-positively invariant.*

Then there exists a family of non-empty compact subsets $\mathcal{K}^\infty = (K_t^\infty)_{t\in\mathbb{R}}$ *contained in* $\mathcal{K}$ *in the sense that* $K_t^\infty \subset K_t$ *for each* $t \in \mathbb{R}$*, which is* Ψ*-strongly invariant. The component sets* K_t^∞ *are given by*

$$K_t^\infty = \bigcap_{t_0 \le t} \Psi(t, t_0, K_{t_0}), \qquad t \in \mathbb{R}.$$

2.2.3 *Attractors of skew product flows*

For complete metric spaces $(\mathfrak{P}, \mathfrak{d}_{\mathfrak{P}})$ and $(\mathfrak{X}, \mathfrak{d}_{\mathfrak{X}})$, let (ϑ, π) be a skew product flow on $\mathfrak{P} \times \mathfrak{X}$.

Similarly to processes we have two types of attractors for skew product flows, pullback and forward attractors.

Definition 2.29. A family $\mathcal{A} = (A_p)_{p\in\mathfrak{P}}$ of π-invariant non-empty compact subsets of $\mathfrak{X}$ is called a *pullback attractor* if it pullback attracts families $\mathcal{B} = (B_p)_{p\in\mathfrak{P}}$ of non-empty bounded subsets of $\mathfrak{X}$, i.e.,

$$\lim_{t\to\infty} \mathrm{dist}_{\mathfrak{X}}\left(\pi(t, \vartheta_{-t}(p), B_{\vartheta_{-t}(p)}), A_p\right) = 0 \qquad \text{for each } p \in \mathfrak{P}.$$

It is called a *forward attractor* if it forward attracts families of non-empty bounded subsets $\mathcal{B} = (B_p)_{p\in\mathfrak{P}}$ of $\mathfrak{X}$, i.e.,

$$\lim_{t\to\infty} \mathrm{dist}_{\mathfrak{X}}\left(\pi(t, p, B_p), A_{\vartheta_t(p)}\right) = 0 \qquad \text{for each } p \in \mathfrak{P}.$$

Also, as for a process, the existence of a pullback attractor for skew product flow is ensured by that of a pullback absorbing family. To handle nonuniformities, as for processes, the following definition similar to Definition 2.20 is needed to ensure that the component sets in the non-autonomous family should not do too quickly.

Definition 2.30. A family $\mathcal{B} = (B_p)_{p\in\mathfrak{P}}$ of non-empty bounded subsets B_p of $\mathfrak{X}$ is said to have *sub-exponential growth* if

$$\limsup_{|t|\to\infty} \|B_{\vartheta_{-t}(p_0)}\| e^{c|t|} = 0 \quad \forall c > 0 \quad \text{where} \quad \|B_{\vartheta_{-t}(p_0)}\| = \sup_{b\in B_{\vartheta_{-t}(p_0)}} \|b\|.$$

In this case it is called a *tempered* family.

Definition 2.31. A family $\mathcal{Q} = (Q_p)_{p\in\mathfrak{P}}$ of non-empty subsets of $\mathfrak{X}$ is called a *pullback absorbing family* for a skew product flow (ϑ, π) on $\mathfrak{P} \times \mathfrak{X}$ if for each $p \in \mathfrak{P}$ and every tempered family $\mathcal{B} = (B_p)_{p\in\mathfrak{P}}$ of non-empty bounded subsets of $\mathfrak{X}$ there exists a $T_{p,\mathcal{B}} \in \mathbb{R}^+$ such that

$$\pi\left(t, \vartheta_{-t}(p), B_{\vartheta_{-t}(p)}\right) \subseteq Q_p \qquad \text{for all } t \ge T_{p,\mathcal{B}}.$$

Definition 2.32. A skew product flow (ϑ, π) on $\mathfrak{P} \times \mathfrak{X}$ is said to be $\mathscr{D}$-pullback asymptotically compact if for any $p \in \mathfrak{P}$ and $\mathcal{D} = (D_t)_{t\in\mathbb{R}} \in \mathscr{D}$, the sequence $\pi(t_n, \vartheta_{-t_n}(p), x_n)$ has a convergence subsequence for any sequences $t_n \to +\infty$ and $x_n \in D_{\vartheta_{-t_n}(p)}$.

The proof of the following theorem here is similar to that of Theorem 2.4.

Theorem 2.7. *(Existence of a pullback attractor) Let $(\mathfrak{P}, \mathfrak{d}_{\mathfrak{P}})$ and $(\mathfrak{X}, \mathfrak{d}_{\mathfrak{X}})$ be complete metric spaces and suppose that a skew product flow (ϑ, π) on $\mathfrak{P} \times \mathfrak{X}$ is pullback asymptotic compact and has a pullback tempered absorbing family $\mathcal{Q} = (Q_p)_{p \in \mathfrak{P}}$ of non-empty closed and bounded sets.*

Then the skew product flow (ϑ, π) has a pullback attractor $\mathcal{A} = (A_p)_{p \in \mathfrak{P}}$ with component subsets determined by

$$A_p = \bigcap_{t \leq 0} \overline{\bigcup_{s \geq t} \pi\left(t, \vartheta_{-t}(p), Q_{\vartheta_{-t}(p)}\right)} \qquad \text{for each } \ p \in \mathfrak{P}.$$

If $\mathcal{Q}$ is π-positively invariant then

$$A_p = \bigcap_{t \leq 0} \pi\left(t, \vartheta_{-t}(p), Q_{\vartheta_{-t}(p)}\right) \qquad \text{for each } \ p \in \mathfrak{P}.$$

Moreover, $\mathcal{A}$ is unique if the components sets are uniformly bounded.

Note that if the pullback attractor is uniformly pullback attracting, i.e., if

$$\lim_{t \to \infty} \sup_{p \in \mathfrak{P}} \operatorname{dist}_{\mathfrak{X}} \left(\pi(t, \vartheta_{-t}(p), Q_{\vartheta_{-t}(p)}), A_p\right) = 0 \qquad \text{for each } p \in \mathfrak{P},$$

then it is uniformly forward attracting, since writing $a = \vartheta_{-t}(p)$,

$$\sup_{p \in \mathfrak{P}} \operatorname{dist}_{\mathfrak{X}} \left(\pi(t, \vartheta_{-t}(p), Q_{\vartheta_{-t}(p)}), A_p\right) = \sup_{a \in \mathfrak{P}} \operatorname{dist}_{\mathfrak{X}} \left(\pi(t, a, Q_a), A_{\vartheta_t(a)}\right).$$

In this case this uniform pullback/forward attractor is called a *uniform (non-autonomous) attractor.*

2.3 Compactness criteria

In a finite dimensional space such as $\mathbb{R}^d$ the compact subsets are the closed and bounded subsets. In an infinite dimensional Banach space $(\mathfrak{E}, \|\cdot\|_{\mathfrak{E}})$ the compact subsets are the closed and totally bounded subsets, i.e., they can be covered by the union of a finite number of balls of arbitrarily small radius.

Equivalently, a subset D of $(\mathfrak{E}, \|\cdot\|_{\mathfrak{E}})$ is compact if it is *sequentially compact*, i.e., if every sequence in D has a convergent subsequence in D.

2.3.1 *Kuratowski measure of non-compactness*

Let $(\mathfrak{E}, \|\cdot\|_{\mathfrak{E}})$ be a Banach space. A mapping $\mathcal{S}$ is called a *κ-contraction* on $\mathfrak{E}$ when it is a contraction with respect the Kuratowski measure of noncompactness of subsets of $\mathfrak{E}$, i.e., if there is a positive number $q < 1$ such that

$$\kappa(\mathcal{S}(D)) < q\kappa(D)$$

for every subset D of $\mathfrak{E}$. The *Kuratowski measure of noncompactness* of a subset D of Banach space $(\mathfrak{E}, \|\cdot\|_{\mathfrak{E}})$ is defined by

$$\kappa(D) = \inf\{d > 0 : \text{there exists an open cover of } \ D \ \text{ with sets of diameter} \leq d\}.$$

The compact sets are the closed subsets D of $\mathfrak{E}$ with $\kappa(D) = 0$.

Basic properties of the Kuratowski measure of noncompactness on a Banach space include:

(i) D is bounded if and only if $\kappa(D) < \infty$.
(ii) $\kappa(\bar{D}) = \kappa(D)$, where $\bar{D}$ denotes the closure of D.
(iii) D is compact if and only if $\kappa(D) = 0$.
(iv) $\kappa(D_1 \cup D_2) = \max(\kappa(D_1), \kappa(D_2))$ for any subsets D_1 and D_2.
(v) κ is continuous with respect to the Hausdorff distance of sets.
(vi) $\kappa(aD) = |a|\kappa(D)$ for any scalar a.
(vii) $\kappa(D_1 + D_2) \leq \kappa(D_1) + \kappa(D_2)$ for any subsets D_1 and D_2.
(viii) $\kappa(\mathrm{conv}D) = \kappa(D)$, where $\mathrm{conv}D$ denotes the convex hull of D.
(ix) if $D_1 \supseteq D_2 \supseteq D_3 \supseteq \cdots$ are non-empty closed subsets of $\mathfrak{E}$ such that $\kappa(D_n) \to 0$ as $n \to \infty$, then $\bigcap_{n\geq 1} D_n$ is non-empty and compact.

2.3.2 *Weak convergence and weak compactness*

Let $(\mathfrak{H}, \|\cdot\|_{\mathfrak{H}}, \langle\cdot,\cdot\rangle_{\mathfrak{H}})$ be a Hilbert space, which will typically be ℓ^2 or ℓ^2_ρ in this book.

Convergence with respect to the norm $\|\cdot\|_{\mathfrak{H}}$ is often called *strong convergence*, i.e., $\boldsymbol{u}_n \to \boldsymbol{u}^*$ strongly if and only if $\|\boldsymbol{u}_n - \boldsymbol{u}^*\|_{\mathfrak{H}} \to 0$ as $n \to \infty$. Another useful convergence is *weak convergence*. A sequence $\{\boldsymbol{u}_n\}_{n\in\mathbb{N}}$ converges weakly to $\boldsymbol{u}^*$ in $\mathfrak{H}$ if and only if

$$\langle h, \boldsymbol{u}_n - \boldsymbol{u}^*\rangle_{\mathfrak{H}} \to 0 \quad \text{as} \quad n \to \infty \quad \text{for all} \quad h \in \mathfrak{H}.$$

Weak convergence is often written as $\boldsymbol{u}_n \rightharpoonup \boldsymbol{u}^*$.

Essentially, weak convergence is with respect to all linear functionals on $\mathfrak{H}$. In general, weak convergence does not imply strong convergence, but the following result holds. See [Banach and Saks (1930); Okada (1984); Partington (1977); Szlenik (1965)].

Theorem 2.8. (Banach-Saks Theorem) *A bounded sequence $\{\boldsymbol{u}_n\}_{n\in\mathbb{N}}$ in a Hilbert $\mathfrak{H}$ contains a subsequence $\{\boldsymbol{u}_{n_k}\}_{k\in\mathbb{N}}$ and a point $\boldsymbol{u}^*$ such that*

$$\frac{1}{N}\sum_{k=1}^{N} \boldsymbol{u}_{n_k} \longrightarrow \boldsymbol{u}^* \quad \textit{strongly} \quad \textit{as} \quad N \to \infty.$$

Definition 2.33. A subset K of a Hilbert space $\mathfrak{H}$ is said to be weakly compact if it is weakly sequentially compact, i.e., if every sequence $\{\boldsymbol{u}_n\}_{n\in\mathbb{N}}$ in K has a weakly convergent subsequence $\boldsymbol{u}_{n_k} \rightharpoonup \boldsymbol{u}^*$ in K.

The following theorem is a special case of a more general result of Kakutani, see Theorem 3.17 in [Brezis (2011)].

Theorem 2.9. *A closed and bounded (in norm) subset D of a Hilbert space $\mathfrak{H}$ is weakly compact.*

A special case of the Banach-Alaoglu theorem is the sequential version of the original theorem.

Theorem 2.10. (Banach-Alaoglu Theorem) *The closed unit ball of the dual space of a separable normed vector space is sequentially compact in the weak*-topology.*

The following result is from [Ülger (1991), Proposition 7], see also [Diestel (1977)].

Lemma 2.3. (Ülger's Lemma) *Let (Ω, Σ, μ) be a probabilistic space, and $\mathfrak{E}$ be an arbitrary Banach space. For any weakly compact subset $K \subset \mathfrak{E}$, the set*

$$\{ f \in \mathcal{L}^1(\mu, \mathfrak{E}) \ : \ f(\omega) \in K \text{ for } \mu\text{-almost every } \omega \in \Omega \}$$

is relatively weakly compact.

The next result is due to [Ülger (1991), Corollary 5].

Lemma 2.4. *Let (Ω, Σ, μ) be a probabilistic space and $\mathfrak{E}$ be a Banach space. Set*

$$U := \{ f \in \mathcal{L}^1(\mu, \mathfrak{E}) \ : \ \|f(\omega)\|_{\mathfrak{E}} \leq 1 \text{ for } \mu - a.e. \omega \in \Omega \}.$$

A sequence $\{f_k(\cdot)\}_{k\in\mathbb{N}}$ in $U \subset \mathcal{L}^1(\mu, \mathfrak{E})$ converges weakly to $f \in \mathcal{L}^1(\mu, \mathfrak{E})$ if and only if for any sub-sequence $\{f_{k_n}(\cdot)\}_{n\in\mathbb{N}}$ given, there exists a sequence $\{g_n(\cdot)\}_{n\in\mathbb{N}}$ with $g_n \in co\{f_{k_n}, f_{k_{(n+1)}}, \dots\}$ such that for μ-a.e. $\omega \in \Omega$,

$$g_n(\omega) \longrightarrow f(\omega) \quad (n \longrightarrow \infty) \qquad \text{weakly in } \mathfrak{E}.$$

2.3.3 *Ascoli-Arzelà Theorem*

The Ascoli-Arzelà Theorem [Green and Valentine (1960/1961)] is a crucial tool in the study of lattice dynamical systems. Let $(\mathfrak{E}, \|\cdot\|_{\mathfrak{E}})$ be a Banach space, let $\mathcal{I}$ be a closed and bounded interval in $\mathbb{R}$ and let $C(\mathcal{I}, \mathfrak{E})$ be the space of all continuous functions $f : \mathcal{I} \to \mathfrak{E}$ with uniform norm $\|f\|_\infty = \max_{t\in\mathcal{I}} \|f(t)\|_{\mathfrak{E}}$.

Definition 2.34. A subset $\mathfrak{S}$ of $C(\mathcal{I}, \mathfrak{E})$ is said to be *equi-continuous* if for every $\varepsilon > 0$ there exists $\delta = \delta(\varepsilon) > 0$ which is independent of $f \in \mathfrak{S}$ such that $\|f(s) - f(t)\|_{\mathfrak{E}} < \varepsilon$ for all s, $t \in \mathcal{I}$ with $|s - t| < \delta$ and all $f \in \mathfrak{S}$.

Theorem 2.11. (Ascoli-Arzelà Theorem). *A subset $\mathfrak{S}$ of $C(\mathcal{I}, \mathfrak{E})$ is relatively compact if and only if $\mathfrak{S}$ is equi-continuous and $\mathfrak{S}(t) := \{f(t) : f \in \mathfrak{S}\}$ is relatively compact in $\mathfrak{E}$ for every $t \in \mathcal{I}$.*

The following consequence of this theorem in a Hilbert space $\mathfrak{H}$, which will typically be ℓ^2 or ℓ^2_ρ in this book, will be used in the sequel. See, e.g. [Lebl (2016)].

Corollary 2.2. *Let $\{f_n(\cdot)\}_{n\in\mathbb{N}}$ be a sequence in $\mathcal{C}([0,T],\mathfrak{H})$, which is uniformly bounded and equi-Lipschitz continuous on $[0,T]$. Then there is an $f^*(\cdot)\in\mathcal{C}([0,T],\mathfrak{H})$ and a convergent subsequence $\{f_{n_k}(\cdot)\}_{k\in\mathbb{N}}$ of $\{f_n(\cdot)\}_{n\in\mathbb{N}}$ such that*

$$f_{n_k}(\cdot)\to f^*(\cdot) \quad \text{strongly in } \mathcal{C}([0,T],\mathfrak{H}) \quad \text{as} \quad n_k\to\infty$$

$$\frac{\mathrm{d}}{\mathrm{d}t}f_{n_k}(\cdot)\rightharpoonup\frac{\mathrm{d}}{\mathrm{d}t}f^*(\cdot) \text{ weakly in } \mathcal{L}^1([0,T],\mathfrak{H}) \quad \text{as} \quad n_k\to\infty.$$

2.3.4 *Asymptotic compactness properties*

Some kind of compactness condition is required to ensure that the omega limit sets defining an attractor are non-empty. For a dynamical system on the finite dimensional state space this is easy since the compact subsets are the closed and bounded subsets. Then, e.g., for an autonomous dynamical system φ with a positively invariant, closed and bounded (hence compact) absorbing set Q, the attractor

$$\mathcal{A}=\bigcap_{t\geq 0}\varphi(t,Q)$$

is the non-empty intersection of the nested compact subsets $\varphi(s,Q)\subset\varphi(t,Q)\subset Q$ for $s>t$, since continuous functions map compact subsets onto compact subsets.

In infinitely dimensional state spaces, closed and bounded subsets need not be compact, so some compactness property must come from the dynamics. A simple property is that the mappings $\varphi(t,\cdot)$ are compact for $t>0$, i.e., map closed and bounded subsets of $\mathfrak{X}$ onto pre-compact subsets of $\mathfrak{X}$. This is usually too strong for most applications, so a weaker asymptotic compactness property is often used.

For specific examples of lattice systems, to show that the system is asymptotic compact, one usually first shows that the lattice dynamical system satisfies an *asymptotic tails property* inside an absorbing set which is positively invariant closed and bounded convex set (such as a ball). In particular, when the state space $\mathfrak{X}$ is a space of bi-infinite real-valued sequences such as ℓ^2 and the set Q is also convex, then it follows that φ is asymptotically compact in Q.

Similar proofs also hold in weighted Hilbert spaces of bi-infinite real-valued sequences such as ℓ^2_ρ.

Assumption 2.1. (Asymptotic tails property: autonomous systems) Let $\varphi=(\varphi_i)_{i\in\mathbb{Z}}$ be an autonomous semi-dynamical system on the Hilbert space $(\ell^2,\|\cdot\|)$ and let B be a positively invariant, closed and bounded subset of ℓ^2, which is φ-positive invariant. Then φ is said to satisfy an *asymptotic tails property* in B if for every $\varepsilon>0$ there exist $T(\varepsilon)>0$ and $I(\varepsilon)\in\mathbb{N}$ such that

$$\sum_{|i|>I(\varepsilon)}|\varphi_i(t,x_0)|^2\leq\varepsilon \quad \forall\, x_0\in B \text{ and } t\geq T(\varepsilon).$$

Lemma 2.5. *Let Assumption 2.1 hold. Then the semi-dynamical system φ is asymptotically compact in B.*

An analogous result also holds for pullback asymptotic compactness of processes and skew product flows. The proof of Lemma 2.5 follows as a simpler case of the proof for Lemma 2.6 below for processes.

Assumption 2.2. (Pullback asymptotic tails property for process) Let $\psi = (\psi_i)_{i\in\mathbb{Z}}$ be a process on the Hilbert space $(\ell^2, \|\cdot\|)$ and let $\mathcal{B} = \{B_t\}_{t\in\mathbb{R}}$ be ψ-positively invariant and consist of closed and bounded subsets of ℓ^2. Then ψ is said to satisfy a *pullback asymptotic tails property* in $\mathcal{B}$ if for every $t \in \mathbb{R}$ and $\varepsilon > 0$ there exist $T(t,\varepsilon) > 0$ and $I(t,\varepsilon) \in \mathbb{N}$ such that

$$\sum_{|i|>I(t,\varepsilon)} |\psi_i(t,t_0,x_0)|^2 \le \varepsilon, \quad \forall\, x_0 \in B_{t_0} \text{ and } t_0 \le t - T(t,\varepsilon).$$

Lemma 2.6. *Let Assumption 2.2 hold. Then the process ψ is pullback asymptotically compact in $\mathcal{B}$.*

Proof. We only need to show that every sequence $\boldsymbol{v}^{(n)} \in \psi(t,t-t_n,B_{t-t_n}) \subset B_t$ with $t_n \to \infty$ as $n \to \infty$ has a converging subsequence in ℓ^2.

For a sequence $\{t_n\}$ with $t_n \to \infty$ as $n \to \infty$, let $\boldsymbol{u}^{(n)} \in B_{t-t_n}$ and

$$\boldsymbol{v}^{(n)} = \psi(t,t-t_n,\boldsymbol{u}^{(n)}) \in B_t, \quad n = 1,2,\cdots.$$

Since B_t is non-empty, closed, and bounded in ℓ^2, it is weakly compact so there is a subsequence of $\{\boldsymbol{v}^{(n)}\}$ (still denoted by $\{\boldsymbol{v}^{(n)}\}$), and $\boldsymbol{v}^* \in B_t$ such that

$$\boldsymbol{v}^{(n)} = \psi(t,t-t_n,\boldsymbol{u}^{(n)}) \rightharpoonup \boldsymbol{v}^* \quad \text{(i.e., weakly in } \ell^2).$$

We now show that this weak convergence is actually strong. Given any $\varepsilon > 0$, by the Assumption 2.2, there exists $I_1(t,\varepsilon) > 0$ and $N_1(t,\varepsilon) > 0$ such that

$$\sum_{|i|\ge I_1(t,\varepsilon)} |\psi_i(t,t-t_n,\boldsymbol{u}_o)|^2 \le \frac{1}{8}\varepsilon, \quad \forall\, n \ge N_1(t,\varepsilon), \tag{2.1}$$

for every $\boldsymbol{u}_o \in B_{t-t_n}$.

Moreover, since $\boldsymbol{v}^* = (v_i^*)_{i\in\mathbb{Z}} \in \ell^2$, there exists an $I_2(\varepsilon) > 0$ such that

$$\sum_{|i|\ge I_2(\varepsilon)} |v_i^*|^2 \le \frac{\varepsilon}{8}. \tag{2.2}$$

Set $I(t,\varepsilon) := \max\{I_1(t,\varepsilon), I_2(\varepsilon)\}$. Since $\psi(t,t-t_n,\boldsymbol{u}^{(n)}) \rightharpoonup \boldsymbol{v}^*$ in ℓ^2, it follows component wise that

$$\psi_i(t,t-t_n,\boldsymbol{u}^{(n)}) \longrightarrow v_i^* \quad \text{for } |i| \le I(t,\varepsilon), \quad \text{as } n \to \infty.$$

Therefore there exists $N_2(t,\varepsilon) > 0$ such that

$$\sum_{|i|\le I(t,\varepsilon)} \left|\psi_i(t,t-t_n,\boldsymbol{u}^{(n)}) - v_i^*\right|^2 \le \frac{1}{2}\varepsilon, \quad \forall\, n \ge N_2(t,\epsilon). \tag{2.3}$$

Set $I(t,\varepsilon) := \max\{I_1(t,\varepsilon), I_2(t,\varepsilon)\}$. Then, using (2.1) – (2.3), for $n \geq I(t,\varepsilon)$ it follows that

$$\begin{aligned}\left\|\psi(t,t-t_n,\boldsymbol{u}^{(n)}) - \boldsymbol{v}^*\right\|^2 &= \sum_{|i|\leq I(t,\varepsilon)} \left|\psi_i(t,t-t_n,\boldsymbol{u}^{(n)}) - v_i^*\right|^2 \\ &\quad + \sum_{|i|>I(t,\varepsilon)} \left|\psi_i(t,t-t_n,\boldsymbol{u}^{(n)}) - v_i^*\right|^2 \\ &\leq \frac{1}{2}\varepsilon + 2\sum_{|i|>I(t,\epsilon)} \left|\psi_i(t,t-t_n,\boldsymbol{u}^{(n)})\right|^2 + |v_i^*|^2 \leq \varepsilon.\end{aligned}$$

Hence $\boldsymbol{v}^{(n)}$ (the subsequence) is strongly convergent in ℓ^2, so ψ is pullback asymptotic compact in $\mathcal{B}$. □

2.4 End notes

There are many classical monographs on autonomous dynamical systems, see, e.g., [Teschl (2012)]. See Mallet-Paret, Wu, Yi & Zhu, [Mallet-Paret *et al.* (2012)] and [Robinson (2001)] for infinite dimensional dynamical systems. For non-autonomous dynamical systems see [Sell (1971)], [Kloeden and Rasmussen (2011)], [Caraballo and Han (2016)], and [Kloeden and Yang (2021)].

Proofs of most of the results on non-autonomous systems stated in this chapter can be found in [Kloeden and Rasmussen (2011)] and [Kloeden and Yang (2021)], with the autonomous counterparts holding as special cases.

See [Ambrosio and Tilli (2004)] for general topics on analysis in metric spaces.

2.5 Problems

Problem 2.1. Consider the attractor A_p of the autonomous scalar ODE

$$\frac{\mathrm{d}x}{\mathrm{d}t} = -x\left(x^4 - 2x^2 + 1 - p\right)$$

with a parameter $p \in P = [-2,2]$. Determine the attractor A_p for each p. Then show that the attractors A_p converge upper semi-continuously to A_{p_0} as $p \to p_0$, but need not converge continuously (in the Hausdorff metric). What properties will ensure that the attractors converge continuously?

Problem 2.2. What is the exact relationship between the asymptotic tails property and total boundedness?

Problem 2.3. Describe the major differences between the process and skew-product flow formulations of non-autonomous dynamical systems. In what scenarios is one more convenient than the other?

PART 2

Laplacian LDS

Chapter 3

Lattice Laplacian models

A lattice reaction-diffusion model is a lattice dynamical system obtained by spatially discretising the Laplacian operator in a parabolic partial differential equation modelling a reaction-diffusion equation such as equation (1.1) in Chapter 1. Bates, Lu & Wang [Bates *et al.* (2001)] investigated dynamical behaviour of the lattice dynamical system (LDS) based on this reaction-diffusion equation and their results, which have profoundly influenced the development of the theory of dissipative LDS, will be presented here.

Consider the autonomous LDS

$$\frac{\mathrm{d}u_i}{\mathrm{d}t} = \nu\left(u_{i-1} - 2u_i + u_{i+1}\right) + f(u_i) + g_i, \quad i \in \mathbb{Z}, \tag{3.1}$$

in the space ℓ^2 (which was defined in Section 1.3), which will be investigated here under the following assumptions.

Assumption 3.1. The function $f : \mathbb{R} \to \mathbb{R}$ is a continuously differentiable function satisfying

$$f(s)s \leq -\alpha s^2 \qquad \forall\, s \in \mathbb{R},$$

for some $\alpha > 0$.

Assumption 3.2. The function $\boldsymbol{g} = (g_i)_{i\in\mathbb{Z}} \in \ell^2$.

Remark 3.1. Since f is smooth, the Assumption 3.1 implies that $f(0) = 0$.

3.1 The discrete Laplace operator

The Laplacian operator on one-dimensional spatial domain is just the second derivative. Using central difference quotient to approximate it leads to the one-dimensional discrete Laplace operator on an appropriate sequence space.

For any $\boldsymbol{u} = (u_i)_{i\in\mathbb{Z}} \in \ell^2$, the discrete Laplace operator Λ is defined from ℓ^2 to ℓ^2 component wise by

$$(\Lambda \boldsymbol{u})_i = u_{i-1} - 2u_i + u_{i+1}, \quad i \in \mathbb{Z}. \tag{3.2}$$

Define the bounded linear operators D^+ and D^- from ℓ^2 to ℓ^2 by

$$(\mathrm{D}^+\boldsymbol{u})_i = u_{i+1} - u_i, \qquad (\mathrm{D}^-\boldsymbol{u})_i = u_{i-1} - u_i. \qquad i \in \mathbb{Z}. \tag{3.3}$$

Note that $-\Lambda = \mathrm{D}^+\mathrm{D}^- = \mathrm{D}^-\mathrm{D}^+$ and

$$\langle \mathrm{D}^-\boldsymbol{u}, \boldsymbol{v}\rangle = \langle \boldsymbol{u}, \mathrm{D}^+\boldsymbol{v}\rangle \qquad \forall \boldsymbol{u}, \boldsymbol{v} \in \ell^2,$$

which imply that

$$\langle \Lambda\boldsymbol{u}, \boldsymbol{u}\rangle = -\|\mathrm{D}^+\boldsymbol{u}\|^2 \le 0 \qquad \forall\ \boldsymbol{u} \in \ell^2.$$

This means that the operator Λ is negative definite in ℓ^2, since $\|\mathrm{D}^+\boldsymbol{u}\| = 0$ implies that the components u_i are identical and hence zero in ℓ^2.

Moreover, since Λ is a bounded linear operator in ℓ^2, it generates a uniformly continuous semi-group in ℓ^2. Throughout the rest of this book, Λ is designated to denote the discrete Laplace operator defined in (3.2), unless otherwise specified.

3.2 The autonomous reaction-diffusion LDS

Under Assumptions 3.1 and 3.2 the LDS (3.1) can be written as the infinite dimensional ordinary differential equation (ODE) in the sequence space ℓ^2,

$$\frac{\mathrm{d}\boldsymbol{u}}{\mathrm{d}t} = \nu\Lambda\boldsymbol{u} + F(\boldsymbol{u}) + \boldsymbol{g}, \tag{3.4}$$

where $\boldsymbol{g} = (g_i)_{i\in\mathbb{Z}}$ and the Nemytskii operator $F : \ell^2 \to \ell^2$ is defined componentwise by

$$F(\boldsymbol{u}) = (f_i(\boldsymbol{u}))_{i\in\mathbb{Z}}, \qquad f_i(\boldsymbol{u}) = f(u_i).$$

Assumption 3.1 ensures that $F : \ell^2 \to \ell^2$ is Lipschitz continuous on bounded sets. Indeed, let $B \subset \ell^2$ be a bounded set contained in a ball $\mathscr{B}_{r_B}(0)$ with radius $r_B > 0$ and center 0 in ℓ^2. Then, since F is smooth, for every $\boldsymbol{u}, \boldsymbol{v} \in B$ we have

$$\|F(\boldsymbol{u}) - F(\boldsymbol{v})\|^2 = \sum_{i\in\mathbb{Z}} |f(u_i) - f(v_i)|^2 = \sum_{i\in\mathbb{Z}} |f'(w_i)|^2 |u_i - v_i|^2,$$

for some w_i with $|w_i| \le |u_i| + |v_i| \le \|\boldsymbol{u}\| + \|\boldsymbol{v}\| \le 2r_B$, $i \in \mathbb{Z}$. Hence,

$$\|F(\boldsymbol{u}) - F(\boldsymbol{v})\| \le L_{r_B}\|\boldsymbol{u} - \boldsymbol{v}\| \tag{3.5}$$

for some constant $L_{r_B} > 0$ depending on r_B. In particular, since $f(0) = 0$ implies that $F(\boldsymbol{0}) = \boldsymbol{0}$, estimate (3.5) with $\boldsymbol{v} = \boldsymbol{0}$ yields

$$\|F(\boldsymbol{u})\| \le \|F(\boldsymbol{u}) - F(\boldsymbol{0})\| + \|F(\boldsymbol{0})\| \le L_{r_B}\|\boldsymbol{u}\|,$$

which shows that $F(\boldsymbol{u}) \in \ell^2$ for every $\boldsymbol{u} \in \ell^2$.

For given initial data $\boldsymbol{u}(0) = \boldsymbol{u}_o \in \ell^2$, the existence and uniqueness of a global solution $\boldsymbol{u}(\,\cdot\,; \boldsymbol{u}_o) \in \mathcal{C}([0;\infty), \ell^2)$ of the ODE system (3.4) follows by standard arguments (e.g., see [Bates *et al.* (2001); Deimling (1977)]). Moreover, the lattice model

(3.1) generates an autonomous semi-dynamical system $\{\varphi(t)\}_{t\geq 0}: \ell^2 \to \ell^2$ defined by

$$\varphi(t, \boldsymbol{u}_o) := \boldsymbol{u}(t; \boldsymbol{u}_o), \qquad t \geq 0,\ \boldsymbol{u}_o \in \ell^2.$$

Theorem 3.1. *Suppose that Assumptions 3.1 and 3.2 are satisfied. The autonomous semi-dynamical system $\{\varphi(t)\}_{t\geq 0}$ generated by the ODE (3.4) on ℓ^2 has a global attractor $\mathcal{A}$ in ℓ^2.*

The proof is given in the following two subsections.

3.2.1 *Existence of an absorbing set*

Taking the inner product of equation (3.4) with $\boldsymbol{u}$ and using Assumption 3.1 gives

$$\begin{aligned}\frac{\mathrm{d}}{\mathrm{d}t}\|\boldsymbol{u}\|^2 &= 2\nu\langle \Lambda \boldsymbol{u}, \boldsymbol{u}\rangle + 2\langle F(\boldsymbol{u}), \boldsymbol{u}\rangle + 2\langle \boldsymbol{g}, \boldsymbol{u}\rangle \\ &\leq 2\sum_{i\in\mathbb{Z}} u_i f(u_i) + 2\sum_{i\in\mathbb{Z}} g_i u_i \leq -\alpha\|\boldsymbol{u}\|^2 + \frac{\|\boldsymbol{g}\|^2}{\alpha},\end{aligned} \tag{3.6}$$

where the last step follows from Young's inequality. Hence, Gronwall's lemma implies that

$$\|\boldsymbol{u}(t)\|^2 \leq \|\boldsymbol{u}_o\|^2 e^{-\alpha t} + \frac{\|\boldsymbol{g}\|^2}{\alpha^2}\left(1 - e^{-\alpha t}\right), \qquad t \geq 0. \tag{3.7}$$

Define the closed ball Q in ℓ^2 by

$$Q := \left\{\boldsymbol{u} \in \ell^2 : \|\boldsymbol{u}\|^2 \leq R^2 := 1 + \frac{\|\boldsymbol{g}\|^2}{\alpha^2}\right\}.$$

The estimate (3.7) then implies that Q is an absorbing set for the autonomous semi-dynamical system φ. In fact, for any $\boldsymbol{u}_o \in B$, which is a bounded set in ℓ^2, it is straightforward to check that

$$\varphi(t, B) \subset Q \qquad \forall t \geq \frac{2}{\alpha}\ln\|B\|,$$

where $\|B\| := \sup_{\boldsymbol{u}\in B}\|\boldsymbol{u}\|$.

Moreover, for every $\boldsymbol{u}_o \in Q$ by estimate (3.7) we have

$$\|\varphi(t, \boldsymbol{u}_o)\|^2 \leq R^2 e^{-\alpha t} + R^2\left(1 - e^{-\alpha t}\right) = R^2,$$

i.e., Q is positive invariant.

3.2.2 *Asymptotic tails property*

In order to establish the asymptotic compactness property in Assumption 2.1 for the autonomous semi-dynamical system $\{\varphi(t)\}_{t\geq 0}$ in ℓ^2, the asymptotic tails estimate needs to be shown to hold.

Consider a smooth function $\xi : \mathbb{R}^+ \to [0,1]$ satisfying

$$\xi(s) = \begin{cases} 0, & 0 \le s \le 1, \\ \in [0,1], & 1 \le s \le 2, \\ 1, & s \ge 2 \end{cases} \tag{3.8}$$

and note that there exists a constant C_0 such that $|\xi'(s)| \le C_0$ for all $s \ge 0$. Notice that this function ξ or a similar function will be used repeatedly throughout this book. Then for a fixed $k \in \mathbb{N}$ (its value will be specified later), define

$$\xi_k(s) = \xi\left(\frac{s}{k}\right) \quad \text{for all} \quad s \in \mathbb{R}.$$

Given $\boldsymbol{u} \in \ell^2$, define $\boldsymbol{v} \in \ell^2$ component wise as

$$v_i := \xi_k(|i|)u_i \quad i \in \mathbb{Z}.$$

Taking the inner product of equation (3.4) with $\boldsymbol{v}$ gives

$$\frac{\mathrm{d}}{\mathrm{d}t}\langle \boldsymbol{u}, \boldsymbol{v}\rangle + \nu\langle \mathrm{D}^+\boldsymbol{u}, \mathrm{D}^+\boldsymbol{v}\rangle = \langle F(\boldsymbol{u}), \boldsymbol{v}\rangle + \langle \boldsymbol{g}, \boldsymbol{v}\rangle,$$

that is

$$\frac{\mathrm{d}}{\mathrm{d}t}\sum_{i\in\mathbb{Z}} \xi_k(|i|)|u_i|^2 + 2\nu\langle \mathrm{D}^+\boldsymbol{u}, \mathrm{D}^+\boldsymbol{v}\rangle = 2\sum_{i\in\mathbb{Z}} \xi_k(|i|)u_i f(u_i) + 2\sum_{i\in\mathbb{Z}} \xi_k(|i|)g_i u_i. \tag{3.9}$$

Each term in equation (3.9) will now be estimated. First,

$$\begin{aligned}
\langle \mathrm{D}^+\boldsymbol{u}, \mathrm{D}^+\boldsymbol{v}\rangle &= \sum_{i\in\mathbb{Z}} (u_{i+1} - u_i)(v_{i+1} - v_i) \\
&= \sum_{i\in\mathbb{Z}} (u_{i+1} - u_i)\left[(\xi_k(|i+1|) - \xi_k(|i|))u_{i+1} + \xi_k(|i|)(u_{i+1} - u_i)\right] \\
&= \sum_{i\in\mathbb{Z}} (\xi_k(|i+1|) - \xi_k(|i|))(u_{i+1} - u_i)u_{i+1} + \sum_{i\in\mathbb{Z}} \xi_k(|i|)(u_{i+1} - u_i)^2 \\
&\ge \sum_{i\in\mathbb{Z}} (\xi_k(|i+1|) - \xi_k(|i|))(u_{i+1} - u_i)u_{i+1}.
\end{aligned}$$

Since

$$\left|\sum_{i\in\mathbb{Z}} (\xi_k(|i+1|) - \xi_k(|i|))(u_{i+1} - u_i)u_{i+1}\right| \le \sum_{i\in\mathbb{Z}} \frac{1}{k}|\xi'(s_i)| \cdot |u_{i+1} - u_i| \cdot |u_{i+1}|,$$

for some s_i between $|i|$ and $|i+1|$, and

$$\sum_{i\in\mathbb{Z}} |\xi'(s_i)||u_{i+1} - u_i||u_{i+1}| \le C_0 \sum_{i\in\mathbb{Z}} \left(|u_{i+1}|^2 + |u_i||u_{i+1}|\right) \le 4C_0\|\boldsymbol{u}\|^2.$$

Then it follows that for all $\boldsymbol{u} \in Q$ and $\boldsymbol{v} \in \ell^2$ defined component wise as $v_i := \xi_k(|i|)u_i$ for $i \in \mathbb{Z}$

$$\langle \mathrm{D}^+\boldsymbol{u}, \mathrm{D}^+\boldsymbol{v}\rangle \ge -\frac{4C_0\|Q\|^2}{k} \tag{3.10}$$

where $\|Q\| := \sup_{\boldsymbol{u}\in Q} \|\boldsymbol{u}\|$.

On the other hand, by Assumption 3.1,

$$2\sum_{i\in\mathbb{Z}} \xi_k(|i|)u_i f(u_i) \le -2\alpha \sum_{i\in\mathbb{Z}} \xi_k(|i|)|u_i|^2$$

and by Young's inequality

$$2\sum_{i\in\mathbb{Z}} \xi_k(|i|)g_i u_i \le \alpha \sum_{i\in\mathbb{Z}} \xi_k(|i|)|u_i|^2 + \frac{1}{\alpha}\sum_{i\in\mathbb{Z}} \xi_k(|i|)|g_i|^2.$$

Thus

$$2\sum_{i\in\mathbb{Z}} \xi_k(|i|)u_i f(u_i) + 2\sum_{i\in\mathbb{Z}} \xi_k(|i|)g_i u_i \le -\alpha \sum_{i\in\mathbb{Z}} \xi_k(|i|)|u_i|^2 + \frac{1}{\alpha}\sum_{|i|\ge k} |g_i|^2. \quad (3.11)$$

Using the estimates (3.10) and (3.11) in equation (3.9) gives

$$\frac{\mathrm{d}}{\mathrm{d}t}\sum_{i\in\mathbb{Z}} \xi_k(|i|)|u_i|^2 + \alpha \sum_{i\in\mathbb{Z}} \xi_k(|i|)|u_i|^2 \le \nu\frac{4C_0\|Q\|^2}{k} + \frac{1}{\alpha}\sum_{|i|\ge k} |g_i|^2. \quad (3.12)$$

Because $\boldsymbol{g} \in \ell^2$ for every $\varepsilon > 0$, there exists $K(\varepsilon)$ such that

$$\nu\frac{4C_0\|Q\|^2}{k} + \frac{1}{\alpha}\sum_{|i|\ge k} |g_i|^2 \le \varepsilon, \quad k \ge K(\varepsilon).$$

The inequality (3.12) along with the relation above give

$$\frac{\mathrm{d}}{\mathrm{d}t}\sum_{i\in\mathbb{Z}} \xi_k(|i|)|u_i|^2 + \alpha \sum_{i\in\mathbb{Z}} \xi_k(|i|)|u_i|^2 \le \varepsilon.$$

Then, Gronwall's lemma implies that

$$\sum_{i\in\mathbb{Z}} \xi_k(|i|)|u_i(t,\boldsymbol{u}_o)|^2 \le e^{-\alpha t}\sum_{i\in\mathbb{Z}} \xi_k(|i|)|u_{o,i}|^2 + \frac{\varepsilon}{\alpha} \le e^{-\alpha t}\|\boldsymbol{u}_o\|^2 + \frac{\varepsilon}{\alpha}.$$

Hence for every $\boldsymbol{u}_o \in Q$,

$$\sum_{i\in\mathbb{Z}} \xi_k(|i|)|u_i(t,\boldsymbol{u}_o)|^2 \le e^{-\alpha t}\|Q\|^2 + \frac{\varepsilon}{\alpha},$$

and therefore

$$\sum_{i\in\mathbb{Z}} \xi_k(|i|)|u_i(t,\boldsymbol{u}_o)|^2 \le \frac{2\varepsilon}{\alpha}, \quad \text{for } t \ge T(\varepsilon) := \frac{1}{\alpha}\ln\frac{\alpha\|Q\|^2}{\varepsilon}.$$

This is the desired asymptotic tails property. The asymptotic compactness in ℓ^2 of φ in the absorbing set Q then follows from Lemma 2.5. This completes the proof of Theorem 3.1.

3.3 Nonautonomous lattice reaction-diffusion LDS

Time variable forcing terms $g_i : \mathbb{R} \to \mathbb{R}$, $i \in \mathbb{Z}$ in the LDS model (3.1) lead to the nonautonomous LDS

$$\frac{\mathrm{d}u_i}{\mathrm{d}t} = \nu\left(u_{i-1} - 2u_i + u_{i+1}\right) - \lambda u_i + f(u_i) + g_i(t) \quad i \in \mathbb{Z}, \quad (3.13)$$

in the space ℓ^2.

As before, the term f is assumed to satisfy Assumption 3.1, but now Assumption 3.2 is replaced by the following:

Assumption 3.3. There exists $M_g > 0$ such that $\boldsymbol{g} := (g_i)_{i\in\mathbb{Z}} \in \mathcal{C}(\mathbb{R}, \ell^2)$ satisfies

$$e^{-\alpha t}\int_{-\infty}^{t} e^{\alpha s}\|\boldsymbol{g}(s)\|^2\,\mathrm{d}s \le M_g \quad \text{for all } t \in \mathbb{R}.$$

The analysis is similar to the autonomous case above, except it now generates a process $\{\psi(t,t_0)\}_{(t,t_0)\in\mathbb{R}^2_{\geq}}$ and the estimates involve the time-dependent norm $\|\boldsymbol{g}(t)\|$. More precisely, the differential inequality (3.6) becomes

$$\frac{\mathrm{d}}{\mathrm{d}t}\|\boldsymbol{u}\|^2 \le -\alpha\|\boldsymbol{u}\|^2 + \frac{1}{\alpha}\|\boldsymbol{g}(t)\|^2,$$

and Gronwall's Lemma implies that

$$\|\boldsymbol{u}(t;t_0,\boldsymbol{u}_o)\|^2 \le \|\boldsymbol{u}_o\|^2 e^{-\alpha(t-t_0)} + \frac{1}{\alpha}e^{-\alpha t}\int_{t_0}^{t} e^{\alpha s}\|\boldsymbol{g}(s)\|^2\,ds. \tag{3.14}$$

Define the closed ball Q in ℓ^2 by

$$Q = \left\{\boldsymbol{u} \in \ell^2 : \|\boldsymbol{u}\|^2 \le 1 + \frac{1}{\alpha}M_g := R^2\right\}.$$

The estimate (3.14) then implies that the closed and bounded subset Q is pullback absorbing for the process $\{\psi(t,t_0)\}_{(t,t_0)\in\mathbb{R}^2_{\geq}}$. Notice that here the absorbing family $\mathcal{Q} = (Q_t)_{t\in\mathbb{R}}$ consists of identical sets $Q_t \equiv Q$. In fact, for any $\boldsymbol{u}_o \in B_{t_0}$ for a tempered family $\mathcal{B} = (B_t)_{t\in\mathbb{R}}$ of bounded sets of ℓ^2, it follows from (3.14) that

$$\psi(t,t_0,B_{t_0}) \subset Q \quad \forall t_0 \le t - \frac{2}{\alpha}\ln\|B_{t_0}\|,$$

since then $\|B_{t_0}\|^2 e^{-\alpha(t-t_0)} \le 1$.

Moreover, for every $\boldsymbol{u}_o \in Q$ by estimate (3.7)

$$\begin{aligned}
\|\psi(t,t_0,\boldsymbol{u}_o)\|^2 &\le R^2 e^{-\alpha(t-t_0)} + \frac{1}{\alpha}e^{-\alpha t}\int_{t_0}^{t} e^{\alpha s}\|\boldsymbol{g}(s)\|^2\,\mathrm{d}s\\
&\le \left(1 + \frac{1}{\alpha}e^{-\alpha t_0}\int_{-\infty}^{t_0} e^{\alpha s}\|\boldsymbol{g}(s)\|^2\,\mathrm{d}s\right)e^{-\alpha(t-t_0)}\\
&\quad + \frac{1}{\alpha}e^{-\alpha t}\int_{t_0}^{t} e^{\alpha s}\|\boldsymbol{g}(s)\|^2\,\mathrm{d}s\\
&= e^{-\alpha(t-t_0)} + \frac{1}{\alpha}e^{-\alpha t}\int_{-\infty}^{t_0} e^{\alpha s}\|\boldsymbol{g}(s)\|^2\,\mathrm{d}s + \frac{1}{\alpha}e^{-\alpha t}\int_{t_0}^{t} e^{\alpha s}\|\boldsymbol{g}(s)\|^2\,\mathrm{d}s\\
&\le 1 + \frac{1}{\alpha}e^{-\alpha t}\int_{-\infty}^{t} e^{\alpha s}\|\boldsymbol{g}(s)\|^2\,\mathrm{d}s \le 1 + \frac{1}{\alpha}M_g := R^2
\end{aligned}$$

i.e., Q is ψ-positive invariant.

The proof that the process ψ has the asymptotic tails property is similar to the autonomous case above, except equation (3.12) is now

$$\frac{\mathrm{d}}{\mathrm{d}t}\sum_{i\in\mathbb{Z}}\xi_k(|i|)|u_i|^2+\alpha\sum_{i\in\mathbb{Z}}\xi_k(|i|)|u_i|^2\le\nu\frac{4C_0\|Q\|^2}{k}+\frac{1}{\alpha}\sum_{|i|\ge k}|g_i(t)|^2,$$

where $\|Q\|:=\sup_{\boldsymbol{u}\in Q}\|\boldsymbol{u}\|\le|R|$ due to Assumption 3.3, and ξ is the smooth cut-off function defined in (3.8). Integrating the above differential inequality gives

$$\begin{aligned}\sum_{i\in\mathbb{Z}}\xi_k(|i|)|u_i(t;t_0,\boldsymbol{u}_o)|^2\le e^{-\alpha(t-t_0)}&\sum_{i\in\mathbb{Z}}\xi_k(|i|)|u_{o,i}|^2\\ &+2\nu\frac{C_0\|Q\|^2}{\alpha k}+\frac{1}{\alpha}e^{-\alpha t}\int_{t_0}^{t}e^{\alpha s}\sum_{|i|\ge k}|g_i(t)|^2\,\mathrm{d}s.\end{aligned}$$

By Assumption 3.3 it follows that

$$\left(\left(e^{-\alpha t}\int_{-\infty}^{t}e^{\alpha s}g_i(s)^2\,\mathrm{d}s\right)^{\frac{1}{2}}\right)_{i\in\mathbb{Z}}\in\ell^2\quad\text{for each}\quad t\in\mathbb{R}.$$

Hence for every $\varepsilon>0$ and $t\in\mathbb{R}$ there exists $K(\varepsilon,t)$ such that

$$2\nu\frac{C_0\|Q\|^2}{\alpha k}+\frac{1}{\alpha}e^{-\alpha t}\int_{-\infty}^{t}e^{\alpha s}\sum_{|i|\ge k}|g_i(s)|^2\,\mathrm{d}s\le\varepsilon,\quad k\ge K(\varepsilon,t),$$

and consequently for every $\boldsymbol{u}_o\in Q$,

$$\sum_{i\in\mathbb{Z}}\xi_k(|i|)|u_i(t;t_0,\boldsymbol{u}_o)|^2\le e^{-\alpha(t-t_0)}\sum_{i\in\mathbb{Z}}\xi_k(|i|)|u_{o,i}|^2+\varepsilon\le e^{-\alpha(t-t_0)}\|Q\|^2+\varepsilon.$$

This implies that $\sum_{i\in\mathbb{Z}}\xi_k(|i|)|u_i(t;t_0,\boldsymbol{u}_o)|^2<\varepsilon$ for $t-t_0$ large enough, which is the desired asymptotic tails property. The asymptotic compactness in ℓ^2 of the process ψ in the pullback absorbing set Q, then follows by Lemma 2.6. Thus the following result holds.

Theorem 3.2. *Suppose that Assumptions 3.1 and 3.3 are satisfied. The process* $\{\psi(t,t_0)\}_{(t,t_0)\in\mathbb{R}^2_{\ge}}$ *generated by the ODE* (3.13) *on* ℓ^2 *has a pullback attractor* $\mathcal{A}=(A_t)_{t\in\mathbb{R}}$ *in* ℓ^2, *which pullback attracts tempered families of bounded subsets of* ℓ^2.

3.4 p-Laplacian reaction-diffusion LDS

p-Laplacian reaction-diffusion equations have been used to model a variety of physical phenomena, such as the motion of non-Newtonian fluid, and the flow in porous media and nonlinear elasticity. A representative one-dimensional p-Laplacian reaction-diffusion equation with $p\ge 2$ is

$$\frac{\partial u}{\partial t}=\frac{\partial}{\partial x}\left(\left|\frac{\partial u}{\partial x}\right|^{p-2}\frac{\partial u}{\partial x}\right)+f(u)+g(x),\qquad x\in\mathbb{R}.$$

Lattice models based on p-Laplacian reaction-diffusion equations were investigated by [Gu and Kloeden (2016)]. A simpler autonomous p-Laplacian lattice model will be considered here:

$$\frac{\mathrm{d}u_i(t)}{\mathrm{d}t} = |u_{i+1} - u_i|^{p-2}(u_{i+1} - u_i) - |u_i - u_{i-1}|^{p-2}(u_i - u_{i-1}) + f(u_i) + g_i, \; i \in \mathbb{Z}. \tag{3.15}$$

It is assumed here that f satisfies Assumption 3.1 and $\boldsymbol{g} = (g_i)_{i\in\mathbb{Z}}$ satisfies Assumption 3.2.

3.4.1 *Discretised p-Laplacian*

The discrete p-Laplacian operator Λ_p is defined componentwise as

$$(\Lambda_p \boldsymbol{u})_i = |u_{i+1} - u_i|^{p-2}(u_{i+1} - u_i) - |u_i - u_{i-1}|^{p-2}(u_i - u_{i-1}),$$

for $\boldsymbol{u} = (u_i)_{i\in\mathbb{Z}}$. It is straightforward to check that

$$\Lambda_p \boldsymbol{u} = -\mathrm{D}^-\left(|(\mathrm{D}^+\boldsymbol{u})|^{p-2} \odot (\mathrm{D}^+\boldsymbol{u})\right)$$

where $a \odot b := (a_i b_i)_{i\in\mathbb{Z}}$, and D^+ and D^- are as defined in (3.3).

Moreover, for $\boldsymbol{u} = (u_i)_{i\in\mathbb{Z}}, \boldsymbol{v} = (v_i)_{i\in\mathbb{Z}} \in \ell^2$,

$$-\langle \Lambda_p \boldsymbol{u}, \boldsymbol{v}\rangle = \langle \mathrm{D}^-\left(|\mathrm{D}^+\boldsymbol{u}|^{p-2} \odot (\mathrm{D}^+\boldsymbol{u})\right), \boldsymbol{v}\rangle = \langle |\mathrm{D}^+\boldsymbol{u}|^{p-2} \odot (\mathrm{D}^+\boldsymbol{u}), \mathrm{D}^+\boldsymbol{v}\rangle. \tag{3.16}$$

In particular,

$$-\langle \Lambda_p \boldsymbol{u}, \boldsymbol{u}\rangle = \langle |\mathrm{D}^+\boldsymbol{u}|^{p-2} \odot (\mathrm{D}^+\boldsymbol{u}), \mathrm{D}^+\boldsymbol{u}\rangle = \|(\mathrm{D}^+\boldsymbol{u})\|_p^p \le 2^p\|\boldsymbol{u}\|^p,$$

which implies that

$$-2^p\|\boldsymbol{u}\|^p \le \langle \Lambda_p \boldsymbol{u}, \boldsymbol{u}\rangle \le 0. \tag{3.17}$$

Thus Λ_p is a bounded negative definite operator from ℓ^2 to ℓ^2.

Furthermore, the nonlinear discrete p-Laplacian operator Λ_p is also locally Lipschitz on ℓ^2, since

$$\begin{aligned}
\|\Lambda_p \boldsymbol{u} - \Lambda_p \boldsymbol{v}\|^2 &= \sum_{i\in\mathbb{Z}} ((\Lambda_p \boldsymbol{u})_i - (\Lambda_p \boldsymbol{v})_i)^2 \\
&= \sum_{i\in\mathbb{Z}} (\mathrm{D}^-(|(\mathrm{D}^+\boldsymbol{u})_i|^{p-2} \odot (\mathrm{D}^+\boldsymbol{u})_i) - \mathrm{D}^-(|(\mathrm{D}^+\boldsymbol{v})_i|^{p-2} \odot (\mathrm{D}^+\boldsymbol{v})_i))^2 \\
&\le C(\|\boldsymbol{u}\| + \|\boldsymbol{v}\|)^{2p-4}\|\boldsymbol{u} - \boldsymbol{v}\|^2.
\end{aligned}$$

3.4.2 *Existence and uniqueness of solutions*

Recall that under the Assumption 3.1 the mapping $F : \ell^2 \to \ell^2$ defined componentwise by

$$F(\boldsymbol{u}) = (F_i(\boldsymbol{u}))_{i\in\mathbb{Z}} := (f(u_i))_{i\in\mathbb{Z}}$$

is locally Lipschitz and satisfies the following dissipativity condition on ℓ^2:

$$\langle F(\boldsymbol{u}), \boldsymbol{u}\rangle \le -\alpha\|\boldsymbol{u}\|^2. \tag{3.18}$$

The LDS (3.15) can be formulated as an ODE on ℓ^2,

$$\frac{\mathrm{d}\boldsymbol{u}}{\mathrm{d}t} = \Lambda_p \boldsymbol{u} + F(\boldsymbol{u}) + \boldsymbol{g}. \tag{3.19}$$

The vector field of (3.19) is locally Lipschitz, and this ensures the existence and uniqueness of a local solution of the lattice ODE (3.19), which is continuous in its initial value.

Taking the inner product of (3.19) with $\boldsymbol{u} \in \ell^2$ and taking into account (3.17) and (3.18) leads to

$$\frac{\mathrm{d}}{\mathrm{d}t}\|\boldsymbol{u}\|^2 \le -\alpha\|\boldsymbol{u}\|^2 + \frac{4\|\boldsymbol{g}\|^2}{\alpha}, \tag{3.20}$$

which implies that the solutions cannot explode at any finite time, and thus exist globally.

Theorem 3.3. *Suppose that Assumptions 3.1 and 3.2 are satisfied. Then the lattice ODE* (3.19) *generates an autonomous semi-dynamical system* $\{\varphi(t)\}_{t\ge0}$ *on* ℓ^2.

3.4.3 *Existence of a global attractor*

Theorem 3.4. *Under Assumptions 3.1 and 3.2 the autonomous semi-dynamical system* $\{\varphi(t)\}_{t\ge0}$ *on* ℓ^2 *has a global attractor in* ℓ^2.

We will first show that there exists a positive invariant closed and bounded absorbing set Q and then verify that φ satisfies the asymptotic tails property in Q, which gives the required asymptotic compactness of Q.

3.4.3.1 *Existence of an absorbing set*

Applying Gronwall's lemma to the estimate (3.20) and integrating from 0 to t gives

$$\|\varphi(t, \boldsymbol{u}_o)\|^2 = \|\boldsymbol{u}(t; \boldsymbol{u}_o)\|^2 \le e^{-\alpha t}\|\boldsymbol{u}_o\|^2 + \frac{4\|\boldsymbol{g}\|^2}{\alpha^2}\left(1 - e^{-\alpha t}\right). \tag{3.21}$$

It follows that the closed and bounded set

$$Q := \left\{\boldsymbol{u} \in \ell^2 : \|\boldsymbol{u}\|^2 \le R^2 := 1 + \frac{4\|\boldsymbol{g}\|^2}{\alpha^2}\right\}$$

is an absorbing set for φ in ℓ^2. Moreover, it is positive invariant since for $\boldsymbol{u}_o \in Q$ the estimate (3.21) gives

$$\begin{aligned}\|\varphi(t,\boldsymbol{u}_o)\|^2 &\le e^{-\alpha t}\|\boldsymbol{u}_o\|^2 + \frac{4\|\boldsymbol{g}\|^2}{\alpha^2}\left(1-e^{-\alpha t}\right)\\ &\le e^{-\alpha t}R^2 + R^2\left(1-e^{-\alpha t}\right) = R^2, \quad \forall\, t \ge 0.\end{aligned}$$

Notice that for $p > 2$, $\|\boldsymbol{u}\|_p \le \|\boldsymbol{u}\|$ for every $\boldsymbol{u} \in \ell^2 \subset \ell^p$, and hence

$$\|\varphi(t,\boldsymbol{u}_o)\|_p^p \le \|\varphi(t,\boldsymbol{u}_o)\|^p \le |R|^p \quad \text{for } t \ge 0, \quad \boldsymbol{u}_o \in Q. \tag{3.22}$$

3.4.3.2 *Asymptotic tails property*

Let ξ be the cut-off function defined in (3.8) with k to be determined later. For $\boldsymbol{u} \in \ell^2$, define $\boldsymbol{v} \in \ell^2$ component wise as $v_i := \xi_k(|i|)u_i$ for $i \in \mathbb{Z}$. Taking the inner product of (3.15) with $\boldsymbol{v}$ in ℓ^2 gives

$$\frac{\mathrm{d}}{\mathrm{d}t}\sum_{i\in\mathbb{Z}}\xi_k(|i|)u_i^2 = 2\langle\Lambda_p\boldsymbol{u},\boldsymbol{v}\rangle + 2\langle F(\boldsymbol{u}),\boldsymbol{v}\rangle + 2\langle \boldsymbol{g},\boldsymbol{v}\rangle. \tag{3.23}$$

Now due to (3.16)

$$\begin{aligned}\langle\Lambda_p\boldsymbol{u},\boldsymbol{v}\rangle &= -\langle|\mathrm{D}^+\boldsymbol{u}|^{p-2}\odot(\mathrm{D}^+\boldsymbol{u}),\mathrm{D}^+\boldsymbol{v})\rangle\\ &\le -\sum_{i\in\mathbb{Z}}(\xi_k(|i+1|)-\xi_k(|i|))|u_{i+1}-u_i|^{p-2}(u_{i+1}-u_i)u_{i+1},\end{aligned}$$

whence

$$\begin{aligned}&\left|\sum_{i\in\mathbb{Z}}\xi_k(|i+1|)-\xi_k(|i|)|u_{i+1}-u_i|^{p-2}(u_{i+1}-u_i)u_{i+1}\right|\\ &\le \frac{C_0}{k}\sum_{i\in,\mathbb{Z}}|u_{i+1}-u_i|^{p-1}|u_{i+1}| \le \frac{C}{k}\|\boldsymbol{u}\|_p^p,\end{aligned} \tag{3.24}$$

for some positive constant C. Also, by Assumptions 3.1 and 3.2

$$2\langle F(\boldsymbol{u}),\boldsymbol{v}\rangle + 2\langle\boldsymbol{g},\boldsymbol{v}\rangle \le -\alpha\|\boldsymbol{v}\|^2 + \frac{4}{\alpha}\sum_{|i|\ge k}\xi_k(|i|)g_i^2. \tag{3.25}$$

Collecting (3.23)–(3.25) gives

$$\frac{\mathrm{d}}{\mathrm{d}t}\sum_{i\in\mathbb{Z}}\xi_k(|i|)u_i^2 + \alpha\sum_{i\in\mathbb{Z}}\xi_k(|i|)u_i^2 \le \frac{C}{k}\|\boldsymbol{u}\|_p^p + \frac{4}{\alpha}\sum_{|i|\ge k}\xi_k(|i|)g_i^2. \tag{3.26}$$

By (3.22) and the fact that $\boldsymbol{g} \in \ell^2$ for $\varepsilon > 0$, there exists $K(\varepsilon)$ such that

$$\frac{C}{k}|R|^p + \frac{4}{\alpha}\sum_{|i|\ge k}g_i \le \frac{\alpha}{2}\varepsilon, \quad k \ge K(\varepsilon).$$

Then applying Gronwall's lemma to (3.26) gives

$$\sum_{i\in\mathbb{Z}} \xi_k(|i|)u_i^2(t) \le e^{-\alpha t}\sum_{i\in\mathbb{Z}} \xi_k(|i|)u_{o,i}^2 + \frac{\varepsilon}{2}$$
$$\le e^{-\alpha t}\|\boldsymbol{u}_o\|^2 + \frac{\varepsilon}{2} \le e^{-\alpha t}R^2 + \frac{\varepsilon}{2} \le \varepsilon,$$

for $\boldsymbol{u}_o \in Q$ and $t \ge T(\varepsilon) := \frac{2}{\alpha}\ln\frac{2|R|}{\varepsilon}$. Consequently,

$$\sum_{|i|\ge 2k} |\varphi(t,\boldsymbol{u}_o)_i|^2 \le \varepsilon, \quad t \ge T(\varepsilon), k \ge K(\varepsilon),$$

which completes the proof of the asymptotic tails property. The asymptotic compactness of the solution mapping in Q then follows by Lemma 2.5. This completes the proof of Theorem 3.4.

Remark 3.2. It would be interesting to extend the above results to apply in the larger space ℓ^p rather than in ℓ^2.

3.4.3.3 *Variable exponent Laplace operators*

Evolution equations with the $p(x)$-Laplacian operator

$$\operatorname{div}\left(|\nabla u|^{p(x)-2}\nabla u\right)$$

have been extensively investigated in the literature, see e.g., [Kloeden and Simsen (2014, 2015)].

Applying the chain rule and the finite difference quotient to the one-dimensional $p(x)$-Laplacian operator gives the one-dimensional discretised $\boldsymbol{p}$-Laplacian operator $\Lambda_{\boldsymbol{p}}$ with variable exponents defined component wise by :

$$(\Lambda_{\boldsymbol{p}}\boldsymbol{u})_i := |\mathrm{D}^- u_i|^{p_i-2}\left[(\mathrm{D}^+p_i)(\mathrm{D}^-u_i)\ln|\mathrm{D}^-u_i| + (p_i-1)\mathrm{D}^+\mathrm{D}^-u_i\right],$$

for variable exponents $\boldsymbol{p} = \{p_i : i \in \mathbb{Z}\}$, where $p_i = p(x_i) \ge 2$, and D^+ and D^- are defined as in (3.3).

To date there has been no investigation of the long term dynamics of lattice dynamical systems with the leading operator $\Lambda_{\boldsymbol{p}}$. This will require the variable exponent bi-infinite real valued sequence spaces

$$\ell^{\boldsymbol{p}} := \left\{\boldsymbol{u} = (u_i)_{i\in\mathbb{Z}} : \sum_{i\in\mathbb{Z}} |u_i|^{p_i} < \infty\right\},$$

which are discrete Musielak-Orlicz spaces of real valued bi-infinite sequence equiped with a norm

$$\|\boldsymbol{u}\|_{\boldsymbol{p}} := \inf\left\{\nu > 0 : \varrho\left(\frac{\boldsymbol{u}}{\nu}\right) \le 1\right\}$$

induced by a semi-modular

$$\varrho(\boldsymbol{u}) := \sum_{i\in\mathbb{Z}} |u_i|^{p_i}.$$

The properties of these spaces are given in Han, Kloeden & Simsen [Han *et al.* (2019)].

3.5 End notes

The results in Section 3.2 in the autonomous case are based on the seminal paper of Bates, Lu & Wang [Bates *et al.* (2001)], while the nonautonomous case in Section 3.3 was first considered by B. Wang [Wang (2007a)]. An alternative compactness argument via the quasi-stability concept of [Chueshov (2015)] was used by [Czaja (2022)].

Some early papers on Laplacian LDS are [Abdallah (2005); Beyn and Pilyugin (2003); Oliveira and Pereira (2010); Van Vleck and Wang (2005); Wang (2006); Zhou (2003, 2004)]. Some other papers on non-autonomous LDS include [Abdallah (2010); Zhao and Zhou (2008, 2009); Zhou and Shi (2006); Zhou *et al.* (2007, 2008); Zhu *et al.* (2022)].

The results for lattice models based on p-Laplacian reaction-diffusion equations in Section 3.4 are due to [Gu and Kloeden (2016)]. Properties of the sequence spaces with variable exponents are investigated in Han, Kloeden & Simsen [Han *et al.* (2019)].

See [Chen and Wang (2022)] for lattice systems with a lattice fractional Laplacian operator.

3.6 Problems

Problem 3.1. Show that the lattice system (3.15) with discretised p-Laplacian has an attractor in the sequence space ℓ^p for $p > 2$.

Problem 3.2. Is ℓ^2 dense in $\ell^{\boldsymbol{p}}$ for variable exponents $\boldsymbol{p} = \{p_i : i \in \mathbb{Z}\}$, where $p_i = p(x_i) \geq 2$?

Problem 3.3. Investigate the existence and uniqueness of solutions of the lattice system (3.15) with the discretised $\boldsymbol{p}$-Laplacian operator (i.e., with variable exponents) instead of the discretised p-Laplacian. Does it have an attractor in the sequence space $\ell^{\boldsymbol{p}}$? Can dissipativity conditions be expressed in terms of the semi-modular rather than the norm $\|\boldsymbol{u}\|_{\boldsymbol{p}}$?

Chapter 4

Approximation of attractors of LDS

Numerical computations of lattice dynamical systems, which are infinite dimensional systems, need finite dimensional approximations. In their seminal paper [Bates *et al.* (2001)], Bates, Lu and Wang considered finite dimensional ODEs approximation the Laplacian LDS

$$\frac{\mathrm{d}u_i}{\mathrm{d}t} = \nu(u_{i-1} - 2u_i + u_{i+1}) + f(u_i) + g_i, \qquad i \in \mathbb{Z}, \tag{4.1}$$

and showed that these finite dimensional ODEs have attractors, which converge upper semi-continuously to the attractor for the infinite lattice system (4.1). These results are presented here. In addition, Euler approximations of an LDS and their finite dimensional approximating ODEs are considered along with the upper semi-continuous convergence on their attractors.

The above equation (4.1) is the same as (3.1). Throughout this chapter it is assumed that Assumptions 3.1 and 3.2 hold.

4.1 Finite dimensional approximations

Bates, Lu & Wang [Bates *et al.* (2001)] showed that the $(2N+1)$-dimensional approximations of the lattice system (4.1) have global attractors $\mathcal{A}_N$ which converge upper semi-continuously to the attractor $\mathcal{A}$ in the Hausdorff semi-distance on ℓ^2, i.e.,

$$\lim_{N\to\infty} \mathrm{dist}_{\ell^2}(\mathcal{A}_N, \mathcal{A}) = 0. \tag{4.2}$$

The proof of upper semi-continuous convergence uses similar ideas to those for the tail estimates above.

Note that they did not simply truncate the LDS (4.1) but assumed that it has periodic boundary conditions, i.e., with $u_N(t) = u_{-N-1}(t)$ and $u_{-N}(t) = u_{N+1}(t)$.

Specifically, they consider the finite dimensional system of ODEs

$$\begin{cases} \frac{\mathrm{d}u_{-N}}{\mathrm{d}t} = \nu\left(u_N - 2u_{-N} + u_{-N+1}\right) + f(u_{-N}) + g_{-N}, \\ \vdots \\ \frac{\mathrm{d}u_i}{\mathrm{d}t} = \nu\left(u_{i-1} - 2u_i + u_{i+1}\right) + f(u_i) + g_i, \quad i = -N+1, \cdots, N-1, \\ \vdots \\ \frac{\mathrm{d}u_N}{\mathrm{d}t} = \nu\left(u_{N-1} - 2u_N + u_{-N}\right) + f(u_N) + g_N. \end{cases} \tag{4.3}$$

Write $\mathfrak{u} = (u_i)_{i=-N,\cdots,N}$. Then the system (4.3) can be reformulated as an ordinary differential equation on $\mathbb{R}^{2N+1}$,

$$\frac{\mathrm{d}\mathfrak{u}}{\mathrm{d}t} = \nu\Lambda_N\mathfrak{u} + \mathfrak{f}_N(\mathfrak{u}) + \mathfrak{g}_N \tag{4.4}$$

with $\mathfrak{f}_N(\mathfrak{u}) = (f(u_{-N}), \cdots f(u_0), \cdots, f(u_N))^T$ and $\mathfrak{g}_N = (g_{-N}, \cdots, g_0, \cdots, g_N)^T$ taking values in $\mathbb{R}^{2N+1}$. Here Λ_N is a $(2N+1)\times(2N+1)$ matrix that can be written as

$$\Lambda_N = -\mathrm{D}_N\mathrm{D}_N^T = -\mathrm{D}_N^T\mathrm{D}_N, \tag{4.5}$$

where

$$\Lambda_N = \begin{bmatrix} -2 & 1 & 0 & \cdots & 0 & 1 \\ 1 & -2 & 1 & \ddots & & 0 \\ 0 & \ddots & \ddots & \ddots & \ddots & \vdots \\ \vdots & \ddots & \ddots & \ddots & \ddots & 0 \\ 0 & & \ddots & 1 & -2 & 1 \\ 1 & 0 & \cdots & 0 & 1 & -2 \end{bmatrix}, \qquad D_N = \begin{bmatrix} -1 & 1 & 0 & \cdots & 0 & 0 \\ 0 & -1 & 1 & \ddots & & 0 \\ 0 & \ddots & \ddots & \ddots & \ddots & \vdots \\ \vdots & \ddots & \ddots & \ddots & \ddots & 0 \\ 0 & & \ddots & 0 & -1 & 1 \\ 1 & 0 & \cdots & 0 & 0 & -1 \end{bmatrix}.$$

Under Assumptions 3.1 and 3.2 the vector field of the ODE (4.4) is locally Lipschitz and satisfies the dissipativity condition

$$\begin{aligned} (\!(\Lambda_N\mathfrak{u} + \mathfrak{f}_N(\mathfrak{u}) + \mathfrak{g}_N, \mathfrak{u})\!) &= \nu(\!(\Lambda_N\mathfrak{u}, \mathfrak{u})\!) + (\!(\mathfrak{f}_N(\mathfrak{u}), \mathfrak{u})\!) + (\!(\mathfrak{g}_N, \mathfrak{u})\!) \\ &\le -\nu|\mathrm{D}_N\mathfrak{u}|^2 - \frac{1}{2}\alpha|\mathfrak{u}|^2 + \frac{2}{\alpha}|\mathfrak{g}_N|^2, \end{aligned} \tag{4.6}$$

where $(\!(\cdot,\cdot)\!)$ and $|\cdot|$ denote the inner product and norm of Euclidean spaces, respectively. It follows from standard existence and uniqueness theorems for finite dimensional ODEs that the ODE (4.4) has a unique global solution and generates an autonomous semi-dynamical system $\{\varphi_N(t)\}_{t\ge 0}$ on $\mathbb{R}^{2N+1}$. Moreover, the inequality (4.6) implies

$$\frac{\mathrm{d}}{\mathrm{d}t}|\mathfrak{u}|^2 \le -\alpha|\mathfrak{u}|^2 + \frac{4}{\alpha}|\mathfrak{g}_N|^2.$$

It then follows by the Gronwall inequality that

$$|\mathfrak{u}(t)|^2 \le |\mathfrak{u}_o|^2 e^{-\alpha t} + \frac{4}{\alpha^2}|\mathfrak{g}_N|^2,$$

which means that the compact set

$$Q_N := \left\{ \mathfrak{u} \in \mathbb{R}^{2N+1} \, : |\mathfrak{u}|^2 \leq 1 + \frac{4}{\alpha^2} |\mathfrak{g}_N|^2 \right\}$$

is an absorbing set for φ_N. Hence φ_N has a global attractor $\mathcal{A}_N$ in $\mathbb{R}^{2N+1}$ which is contained in Q_N.

Remark 4.1. Every vector $\mathfrak{u} = (u_i)_{|i|\leq N} \in \mathbb{R}^{2N+1}$ can be naturally extended to an element $\boldsymbol{u}^{(N)} = (u_i)_{i\in\mathbb{Z}}$ in ℓ^2, by setting $u_i = 0$ for all $|i| > N$. In this sense the above attractors $\mathcal{A}_N$ in $\mathbb{R}^{2N+1}$ can be naturally embedded into ℓ^2. The vector $\mathfrak{g}_N$ can also be extended in this way. Since $|\mathfrak{g}_N| \leq \|\boldsymbol{g}\|$, it follows that the natural embedding of the absorbing sets Q_N into ℓ^2, denoted as $\tilde{Q}_N$, are subsets of the absorbing set

$$Q := \left\{ \boldsymbol{u} \in \ell^2 \, :\in \|\boldsymbol{u}\|^2 \leq R^2 := 1 + \frac{4}{\alpha^2} \|\boldsymbol{g}\|^2 \right\}$$

of the lattice system (4.1).

4.2 Upper semi-continuous convergence of the finite dimensional attractors

The first step to prove the upper semi-continuous convergence (4.2) is to establish an asymptotic tails property.

Lemma 4.1. *Suppose that Assumptions 3.1 and 3.2 hold and let $\mathfrak{u}^{(N)}(t)$ be the solution of ODE (4.4) for the initial value $\mathfrak{u}_o^{(N)} \in Q_N$. Then for every $\varepsilon > 0$ there exist $K(\varepsilon) \in \mathbb{N}$ and $T(\varepsilon) > 0$ such that*

$$\sum_{K(\varepsilon)\leq|i|\leq N} \left| \mathfrak{u}^{(N)}(t) \right|^2 \leq \varepsilon, \qquad t \geq T(\varepsilon),$$

for every $N \geq K(\varepsilon)$.

Proof. The proof is similar to that in Chapter 3, so only the main points will be given here. For convenience the superscript (N) will be omitted when the context is clear. Given $\mathfrak{u} \in \mathbb{R}^{2N+1}$, define $\mathfrak{v} = (v_i)_{i\in\mathbb{Z}} \in \mathbb{R}^{2N+1}$ component wise as $v_i := \xi_k(|i|)u_i$ for $i \in \mathbb{Z}$, where ξ is the cut-off function defined in (3.8) and $\xi_k(s) = \xi(\frac{s}{k})$ for all $s \in \mathbb{R}$ for some $k \in \mathbb{N}$. Notice that here $k \leq N$, otherwise the sums are vacuous.

Taking the inner product of equation (4.4) with $\mathfrak{v}$ in $\mathbb{R}^{2N+1}$ and using (4.5) gives

$$\frac{\mathrm{d}}{\mathrm{d}t} (\!| \mathfrak{u}, \mathfrak{v} |\!) + \nu (\!| \mathrm{D}_N \mathfrak{u}, \mathrm{D}_N \mathfrak{v} |\!) = (\!| \mathfrak{f}_N(\mathfrak{u}), \mathfrak{v} |\!) + (\!| \mathfrak{g}_N, \mathfrak{v} |\!),$$

which is equivalent to

$$\frac{\mathrm{d}}{\mathrm{d}t} \sum_{|i|\leq N} \xi_k(|i|)|u_i|^2 + 2\nu (\!| \mathrm{D}_N \mathfrak{u}, \mathrm{D}_N \mathfrak{v} |\!) = 2 \sum_{|i|\leq N} \xi_k(|i|) u_i f(u_i) + 2 \sum_{|i|\leq N} \xi_k(|i|) g_i u_i.$$

Estimating each term in the equation above and using $|\xi'(s)| \le C_0$ leads to

$$\frac{\mathrm{d}}{\mathrm{d}t}\sum_{|i|\le N}\xi_k(|i|)|u_i|^2 + \alpha\sum_{|i|\le N}\xi_k(|i|)|u_i|^2 \le \nu\frac{4C_0\|Q\|^2}{k} + \frac{1}{\alpha}\sum_{k\le|i|\le N}|g_i|^2,$$

since for all $\mathfrak{u} \in Q_N$ with its natural ℓ^2 extension $\boldsymbol{u}^{(N)} \in Q$ we have

$$(\!(\mathrm{D}_N\mathfrak{u}, \mathrm{D}_N\mathfrak{v})\!) \ge -\frac{C_0}{k}\sum_{i=-N}^{N}\left(|u_{i+1}|^2 + |u_i||u_{i+1}|\right) \ge -\frac{2C_0\|Q_N\|^2}{k} \ge -\frac{2C_0\|Q\|^2}{k}$$

where $\|Q_N\| := \sup_{\mathfrak{u}\in Q_N}\|\mathfrak{u}\|$ and $\|Q\| := \sup_{\boldsymbol{u}\in Q}\|\boldsymbol{u}\|$.

Since $\boldsymbol{g} \in \ell^2$, then for every $\varepsilon > 0$ there exists $K(\varepsilon)$ such that

$$\nu\frac{4C_0\|Q\|^2}{k} + \frac{1}{\alpha}\sum_{|i|\ge k}|g_i|^2 \le \varepsilon, \quad k \ge K(\varepsilon),$$

and thus

$$\frac{\mathrm{d}}{\mathrm{d}t}\sum_{|i|\le N}\xi_k(|i|)|u_i|^2 + \alpha\sum_{|i|\le N}\xi_k(|i|)|u_i|^2 \le \varepsilon.$$

Then, Gronwall's Lemma implies that

$$\sum_{|i|\le N}\xi_k(|i|)|u_i(t,\boldsymbol{u}_o)|^2 \le e^{-\alpha t}\sum_{|i|\le N}\xi_k(|i|)|u_{o,i}|^2 + \frac{\varepsilon}{\alpha} \le e^{-\alpha t}\|\boldsymbol{u}_o\|^2 + \frac{\varepsilon}{\alpha}.$$

Hence, for $\mathfrak{u}_o \in Q_N$,

$$\sum_{|i|\le N}\xi_k(|i|)|u_i(t,\mathfrak{u}_o)|^2 \le e^{-\alpha t}\|Q_N\|^2 + \frac{\varepsilon}{\alpha},$$

so

$$\sum_{K(\varepsilon)\le|i|\le N}|u_i(t,\mathfrak{u}_o)|^2 \le \frac{2\varepsilon}{\alpha}$$

for $t \ge T(\varepsilon) := \frac{1}{\alpha}\ln\frac{\alpha\|Q_N\|^2}{\varepsilon}$. □

In the following lemma and its proof, we call the element in ℓ^2 obtained from expanding $\mathfrak{u} \in \mathbb{R}^{2N+1}$ with zero components for $|i| > N$ the natural extension of $\mathfrak{u}$ in ℓ^2, denoted by $\boldsymbol{u}^{(N)}$.

Lemma 4.2. *Suppose that Assumptions 3.1 and 3.2 hold. For $N \in \mathbb{N}$, let $\mathfrak{u}_o^{(N)} \in \mathcal{A}_N$ and $\boldsymbol{u}_o^{(N)}$ be the natural extension of $\mathfrak{u}_o^{(N)}$ in ℓ^2. Then there exists a subsequence $\{\boldsymbol{u}_o^{(N_j)}\}$ of $\{\boldsymbol{u}_o^{(N)}\}$ and an element $\boldsymbol{u}_o^* \in \mathcal{A}$ such that $\{\boldsymbol{u}_o^{(N_j)}\}$ converges to $\boldsymbol{u}_o^*$ in ℓ^2 as $j \to \infty$.*

Proof. Let $\mathfrak{u}^{(N)}(t) = \varphi_N(t, \mathfrak{u}_o^{(N)})$ be the entire solution of the ODE (4.4) with $\mathfrak{u}^{(N)}(0) = \mathfrak{u}_o^{(N)}$, and let $\boldsymbol{u}^{(N)}(t)$ be the natural extension of $\mathfrak{u}^{(N)}(t)$ into ℓ^2. Since $\mathfrak{u}^{(N)}(t) \in Q_N$ for all $t \in \mathbb{R}$, then by Remark 4.1, $\boldsymbol{u}^{(N)}(t) \subset \tilde{Q}_N \subset Q$ for all $t \in \mathbb{R}$ and hence

$$\|\boldsymbol{u}^{(N)}(t)\| \le \|Q\| \quad \forall t \in \mathbb{R},\ N \in \mathbb{N}, \tag{4.7}$$

where $\|Q\|$ is the radius of the set Q which is independent of N.

Let $\tilde{\Lambda}_N \boldsymbol{u}^{(N)}(t)$, $\tilde{\mathfrak{f}}_N(\boldsymbol{u}^{(N)}(t))$, and $\tilde{\mathfrak{g}}_N$ be the natural extensions of $\Lambda_N \mathrm{u}^{(N)}(t)$, $\mathfrak{f}_N(\mathrm{u}^{(N)}(t))$ and $\mathfrak{g}_N$ into ℓ^2, respectively. Then it follows from the ODE (4.4) that

$$\left\| \frac{\mathrm{d}}{\mathrm{d}t} \boldsymbol{u}^{(N)}(t) \right\| \leq \nu \left\| \tilde{\Lambda}_N \boldsymbol{u}^{(N)}(t) \right\| + \left\| \tilde{\mathfrak{f}}_N(\boldsymbol{u}^{(N)}(t)) \right\| + \|\tilde{\mathfrak{g}}_N\| . \tag{4.8}$$

Notice that

$$\left\| \tilde{\Lambda}_N \boldsymbol{u}^{(N)} \right\| = \sqrt{\|\mathrm{D}^+ \boldsymbol{u}^{(N)}\| \, \|\mathrm{D}^- \boldsymbol{u}^{(N)}\|} \leq 4 \|\boldsymbol{u}^{(N)}\|, \tag{4.9}$$

and that $\mathfrak{f}_N$ is locally Lipschitz with

$$\left\| \mathfrak{f}_N(\boldsymbol{u}^{(N)}) - \mathfrak{f}_N(\boldsymbol{0}) \right\| \leq L_Q \left\| \boldsymbol{u}^{(N)} \right\| \quad \text{for } \boldsymbol{u}^{(N)} \in Q. \tag{4.10}$$

It then follows from (4.8)–(4.10) that

$$\left\| \frac{\mathrm{d}}{\mathrm{d}t} \boldsymbol{u}^{(N)}(t) \right\| \leq (4\nu + L_Q)\|Q\| + \|\boldsymbol{g}\| := R, \quad \forall t \in \mathbb{R},\, N \in \mathbb{N}. \tag{4.11}$$

Then on each time interval $J_k = [-k, k]$,

$$\|\boldsymbol{u}^{(N)}(t) - \boldsymbol{u}^{(N)}(s)\| \leq \|\dot{\boldsymbol{u}}^{(N)}(\zeta_{t,s})\| \cdot |t - s| \leq R|t - s|, \quad t, s \in J_k,$$

which implies that the sequence $\{\boldsymbol{u}^{(N)}(t)\}$ is equi-continuous on $\mathbb{R} = \cup_{k \in \mathbb{N}} J_k$.

Now by the bound (4.7), for each $t \in J_k$ there exists a subsequence $\{\boldsymbol{u}^{(N_j)}(t)\}$ and an element $\boldsymbol{v}(t) \in \ell^2$ such that $\{\boldsymbol{u}^{(N_j)}(t)\}$ converges weakly to $\boldsymbol{v}(t)$ in ℓ^2. Similar to the proof of Lemma 2.6 it can be shown that this convergence is in fact strong convergence in ℓ^2. By the Ascoli-Arzelà Theorem the subsequence $\{\boldsymbol{u}^{(N_j)}(\cdot)\}$ in $\mathcal{C}(J_k, \ell^2)$ and $\boldsymbol{v}(\cdot) \in \mathcal{C}(J_k, \ell^2)$ such that $\{\boldsymbol{u}^{(N_j)}(\cdot)\}$ converges strongly to $\boldsymbol{v}(\cdot)$ in $\mathcal{C}(J_k, \ell^2)$ as $N_j \to \infty$. Since this can be done on each J_k, a diagonal subsequence argument can be used to find a common convergent subsequence $\{\boldsymbol{u}^{(N_j)}(\cdot)\}$ converging strongly to $\boldsymbol{v}(\cdot)$ in $\mathcal{C}(J, \ell^2)$ for any compact interval J in $\mathbb{R}$. Since $\{\boldsymbol{u}^{(N_j)}(t)\}$ takes values in the closed and bounded subset Q of ℓ^2 so does the limit $\boldsymbol{v}(t)$. Hence

$$\|\boldsymbol{v}(t)\| \leq \|Q\| \quad \text{for all } t \in \mathbb{R}.$$

In order to show that $\boldsymbol{v}(0) \in \mathcal{A}$ it needs to be shown that $\boldsymbol{v}(t)$ is a solution of the lattice system (4.1) in ℓ^2. For simplicity the convergent subsequence $\{\boldsymbol{u}^{(N_j)}(t)\}$ will be denoted by $\{\boldsymbol{u}^{(N)}(t)\}$. It follows from the uniform bound (4.11) that

$$\dot{\boldsymbol{u}}^{(N)} \to \dot{\boldsymbol{v}} \quad \text{in } \mathcal{L}^\infty(\mathbb{R}, \ell^2) \quad \text{(weak star convergence)}.$$

Fix $i \in \mathbb{Z}$ with $|i| < N$. Then

$$\frac{\mathrm{d}u_i^{(N)}}{\mathrm{d}t} = \nu(u_{i-1}^{(N)} - 2u_i^{(N)} + u_{i+1}^{(N)}) + f(u_i^{(N)}) + g_i, \qquad t \in \mathbb{R}$$

since $\boldsymbol{u}^{(N)}(t)$ is a solution of the system of ODEs (4.3). Thus for any $\phi \in \mathcal{C}(J, \mathbb{R})$

$$\int_J \dot{u}_i^{(N)}(t)\phi(t)\mathrm{d}t = \nu \int_J \left(u_{i-1}^{(N)}(t) - 2u_i^{(N)}(t) + u_{i+1}^{(N)}(t) \right) \phi(t)\mathrm{d}t$$
$$+ \int_J f(u_i^{(N)}(t))\phi(t)\mathrm{d}t + \int_J g_i \phi(t)\mathrm{d}t.$$

Note in particular that

$$\left|\int_J f(u_i^{(N)}(t))\phi(t) - f(v_i(t))\phi(t)\mathrm{d}t\right| \leq L_Q \int_J \left|u_i^{(N)}(t) - v_i(t)\right| |\phi(t)| \,\mathrm{d}t$$
$$\leq L_Q \sup_{t\in J}\left|u_i^{(N)}(t) - v_i(t)\right| \int_J |\phi(t)| \,\mathrm{d}t \to 0.$$

Hence in the limit

$$\frac{\mathrm{d}v_i}{\mathrm{d}t} = \nu(v_{i-1} - 2v_i + v_{i+1}) + f(v_i) + g_i, \quad i \in \mathbb{Z},\ t \in J,$$

so $\boldsymbol{v} = (v_i)_{i\in\mathbb{Z}}$ is a solution of the lattice system (4.1). Moreover, this limit holds on any compact interval J, thus on $\mathbb{R}$, so $\boldsymbol{v}$ is an entire solution of the lattice system (4.1). Therefore $\boldsymbol{v}(t) \in \mathcal{A}$ for all $t \in \mathbb{R}$. In particular, $(u_i^{(N)}(0))_{i\in\mathbb{Z}} = (u_{o,i}^{(N)})_{i\in\mathbb{Z}}$ converges to $\boldsymbol{v}(0) = \boldsymbol{u}^*$ *in* $\mathcal{A}$. □

Theorem 4.1. *Suppose that Assumptions 3.1 and 3.2 hold. Let $\tilde{\mathcal{A}}_N$ be the natural embedding of $\mathcal{A}_N$ in ℓ^2. Then*

$$\lim_{N\to\infty} \mathrm{dist}_{\ell^2}\left(\tilde{\mathcal{A}}_N, \mathcal{A}\right) = 0.$$

Proof. The proof is by contradiction. Suppose that there is a sequence $\boldsymbol{a}^{(N)} \in \tilde{\mathcal{A}}_N$ and an $\varepsilon_0 > 0$ such that

$$\mathrm{dist}_{\ell^2}\left(\boldsymbol{a}^{(N)}, \mathcal{A}\right) \geq \varepsilon_0. \tag{4.12}$$

By Lemma 4.2 there exists a convergence subsequence $\{\boldsymbol{a}^{(N_j)}\}$ with $\mathrm{dist}_{\ell^2}\left(\boldsymbol{a}^{(N_j)}, \mathcal{A}\right) \to 0$. This contradicts the inequality (4.12). □

4.3 Numerical approximation of lattice attractors

Han, Kloeden & Sonner [Han *et al.* (2020a)] considered the numerical approximation of the attractor of the LDS (4.1). They focused on the implicit Euler scheme (IES) (4.13) below and first apply it to lattice ODE (3.4) in the space ℓ^2, where it has the form

$$\boldsymbol{u}_{n+1}^{(h)} = \boldsymbol{u}_n^{(h)} + h\nu\Lambda\boldsymbol{u}_{n+1}^{(h)} + hF\left(\boldsymbol{u}_{n+1}^{(h)}\right) + h\boldsymbol{g}, \qquad n \in \mathbb{N}_0, \tag{4.13}$$

for a constant time step $h > 0$. Here the subscript n denotes the time step, while the subscript i corresponds to the point in the lattice. More precisely,

$$\boldsymbol{u}_n^{(h)} = \left(u_{n,i}^{(h)}\right)_{i\in\mathbb{Z}} \in \ell^2 \quad \text{for all} \quad n \in \mathbb{N}_0.$$

The IES (4.13) is well defined, i.e., all terms belong to ℓ^2, and, restricted to bounded sets in ℓ^2, is uniquely solvable for sufficiently small step sizes h (see Lemma 4.4 below).

The aim here is to construct approximations for the global attractor $\mathcal{A}$ of the LDS (4.1), which is contained in the closed and bounded, positively invariant absorbing set

$$Q := \left\{ \boldsymbol{u} \in \ell^2 : \|\boldsymbol{u}\|^2 \leq R^2 := 1 + \frac{\|g\|^2}{\alpha} \right\}.$$

Attention will henceforth be restricted to the IES in Q and to show that it generates a discrete semi-dynamical system in Q.

Lemma 4.3. *Let Assumptions 3.1 and 3.2 hold. Then, for any $h > 0$ and $n \in \mathbb{N}_0$, the solutions of* (4.13) *satisfy*

$$\left\| \boldsymbol{u}_{n+1}^{(h)} \right\|^2 \leq \frac{1}{1 + h\alpha} \left\| \boldsymbol{u}_n^{(h)} \right\|^2 + \frac{h}{1 + h\alpha} \frac{\|\boldsymbol{g}\|^2}{\alpha}.$$

In particular, if $\boldsymbol{u}_o \in Q$, then $\boldsymbol{u}_n^{(h)} \in Q$ for all $n \in \mathbb{N}_0$, i.e., the bounded positive invariant absorbing set Q for the lattice system (4.1) *is also positive invariant for the IES* (4.13).

Proof. Taking the inner product of (4.13) with $\boldsymbol{u}_{n+1}^{(h)}$ and using Assumption 3.1 gives

$$\begin{aligned}
\|\boldsymbol{u}_{n+1}^{(h)}\|^2 &= \langle \boldsymbol{u}_n^{(h)}, \boldsymbol{u}_{n+1}^{(h)} \rangle + h\nu \langle \Lambda \boldsymbol{u}_{n+1}^{(h)}, \boldsymbol{u}_{n+1}^{(h)} \rangle + h \langle F(\boldsymbol{u}_{n+1}^{(h)}) + \boldsymbol{g}, \boldsymbol{u}_{n+1}^{(h)} \rangle \\
&\leq \frac{1}{2} \|\boldsymbol{u}_{n+1}^{(h)}\|^2 + \frac{1}{2} \|\boldsymbol{u}_n^{(h)}\|^2 + \frac{h}{2} \left(-\alpha \|\boldsymbol{u}_{n+1}^{(h)}\|^2 + \frac{1}{\alpha} \|\boldsymbol{g}\|^2 \right) \\
&= \left(\frac{1 - h\alpha}{2} \right) \|\boldsymbol{u}_{n+1}^{(h)}\|^2 + \frac{1}{2} \|\boldsymbol{u}_n^{(h)}\|^2 + \frac{h}{2\alpha} \|\boldsymbol{g}\|^2,
\end{aligned}$$

which implies the desired estimate.

Moreover, if $\boldsymbol{u}_n^{(h)} \in Q$, then $\|\boldsymbol{u}_n^{(h)}\|^2 \leq R^2 = \frac{2\|g\|^2}{\alpha^2} + 1$. Hence, it follows that

$$\begin{aligned}
\|\boldsymbol{u}_{n+1}^{(h)}\|^2 &\leq \frac{1}{1 + h\alpha} \left(\frac{2\|\boldsymbol{g}\|}{\alpha^2} + 1 \right) + \frac{h}{1 + h\alpha} \frac{\|\boldsymbol{g}\|^2}{\alpha} \\
&\leq \frac{1}{1 + h\alpha} \frac{1}{\alpha^2} \left((2 + h\alpha) \|\boldsymbol{g}\|^2 + 1 \right) \leq R^2,
\end{aligned}$$

which proves the positive invariance of Q. □

Lemma 4.4. *Let Assumptions 3.1 and 3.2 hold. Then, there exists $h^* > 0$ such that the IES* (4.13) *is uniquely solvable for every $\boldsymbol{u}_o \in Q$ and $h \in (0, h^*)$.*

Proof. For $h > 0$ and given $\boldsymbol{u} \in Q$ define the mapping $\mathfrak{F}_h : Q \to Q$ by

$$\mathfrak{F}_h(\boldsymbol{u}) := \boldsymbol{u} + h\nu \Lambda \boldsymbol{u} + hF(\boldsymbol{u}) + h\boldsymbol{g}, \qquad \boldsymbol{u} \in Q.$$

The positive invariance of Q was shown in Lemma 4.3 and implies that $\mathfrak{F}_h(\boldsymbol{u}) \in Q$.

For every $\boldsymbol{u}, \boldsymbol{v} \in Q$ the difference between $\mathfrak{F}_h(\boldsymbol{u})$ and $\mathfrak{F}_h(\boldsymbol{v})$ satisfies

$$\|\mathfrak{F}_h(\boldsymbol{u}) - \mathfrak{F}_h(\boldsymbol{v})\| \leq h\nu \|\Lambda(\boldsymbol{u} - \boldsymbol{v})\| + h\|F(\boldsymbol{u}) - F(\boldsymbol{v})\|. \tag{4.14}$$

Recall that the discrete Laplacian Λ satisfies $\|\Lambda \boldsymbol{u}\| \leq 4\|\boldsymbol{u}\|$ for every $\boldsymbol{u} \in \ell^2$, hence

$$\|\Lambda(\boldsymbol{u}-\boldsymbol{v})\| \leq 4\|\boldsymbol{u}-\boldsymbol{v}\|.$$

Moreover, since f is smooth it is locally Lipschitz on Q, so

$$\|F(\boldsymbol{u})-F(\boldsymbol{v})\| \leq L_Q\|\boldsymbol{u}-\boldsymbol{v}\|, \quad \text{with} \quad L_Q = \sup_{r\in[-2R,2R]} |f'(r)|.$$

Inserting the above two inequalities in (4.14) yields the estimate

$$\|\mathfrak{F}_h(\boldsymbol{u})-\mathfrak{F}_h(\boldsymbol{v})\| \leq h(4\nu+L_Q)\|\boldsymbol{u}-\boldsymbol{v}\|.$$

Define

$$h^* = \frac{1}{4\nu+L_Q}.$$

Then for all $h < h^*$, by the contraction mapping principle it follows that $\mathfrak{F}_h$ possesses a unique fixed point in ℓ^2, which implies that the IES (4.13) is uniquely solvable for all $n \in \mathbb{N}_0$. □

By Lemma 4.4, for all $h \in (0, h^*)$ the IES (4.13) can be written in an explicit form $\boldsymbol{u}_{n+1}^{(h)} = \mathcal{S}^{(h)}\left(\boldsymbol{u}_n^{(h)}\right)$ for an appropriate mapping $\mathcal{S}^{(h)} : Q \to Q$.

Lemma 4.5. *Let Assumptions 3.1 and 3.2 hold and $h \in (0, h^*)$. Then, $\mathcal{S}^{(h)} : Q \to Q$ is Lipschitz continuous.*

Proof. For any $\boldsymbol{u}_n^{(h)}, \boldsymbol{v}_n^{(h)} \in Q$, the next iterates $\boldsymbol{u}_{n+1}^{(h)}, \boldsymbol{v}_{n+1}^{(h)} \in Q$ by Lemma 4.3. Their difference satisfies

$$\begin{aligned}\left\|\boldsymbol{u}_{n+1}^{(h)}-\boldsymbol{v}_{n+1}^{(h)}\right\| \leq \left\|\boldsymbol{u}_n^{(h)}-\boldsymbol{v}_n^{(h)}\right\| + h\nu\left\|\Lambda(\boldsymbol{u}_{n+1}^{(h)}-\boldsymbol{v}_{n+1}^{(h)})\right\| \\ +h\left\|F\left(\boldsymbol{u}_{n+1}^{(h)}\right)-F\left(\boldsymbol{v}_{n+1}^{(h)}\right)\right\|.\end{aligned}$$

Moreover, the estimates in the proof of Lemma 4.4 imply that

$$\left\|\boldsymbol{u}_{n+1}^{(h)}-\boldsymbol{v}_{n+1}^{(h)}\right\| \leq \left\|\boldsymbol{u}_n^{(h)}-\boldsymbol{v}_n^{(h)}\right\| + h(4\nu+L_Q)\left\|\boldsymbol{u}_{n+1}^{(h)}-\boldsymbol{v}_{n+1}^{(h)}\right\|.$$

Consequently,

$$\left\|\boldsymbol{u}_{n+1}^{(h)}-\boldsymbol{v}_{n+1}^{(h)}\right\| \leq \frac{1}{1-h(4\nu+L_Q)}\left\|\boldsymbol{u}_n^{(h)}-\boldsymbol{v}_n^{(h)}\right\|,$$

provided that $h < h^*$. □

The following lemma on the order of global discretisation error of IES on ℓ^2 was proved in [Han *et al.* (2020a)].

Lemma 4.6. *Let Assumptions 3.1 and 3.2 hold. Then, the IES* (4.13) *on ℓ^2 has global discretisation error of order one in the absorbing set Q, i.e.,*

$$\|\mathcal{E}(h)\| = \|\boldsymbol{u}(nh;\boldsymbol{u}_o)-\boldsymbol{u}_n^{(h)}\| \leq C_{T,B}h \qquad 0 \leq nh \leq T,$$

for some constant $C_{T,B} > 0$.

Existence of the numerical attractor

First, note that a numerical scheme needs not possess a globally attracting set even if the original attractor is globally attracting; see [Han and Kloeden (2017b), Example 2.13].

Due to Lemmas 4.3, 4.4 and 4.5, for every $h \in (0, h^*)$ the IES (4.13), restricted to Q, generates a discrete semi-dynamical system $\varphi_n^{(h)} : Q \to Q$, $n \in \mathbb{N}_0$, defined by

$$\varphi_n^{(h)}(\boldsymbol{u}_o) = \boldsymbol{u}_n^{(h)}(\boldsymbol{u}_o), \qquad \boldsymbol{u}_o \in Q,$$

where $\boldsymbol{u}_n^{(h)}(\boldsymbol{u}_o) = \boldsymbol{u}_n^{(h)}$ denotes the unique solution of (4.13) with initial data $\boldsymbol{u}_o \in Q$.

Remark 4.2. For the following analysis it is important that Q is a common absorbing set for all step sizes $h \in (0, h^*)$ under consideration.

Next, it will be shown that $\{\varphi_n^{(h)}\}_{n\in\mathbb{N}_0}$ possesses an attractor $\mathcal{A}^{(h)}$ in ℓ^2, namely, the *numerical attractor* corresponding to the IES applied to the lattice ODE (4.1). Since the restricted state space Q is bounded, it suffices to prove that the semi-dynamical system $\{\varphi_n^{(h)}\}_{n\in\mathbb{N}_0}$ is asymptotically compact. To this end, in the following uniform bounds for the tail ends of solutions are needed. Such estimates were obtained in the continuous time setting in Chapter 3.

Lemma 4.7. *Assume that Assumptions 3.1 and 3.2 hold, let $h \in (0, h^*)$ and $\boldsymbol{u}_o \in Q$. Then, for every $\varepsilon > 0$ there exist $N^{(h)}(\varepsilon)$ and $I(\varepsilon)$ in $\mathbb{N}$ such that the solution of the IES (4.13) satisfies*

$$\sum_{|i|>I(\varepsilon)} \left|u_{n,i}^{(h)}\right|^2 \le \varepsilon \quad \textit{for all } n \ge N^{(h)}(\varepsilon).$$

Proof. Let ξ be the cut-off function defined in (3.8), and for $k \in \mathbb{N}$ define $\xi_k(s) = \xi(s/k)$. Recall that there exists a constant C_0 such that $|\xi'(s)| \le C_0$ for all $s \ge 0$.

For any $\boldsymbol{u}_n^{(h)} = (u_{n,i}^{(h)})_{i\in\mathbb{Z}}$ set $\boldsymbol{v}_n^{(h)} = (v_{n,i}^{(h)})_{i\in\mathbb{Z}}$ with $v_{n,i}^{(h)} := \xi_k(|i|)u_{n,i}^{(h)}$. Taking the inner product of (4.13) with $\boldsymbol{v}_{n+1}^{(h)}$ in ℓ^2 results in

$$\begin{aligned}\sum_{i\in\mathbb{Z}} \xi_k(|i|)|u_{n+1,i}^{(h)}|^2 = &\sum_{i\in\mathbb{Z}} \xi_k(|i|)u_{n,i}^{(h)}u_{n+1,i}^{(h)} + h\nu\langle \Lambda \boldsymbol{u}_{n+1}^{(h)}, \boldsymbol{v}_{n+1}^{(h)}\rangle \\ &+ h\langle F(\boldsymbol{u}_{n+1}^{(h)}) + \boldsymbol{g}, \boldsymbol{v}_{n+1}^{(h)}\rangle.\end{aligned} \tag{4.15}$$

The terms on the right hand side of the equation are estimated as follows. First, by the triangle inequality, it follows that

$$\sum_{i\in\mathbb{Z}} \xi_k(|i|)u_{n,i}^{(h)}u_{n+1,i}^{(h)} \le \frac{1}{2}\sum_{i\in\mathbb{Z}} \xi_k(|i|)|u_{n+1,i}^{(h)}|^2 + \frac{1}{2}\sum_{i\in\mathbb{Z}} \xi_k(|i|)|u_{n,i}^{(h)}|^2.$$

Second, since $\langle \Lambda \boldsymbol{u}, \boldsymbol{v}\rangle = -\langle \mathrm{D}^+\boldsymbol{u}, \mathrm{D}^+\boldsymbol{v}\rangle$ for all $\boldsymbol{u}, \boldsymbol{v} \in \ell^2$, it follows that

$$\begin{aligned}\langle \Lambda \boldsymbol{u}_{n+1}^{(h)}, \boldsymbol{v}_{n+1}^{(h)}\rangle &= -\langle \mathrm{D}^+\boldsymbol{u}_{n+1}^{(h)}, \mathrm{D}^+\boldsymbol{v}_{n+1}^{(h)}\rangle \\ &= -\sum_{i\in\mathbb{Z}} \left(u_{n+1,i+1}^{(h)} - u_{n+1,i}^{(h)}\right) \Big(\xi_k(|i|)(u_{n+1,i+1}^{(h)} - u_{n+1,i}^{(h)}) \\ &\qquad + (\xi_k(|i+1|) - \xi_k(|i|))u_{n+1,i+1}^{(h)}\Big) \\ &\le -\sum_{i\in\mathbb{Z}} (\xi_k(|i+1|) - \xi_k(|i|)) \left(u_{n+1,i+1}^{(h)} - u_{n+1,i}^{(h)}\right) u_{n+1,i+1}^{(h)}.\end{aligned}$$

Then, by the smoothness of ξ, there exists $s_i \in \left(\frac{|i|}{k}, \frac{|i+1|}{k}\right)$ such that

$$\begin{aligned}&\left|\sum_{i\in\mathbb{Z}} (\xi_k(|i+1|) - \xi_k(|i|)) \left(u_{n+1,i+1}^{(h)} - u_{n+1,i}^{(h)}\right) u_{n+1,i+1}^{(h)}\right| \\ &\quad\le \sum_{i\in\mathbb{Z}} \frac{1}{k}|\xi'(s_i)| \left|u_{n+1,i+1}^{(h)} - u_{n+1,i}^{(h)}\right| \left|u_{n+1,i+1}^{(h)}\right| \\ &\quad\le \frac{C_0}{k}\sum_{i\in\mathbb{Z}} \left(\left|u_{n+1,i+1}^{(h)}\right|^2 + \left|u_{n+1,i}^{(h)}\right|\left|u_{n+1,i+1}^{(h)}\right|\right) \le \frac{2C_0\|Q\|^2}{k},\end{aligned}$$

due to the positive invariance of Q which has radius $\|Q\|$.

Moreover, to estimate the last term in (4.15), Assumption 3.1 and Young's inequality yield

$$\begin{aligned}\langle F(\boldsymbol{u}_{n+1}^{(h)}) + \boldsymbol{g}, \boldsymbol{v}_{n+1}^{(h)}\rangle &= \sum_{i\in\mathbb{Z}} \xi_k(|i|) \left(f(u_{n+1,i}^{(h)})u_{n+1,i}^{(h)} + g_i u_{n+1,i}^{(h)}\right) \\ &\le -\frac{\alpha}{2}\sum_{i\in\mathbb{Z}} |\xi'(s_i)||u_{n+1,i}^{(h)}|^2 + \frac{1}{2\alpha}\sum_{i\in\mathbb{Z}} |\xi'(s_i)|g_i^2.\end{aligned}$$

Inserting these inequalities into (4.15) then gives

$$(1+\alpha h)\sum_{i\in\mathbb{Z}} |\xi'(s_i)||u_{n+1,i}^{(h)}|^2 \le \sum_{i\in\mathbb{Z}} |\xi'(s_i)||u_{n,i}^{(h)}|^2 + h\left(\frac{4\nu C_0\|Q\|^2}{k} + \frac{1}{\alpha}\sum_{i\in\mathbb{Z}} |\xi'(s_i)|g_i^2\right). \tag{4.16}$$

Since $\boldsymbol{g} = (g_i)_{i\in\mathbb{Z}} \in \ell^2$, for every $\varepsilon > 0$ there exists $I(\varepsilon) > 0$ such that

$$\frac{4\nu C_0\|Q\|^2}{k} + \frac{1}{\alpha}\sum_{i\in\mathbb{Z}} |\xi'(s_i)|g_i^2 < \frac{\varepsilon\alpha}{2} \qquad \forall\, k \ge I(\varepsilon).$$

Hence, by iterating estimate (4.16) it follows that for every $\boldsymbol{u}_o \in Q$,

$$\begin{aligned}\sum_{i\in\mathbb{Z}} |\xi'(s_i)||u_{n,i}^{(h)}|^2 &\le \frac{1}{1+\alpha h}\sum_{i\in\mathbb{Z}} |\xi'(s_i)||u_{n-1,i}^{(h)}|^2 + \frac{\varepsilon}{2}\frac{\alpha h}{1+\alpha h} \\ &\le \frac{1}{(1+\alpha h)}\sum_{i\in\mathbb{Z}} |\xi'(s_i)||u_{0,i}^{(h)}|^2 + \frac{\varepsilon}{2}\frac{h\alpha}{1+h\alpha}\sum_{j=0}^{n-1}\frac{1}{(1+h\alpha)^j} \\ &\le \frac{\|\boldsymbol{u}_o\|^2}{(1+\alpha h)^n} + \frac{\varepsilon}{2} \le \frac{\|Q\|^2}{(1+\alpha h)^n} + \frac{\varepsilon}{2},\end{aligned}$$

Finally, let

$$N^{(h)}(\varepsilon) := \log_{1+h\alpha}\left(\frac{2\|Q\|^2}{\varepsilon}\right).$$

Then, we have for all $n \geq N^{(h)}(\varepsilon)$

$$\sum_{|i|\geq 2k} |u_{n,i}|^2 \leq \sum_{i\in\mathbb{Z}} \xi_k(|i|)|u_{n,i}^{(h)}|^2 < \varepsilon \quad \text{for all } k \geq I(\varepsilon),$$

which implies the stated assertion. □

Theorem 4.2. *For all step sizes* $h \in (0, h^*)$ *the discrete semi-dynamical system* $\{\varphi_n^{(h)}\}_{n\in\mathbb{N}}$ *generated by the implicit Euler scheme* (4.13) *possesses an attractor* $\mathcal{A}^{(h)} \subset Q$ *in* ℓ^2.

Proof. The restricted state space Q is bounded in ℓ^2 and consequently, it suffices to verify that the discrete time semi-dynamical system $\varphi^{(h)}$ is asymptotically compact.

To this end let $\{\boldsymbol{u}_k\}_{k\in\mathbb{N}}$ be a sequence in Q and let $\{n_k\}_{k\in\mathbb{N}}$ be a sequence in $\mathbb{N}$ converging to ∞. Since Q is bounded, closed and convex as well positively invariant, there exists $\hat{\boldsymbol{u}} \in \ell^2$ and a subsequence of $\{\varphi_{n_k}^{(h)}(\boldsymbol{u}_k)\}_{k\in\mathbb{N}}$, again denoted by $\{\varphi_{n_k}^{(h)}(\boldsymbol{u}_k)\}_{k\in\mathbb{N}}$, such that

$$\varphi_{n_k}^{(h)}(\boldsymbol{u}_k) \rightharpoonup \hat{\boldsymbol{u}} \quad \text{weakly in } \ell^2 \text{ as } k \to \infty. \tag{4.17}$$

It will be shown that the convergence above is, in fact, strong in ℓ^2.

Let $\varepsilon > 0$ and denote by $\varphi_{n_k,i}^{(h)}(\boldsymbol{u}_k)$ the ith component of $\varphi_{n_k}^{(h)}(\boldsymbol{u}_k)$. By Lemma 4.7 there exist $I_1(\varepsilon)$ and $N_1^{(h)}(\varepsilon)$ in $\mathbb{N}$ such that

$$\sum_{|i|>I_1(\varepsilon)} \left(\varphi_{n_k,i}^{(h)}(\boldsymbol{u}_k)\right)^2 < \frac{\varepsilon^2}{8} \quad \text{for all } k \geq N_1^{(h)}(\varepsilon). \tag{4.18}$$

Moreover, since $\hat{\boldsymbol{u}} \in \ell^2$, there exists $I_2(\varepsilon) \in \mathbb{N}$ such that

$$\sum_{|i|>I_2(\varepsilon)} \hat{u}_i^2 < \frac{\varepsilon^2}{8}. \tag{4.19}$$

On the other hand, due to the weak convergence (4.17), it follows that

$$\left(\varphi_{n_k,i}^{(h)}(\boldsymbol{u}_k)\right)_{|i|\leq I(\varepsilon)} \to \left(\hat{u}_i\right)_{|i|\leq I(\varepsilon)} \quad \text{in } \mathbb{R}^{2I(\varepsilon)+1} \text{ as } k \to \infty,$$

where $I(\varepsilon) = \max\{I_1(\varepsilon), I_2(\varepsilon)\}$. Hence, there exists $N_2^{(h)}(\varepsilon) \in \mathbb{N}$ such that

$$\sum_{|i|\leq I(\varepsilon)} \left(\varphi_{n_k,i}^{(h)} u_{k,i} - \hat{u}\right)^2 < \frac{\varepsilon^2}{2} \quad \text{for all } k \geq N_2^{(h)}(\varepsilon). \tag{4.20}$$

Finally, for $k \geq \max\left\{N_1^{(h)}(\varepsilon), N_2^{(h)}(\varepsilon)\right\}$ it follows from (4.18)–(4.20) that

$$\begin{aligned}\|\varphi_{n_k}^{(h)}(\boldsymbol{u}_k) - \hat{\boldsymbol{u}}\|^2 &= \sum_{|i|\leq I(\varepsilon)} \left(\varphi_{n_k,i}^{(h)}(\boldsymbol{u}_k) - \hat{u}_i\right)^2 + \sum_{|i|>I(\varepsilon)} \left(\varphi_{n_k,i}^{(h)}(\boldsymbol{u}_k) - \hat{u}_i\right)^2 \\ &< \frac{\varepsilon^2}{2} + 2\sum_{|i|>I(\varepsilon)} \left(\left(\varphi_{n_k,i}^{(h)}(\hat{\boldsymbol{u}})\right)^2 + \hat{u}_i^2\right) < \varepsilon^2,\end{aligned}$$

which implies that $\left\{\varphi_{n_k}^{(h)}(\boldsymbol{u}_k)\right\}_{k\in\mathbb{N}}$ converges strongly to $\hat{\boldsymbol{u}}$ in ℓ^2 as $k \to \infty$. □

The next theorem establishes the upper semi-continuous convergence of the numerical attractors $\mathcal{A}^{(h)}$ to the global attractor $\mathcal{A}$ of the lattice system as $h \to 0^+$.

Theorem 4.3. *Let Assumptions 3.1 and 3.2 hold. Then,*

$$dist_{\ell^2}\left(\mathcal{A}^{(h)}, \mathcal{A}\right) \to 0 \quad \text{as } h \to 0^+,$$

where $dist_{\ell^2}(\cdot,\cdot)$ denotes the Hausdorff semi distance in ℓ^2.

Proof. Suppose that the assertion is false. Then, there exist $\varepsilon_0 > 0$ and a sequence $h_j \to 0^+$ as $j \to \infty$ such that

$$\mathrm{dist}_{\ell^2}(\mathcal{A}^{(h_j)}, \mathcal{A}) \geq \varepsilon_0 \quad \text{for all } j \in \mathbb{N}.$$

Since $\mathcal{A}^{(h_j)}$ and $\mathcal{A}$ are compact, for every $j \in \mathbb{N}$ there exists $\boldsymbol{a}^j \in \mathcal{A}^{(h_j)}$ such that

$$\mathrm{dist}_{\ell^2}(\boldsymbol{a}^j, \mathcal{A}) = \mathrm{dist}_{\ell^2}(\mathcal{A}^{(h_j)}, \mathcal{A}) \geq \varepsilon_0.$$

On the other hand, due to the invariance of $\mathcal{A}^{(h_j)}$, there exists $\boldsymbol{b}^j \in \mathcal{A}^{(h_j)}$ and $N_j \in \mathbb{N}$ such that the N_jth iterate of the IES (4.13) starting at $\boldsymbol{b}^j$ with step size h_j equals $\boldsymbol{a}^j$, i.e., $\boldsymbol{u}_{N_j}^{(h_j)}(\boldsymbol{b}^j) = \boldsymbol{a}^j$. Denote by $\boldsymbol{u}(N_j h_j; \boldsymbol{b}^j)$ the solution of the lattice system starting at $\boldsymbol{b}^j$, evaluated at $T_j := N_j h_j$. Then,

$$\begin{aligned}\mathrm{dist}_{\ell^2}(\mathcal{A}^{(h_j)}, \mathcal{A}) &= \mathrm{dist}_{\ell^2}(\boldsymbol{a}^j, \mathcal{A}) = \mathrm{dist}_{\ell^2}\left(\boldsymbol{u}_{N_j}^{(h_j)}(\boldsymbol{b}^j, \mathcal{A})\right) \\ &\leq \left\|\boldsymbol{u}_{N_j}^{(h_j)}(\boldsymbol{b}^j) - \boldsymbol{u}(N_j h_j; \boldsymbol{b}^j)\right\| + \mathrm{dist}_{\ell^2}\left(\boldsymbol{u}(N_h h_j; \boldsymbol{b}^j), \mathcal{A}\right). \qquad (4.21)\end{aligned}$$

Since $\mathcal{A}$ is the global attractor for the lattice system, there exists $T(\varepsilon_0) > 0$ and for each $j \in \mathbb{N}$ an $M_j \in \mathbb{N}$ such that $T(\varepsilon_0) \leq M_j h_j < T(\varepsilon_0) + 1$ and

$$\mathrm{dist}_{\ell^2}\left(\boldsymbol{u}(M_j h_j; \boldsymbol{b}^j), \mathcal{A}\right) \leq \frac{\varepsilon_0}{4}. \qquad (4.22)$$

Due to Lemma 4.6 and the fact that $\boldsymbol{b}^j \in \mathcal{A}^{(h_j)} \subset Q$, we have

$$\left\|u_{M_j}^{(h_j)}(\boldsymbol{b}^j) - u(M_j h_j; \boldsymbol{b}^j)\right\| \leq C_{T(\varepsilon_0)+1,Q} h_j \leq \frac{\varepsilon_0}{4}, \qquad (4.23)$$

if j is sufficiently large.

Finally, inserting (4.22) and (4.23) into (4.21) immediately implies that there exists $j \in \mathbb{N}$ such that

$$\mathrm{dist}_{\ell^2}(\mathcal{A}^{(h_j)}, \mathcal{A}) \leq \frac{\varepsilon_0}{2},$$

which contradicts the hypothesis and completes the proof. □

4.4 Finite dimensional approximations of the IES

Numerical computations are only possible in finite dimensional spaces. For $N \in \mathbb{N}$ the $(2N+1)$-dimensional implicit system of difference equations

$$
\begin{aligned}
u_{n,-N}^{(h)} &= u_{n-1,-N}^{(h)} + h\nu(u_{n,N}^{(h)} - 2u_{n,-N}^{(h)} + u_{n,-N+1}^{(h)}) + hf(u_{n,-N}^{(h)}) + hg_{-N} \\
u_{n,-N+1}^{(h)} &= u_{n-1,-N+1}^{(h)} + h\nu(u_{n,-N}^{(h)} - 2u_{n,-N+1}^{(h)} + u_{n,-N+2}^{(h)}) + hf(u_{n,-N+1}^{(h)}) + hg_{-N+1} \\
&\ \vdots \\
u_{n,N-1}^{(h)} &= u_{n-1,N-1}^{(h)} + h\nu(u_{n,N}^{(h)} - 2u_{n,N-1}^{(h)} + u_{n,-N-2}^{(h)}) + hf(u_{n,N-1}^{(h)}) + hg_{N-1} \\
u_{n,N}^{(h)} &= u_{n-1,N}^{(h)} + h\nu(u_{n,N-1}^{(h)} - 2u_{n,N}^{(h)} + u_{n,-N}^{(h)}) + hf(u_{n,N}^{(h)}) + hg_N
\end{aligned}
$$

with initial data $(u_{-N}^{(h)}, \ldots, u_N^{(h)})(0) = (u_{o,-N}, \ldots, u_{o,N}) \in \mathbb{R}^{2N+1}$ provides a finite dimensional approximations for the Implicit Euler Scheme (IES) (4.13).

This corresponds to the Implicit Euler Scheme applied to the $(2N+1)$ dimensional system of ODEs with periodic boundary conditions, which was used to approximate the LDS.

In order to simplify the notation, $\mathfrak{u}_n^{(h)} \in \mathbb{R}^{2N+1}$ will be used to denote the $(2N+1)$-dimensional truncation of $\boldsymbol{u}_n^{(h)} \in \ell^2$. Notice that $\mathfrak{u}_n^{(h)}$ actually depends on N, so when the dependence on N is explicitly needed $\mathfrak{u}_n^{(h)} = \mathfrak{u}_n^{(h,N)}$ will be used. As before, every $\mathfrak{u} = (u_i)_{|i|\le N} \in \mathbb{R}^{2N+1}$ can be naturally extended to an element $\boldsymbol{u} = (u_i)_{i\in\mathbb{Z}}$ in ℓ^2, by setting $u_i = 0$ for all $|i| > N$.

The finite dimensional IES can be rewritten compactly in $\mathbb{R}^{2N+1}$ as

$$
\begin{cases}
\mathfrak{u}_{n+1}^{(h)} = \mathfrak{u}_n^{(h)} + h\nu\Lambda_N \mathfrak{u}_{n+1}^{(h)} + h\mathfrak{f}_N(\mathfrak{u}_{n+1}^{(h)}) + h\mathfrak{g}_N, \qquad N \in \mathbb{N}, \\
\mathfrak{u}_0^{(h)} \ = (u_{o,i})_{|i|\le N} \in \mathbb{R}^{2N+1}.
\end{cases}
\tag{4.24}
$$

where $\mathfrak{f}_N(\mathfrak{u}_n^{(h)}) = (f(u_{n,i}^{(h)}))_{|i|\le N}$, $\mathfrak{g}_N = (g_i)_{|i|\le N}$, and Λ_N is defined as in (4.5).

4.4.1 *Finite dimensional numerical attractors $\mathcal{A}_N^{(h)}$*

As in the continuous time case, we restrict the truncated system (4.24) to a suitable bounded, positive invariant ball $Q_N \subset \mathbb{R}^{2N+1}$, namely,

$$
Q_N = \left\{ \mathfrak{u} \in \mathbb{R}^{2N+1} : |\mathfrak{u}|^2 \le R^2 = \frac{2\|\boldsymbol{g}\|^2}{\alpha^2} + 1 \right\}.
$$

Here and in the sequel, the Euclidean norm in $\mathbb{R}^{2N+1}$ is denoted by $|\cdot|$, while $\|\cdot\|$ denotes the ℓ^2-norm. The inner product in Euclidean space is denoted by $(\!|\cdot,\cdot|\!)$.

One can show that for every $h \in (0, h^*)$ and $\mathfrak{u}_o^{(h)} \in Q_N$ the system(4.24) is globally well-posed, and the solution takes values in Q_N. Hence, it generates a discrete semi-dynamical system $\varphi_N^{(h)}(n) : Q_N \to Q_N$, $n \in \mathbb{N}$, defined by

$$
\varphi_N^{(h)}(n, \mathfrak{u}_o^{(h)}) = \mathfrak{u}_N^{(h)}(nh; \mathfrak{u}_o^{(h)}), \qquad \mathfrak{u}_o^{(h)} \in Q_N.
$$

Lemma 4.8. *Let Assumptions 3.1 and 3.2 hold. Then, for every $h \in (0, h^*)$ and $\mathrm{u}_0^{(h)} \in Q_N$ the solution of (4.24) satisfies*

$$\left|\mathrm{u}_{n+1}^{(h)}\right|^2 \le \frac{1}{1+h\alpha}|\mathrm{u}_n^{(h)}|^2 + \frac{h}{1+h\alpha}\frac{\|g\|^2}{\alpha} \quad \textit{for all } n \in \mathbb{N}_0.$$

In particular, the set Q_N is positive invariant for $\varphi_N^{(h)}$.

Proof. Recall that $\Lambda_N = -\mathrm{D}_N(\mathrm{D}_N)^T = -(\mathrm{D}_N)^T\mathrm{D}_N$. Therefore, it follows that

$$⦇ \Lambda_N \mathrm{u}_{n+1}^{(h)}, \mathrm{u}_{n+1}^{(h)} ⦈ = -\left|\mathrm{D}_N \mathrm{u}_{n+1}^{(h)}\right|^2 \le 0.$$

As in the proof of Lemma 4.6 we take the inner product in $\mathbb{R}^{2N+1}$ of (4.24) with $\mathrm{u}_{n+1}^{(h)}$ and use (3.1) to obtain

$$\begin{aligned}
\left|\mathrm{u}_{n+1}^{(h)}\right|^2 &= ⦇ \mathrm{u}_n^{(h)}, \mathrm{u}_{n+1}^{(h)} ⦈ + h\nu ⦇ \Lambda_N \mathrm{u}_{n+1}^{(h)}, \mathrm{u}_{n+1}^{(h)} ⦈ \mathrm{u}_{n+1}^{(h)} + h ⦇ \mathrm{f}_N(\mathrm{u}_{+1}^{(h)}) + \mathrm{g}_N, \mathrm{u}_{n+1}^{(h)} ⦈ \\
&\le \frac{1}{2}\left(|\mathrm{u}_n^{(h)}|^2 + |\mathrm{u}_{n+1}^{(h)}|^2\right) + h \sum_{|i|\le N} f(\mathrm{u}_{n+1,i}^{(h)})\mathrm{u}_{n+1,i}^{(h)} + \frac{h}{2}\left(- \alpha|\mathrm{u}_{n+1}^{(h)}|^2 + \frac{1}{\alpha}|g_N|^2\right) \\
&\le \frac{1-h\alpha}{2}|\mathrm{u}_{n+1}^{(h)}|^2 + \frac{1}{2}|\mathrm{u}_n^{(h)}|^2 + \frac{1}{2\alpha}\|g\|^2,
\end{aligned}$$

which implies the desired estimate.

Moreover, if $\mathrm{u}_n^{(h)} \in Q_N$, i.e. $|\mathrm{u}_n^{(h)}|^2 \le R^2 = \frac{2\|g\|^2}{\alpha^2} + 1$, then

$$\left|\mathrm{u}_{n+1}^{(h)}\right|^2 \le \frac{1}{1+h\alpha}\left(\frac{2\|g\|^2}{\alpha^2} + 1\right) + \frac{h}{1+h\alpha}\frac{\|g\|^2}{\alpha} \le R^2,$$

which proves the positive invariance of Q_N. □

Since the phase space $Q_N \subset \mathbb{R}^{2N+1}$ is compact, it immediately follows that the truncated IES possesses an attractor $\mathcal{A}_N^{(h)} \subset Q_N$.

Lemma 4.9. *Let Assumptions 3.1 and 3.2 hold and $h \in (0, h^*)$. Then, the discrete semi-dynamical system $\{\varphi_N^{(h)}(n)\}_{n\in\mathbb{N}}$ generated by the finite dimensional IES (4.24) possesses an attractor $\mathcal{A}_N^{(h)} \subset Q_N$.*

4.4.2 *Upper semi continuous convergence*

For any fixed $h \in (0, h^*)$, denote by $\tilde{\mathcal{A}}_N^{(h)}$ the natural embedding of the finite dimensional attractors $\mathcal{A}_N^{(h)}$ into ℓ^2. It will be shown here that they converge upper semi-continuously to the numerical attractor $\mathcal{A}^{(h)}$ in ℓ^2 as $N \to \infty$. This requires the following asymptotic tails estimate.

Lemma 4.10. *Let Assumptions 3.1 and 3.2 hold and $h \in (0, h^*)$. Then, for every $\varepsilon > 0$ there exists $I(\varepsilon) \in \mathbb{N}$ such that*

$$\sum_{I(\varepsilon)\le|i|\le N} (a_i)^2 \le \varepsilon \quad \textit{for all } \boldsymbol{a} = (a_i)_{|i|\le N} \in \mathcal{A}_N^{(h)}.$$

Proof. Let ξ be the smooth cut-off function defined in (3.8), and let $\xi_k(s) = \xi(s/k)$ for $s \in \mathbb{R}^+$, where $k \leq N$ is a fixed integer to be determined. For $\mathfrak{u}_n^{(h)} = (u_{n,i}^{(h)})_{|i|\leq N}$ set $\mathfrak{v}_n^{(h)} = \left(\xi_k(|i|)u_{n,i}^{(h)}\right)_{|i|\leq N}$ for each $n \in \mathbb{N}$.

Taking the inner product of equation (4.24) with $\mathfrak{v}_n^{(h)}$ in $\mathbb{R}^{2N+1}$, using the fact that $(\!|\Lambda_N \mathfrak{u}_n^{(h)}, \mathfrak{v}_n^{(h)}|\!) = -(\!|D_N \mathfrak{u}_n^{(h)}, D_N \mathfrak{v}_n^{(h)}|\!)$, and then following exactly the same arguments as in the proof of Lemma 4.7, it follows that for every $\varepsilon > 0$ there exist $I(\varepsilon)$ and $N^{(h)}(\varepsilon)$ in $\mathbb{N}$ such that for $k \geq I(\varepsilon)$

$$\sum_{2k\leq|i|\leq N} \left(u_{n,i}^{(h)}\right)^2 \leq \sum_{|i|\leq N} \xi_k(|i|)\left(u_{n,i}^{(h)}\right)^2 \leq \varepsilon \quad \text{for all } n \geq N^{(h)}(\varepsilon).$$

Since the global attractor $\mathcal{A}_N^{(h)}$ is invariant under $\varphi_N^{(h)}$, for any $\boldsymbol{a} \in \mathcal{A}_N^{(h)}$ and $N \in \mathbb{N}$ there exists $\mathfrak{u}_o \in \mathcal{A}_N^{(h)}$ such that $\boldsymbol{a} = \mathfrak{u}_N^{(h)}(\mathfrak{u}_o) = \left(u_{N,i}^{(h)}\right)_{|i|\leq N}$. This implies that

$$\sum_{I(\varepsilon)\leq|i|\leq N} a_i^2 = \sum_{I(\varepsilon)\leq|i|\leq N} \left(u_{N,i}^{(h)}\right)^2 \leq \varepsilon \quad \text{for all } N \geq N^{(h)}(\varepsilon),$$

which completes the proof. □

Lemma 4.11. *Let Assumptions 3.1 and 3.2 hold, $h \in (0, h^*)$ be fixed and $\boldsymbol{a}^{(N)} \in \mathcal{A}_N^{(h)}$ for $N \in \mathbb{N}$. Then, there exists a subsequence of the natural extension of $\boldsymbol{a}^{(N)}$ in ℓ^2, denoted by $(\tilde{\boldsymbol{a}}^{(N_j)})_{j\in\mathbb{N}}$ and an element $\boldsymbol{a}^* \in \mathcal{A}^{(h)}$ such that $\tilde{\boldsymbol{a}}^{(N_j)}$ converges to $\boldsymbol{a}^*$ in ℓ^2 as $j \to \infty$.*

Proof. Since h is fixed and the explicit dependence on N is important in this lemma, $\mathfrak{u}_n^{(h,N)}$ will be written as $\mathfrak{u}_n^{(N)}$ for every $\mathfrak{u}_n^{(h,N_j)} = (u_{n,i}^{(h)})_{|i|\leq N}$ and all $j \in \mathbb{N}_0$.

Given $\boldsymbol{a}^{(N)} \in \mathcal{A}_N^{(h)}$, let

$$\mathfrak{u}^{(N)} = \varphi_N^{(h)}(\boldsymbol{a}^{(N)})$$

be the solution of the truncated IES (4.24) with initial data $\mathfrak{u}_o^{(h)} = (u_{o,i})_{|i|\leq N} = \boldsymbol{a}^{(N)}$. Since $\boldsymbol{a}^{(N)} \in \mathcal{A}_N^{(h)}$, it follows that $\mathfrak{u}_n^{(N)} \in \mathcal{A}_N^{(h)}$, and it is an entire solution, i.e., it is defined for all $n \in \mathbb{Z}$.

For each $N \in \mathbb{N}$ and $n \in \mathbb{Z}$ let $\boldsymbol{u}_n^{(N)}$ be the natural extension of $\mathfrak{u}_n^{(N)}$ in ℓ^2, i.e.,

$$\boldsymbol{u}_n^{(N)} = \left(\ldots, 0, \ldots, 0, u_{n,-N}^{(N)}, \ldots, u_{n,0}^{(N)}, \ldots, u_{n,N}^{(N)}, 0, \ldots, 0, \ldots\right).$$

Denote by $\widetilde{Q}_N$ the natural embedding of $Q_N \subset \mathbb{R}^{2N+1}$ in ℓ^2 and observe that $\widetilde{Q}_N \subset Q$ for all $N \in \mathbb{N}$, and hence, $\left(\boldsymbol{u}_n^{(N)}\right)_{N\in\mathbb{N}}$ is uniformly bounded in ℓ^2. Consequently, by a diagonal argument there exists $\boldsymbol{u}_n^* = (u_{n,i}^*)_{i\in\mathbb{Z}} \in \ell^2$ and a subsequence $(u_n^{(N_j)})_{j\in\mathbb{N}}$ converging weakly to $\boldsymbol{u}_n^*$ in ℓ^2,

$$\boldsymbol{u}_n^{(N_j)} \rightharpoonup \boldsymbol{u}_n^* \qquad \text{as } j \to \infty \quad \text{for all } n \in \mathbb{Z}.$$

Similarly, as in the proof of Theorem 4.2, it can be shown that the weak convergence is, in fact, strong, i.e., for all $n \in \mathbb{Z}$,

$$\|\boldsymbol{u}_n^{(N_j)} - \boldsymbol{u}_n^*\| \to 0 \qquad \text{as } j \to \infty, \tag{4.25}$$

and consequently, $\boldsymbol{u}_n^* \in Q$ for all $n \in \mathbb{Z}$.

Next, it will be shown that $\boldsymbol{u}_n^*$ is an entire solution of the infinite dimensional IES (4.13) with $N = N_j$ for all $n \in \mathbb{Z}$. For fixed $k \in \mathbb{Z}$ let $N_j > |i|$. Since $\boldsymbol{u}_n^{(N_j)}$ is a solution of the truncated IES (4.24), it follows that

$$u_{n,i}^{(N_j)} = u_{n-1,i}^{(N_j)} + h\nu\left(u_{n,i-1}^{(N_j)} - 2u_{n,i}^{(N_j)} + u_{n,i+1}^{(N_j)}\right) + hf\left(u_{n,i}^{(N_j)}\right) + hg_i, \quad i \in \mathbb{Z}. \tag{4.26}$$

By estimate (3.5) it follows that

$$\|F(\boldsymbol{u}_n^{(N_j)}) - F(\boldsymbol{u}_n^*)\| \le L_Q \|\boldsymbol{u}_n^{(N_j)} - \boldsymbol{u}_n^*\|.$$

Hence, taking the limit $j \to \infty$ in equation (4.26) and using (4.25) yields

$$u_{n,i}^* = u_{n-1,i}^* + h\nu\left(u_{n,i-1}^* - 2u_{n,i}^* + u_{n,i+1}^*\right) + hf(u_{n,i}^*) + hg_i, \quad i \in \mathbb{Z}.$$

Consequently, $\{\boldsymbol{u}_n^*\}_{n\in\mathbb{Z}}$ is a bounded, entire solution of (4.13) in ℓ^2, which implies that it is contained in the attractor $\mathcal{A}^{(h)}$. In particular, $\boldsymbol{u}_o^* \in \mathcal{A}^{(h)}$, and by (4.25) it follows that

$$\|\boldsymbol{u}_o^{(N_j)} - \boldsymbol{u}_o^*\| = \|\boldsymbol{a}^{(N_j)} - \boldsymbol{u}_o^*\| \to 0 \qquad \text{as } j \to \infty,$$

which concludes the proof of the lemma. □

Theorem 4.4. *Let Assumptions 3.1 and 3.2 hold and $h \in (0, h^*)$ be fixed. Then, the attractors $\mathcal{A}_N^{(h)}$ of the trunctated IES* (4.24) *converge upper semi-continuously to the attractor $\mathcal{A}^{(h)}$ of the IES* (4.13),

$$\lim_{N\to\infty} \operatorname{dist}_{\ell^2}(\mathcal{A}_N^{(h)}, \mathcal{A}^{(h)}) = 0.$$

Proof. Suppose that the assertion is false. Then, there exist $\varepsilon_0 > 0$ and sequences $\boldsymbol{a}^{(N_j)} \in \mathcal{A}_{n_j}^{(h)}$ and $N_j \to \infty$ such that

$$\operatorname{dist}_{\ell^2}\left(\boldsymbol{a}^{(N_j)}, \mathcal{A}^{(h)}\right) \ge \varepsilon_0 \quad \text{for all } j \in \mathbb{N}. \tag{4.27}$$

However, since $\boldsymbol{a}^{(N_j)} \in \mathcal{A}_{N_j}^{(h)}$ and by Lemma 4.11, there exists a subsequence $(\boldsymbol{a}^{(N_{j_k})})_{k\in\mathbb{N}}$ of $(\boldsymbol{a}^{(N_j)})_{j\in\mathbb{N}}$ such that

$$\operatorname{dist}_{\ell^2}\left(\boldsymbol{a}^{(N_{j_k})}, \mathcal{A}^{(h)}\right) \to 0 \quad \text{as} \quad N_{j_k} \to \infty.$$

This contradicts (4.27) and completes the proof. □

4.4.3 *Convergence of numerical attractors*

Combining Theorem 4.3 and Theorem 4.4 the convergence of the approximated numerical attractors $\mathcal{A}_N^{(h)}$ to the attractor $\mathcal{A}$ of the lattice system follows immediately.

Corollary 4.1. *The finite dimensional numerical attractors* $\mathcal{A}_N^{(h)}$ *of* (4.24) *converge upper semi-continuously to the global attractor* $\mathcal{A}$ *of the system* (4.1),

$$\lim_{h\to 0^+}\lim_{N\to\infty}\mathrm{dist}_{\ell^2}(\mathcal{A}_N^{(h)},\mathcal{A})=0.$$

A general theorem in [Kloeden and Lorenz (1986)] can be applied to conclude that a one-step numerical scheme (such as an Euler or Runge-Kutta scheme) with constant time step h applied to the ODE system (4.3) has an attractor $\mathcal{A}_N^{(h)}$, which converges upper semi continuously to $\mathcal{A}_N$ for each N, i.e.,

$$\lim_{h\to 0+}\mathrm{dist}_{\mathbb{R}^{2N+1}}\left(\mathcal{A}_N^{(h)},\mathcal{A}_N\right)=0. \tag{4.28}$$

Thus, combining the convergences (4.2) and (4.28), shows that $\mathcal{A}_N^{(h)}$ can also be used as an approximation for the lattice attractor $\mathcal{A}$ for the LDS (4.1) when h is small enough and N is large enough.

There are thus two approaches to approximating the lattice attractor $\mathcal{A}$ numerically, either by applying a one-step numerical scheme to a finite dimensional system of ODEs approximating the lattice system (4.1) or by applying a one-step numerical scheme such as the IES to (4.1) in ℓ^2 and then approximating it by a finite dimensional counterpart.

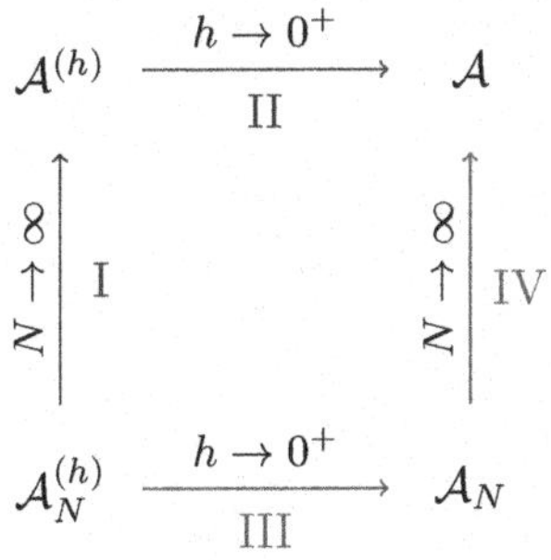

Fig. 4.1 Convergence paths for the approximated numerical attractor to the analytical attractor.

Two different paths for the convergence of the finite dimensional numerical attractors $\mathcal{A}_N^{(h)}$ to the analytical attractor $\mathcal{A}$ are illustrated in Figure 4.1. Bates, Lu & Wang proved the convergence IV in [Bates *et al.* (2001)] (presented in the first part of this chapter), while the convergence III follows by a general result of Kloeden & Lorenz in [Kloeden and Lorenz (1986)] on the discretisation of attractors of ordinary differential equations (see also [Han and Kloeden (2017b)], Chapter 9). Han, Kloeden & Sonner [Han *et al.* (2020a)] proved the convergence via paths I and II.

4.5 End notes

The first part of this chapter is taken from Bates, Lu & Wang [Bates *et al.* (2001)], while the second part is based on Han, Kloeden & Sonner [Han *et al.* (2020a)]. Approximation results are also given in [Zhou (2004)].

The discretisation of attractors in ODEs (rather than LDS) is discussed in the monographs [Han and Kloeden (2017b); Stuart and Humphries (1996)] on numerical dynamics.

4.6 Problems

Problem 4.1. Determine conditions to ensure that the attractors of the finite dimensional approximating systems converge to the attractor of the lattice system not just upper semi-continuously, but in the Hausdorff metric.

Problem 4.2. Investigate the technical difficulties that arise when the explicit Euler scheme is used in Section 4.3 instead of the implicit Euler scheme.

Chapter 5

Non-autonomous Laplacian lattice systems in weighted sequence spaces

Lattice dynamical systems with the discrete Laplacian operator are considerably more difficult to analyse in sequence spaces ℓ_ρ^2 with weighted norms than in the space ℓ^2 that was used in Chapter 3. Such spaces are important in applications as they include travelling wave solutions which do not exist in the space ℓ^2 since the components of its sequence elements become vanishingly small as the index increases in magnitude.

For appropriately chosen weights, the space ℓ^2 is densely contained in ℓ_ρ^2. Bixiang Wang [Wang (2006)] used this to determine first the existence and uniqueness of solutions of an autonomous system in ℓ^2 and then to extend the solution mapping to a semi-group on the larger space ℓ_ρ^2.

The aim here is to show how this approach can be applied to a non-autonomous lattice system, to prove that it generates a non-autonomous process or two-parameter semi-group on ℓ_ρ^2 and to establish the existence of a non-autonomous attractor. A simple non-autonomous lattice system with a discrete Laplacian is chosen as an illustrative example here, so attention can be focused on ideas and techniques needed to use this extension.

Consider the non-autonomous lattice system

$$\frac{du_i}{dt} = \nu\left(u_{i-1} - 2u_i + u_{i+1}\right) + f(u_i) + g_i(t), \quad i \in \mathbb{Z}. \tag{5.1}$$

The LDS (5.1) here differs from the LDS (3.1) in Chapter 3 by the time dependence of the forcing term. The following Assumptions will be used.

Assumption 5.1. $f : \mathbb{R} \to \mathbb{R}$ is a continuously differentiable function satisfying

$$f(0) = 0, \quad f(x)x \le -\alpha x^2 + \beta, \quad f'(x) \le D_f \qquad \forall\, x \in \mathbb{R},$$

for positive constants α, β, $D_f > 0$.

Assumption 5.2. $\boldsymbol{g} = (g_i)_{i\in\mathbb{Z}} : \mathbb{R} \to \ell^2$ is continuous and uniformly bounded with $\sup_{t\in\mathbb{R}} \|\boldsymbol{g}(t)\| = M_{\boldsymbol{g}} < \infty$.

Remark 5.1. The uniformly boundedness of $\boldsymbol{g}$ can be weakened as in Chapter 3. It is used here to simplify various estimates. It nevertheless includes important situations with periodic, almost periodic forcing or even more general forcing, see [Kloeden and Rodrigues (2011)].

5.1 The discrete Laplacian on weighted sequence spaces

Given a sequence of positive weights $(\rho_i)_{i\in\mathbb{Z}}$, we consider the separable Hilbert space

$$\ell^2_\rho := \left\{\boldsymbol{u} = (u_i)_{i\in\mathbb{Z}} : \sum_{i\in\mathbb{Z}} \rho_i u_i^2 < \infty\right\}$$

with the inner product

$$\langle \boldsymbol{u}, \boldsymbol{v}\rangle_\rho := \sum_{i\in\mathbb{Z}} \rho_i u_i v_i \quad \text{for } \boldsymbol{u} = (u_i)_{i\in\mathbb{Z}}, \boldsymbol{v} = (v_i)_{i\in\mathbb{Z}} \in \ell^2_\rho$$

and the norm

$$\|\boldsymbol{u}\|_\rho := \sqrt{\sum_{i\in\mathbb{Z}} \rho_i u_i^2} \quad \text{for } \boldsymbol{u} = (u_i)_{i\in\mathbb{Z}} \in \ell^2_\rho.$$

The weights ρ_i are often assumed to satisfy the Assumptions 1.1 and 1.2, which are restated below for the reader's convenience.

Assumption 5.3. $\rho_i > 0$ for all $i \in \mathbb{Z}$ and $\rho_\Sigma := \sum_{i\in\mathbb{Z}} \rho_i < \infty$.

Assumption 5.4. There exist positive constants γ_0 and γ_1 such that

$$\rho_{i\pm1} \le \gamma_0 \rho_i, \qquad |\rho_{i\pm1} - \rho_i| \le \gamma_1 \rho_i \quad \text{for all } i \in \mathbb{Z}.$$

The following lemma is based on Lemma 5.1 in [Zhao and Zhou (2008)].

Lemma 5.1. *Let Assumption 5.3 hold. Then ℓ^2 is dense in ℓ^2_ρ.*

Proof. The inclusion $\ell^2 \subset \ell^2_\rho$ follows from Lemma 1.1, so it remains to show that the inclusion is dense. To that end, let $\boldsymbol{u} = (u_i)_{i\in\mathbb{Z}} \in \ell^2_\rho$ be a given element of ℓ^2_ρ. Then for any $\varepsilon > 0$ there exists $I(\varepsilon) \in \mathbb{N}$ such that $\sum_{|i|>I(\varepsilon)} \rho_i u_i^2 < \varepsilon^2$.

Choose $\tilde{\boldsymbol{u}} = (\tilde{u}_i)_{i\in\mathbb{Z}}$ such that $\tilde{u}_i = u_i$ for $|i| \le I(\varepsilon)$ and $\tilde{u}_i = 0$ for all $|i| > I(\varepsilon)$. Then

$$\|\tilde{\boldsymbol{u}}\|^2 = \sum_{i\in\mathbb{Z}} \tilde{u}_i^2 = \sum_{|i|\le I(\varepsilon)} u_i^2 \le \frac{1}{\min_{|i|\le I(\varepsilon)} \rho_i} \sum_{|i|\le I(\varepsilon)} \rho_i u_i^2 \le \frac{1}{\min_{|i|\le I(\varepsilon)} \rho_i} \|\boldsymbol{u}\|^2_\rho,$$

that is, $\tilde{\boldsymbol{u}} = (\tilde{u}_i)_{i\in\mathbb{Z}} \in \ell^2$. Moreover, $\|\boldsymbol{u} - \tilde{\boldsymbol{u}}\|^2_\rho = \sum_{|i|>I(\varepsilon)} \rho_i u_i^2 < \varepsilon^2$. This implies the density of ℓ^2 in ℓ^2_ρ. □

Properties of the discrete Laplacian operator in ℓ^2_ρ

For any $\boldsymbol{u} = (u_i)_{i\in\mathbb{Z}} \in \ell^2_\rho$, the discrete Laplace operator Λ and linear operators D^+ and D^- are defined in the same way as in Chapter 3, i.e., component wise by

$$(\Lambda \boldsymbol{u})_i = u_{i-1} - 2u_i + u_{i+1}, \quad (\mathrm{D}^+\boldsymbol{u})_i = u_{i+1} - u_i, \quad (\mathrm{D}^-\boldsymbol{u})_i = u_{i-1} - u_i, \quad i \in \mathbb{Z}.$$

Recall that $-\Lambda = \mathrm{D}^+\mathrm{D}^- = \mathrm{D}^-\mathrm{D}^+$.

Lemma 5.2. *Let Assumption 5.4 hold. Then the operators Λ, D^+ and D^- are bounded from ℓ^2_ρ into ℓ^2_ρ.*

Proof. For each $\boldsymbol{u} \in \ell^2_\rho$

$$\begin{aligned}
\|\mathrm{D}^+\boldsymbol{u}\|^2_\rho &= \sum_{i\in\mathbb{Z}} \rho_i \left(\mathrm{D}^+\boldsymbol{u}\right)^2_i = \sum_{i\in\mathbb{Z}} \rho_i \left(u_{i+1} - u_i\right)^2 \\
&\le 2\sum_{i\in\mathbb{Z}} \rho_i u^2_{i+1} + 2\sum_{i\in\mathbb{Z}} \rho_i u^2_i \le 2\gamma_0 \sum_{i\in\mathbb{Z}} \rho_{i+1} u^2_{i+1} + 2\sum_{i\in\mathbb{Z}} \rho_i u^2_i \\
&\le 2(\gamma_0 + 1)\|\boldsymbol{u}\|^2_\rho,
\end{aligned}$$

and similarly for D^-. Then

$$\|\Lambda \boldsymbol{u}\|^2_\rho = \sum_{i\in\mathbb{Z}} \rho_i \left(\mathrm{D}^+\mathrm{D}^-\boldsymbol{u}\right)^2_i \le 2(\gamma_0 + 1)\|\mathrm{D}^-\boldsymbol{u}\|^2_\rho \le 4(\gamma_0 + 1)^2\|\boldsymbol{u}\|^2_\rho.$$

□

Lemma 5.3. *Let Assumption 5.4 hold. Then for all $\boldsymbol{u}, \boldsymbol{v} \in \ell^2_\rho$*

$$\left\langle \mathrm{D}^-\boldsymbol{u}, \boldsymbol{v}\right\rangle_\rho = \sum_{i\in\mathbb{Z}} u_i(\mathrm{D}^+(\rho \boldsymbol{v}))_i, \qquad \left\langle \boldsymbol{u}, \mathrm{D}^+\boldsymbol{v}\right\rangle_\rho = \sum_{i\in\mathbb{Z}} v_i(\mathrm{D}^-(\rho \boldsymbol{u}))_i.$$

Proof. Given any $\boldsymbol{u}, \boldsymbol{v} \in \ell^2_\rho$

$$\begin{aligned}
\left\langle \mathrm{D}^-\boldsymbol{u}, \boldsymbol{v}\right\rangle_\rho = \sum_{i\in\mathbb{Z}} \rho_i v_i \left(u_{i-1} - u_i\right) &= \sum_{i\in\mathbb{Z}} \rho_i v_i u_{i-1} - \sum_{i\in\mathbb{Z}} \rho_i v_i u_i \\
&= \sum_{j\in\mathbb{Z}} \rho_{j+1} v_{j+1} u_j - \sum_{i\in\mathbb{Z}} \rho_i v_i u_i \\
&= \sum_{i\in\mathbb{Z}} \left(\rho_{i+1} v_{i+1} - \rho_i v_i\right) u_i = \sum_{i\in\mathbb{Z}} u_i(\mathrm{D}^+(\rho \boldsymbol{v}))_i,
\end{aligned}$$

and similarly for the other expression. □

Lemma 5.4. *Let Assumption 5.4 hold. Then for all $\boldsymbol{u}, \boldsymbol{v} \in \ell^2_\rho$*

$$\langle \Lambda \boldsymbol{u}, \boldsymbol{v}\rangle_\rho = -\langle \mathrm{D}^+\boldsymbol{u}, \mathrm{D}^+\boldsymbol{v}\rangle_\rho + \sum_{i\in\mathbb{Z}} (D^+\boldsymbol{u})_i \left(\rho_i(D^+\boldsymbol{v})_i - (D^+(\rho \boldsymbol{v}))_i\right).$$

Proof. By Lemma 5.3

$$\langle \Lambda \boldsymbol{u}, \boldsymbol{v} \rangle_\rho = \left\langle -(\mathrm{D}^-\mathrm{D}^+\boldsymbol{u}), \boldsymbol{v} \right\rangle_\rho = -\sum_{i\in\mathbb{Z}} (\mathrm{D}^+\boldsymbol{u})_i (\mathrm{D}^+(\rho\boldsymbol{v}))_i$$

Then use

$$-(\mathrm{D}^+(\rho\boldsymbol{v}))_i = -\rho_i(\mathrm{D}^+\boldsymbol{v})_i + \rho_i(\mathrm{D}^+\boldsymbol{v})_i - (\mathrm{D}^+(\rho\boldsymbol{v}))_i$$

to obtain the result. □

Expanding component wise gives

$$\rho_i(\mathrm{D}^+\boldsymbol{v})_i - (\mathrm{D}^+(\rho\boldsymbol{v}))_i = -\left(\rho_{i+1} - \rho_i\right) v_{i+1}$$

which implies that

$$\left|\rho_i(\mathrm{D}^+\boldsymbol{v})_i - (\mathrm{D}^+(\rho\boldsymbol{v}))_i\right| = |\rho_{i+1} - \rho_i|\,|v_{i+1}| \le \gamma_1 \rho_i |v_{i+1}|.$$

Consequently, we have

$$\begin{aligned}
\left|\sum_{i\in\mathbb{Z}} (\mathrm{D}^+\boldsymbol{u})_i \left((\mathrm{D}^+(\rho\boldsymbol{v}))_i - \rho_i(D^+\boldsymbol{v})_i\right)\right| &\le \gamma_1 \sum_{i\in\mathbb{Z}} \rho_i |(\mathrm{D}^+\boldsymbol{u})_i||v_{i+1}| \\
&\le \frac{1}{2}\sum_{i\in\mathbb{Z}} \left(\rho_i (\mathrm{D}^+\boldsymbol{u})_i^2 + \gamma_1^2 \rho_i |v_{i+1}|^2\right) \\
&\le \frac{1}{2}\left\|\mathrm{D}^+\boldsymbol{u}\right\|_\rho^2 + \frac{1}{2}\gamma_0\gamma_1^2 \sum_{i\in\mathbb{Z}} \rho_{i+1} v_{i+1}^2 \\
&\le \frac{1}{2}\left\|\mathrm{D}^+\boldsymbol{u}\right\|_\rho^2 + \frac{1}{2}\gamma_0\gamma_1^2 \left\|\boldsymbol{v}\right\|_\rho^2.
\end{aligned}$$

Hence the following lemma holds.

Lemma 5.5. *Let Assumption 5.4 hold. Then for all $\boldsymbol{u} \in \ell^2_\rho$*

$$\langle \Lambda \boldsymbol{u}, \boldsymbol{u} \rangle_\rho \le -\frac{1}{2}\left\|\mathrm{D}^+\boldsymbol{u}\right\|_\rho + \frac{1}{2}\gamma_0\gamma_1^2 \left\|\boldsymbol{u}\right\|_\rho^2.$$

5.2 Generation of a non-autonomous dynamical system on ℓ^2_ρ

Under Assumptions 5.1 and 5.2 the LDS (5.1) can be written as the infinite dimensional ordinary differential equation (ODE) in the sequence space ℓ^2

$$\frac{\mathrm{d}\boldsymbol{u}}{\mathrm{d}t} = \nu\Lambda\boldsymbol{u} + F(\boldsymbol{u}) + \boldsymbol{g}(t) =: \mathfrak{F}(\boldsymbol{u}, t), \tag{5.2}$$

where $\boldsymbol{g}(t) = (g_i(t))_{i\in\mathbb{Z}}$ is continuous and the Nemytskii operator $F : \ell^2 \to \ell^2$ is defined componentwise by

$$F(\boldsymbol{u}) = (F_i(\boldsymbol{u}))_{i\in\mathbb{Z}} \quad \text{with} \quad F_i(\boldsymbol{u}) = f(u_i), \quad i \in \mathbb{Z}.$$

Theorem 5.1. *Under Assumptions 5.1–5.4 one can associate the initial value problem for the lattice system (5.2) with a two-parameter semi-group or process $\{\psi(t,t_0)\}_{(t,t_0)\in\mathbb{R}^2_\ge}$ in ℓ^2_ρ such that $\psi(t,t_0,\boldsymbol{u}_o)$ is the unique solution of (5.1) with the initial value $\boldsymbol{u}_o \in \ell^2$.*

The proof consists of several parts. Firstly the existence and uniqueness of solutions of (5.2) in ℓ^2 will be established and such solutions for fixed initial time t_0, which can be considered as functions in ℓ^2_ρ since ℓ^2 is contained in ℓ^2_ρ, will be shown to be globally Lipschitz in the norm of ℓ^2_ρ. This allows the solution mapping to be extended from $\mathcal{C}\left([t_0,T],\ell^2\right)$ to $\mathcal{C}\left([t_0,T],\ell^2_\rho\right)$. Since the solutions are also continuous in the initial times and satisfy a two-parameter semi-group property, these extended functions in ℓ^2_ρ then form a two-parameter semi-group on the space ℓ^2_ρ.

The proofs below require the fact that the mapping F satisfies a one-sided Lipschitz condition in ℓ^2 as well as in ℓ^2_ρ, which follows from Assumption 5.1.

Lemma 5.6. *Let Assumptions 5.1 hold. Then*

$$\langle F(\boldsymbol{u}) - F(\boldsymbol{v}), \boldsymbol{u} - \boldsymbol{v}\rangle_\rho \leq D_f \left\|\boldsymbol{u} - \boldsymbol{v}\right\|_\rho^2, \quad \forall\, \boldsymbol{u} = (u_i)_{i\in\mathbb{Z}},\ \boldsymbol{v} = (v_i)_{i\in\mathbb{Z}} \in \ell^2_\rho.$$

Proof. By the mean value theorem for derivatives there exists ζ_{xy} between x and y in $\mathbb{R}$, such that

$$f(x) - f(y) = f'(\zeta_{xy})\,(x-y)$$

and thus, by Assumption 5.1,

$$(f(x) - f(y))\,(x-y) = f'(\zeta_{xy})\,(x-y)^2 \leq D_f\,(x-y)^2.$$

Hence for $\boldsymbol{u} = (u_i)_{i\in\mathbb{Z}}$ and $\boldsymbol{v} = (v_i)_{i\in\mathbb{Z}} \in \ell^2_\rho$,

$$\rho_i\,(f(u_i) - f(v_i))\,(u_i - v_i) \leq D_f\ \rho_i\,(u_i - v_i)^2, \quad i \in \mathbb{Z}.$$

Summing the above inequality over $i \in \mathbb{Z}$ gives the result. A similar inequality holds in ℓ^2. □

Remark 5.2. Since $f(0) = 0$, then with $\boldsymbol{v} = \boldsymbol{0}$ Lemma 5.6 reduces to

$$\langle F(\boldsymbol{u}), \boldsymbol{u}\rangle_\rho \leq D_f \left\|\boldsymbol{u}\right\|_\rho^2.$$

5.2.1 *Existence and uniqueness of solutions in* ℓ^2

As seen in Chapter 3, Assumption 5.1 ensures that $F : \ell^2 \to \ell^2$ satisfies $F(\boldsymbol{0}) = \boldsymbol{0}$ and is Lipschitz on any bounded set $B \subset \ell^2$ with

$$\|F(\boldsymbol{u}) - F(\boldsymbol{v})\| \leq L_B \|u - v\|$$

for the local Lipschitz constant $L_B > 0$.

Hence the mapping $\mathfrak{F}(\boldsymbol{u}, t) : \ell^2 \times \mathbb{R} \to \ell^2$ is continuous in both variables and locally Lipschitz continuous in its first variable. Moreover, the solutions are bounded on bounded time intervals. From Remark 5.2 and analysis similar to Section 3.2.1

$$\frac{\mathrm{d}}{\mathrm{d}t}\|\boldsymbol{u}\|^2 \leq (2D_f + \alpha)\,\|\boldsymbol{u}\|^2 + \frac{1}{\alpha}\|\boldsymbol{g}(t)\|^2, \quad t \geq t_0,$$

and hence by Gronwall's inequality

$$\|\boldsymbol{u}(t)\|^2 \leq \|\boldsymbol{u}(t_0)\|^2 e^{(2D_f+\alpha)(t-t_0)} + \frac{1}{\alpha\,(2D_f+\alpha)} M_{\boldsymbol{g}}^2 \left(e^{(2D_f+\alpha)(t-t_0)} - 1\right), \quad t \geq t_0.$$

Given $t_0 \in \mathbb{R}$ and initial data $\boldsymbol{u}(t_0) = \boldsymbol{u}_o \in \ell^2$, the existence and uniqueness of a global solution $\boldsymbol{u}(\,\cdot\,; t_0, \boldsymbol{u}_o) \in \mathcal{C}([t_0;\infty), \ell^2)$ of the ODE system (5.2) in the Hilbert space ℓ^2 follows by standard arguments. Moreover, these solutions are continuous in their initial data and by the uniqueness of solutions satisfy the two-parameter semi-group property:

$$\boldsymbol{u}(t_2; t_0, \boldsymbol{u}_o) = \boldsymbol{u}(t_2; t_1, \boldsymbol{u}(t_1; t_0, \boldsymbol{u}_o)), \qquad t_0 \leq t_1 \leq t_2.$$

5.2.2 *Lipschitz continuity of solutions in initial data in the ℓ^2_ρ norm*

The next lemma establishes the Lipschitz continuity of these solutions in ℓ^2_ρ which is required for the extension of the solution operator from ℓ^2 to ℓ^2_ρ.

Lemma 5.7. *Let Assumptions 5.1–5.4 hold and let $\boldsymbol{u}(t) = \boldsymbol{u}(t; t_0, \boldsymbol{u}_o)$ and $\boldsymbol{v}(t) = \boldsymbol{v}(t; t_0, \boldsymbol{v}_o)$ be two solutions of* (5.2) *for the initial values $\boldsymbol{u}_o = \boldsymbol{u}(t_0)$ and $\boldsymbol{v}_o = \boldsymbol{v}(t_0)$ in ℓ^2, respectively, at time $t_0 \in \mathbb{R}$. Then for each $T > 0$ there exists a constant K_T depending on T such that*

$$\|\boldsymbol{u}(t; t_0, \boldsymbol{u}_o) - \boldsymbol{v}(t; t_0, \boldsymbol{v}_o)\|_\rho \leq K_T \|\boldsymbol{u}_o - \boldsymbol{v}_o\|_\rho \qquad t_0 \leq t \leq T.$$

Proof. Define $\boldsymbol{w}(t) = \boldsymbol{u}(t; t_0, \boldsymbol{u}_o) - \boldsymbol{v}(t; t_0, \boldsymbol{v}_o)$. Then, from (5.2)

$$\frac{\mathrm{d}\boldsymbol{w}}{\mathrm{d}t} = \nu\Lambda\boldsymbol{w} + F(\boldsymbol{u}) - F(\boldsymbol{v}).$$

Taking the inner product in ℓ^2_ρ with $\boldsymbol{w}$ gives

$$\frac{1}{2}\frac{\mathrm{d}}{\mathrm{d}t}\|\boldsymbol{w}\|_\rho^2 = \nu\left\langle \Lambda\boldsymbol{w}, \boldsymbol{w}\right\rangle_\rho + \left\langle F(\boldsymbol{u}) - F(\boldsymbol{v}), \boldsymbol{w}\right\rangle_\rho \leq \left\langle \Lambda\boldsymbol{w}, \boldsymbol{w}\right\rangle_\rho + D_f\|\boldsymbol{w}\|_\rho^2,$$

using Lemma 5.6. Then by Lemma 5.5

$$\begin{aligned}\frac{1}{2}\frac{\mathrm{d}}{\mathrm{d}t}\|\boldsymbol{w}\|_\rho^2 &\leq \nu\left\langle \Lambda\boldsymbol{w}, \boldsymbol{w}\right\rangle_\rho + D_f\|\boldsymbol{w}\|_\rho^2 \\ &\leq -\frac{\nu}{2}\left\|\mathrm{D}^+\boldsymbol{w}\right\|_\rho + \frac{\nu}{2}\gamma_0\gamma_1^2\|\boldsymbol{w}\|_\rho^2 + D_f\|\boldsymbol{w}\|_\rho^2,\end{aligned}$$

from which follows

$$\frac{\mathrm{d}}{\mathrm{d}t}\|\boldsymbol{w}\|_\rho^2 \leq \left(\nu\gamma_0\gamma_1^2 + 2D_f\right)\|\boldsymbol{w}\|_\rho^2.$$

Integrating the differential inequality above gives

$$\|\boldsymbol{w}(t)\|_\rho^2 \leq e^{\left(\nu\gamma_0\gamma_1^2 + 2D_f\right)(t-t_0)}\|\boldsymbol{w}(t_0)\|_\rho^2, \qquad t_0 \leq t \leq T,$$

and thus

$$\|\boldsymbol{u}(t; t_0, \boldsymbol{u}_o) - \boldsymbol{v}(t; t_0, \boldsymbol{v}_o)\|_\rho \leq K_T\|\boldsymbol{u}_o - \boldsymbol{v}_o\|_\rho, \qquad t_0 \leq t \leq T,$$

where $K_T = e^{\frac{1}{2}(\nu\gamma_0\gamma_1^2 + 2D_f)T}$. □

Lemma 5.8. *Let $\boldsymbol{u}(\cdot;t_0,\boldsymbol{u}_o)$ be a solution of (5.2) for the initial value $\boldsymbol{u}_o \in \ell^2$ at time $t_0 \in \mathbb{R}$. Then there exist constants K_1 and K_2 with K_2 depending on $\boldsymbol{u}_o$ such that*

$$\|\boldsymbol{u}(t;t_0,\boldsymbol{u}_o) - \boldsymbol{u}_o\|_\rho \le K_2 e^{K_1(t-t_0)} \quad \forall\, t > t_0.$$

Proof. Suppose that $t > t_0$ and define $\boldsymbol{w}(t) = \boldsymbol{u}(t;t_0,\boldsymbol{u}_o) - \boldsymbol{u}_o$. Then $\boldsymbol{w}(t)$ satisfies

$$\frac{\mathrm{d}\boldsymbol{w}}{\mathrm{d}t} = \nu\Lambda\boldsymbol{w} + \nu\Lambda\boldsymbol{u}_o + F(\boldsymbol{u}) + \boldsymbol{g}(t).$$

Taking inner product $\langle\cdot,\cdot\rangle_\rho$ of the above equation with $\boldsymbol{w}(t)$ gives

$$\begin{aligned}\frac{1}{2}\frac{\mathrm{d}}{\mathrm{d}t}\|\boldsymbol{w}\|_\rho^2 &= \nu\left\langle \Lambda\boldsymbol{w},\boldsymbol{w}\right\rangle_\rho + \nu\left\langle \Lambda\boldsymbol{u}_o,\boldsymbol{w}\right\rangle_\rho \\ &\quad + \left\langle F(\boldsymbol{u}) - F(\boldsymbol{u}_o),\boldsymbol{w}\right\rangle_\rho + \left\langle F(\boldsymbol{u}_o),\boldsymbol{w}\right\rangle_\rho + \left\langle \boldsymbol{g}(t),\boldsymbol{w}\right\rangle_\rho . \end{aligned} \tag{5.3}$$

By Assumptions 5.2–5.3, Hölder inequality and Lemma 5.2 we have

$$\begin{aligned}\left\langle \Lambda\boldsymbol{u}_o,\boldsymbol{w}\right\rangle_\rho &\le \|\Lambda\boldsymbol{u}_o\|_\rho \cdot \|\boldsymbol{w}\|_\rho \le 2(\gamma_0+1)\|\boldsymbol{u}_o\|_\rho \cdot \|\boldsymbol{w}\|_\rho \\ &\le (\gamma_0+1)\|\boldsymbol{u}_o\|_\rho^2 + (\gamma_0+1)\|\boldsymbol{w}\|_\rho^2, \end{aligned} \tag{5.4}$$

$$\left\langle F(\boldsymbol{u}) - F(\boldsymbol{u}_o),\boldsymbol{w}\right\rangle_\rho \le D_f\|\boldsymbol{w}\|_\rho^2, \tag{5.5}$$

$$\left\langle F(\boldsymbol{u}_o),\boldsymbol{w}\right\rangle_\rho \le \|F(\boldsymbol{u}_o)\|_\rho \cdot \|\boldsymbol{w}\|_\rho \le \frac{1}{2}\big(\|F(\boldsymbol{u}_o)\|_\rho^2 + \|\boldsymbol{w}\|_\rho^2\big), \tag{5.6}$$

$$\left\langle \boldsymbol{g}(t),\boldsymbol{w}\right\rangle_\rho \le \|\boldsymbol{g}(t)\|_\rho \cdot \|\boldsymbol{w}\|_\rho \le \frac{1}{2}\big(\|\boldsymbol{w}\|_\rho^2 + M_{\boldsymbol{g}}^2\rho_\Sigma\big). \tag{5.7}$$

Inserting (5.4)–(5.7) into (5.3) and using Lemma 5.4 we obtain

$$\begin{aligned}\frac{\mathrm{d}}{\mathrm{d}t}\|\boldsymbol{w}\|_\rho^2 &\le 2\nu\left\langle \Lambda\boldsymbol{w},\boldsymbol{w}\right\rangle_\rho + 2\nu(\gamma_0+1)\left(\|\boldsymbol{u}_o\|_\rho^2 + \|\boldsymbol{w}\|_\rho^2\right) + 2D_f\|\boldsymbol{w}\|_\rho^2 \\ &\quad + \|F(\boldsymbol{u}_o)\|_\rho^2 + 2\|\boldsymbol{w}\|_\rho^2 + M_{\boldsymbol{g}}^2\rho_\Sigma \\ &\le -\nu\left\|\mathrm{D}^+\boldsymbol{w}\right\|_\rho^2 + \left(\nu\gamma_0\gamma_1^2 + 2\nu(\gamma_0+1) + 2D_f + 2\right)\|\boldsymbol{w}\|_\rho^2 \\ &\quad + 2\nu(\gamma_0+1)\|\boldsymbol{u}_o\|_\rho^2 + \|F(\boldsymbol{u}_o)\|_\rho^2 + M_{\boldsymbol{g}}^2\rho_\Sigma \\ &\le K_1\|\boldsymbol{w}\|_\rho^2 + K_1K_2 \end{aligned} \tag{5.8}$$

with

$$K_1 := \nu\gamma_0\gamma_1^2 + 2\nu(\gamma_0+1) + 2D_f + 2,$$

$$K_2 := \frac{2\nu(\gamma_0+1)\|\boldsymbol{u}_o\|_\rho^2 + \|F(\boldsymbol{u}_o)\|_\rho^2 + M_{\boldsymbol{g}}^2\rho_\Sigma}{\nu\gamma_0\gamma_1^2 + 2\nu(\gamma_0+1) + 2D_f + 2}.$$

Integrating (5.8) from t_0 to t and using $\boldsymbol{w}(t_0) = \mathbf{0}$ then results in

$$\|\boldsymbol{u}(t;t_0,\boldsymbol{u}_o) - \boldsymbol{u}_o\|_\rho^2 \le K_2 e^{K_1(t-t_0)}.$$

The proof is complete. □

It follows from the above lemmas that

$$\begin{aligned}\|\boldsymbol{u}(t;t_0,\boldsymbol{u}_o)-\boldsymbol{v}(t;s,\boldsymbol{v}_o)\|_\rho &\le K_T\,\|\boldsymbol{u}(s;t_0,\boldsymbol{u}_o)-\boldsymbol{v}_o\|_\rho\\ &\le K_T\left(\|\boldsymbol{u}(s;t_0,\boldsymbol{u}_o)-\boldsymbol{u}_o\|_\rho+\|\boldsymbol{u}_o-\boldsymbol{v}_o\|_\rho\right),\\ &\le K_T\left(K_2^{\frac{1}{2}}e^{\frac{1}{2}K_1(s-t_0)}+\|\boldsymbol{u}_o-\boldsymbol{v}_o\|_\rho\right),\quad t_0\le s\le t\le T.\end{aligned}$$

Thus the solution mapping is continuous in its initial data.

5.2.3 *Generation of semi-group on ℓ_p^2*

Fix $t_0 < T$ in $\mathbb{R}$. Then by Lemma 5.7 there exists a mapping $\mathcal{S}$ from ℓ^2 into $\mathcal{C}\left([t_0,T],\ell_\rho^2\right)$ such that $\mathcal{S}(\boldsymbol{u}_o)$ for each $\boldsymbol{u}_o\in\ell^2$ is the unique solution $\boldsymbol{u}(t;t_0,\boldsymbol{u}_o)$ of the initial value problem for the system (5.2) with initial time t_0. Further, $\mathcal{S}$ is continuous in initial data from $\ell^2\subset\ell_\rho^2$ into $\mathcal{C}\left([t_0,T],\ell_\rho^2\right)$. Since ℓ^2 is dense ℓ_ρ^2, following Bixiang Wang [Wang (2006)], the mapping $\mathcal{S}$ can be extended uniquely to a mapping $\widehat{\mathcal{S}}$ from ℓ_ρ^2 into $\mathcal{C}\left([t_0,T],\ell_\rho^2\right)$.

Now for fixed t_0 and $t\ge t_0$ define mappings $\psi(t,t_0,\cdot):\ell_\rho^2\to\ell_\rho^2$ such that $\psi(t,t_0,\boldsymbol{u}_o)=\widehat{\mathcal{S}}(\boldsymbol{u}_o)(t)$ for every $t\in[t_0,T]$ and $\boldsymbol{u}_o\in\ell_\rho^2$. By Lemmas 5.7 and 5.8, the mapping ψ is continuous in both t_0 and $\boldsymbol{u}_o$.

Finally, $\{\psi(t,t_0)\}_{(t,t_0)\in\mathbb{R}^2_\ge}$ inherits the two-parameter semi-group property of the solutions of the system (5.2) in ℓ^2. Hence ψ is a non-autonomous process or two-parameter semi-group on ℓ_ρ^2. The proof of Theorem 5.1 is complete. □

5.3 Existence of pullback attractors

From now on, this process $\{\psi(t,t_0)\}_{(t,t_0)\in\mathbb{R}^2_\ge}$ defined in Section 5.2.3 is referred to as the process generated by the non-autonomous system (5.2). The main result on the existence of pullback attractors is stated in the following theorem.

Theorem 5.2. *Suppose that Assumptions 5.1–5.4 are satisfied and that*

$$\alpha>\nu\gamma_1\max\left\{\gamma_0\gamma_1,\gamma_0+1\right\}. \tag{5.9}$$

Then the process $\{\psi(t,t_0)\}_{(t,t_0)\in\mathbb{R}^2_\ge}$ generated by the system (5.2) on ℓ_ρ^2 has a pullback attractor $\mathcal{A}=(A_t)_{t\in\mathbb{R}}$ in ℓ_ρ^2, which pullback attracts families of bounded subsets of ℓ_ρ^2.

5.3.1 *Existence of an absorbing set*

Lemma 5.9. *Suppose that $\alpha>\nu\gamma_0\gamma_1^2$. Then the process $\{\psi(t,t_0)\}_{(t,t_0)\in\mathbb{R}^2_\ge}$ on ℓ_ρ^2 generated by the system (5.2) has a closed and bounded subset of ℓ_ρ^2 which is positively invariant and absorbing in both the forward and pullback senses.*

Proof. Taking the inner product in ℓ_ρ^2 of the equation (5.2) with $\boldsymbol{u} = \boldsymbol{u}(t; t_0, \boldsymbol{u}_o)$ and using Lemma 5.5 gives

$$\begin{aligned}\frac{\mathrm{d}}{\mathrm{d}t}\|\boldsymbol{u}\|_\rho^2 &= 2\nu\left\langle \Lambda\boldsymbol{u}, \boldsymbol{u}\right\rangle_\rho + 2\langle F(\boldsymbol{u}), \boldsymbol{u}\rangle_\rho + 2\langle \boldsymbol{g}(t), \boldsymbol{u}\rangle_\rho \\ &\leq -\nu\left\|\mathrm{D}^+\boldsymbol{u}\right\|_\rho + \nu\gamma_0\gamma_1^2\left\|\boldsymbol{u}\right\|_\rho^2 - 2\alpha\left\|\boldsymbol{u}\right\|_\rho^2 + 2\beta\rho_\Sigma + 2\langle \boldsymbol{g}(t), \boldsymbol{u}\rangle_\rho \\ &\leq \nu\gamma_0\gamma_1^2\left\|\boldsymbol{u}\right\|_\rho^2 - \alpha\left\|\boldsymbol{u}\right\|_\rho^2 + 2\beta\rho_\Sigma + \frac{4}{\alpha}\|\boldsymbol{g}(t)\|_\rho^2\end{aligned}$$

and hence

$$\frac{\mathrm{d}}{\mathrm{d}t}\|\boldsymbol{u}\|_\rho^2 \leq -K_1\|\boldsymbol{u}\|_\rho^2 + 2\left(\beta + \frac{2}{\alpha}M_{\boldsymbol{g}}^2\right)\rho_\Sigma,$$

where $K_1 = \alpha - \nu\gamma_0\gamma_1^2 > 0$. The Gronwall inequality then gives

$$\|\boldsymbol{u}(t; t_0, \boldsymbol{u}_o)\|_\rho^2 \leq \|\boldsymbol{u}_o\|_\rho^2 e^{-K_1(t-t_0)} + 2\left(\beta + \frac{2}{\alpha}M_{\boldsymbol{g}}^2\right)\rho_\Sigma.$$

Hence the closed and bounded subset of ℓ_ρ^2

$$Q := \left\{\boldsymbol{u} \in \ell_\rho^2 : \|\boldsymbol{u}\|_\rho^2 \leq 1 + \frac{2\left(\alpha\beta + 2L_{\boldsymbol{g}}^2\right)\rho_\Sigma}{\alpha(\alpha - \nu\gamma_0\gamma_1^2)} := R^2\right\} \tag{5.10}$$

is an absorbing set for the process $\{\psi(t, t_0)\}_{(t,t_0)\in\mathbb{R}_\geq^2}$ on ℓ_ρ^2 in both the forward and pullback senses. Moreover it is positively invariant. □

5.3.2 *Asymptotic tails and asymptotic compactness*

The next step of the proof is to derive an asymptotic tails estimate for the process $\{\psi(t, t_0)\}_{(t,t_0)\in\mathbb{R}_\geq^2}$ on ℓ_ρ^2 in the positive invariant absorbing set Q.

Lemma 5.10. *For every $\varepsilon > 0$ there exist $T(\varepsilon) > 0$ and $I(\varepsilon) \in \mathbb{N}$ such that*

$$\sum_{|i|>I(\varepsilon)} \rho_i \left|u_i(t; t_0, \boldsymbol{u}_o)\right|^2 \leq \varepsilon$$

for all $\boldsymbol{u}_o \in Q$ and $t_0 \leq t - T(\varepsilon)$.

Proof. Consider the smooth function $\xi : \mathbb{R} \to [0, 1]$ defined in (3.8), and recall that there exists a constant C_0 such that $|\xi'(s)| \leq C_0$ for all $s \geq 0$. Then define $\xi_k(s) = \xi(\frac{s}{k})$ for all $s \in \mathbb{R}$ and a fixed $k \in \mathbb{N}$ (its value will be specified later). Given $\boldsymbol{u} \in \ell_\rho^2$, define $\boldsymbol{v} \in \ell_\rho^2$ component wise as $v_i := \xi_k(|i|)u_i$ for $i \in \mathbb{Z}$.

Taking the inner product in ℓ_ρ^2 of equation (5.2) with $\boldsymbol{v}$ gives

$$\frac{\mathrm{d}}{\mathrm{d}t}\langle \boldsymbol{u}, \boldsymbol{v}\rangle_\rho = \nu\left\langle \Lambda\boldsymbol{u}, \boldsymbol{v}\right\rangle_\rho + \langle F(\boldsymbol{u}), \boldsymbol{v}\rangle_\rho + \langle \boldsymbol{g}(t), \boldsymbol{v}\rangle_\rho,$$

that is

$$\frac{\mathrm{d}}{\mathrm{d}t}\sum_{i\in\mathbb{Z}} \xi_k(|i|)\rho_i|u_i|^2 = 2\nu\left\langle \Lambda\boldsymbol{u}, \boldsymbol{v}\right\rangle_\rho + 2\sum_{i\in\mathbb{Z}} \xi_k(|i|)\rho_i u_i f(u_i) + 2\sum_{i\in\mathbb{Z}} \xi_k(|i|)\rho_i g_i(t)u_i \tag{5.11}$$

By Lemma 5.4

$$\langle \Lambda \boldsymbol{u}, \boldsymbol{v}\rangle_\rho = -\left\langle \mathrm{D}^+\boldsymbol{u}, \mathrm{D}^+\boldsymbol{v}\right\rangle_\rho + \sum_{i\in\mathbb{Z}} (\mathrm{D}^+\boldsymbol{u})_i \left(\rho_i(\mathrm{D}^+\boldsymbol{v})_i - (\mathrm{D}^+(\rho\boldsymbol{v}))_i\right). \tag{5.12}$$

Now each term in (5.12) will be estimated. First,

$$\begin{aligned}
\langle \mathrm{D}^+\boldsymbol{u}, \mathrm{D}^+\boldsymbol{v}\rangle_\rho &= \sum_{i\in\mathbb{Z}} \rho_i(u_{i+1}-u_i)(v_{i+1}-v_i)\\
&= \sum_{i\in\mathbb{Z}} \rho_i(u_{i+1}-u_i)\left((\xi_k(|i+1|)-\xi_k(|i|))u_{i+1} + \xi_k(|i|)(u_{i+1}-u_i)\right)\\
&= \sum_{i\in\mathbb{Z}} \rho_i(\xi_k(|i+1|)-\xi_k(|i|))(u_{i+1}-u_i)u_{i+1} + \sum_{i\in\mathbb{Z}} \rho_i\xi_k(|i|)(u_{i+1}-u_i)^2\\
&\geq \sum_{i\in\mathbb{Z}} \rho_i(\xi_k(|i+1|)-\xi_k(|i|))(u_{i+1}-u_i)u_{i+1}.
\end{aligned}$$

Since

$$\left|\sum_{i\in\mathbb{Z}} \rho_i(\xi_k(|i+1|)-\xi_k(|i|))(u_{i+1}-u_i)u_{i+1}\right| \leq \sum_{i\in\mathbb{Z}} \rho_i \frac{1}{k}|\xi'(s_i)||u_{i+1}-u_i||u_{i+1}|$$

by the Mean Value Theorem and

$$\begin{aligned}
\sum_{i\in\mathbb{Z}} \rho_i \frac{1}{k}|\xi'(s_i)||u_{i+1}-u_i||u_{i+1}| &\leq \frac{C_0}{k}\sum_{i\in\mathbb{Z}} \rho_i\left(|u_{i+1}|^2 + |u_i|^2\right)\\
&\leq \frac{C_0}{k}\left(\gamma_0 \sum_{i\in\mathbb{Z}} \rho_{i+1}|u_{i+1}|^2 + \sum_{i\in\mathbb{Z}} \rho_i|u_i|^2\right)\\
&= \frac{C_0(\gamma_0+1)}{k}\|\boldsymbol{u}\|_\rho^2 \leq \frac{C_0(\gamma_0+1)\|Q\|^2}{k}
\end{aligned}$$

for all $\boldsymbol{u} \in Q$. It follows immediately

$$\langle \mathrm{D}^+\boldsymbol{u}, \mathrm{D}^+\boldsymbol{v}\rangle_\rho \geq -\frac{C_0(\gamma_0+1)\|Q\|^2}{k}, \quad \forall\, \boldsymbol{u}\in Q,\ \boldsymbol{v} = (\xi_k(|i|)u_i)_{i\in\mathbb{Z}}.$$

In addition,

$$\begin{aligned}
\left|\sum_{i\in\mathbb{Z}} (\mathrm{D}^+\boldsymbol{u})_i((\mathrm{D}^+(\rho\boldsymbol{v}))_i - \rho_i(D^+\boldsymbol{v})_i)\right| &\leq \gamma_1 \sum_{i\in\mathbb{Z}} \rho_i|(\mathrm{D}^+\boldsymbol{u})_i||v_{i+1}|\\
&= \gamma_1 \sum_{i\in\mathbb{Z}} \xi_k(|i+1|)\rho_i|u_{i+1}-u_i||u_{i+1}|\\
&\leq \gamma_1 \sum_{i\in\mathbb{Z}} \xi_k(|i+1|)\rho_i(u_i^2 + u_{i+1}^2)\\
&\leq \gamma_0\gamma_1 \sum_{i\in\mathbb{Z}} \xi_k(|i|)\rho_i u_i^2 + \gamma_1 \sum_{i\in\mathbb{Z}} \xi_k(|i+1|)\rho_i u_i^2
\end{aligned}$$

where

$$\begin{aligned}
\sum_{i\in\mathbb{Z}} \xi_k(|i+1|)\rho_i u_i^2 &\leq \sum_{i\in\mathbb{Z}} \left(\xi_k(|i+1|)-\xi_k(|i|)\right)\rho_i u_i^2 + \sum_{i\in\mathbb{Z}} \xi_k(|i|)\rho_i u_i^2\\
&\leq \frac{C_0(\gamma_0+1)}{k}\|\boldsymbol{u}\|_\rho^2 + \sum_{i\in\mathbb{Z}} \xi_k(|i|)\rho_i u_i^2,
\end{aligned}$$

and thus

$$\gamma_1 \Big| \sum_{i\in\mathbb{Z}} (\mathrm{D}^+\boldsymbol{u})_i \big((\mathrm{D}^+(\rho \boldsymbol{v}))_i - \rho_i (\mathrm{D}^+\boldsymbol{v})_i \big) \Big|$$
$$\leq (\gamma_0+1)\gamma_1 \sum_{i\in\mathbb{Z}} \xi_k(|i|)\rho_i u_i^2 + \frac{C_0\gamma_1(\gamma_0+1)}{k}\|\boldsymbol{u}\|_\rho^2.$$

Combining these estimates and inserting into (5.12) gives

$$\langle \Lambda\boldsymbol{u}, \boldsymbol{v}\rangle_\rho \leq (\gamma_0+1)\gamma_1 \sum_{j\in\mathbb{Z}} \xi_k(|i|)\rho_i u_i^2 + \frac{C_0(\gamma_0+1)(\gamma_1+1)}{k}\|Q\|^2,$$

since $\boldsymbol{u}_o \in Q$.

On the other hand, by Assumption 5.1,

$$2\sum_{i\in\mathbb{Z}} \xi_k(|i|)\rho_i u_i f(u_i) \leq -2\alpha \sum_{i\in\mathbb{Z}} \xi_k(|i|)\rho_i|u_i|^2 + 2\beta \sum_{i\in\mathbb{Z}} \xi_k(|i|)\rho_i$$

and by Young's inequality

$$2\sum_{i\in\mathbb{Z}} \xi_k(|i|)\rho_i g_i(t) u_i = \alpha \sum_{i\in\mathbb{Z}} \xi_k(|i|)\rho_i|u_i|^2 + \frac{1}{\alpha}\sum_{i\in\mathbb{Z}} \xi_k(|i|)\rho_i|g_i(t)|^2,$$

so

$$2\sum_{i\in\mathbb{Z}} \xi_k(|i|) u_i f(u_i) + 2\sum_{i\in\mathbb{Z}} \xi_k(|i|)\rho_i g_i(t) u_i \leq -\alpha \sum_{i\in\mathbb{Z}} \xi_k(|i|)\rho_i|u_i|^2 + \frac{M_{\boldsymbol{g}}^2}{\alpha}\sum_{i\in\mathbb{Z}} \xi_k(|i|)\rho_i$$

Using the above estimates in equation (5.11) gives

$$\frac{\mathrm{d}}{\mathrm{d}t}\sum_{|i|>k} \rho_i|u_i|^2 + \gamma_2 \sum_{|i|>k} \rho_i|u_i|^2 \leq \frac{\nu C_0(\gamma_0+1)(\gamma_1+1)}{k}\|Q\|^2 + \frac{2\alpha\beta + M_{\boldsymbol{g}}^2}{\alpha}\sum_{|i|>k}\rho_i,$$

where

$$\gamma_2 := \alpha - \nu(\gamma_0+1)\gamma_1 > 0.$$

Since the sum of weights is a convergent series by Assumption 5.3 for every $\varepsilon > 0$ there exists $K(\varepsilon)$ such that

$$\frac{\nu C_0(\gamma_0+1)(\gamma_1+1)}{k}\|Q\|^2 + \frac{M_{\boldsymbol{g}}^2}{\alpha}\sum_{|i|>k}\rho_i \leq \varepsilon, \quad k \geq K(\varepsilon),$$

and thus

$$\frac{\mathrm{d}}{\mathrm{d}t}\sum_{|i|>k} \rho_i|u_i|^2 + \gamma_2 \sum_{|i|>k} \rho_i|u_i|^2 \leq \varepsilon.$$

Then, Gronwall's lemma implies that

$$\sum_{|i|>k} \rho_i|\boldsymbol{u}(t;t_0,\boldsymbol{u}_o)_i|^2 \leq e^{-\gamma_2(t-t_0)}\sum_{|i|>k} \rho_i|u_{o,i}|^2 + \frac{\varepsilon}{\gamma_2} \leq e^{-\gamma_2 t}\|\boldsymbol{u}_o\|^2 + \frac{\varepsilon}{\gamma_2}.$$

Hence, for $\boldsymbol{u}_o \in Q$,

$$\sum_{|i|>k} \rho_i |\boldsymbol{u}(t;t_0,\boldsymbol{u}_o)_i|^2 \le e^{-\gamma_2(t-t_0)}\|Q\|^2 + \frac{\varepsilon}{\gamma_2}, \tag{5.13}$$

which implies that

$$\sum_{|i|>k} \rho_i |\boldsymbol{u}(t;t_0,\boldsymbol{u}_o)_i|^2 \le \frac{2\varepsilon}{\gamma_2}, \quad \text{for} \quad 0 \le T(\varepsilon) := \frac{1}{\gamma_2}\ln\frac{\gamma_2\|Q\|^2}{\varepsilon}.$$

This is the desired pullback asymptotic tails property. The asymptotic compactness in ℓ^2_ρ of $\{\psi(t,t_0)\}_{(t,t_0)\in\mathbb{R}^2_\geq}$ in the absorbing set Q then follows by Lemma 2.5. This completes the proof of Lemma 5.10 and hence of Theorem 5.2. □

5.4 Uniformly strictly contracting Laplacian lattice systems

The estimates in Lemmas 5.9 and 5.10 depend only on the elapsed time $t - t_0$, so the results also apply in the forward as well as pullback senses. However, they do not suffice to show that the family $\mathcal{A} = (A_t)_{t\in\mathbb{R}}$ is a forward attractor, see [Kloeden and Lorenz (2016)].

If the dissipativity condition in Assumption 5.1 is replaced by strictly dissipative one-sided Lipschitz condition such as in

Assumption 5.5. $f : \mathbb{R} \to \mathbb{R}$ is a continuously differentiable function satisfying

$$f(0) = 0, \quad (f(x) - f(y))(x-y) \le -\alpha(x-y)^2, \quad f'(x) \le D_f \quad \forall\, x, y \in \mathbb{R},$$

for some $\alpha > 0$,

then the process $\{\psi(t,t_0)\}_{(t,t_0)\in\mathbb{R}^2_\geq}$ is uniformly strictly contracting. It can then be shown that the process $\{\psi(t,t_0)\}_{(t,t_0)\in\mathbb{R}^2_\geq}$ generated by the lattice system (5.2) on ℓ^2_ρ has a pullback attractor $\mathcal{A} = (A_t)_{t\in\mathbb{R}}$ in ℓ^2_ρ, which is not only a forward attractor, but consists of singleton component sets $A_t = \{\mathfrak{e}(t)\}$ for $t \in \mathbb{R}$, where $\mathfrak{e}$ is an entire trajectory of the process, i.e., satisfies $\mathfrak{e}(t) = \psi(t,s,\mathfrak{e}(s))$ for all $s \le t$ in $\mathbb{R}$.

Theorem 5.3. *Suppose that Assumptions 5.2–5.5 and the condition* (5.9) *are satisfied. Then the pullback attractor $\mathcal{A}$ is also a forward attractor for the process $\{\psi(t,t_0)\}_{(t,t_0)\in\mathbb{R}^2_\geq}$, which consists of a single entire trajectory $\mathfrak{e}(t) = \psi(t,s,\mathfrak{e}(s))$ for all $s \le t$ in $\mathbb{R}$.*

Proof. Given any two initial value $\boldsymbol{u}_o$ and $\boldsymbol{v}_o$, define $\boldsymbol{w}(t) = \boldsymbol{u}(t;t_0,\boldsymbol{u}_o) - \boldsymbol{v}(t;t_0,\boldsymbol{v}_o)$. Then, from (5.2),

$$\frac{\mathrm{d}\boldsymbol{w}}{\mathrm{d}t} = \nu\Lambda\boldsymbol{w} + F(\boldsymbol{u}) - F(\boldsymbol{v}).$$

Taking the inner product of the equation above in ℓ^2_ρ with $\boldsymbol{w}$ gives

$$\frac{1}{2}\frac{\mathrm{d}}{\mathrm{d}t}\|\boldsymbol{w}\|^2_\rho = \nu\langle\Lambda\boldsymbol{w},\boldsymbol{w}\rangle_\rho + \langle F(\boldsymbol{u}) - F(\boldsymbol{v}),\boldsymbol{w}\rangle_\rho \le \langle\Lambda\boldsymbol{w},\boldsymbol{w}\rangle_\rho - \alpha\|\boldsymbol{w}\|^2_\rho,$$

using the dissipative one-sided Lipschitz property of the mapping F in Assumption 5.5. Then by Lemma 5.5

$$\frac{1}{2}\frac{\mathrm{d}}{\mathrm{d}t}\|\boldsymbol{w}\|_\rho^2 \le \nu\left\langle \Lambda \boldsymbol{w}, \boldsymbol{w}\right\rangle_\rho - \alpha\|\boldsymbol{w}\|_\rho^2 \le -\frac{\nu}{2}\left\|\mathrm{D}^+\boldsymbol{w}\right\|_\rho + \frac{\nu}{2}\gamma_0\gamma_1^2\left\|\boldsymbol{w}\right\|_\rho^2 - \alpha\|\boldsymbol{w}\|_\rho^2,$$

from which it follows

$$\frac{\mathrm{d}}{\mathrm{d}t}\|\boldsymbol{w}\|_\rho^2 \le -(2\alpha - \nu\gamma_0\gamma_1^2)\|\boldsymbol{w}\|_\rho^2.$$

Integrating the above differential inequality gives

$$\|\boldsymbol{w}(t)\|_\rho^2 \le e^{-\left(2\alpha-\nu\gamma_0\gamma_1^2\right)(t-t_0)}\|\boldsymbol{w}(t_0)\|_\rho^2,$$

and thus

$$\|\boldsymbol{u}(t;t_0,\boldsymbol{u}_o) - \boldsymbol{v}(t;t_0,\boldsymbol{v}_o)\|_\rho \le e^{-\left(\alpha-\frac{1}{2}\nu\gamma_0\gamma_1^2\right)(t-t_0)}\left\|\boldsymbol{u}_o - \boldsymbol{v}_o\right\|_\rho, \qquad t \ge t_0. \quad (5.14)$$

This is the uniformly strict contraction property and holds if $\alpha > \frac{1}{2}\nu\gamma_0\gamma_1^2$, which is ensured by the condition (5.9).

The other assumptions in (5.9) ensures that there exists a pullback attractor $\mathcal{A} = (A_t)_{t\in\mathbb{R}}$ in ℓ_ρ^2. Suppose that the pullback attractor does not consist of singleton component sets. Then there are $\boldsymbol{u}$, $\boldsymbol{v} \in A_\tau$ for some $\tau \in \mathbb{R}$ with $\|\boldsymbol{u}-\boldsymbol{v}\|_\rho \ge \varepsilon_0$ for some $\varepsilon_0 > 0$.

Since the pullback attractor $\mathcal{A}$ is invariant for the process ψ generated by equation (5.2), then for every $t_0 < \tau$ there exist points $\boldsymbol{u}(t_0)$, $\boldsymbol{v}(t_0) \in A_{t_0}$ such that

$$\psi\left(\tau, t_0, \boldsymbol{u}(t_0)\right) = \boldsymbol{u}, \qquad \psi\left(\tau, t_0, \boldsymbol{v}(t_0)\right) = \boldsymbol{v}.$$

By the uniformly strict contractivity condition (5.14),

$$\begin{aligned}\varepsilon_0^2 \le \|\boldsymbol{u}-\boldsymbol{v}\|_\rho^2 &= \left\|\psi\left(\tau,t_0,\boldsymbol{u}(t_0)\right) - \psi\left(\tau,t_0,\boldsymbol{v}(t_0)\right)\right\|_\rho^2 \\ &\le 2\|Q\|^2 e^{-\left(2\alpha-\gamma_0\gamma_1^2\right)(\tau-t_0)}\end{aligned}$$

since $\boldsymbol{u}$, $\boldsymbol{v} \in Q$. On the other hand, $2\|Q\|^2 e^{-\left(2\alpha-\gamma_0\gamma_1^2\right)(\tau-t_0)} \le \frac{1}{2}\varepsilon_0^2$ for

$$t_0 \le \tau - \frac{1}{2\left(2\alpha - \nu\gamma_0\gamma_1^2\right)}\ln(4\|Q\|^2/\varepsilon_0^2),$$

so

$$\varepsilon_0^2 \le \|\boldsymbol{u}-\boldsymbol{v}\|_\rho^2\varepsilon_0^2 \le \frac{1}{2}\varepsilon_0^2.$$

This is a contradiction, so $\boldsymbol{u} = \boldsymbol{v}$, i.e., the components A_t are singleton sets and hence the pullback attractor consists of a single entire trajectory.

Let this entire solution be denoted by $\mathfrak{e}(t)$, so the components sets $A_t = \{\mathfrak{e}(t)\}$ for all $t \in \mathbb{R}$. Then, for any other solution $\boldsymbol{u}$ of equation (5.2) the uniformly strict contractivity condition (5.14) gives

$$\|\boldsymbol{u}(t) - \mathfrak{e}(t)\|_\rho \le \|\boldsymbol{u}(t_0) - \mathfrak{e}(t_0)\|_\rho e^{-\left(\alpha-\frac{1}{2}\nu\gamma_0\gamma_1^2\right)(t-t_0)} \to 0 \quad \text{as } t \to \infty.$$

Thus the family of sets $(A_t)_{t\in\mathbb{R}}$ given by $A_t = \{\mathfrak{e}(t)\}$ is also a forward attractor. □

Remark 5.3. It can be shown directly that there exists a pullback attractor $\mathscr{A} = (A_t)_{t\in\mathbb{R}}$ in ℓ^2_ρ using the strict contraction property and a Cauchy sequence argument rather than asymptotic compactness, see [Caraballo, Kloeden and Schmalfuß (2004)] and [Kloeden and Yang (2021), Theorem 7.2].

5.5 Forward dynamics

Under the Assumptions 5.1–5.4 and the inequality (5.9) in Theorem 5.2 the family $\mathscr{A} = (A_t)_{t\in\mathbb{R}}$ forming the pullback attractor exists, but need not be a forward attractor. The reason is that there may be omega-limit points starting in the absorbing set that are not omega-limit points starting within the corresponding sets A_t, see [Kloeden and Lorenz (2016); Kloeden and Yang (2021)]. In this case these omega-limit sets characterise the future asymptotic dynamics of the system.

Let $\{\psi(t,t_0)\}_{(t,t_0)\in\mathbb{R}^2_\geq}$ be the process on ℓ^2_ρ associated with the system (5.2) and let Q be the closed and bounded ball in ℓ^2_ρ defined by (5.10), which is positive invariant under ψ. Since the tails estimate (5.13) depends only on the elapsed time $t - t_0$, the set Q is forward as well as pullback absorbing. The asymptotic compactness property thus also holds in the forward sense, so the omega-limit set starting in Q at time t_0, defined as

$$\omega_Q(t_0) := \bigcap_{s\geq t_0} \overline{\bigcup_{t\geq s} \psi(t,t_0,Q)},$$

is non-empty and compact. Moreover,

$$\lim_{t\to\infty} \operatorname{dist}_{\ell^2_\rho}\left(\psi(t,t_0,Q), \omega_Q(t_0)\right) = 0, \qquad \forall t_0 \in \mathbb{R}. \tag{5.15}$$

Notice that $\omega_Q(t_0) \subset \omega_Q(s_0)$ if $t_0 \leq s_0$. Hence the totality of omega-limit points

$$\omega_Q^\infty := \overline{\bigcup_{t_0\in\mathbb{R}} \omega_Q(t_0)}$$

is nonempty and closed in ℓ^2_ρ. It is the union of increasing compact sets, so does not need to be compact. It follows from (5.15) that

$$\lim_{t\to\infty} \operatorname{dist}_{\ell^2_\rho}\left(\psi(t,t_0,Q), \omega_Q^\infty\right) = 0, \qquad \forall t_0 \in \mathbb{R}.$$

The following result from [Crauel and Kloeden (2015), Proposition 4] characterises the sets $\omega_Q(t_0)$ and ω_Q^∞.

Proposition 5.1. *Given any point $\boldsymbol{u} \in \ell^2_\rho$:*

(i) $\boldsymbol{u} \in \omega_Q(t_0)$ if and only if there exist sequences $t_n \to \infty$ and $\{\boldsymbol{u}_o^{(n)}\}_{n\in\mathbb{N}} \subset Q$ such that $\psi(t_n, t_0, \boldsymbol{u}_o^{(n)}) \to \boldsymbol{u}$.

(ii) $\boldsymbol{u} \in \omega_Q^\infty$ if and only if there exist sequences $t_n \to \infty$, $\tau_n \to \infty$ and $\{\boldsymbol{u}_o^{(n)}\}_{n\in\mathbb{N}} \subset Q$ with $t_n - \tau_n \to \infty$ such that $\psi(t_n, \tau_n, \boldsymbol{u}_o^{(n)}) \to \boldsymbol{u}$.

The set ω_Q^∞ is usually not invariant, though under additional assumptions can be shown to be positively and even negatively asymptotically invariant, see [Crauel and Kloeden (2015); Kloeden and Yang (2021)]. For this reason it was called a *forward attracting set* in [Kloeden and Yang (2021)] rather than a forward attractor. Nevertheless it suffices to characterise all of the future asymptotic behaviour of the lattice system considered here.

Note that the behaviour of the system in the distant future does not require knowledge of its behaviour in the distant past. It does not even require the process ψ to be defined for all $t \in \mathbb{R}$, just for $t \geq T^*$ for some finite $T^* \in \mathbb{R}$. In fact,

$$\omega_Q^\infty = \overline{\bigcup_{t_0 \geq T^*} \omega_Q(t_0)} \equiv \overline{\bigcup_{t_0 \in \mathbb{R}} \omega_Q(t_0)},$$

when the process is defined for all $t_0 \in \mathbb{R}$, so nothing is lost in considering the process only after time T^*. This is more realistic in many biological applications.

5.6 End notes

The chapter is based on [Han and Kloeden (2022)], which itself is loosely based on [Wang (2006)] by Bixiang Wang, who was one of the first to investigate LDS with the discrete Laplacian operator with weighted norms. Here a general class of weights with appropriate properties are used rather than a particular weight in [Wang (2006)], due to which the proofs here do not need the auxiliary function introduced there.

An alternative compactness argument via the quasi-stability concept in [Chueshov (2015)] was used by [Czaja (2022)] in the sequence space ℓ^2.

5.7 Problems

Problem 5.1. Formulate the d-dimensional discretised Laplacian operator in terms of appropriate vector-valued indices and difference operators.

Problem 5.2. In [Wang (2006)] the function $\phi(x) = (1 + x^2)^{-\sigma}$ for $x \in \mathbb{R}$, with $\sigma > \frac{1}{2}$, was used to define weights $\sigma_i = (1 + i^2)^{-\sigma}$ for $i \in \mathbb{Z}$. The properties of an auxiliary function $\phi_\delta(x) = (1 + |\delta x|^2)^{-\sigma}$ for an appropriate $\delta > 0$ were then used to establish the properties of the weights. Show that these weights satisfy the assumptions in this chapter. Compare with the derivations in [Wang (2006)].

Problem 5.3. Show that the semi-group property in ℓ^2 is extended uniquely to one in ℓ^2_ρ as used in Subsection 5.2.3.

PART 3

A selection of lattice models

Chapter 6

Lattice dynamical systems with delays

The presence of time delays is well justified in many applications, including lattice models. In this chapter we consider a Laplacian lattice system with finite delay terms

$$\begin{cases} \dfrac{\mathrm{d}u_i}{\mathrm{d}t} = \nu\left(u_{i-1} - 2u_i + u_{i+1}\right) - \lambda u_i + f_i\left(u_i\left(t - \theta_i\right)\right) + g_i, & i \in \mathbb{Z}, \\ u_i\left(s\right) = \phi_i\left(s\right), \text{ for all } s \in [-\theta_i, 0], & i \in \mathbb{Z}, \end{cases} \tag{6.1}$$

where λ is a positive real number and the delays $\theta_i > 0$ are assumed to be uniformly bounded, i.e., satisfy

Assumption 6.1. There exists a constant $h > 0$ that $0 < \theta_i \leq h$ for all $i \in \mathbb{Z}$.

The appropriate function space for the solutions of the lattice system with delays (6.1) is the Banach space $\mathcal{C}([-h, 0], \ell^2)$ of continuous functions $\mathcal{U} : [-h, 0] \to \ell^2$ with the norm

$$\|\mathcal{U}\|_{\mathcal{C}([-h,0],\ell^2)} = \max_{s\in[-h,0]} \|\mathcal{U}(s)\|_{\ell^2}.$$

The initial condition in (6.1) must then satisfy $(\phi_i(\cdot))_{i\in\mathbb{Z}} \in \mathcal{C}([-h, 0], \ell^2)$.

6.1 The coefficient terms

Let the discrete Laplace operator Λ be defined as in Chapter 3. Recall that Λ is a negative definite bounded linear operator on ℓ^2 with

$$\|\Lambda \boldsymbol{u}\| \leq \left(\sum_{i\in\mathbb{Z}} (4u_{i-1}^2 + 8u_i^2 + 4u_{i+1}^2)\right)^{1/2} = 4\|\boldsymbol{u}\| \qquad \forall\, \boldsymbol{u} \in \ell^2. \tag{6.2}$$

The reaction terms f_i are assumed to satisfy

Assumption 6.2. The functions $f_i : \mathbb{R} \to \mathbb{R}$ are equi-globally Lipschitz continuous, i.e., there exists a common positive constant L_f such that

$$|f_i(x) - f_i(y)| \leq L_f\, |x - y|, \quad \forall i \in \mathbb{Z}, x, y \in \mathbb{R}.$$

Assumption 6.3. $f_i(0) = 0$ for all $i \in \mathbb{Z}$.

Given any $\mathcal{U} = (\mathcal{U}_i)_{i\in\mathbb{Z}} \in \mathcal{C}([-h,0],\ell^2)$ define the delay operator term F on $\mathcal{C}([-h,0],\ell^2)$ by

$$F(\mathcal{U}) := (F_i(\mathcal{U}))_{i\in\mathbb{Z}} \quad \text{with} \quad F_i(\mathcal{U}) = f_i(\mathcal{U}_i(-\theta_i)).$$

Lemma 6.1. *The operator F maps $\mathcal{C}([-h,0],\ell^2)$ to ℓ^2 and is globally Lipschitz continuous with Lipschitz constant L_f.*

Proof. Let $\mathcal{U}$, $\mathcal{V} \in \mathcal{C}([-h,0],\ell^2)$. Since each function f_i is globally Lipschitz with Lipschitz constant L_f, then

$$\sum_{i\in\mathbb{Z}} |f_i(\mathcal{U}_i(-\theta_i)) - f_i(\mathcal{V}_i(-\theta_i))|^2 \le L_f^2 \sum_{i\in\mathbb{Z}} |\mathcal{U}_i(-\theta_i) - \mathcal{V}_i(-\theta_i)|^2.$$

Since $F_i(\mathcal{U}) = f_i(\mathcal{U}_i(-\theta_i))$ this means that

$$\sum_{i\in\mathbb{Z}} |F_i(\mathcal{U}) - F_i(\mathcal{V})|^2 \le L_f^2 \sum_{i\in\mathbb{Z}} \max_{-h\le s\le 0} |\mathcal{U}_i(s) - \mathcal{V}_i(s)|^2, \tag{6.3}$$

that is

$$\|F(\mathcal{U}) - F(\mathcal{V})\|_{\ell^2} \le L_f \|\mathcal{U} - \mathcal{V}\|_{\mathcal{C}([-h,0],\ell^2)}.$$

Notice that $F(0) \equiv \mathbf{0}$ in $\mathcal{C}([-h,0],\ell^2)$ due to Assumption 6.3, so

$$\|F(\mathcal{U})\|_{\ell^2} \le L_f \|\mathcal{U}\|_{\mathcal{C}([-h,0],\ell^2)}.$$

Hence $F(\mathcal{U})$ takes its value in ℓ^2 as claimed. □

Finally, we suppose that the constant forcing term $\boldsymbol{g} := (g_i)_{i\in\mathbb{Z}}$ satisfies the following assumption.

Assumption 6.4. $\boldsymbol{g} \in \ell^2$.

6.2 Existence and uniqueness of solutions

For a continuous function $\boldsymbol{u} = (u_i)_{i\in\mathbb{Z}} : [-h,T] \to \ell^2$, denote by $\boldsymbol{u}_t$ the segment of $\boldsymbol{u}$ in $\mathcal{C}([-h,0],\ell^2)$ defined by

$$\boldsymbol{u}_t(s) = \boldsymbol{u}(t+s) \quad \text{for each} \quad s \in [-h,0].$$

The lattice differential equation (6.1) can be rewritten as an infinite dimensional delay differential equation on ℓ^2,

$$\frac{\mathrm{d}}{\mathrm{d}t}\boldsymbol{u}(t) = \nu\Lambda\boldsymbol{u}_t(0) - \lambda\boldsymbol{u}_t(0) + F(\boldsymbol{u}_t) + \boldsymbol{g} := \mathfrak{F}(\boldsymbol{u}_t), \tag{6.4}$$

where

$$\mathfrak{F}(\mathcal{U}) := \nu\Lambda\mathcal{U}(0) - \lambda\mathcal{U}(0) + F(\mathcal{U}) + \boldsymbol{g}, \quad \mathcal{U} \in \mathcal{C}([-h,0],\ell^2).$$

Lemma 6.2. *The mapping $\mathcal{U} \mapsto \mathfrak{F}(\mathcal{U})$ maps $\mathcal{C}([-h,0],\ell^2)$ to ℓ^2 and is globally Lipschitz continuous with Lipschitz constant $4\nu + \lambda + L_f$.*

Proof. For any $\mathcal{U}$, $\mathcal{V} \in \mathcal{C}([-h,0],\ell^2)$, by (6.2)

$$\begin{aligned}\|\mathfrak{F}(\mathcal{U}) - \mathfrak{F}(\mathcal{V})\|_{\ell^2} &\leq \nu\|\Lambda(\mathcal{U}(0) - \mathcal{V}(0))\|_{\ell^2} + \lambda\|\mathcal{U}(0) - \mathcal{V}(0)\|_{\ell^2} + \|F(\mathcal{U}) - F(\mathcal{V})\|_{\ell^2} \\ &\leq (4\nu + \lambda)\|\mathcal{U}(0) - \mathcal{V}(0)\|_{\ell^2} + L_f\|\mathcal{U} - \mathcal{V}\|_{\mathcal{C}([-h,0],\ell^2)} \\ &\leq (4\nu + \lambda + L_f)\|\mathcal{U} - \mathcal{V}\|_{\mathcal{C}([-h,0],\ell^2)}.\end{aligned}$$

Taking $\mathcal{V} \equiv \mathbf{0}$ in $\mathcal{C}([-h,0],\ell^2)$ then gives

$$\|\mathfrak{F}(\mathcal{U})\|_{\ell^2} \leq (4\nu + \lambda + L_f)\|\mathcal{U}\|_{\mathcal{C}([-h,0],\ell^2)},$$

so $\mathfrak{F}(\mathcal{U})$ takes its value in ℓ^2 as claimed. □

Write the ith component of $\mathfrak{F}(\boldsymbol{u}_t)$ as

$$\mathfrak{F}_i(\boldsymbol{u}_t) := \nu(\Lambda\boldsymbol{u}(t))_i - \lambda u_i(t) + f_i(u_i(t-\theta_i)) + g_i.$$

Lemma 6.3. *For $I \in \mathbb{N}$ there exists R_I with $R_I \to 0^+$ as $I \to \infty$ such that*

$$\sum_{|i|\geq I} |\mathfrak{F}_i(\boldsymbol{u}_t)|^2 \leq 4(16\nu + \lambda^2 + L_f^2) \max_{s\in[-h,0]} \sum_{|i|\geq I} |u_i(t+s)|^2 + R_I.$$

Proof. Using the inequalities $(a+b+c+d)^2 \leq 4(a^2+b^2+c^2+d^2)$ and (6.2) gives

$$\begin{aligned}\sum_{|i|\geq I} |\mathfrak{F}_i(\boldsymbol{u}_t)|^2 &\leq 4\nu \sum_{|i|\geq I} (\Lambda\boldsymbol{u}(t))_i^2 + 4\lambda^2 \sum_{|i|\geq I} |u_i(t)|^2 \\ &\quad + 4\sum_{|i|\geq I} |f_i(u_i(t-\theta_i))|^2 + 4\sum_{|i|\geq I} |g_i|^2 \\ &\leq (64\nu + 4\lambda^2) \sum_{|i|\geq I} |u_i(t)|^2 + 4L_f^2 \sum_{|i|\geq I} |u_i(t-\theta_i))|^2 + R_I \\ &\leq (64\nu + 4\lambda^2 + 4L_f^2) \max_{s\in[-h,0]} \sum_{|i|\geq I} |u_i(t+s)|^2 + R_I,\end{aligned}$$

where $R_I = 4\sum_{|i|\geq I} |g_i|^2 \to 0^+$ as $I \to \infty$. □

6.2.1 *Existence of solutions*

Theorem 6.1. *Suppose that Assumptions 6.1–6.4 hold. Then for each $r > 0$ there exists $T(r) > 0$ such that for every $\phi = (\phi_i)_{i\in\mathbb{Z}} \in \mathcal{C}([-h,0],\ell^2)$ satisfying $\|\phi\|_{\mathcal{C}([-h,0],\ell^2)} \leq r$, the lattice delay equation (6.4) has at least one solution defined on $[0,T(r)]$. Moreover, the solution $\boldsymbol{u}(\cdot) \in \mathcal{C}^1([0,T(r)],\ell^2)$.*

Proof. It follows by Lemma 6.2 that $\mathfrak{F} : \mathcal{C}([-h,0],\ell^2) \to \ell^2$ is well defined, bounded and globally Lipschitz continuous.

These are the conditions (H1)–(H3) in Caraballo, Morillas & Valero [Caraballo *et al.* (2014)]. In fact, a similar estimate to (6.3) with a cutoff function shows that the components of $\mathfrak{F}(\boldsymbol{u}_t)$ satisfy the tails estimate in (H3). Hence Theorem 10 and Corollary 13 there can then be applied to give the existence of a local solution $\boldsymbol{u}(\cdot) \in \mathcal{C}^1([0,T(r)],\ell^2)$. □

6.2.2 *An a prior estimate of solutions*

Here we will establish some estimates of the solutions, which imply that the solutions are bounded uniformly with respect to bounded sets of initial conditions and all positive values of time. To that end, the following lemma is needed.

Lemma 6.4. *Let $\boldsymbol{u}_t(\boldsymbol{\phi}) = (u_{t,i}(\boldsymbol{\phi}))_{i\in\mathbb{Z}} \in \mathcal{C}([-h,0],\ell^2)$ be the solution of the lattice delay system* (6.1) *with the initial value $\boldsymbol{\phi} \in (\phi_i)_{i\in\mathbb{Z}} \in \mathcal{C}([-h,0],\ell^2)$. Then for every $\epsilon > 0$ there holds*

$$\int_0^t e^{\epsilon s}\sum_{i\in\mathbb{Z}}|u_i(s-\theta_i)|^2\,\mathrm{d}s \le \frac{e^{\epsilon h}}{\epsilon}\|\boldsymbol{\phi}\|^2_{\mathcal{C}([-h,0],\ell^2)}\left(1-e^{\epsilon h}\right) + e^{-\epsilon h}\int_0^t e^{\epsilon s}\|\boldsymbol{u}(s)\|^2_{\ell^2}\,\mathrm{d}s.$$

Proof. For each $i \in \mathbb{Z}$, we have $\phi_i \in \mathcal{C}([-h,0],\mathbb{R})$ and

$$\begin{aligned}\int_0^t e^{\epsilon s}|u_i(s-\theta_i)|^2\,\mathrm{d}s &= e^{\epsilon\theta_i}\int_{-\theta_i}^{t-\theta_i} e^{\epsilon s}|u_i(s)|^2\,\mathrm{d}s\\ &\le e^{\epsilon\theta_i}\int_{-\theta_i}^{0} e^{\epsilon s}|\phi_i(s)|^2\,\mathrm{d}s + e^{\epsilon\theta_i}\int_0^t e^{\epsilon s}|u_i(s)|^2\,\mathrm{d}s\\ &\le e^{\epsilon h}\int_{-h}^{0} e^{\epsilon s}|\phi_i(s)|^2\,\mathrm{d}s + e^{\epsilon h}\int_0^t e^{\epsilon s}|u_i(s)|^2\,\mathrm{d}s\\ &\le \frac{e^{\epsilon h}}{\epsilon}\|\phi_i\|^2_{\mathcal{C}([-h,0],\mathbb{R})}\left(1-e^{-\epsilon h}\right) + e^{\epsilon h}\int_0^t e^{\epsilon r}|u_i(s)|^2\,\mathrm{d}s\\ &\le \frac{e^{\epsilon h}}{\epsilon}\|\phi_i\|^2_{\mathcal{C}([-h,0],\mathbb{R})}\left(1-e^{-\epsilon h}\right) + e^{\epsilon h}\int_0^t e^{\epsilon r}|u_i(s)|^2\,\mathrm{d}s.\end{aligned}$$

Summing over $i \in \mathbb{Z}$ gives

$$\begin{aligned}\int_0^t e^{\epsilon s}\sum_{i\in\mathbb{Z}}|u_i(s-\theta_i)|^2\,\mathrm{d}s &\le \frac{e^{\epsilon h}}{\epsilon}\sum_{i\in\mathbb{Z}}\|\phi_i\|^2_{\mathcal{C}([-h,0],\mathbb{R})}\left(1-e^{-\epsilon h}\right) + e^{\epsilon h}\int_0^t e^{\epsilon s}\sum_{i\in\mathbb{Z}}|u_i(s)|^2\,\mathrm{d}s.\\ &\le \frac{e^{\epsilon h}}{\epsilon}\|\boldsymbol{\phi}\|^2_{\mathcal{C}([-h,0],\ell^2)}\left(1-e^{-\epsilon h}\right) + e^{\epsilon h}\int_0^t e^{\epsilon s}\|\boldsymbol{u}(s)\|^2_{\ell^2}\,\mathrm{d}s.\end{aligned}$$ □

The following assumption on the positive parameter λ is needed to ensure the dissipativity of the lattice system.

Assumption 6.5. There exists $\epsilon_0 > 0$ such that

$$\lambda > \epsilon_0 > \frac{4L_f^2 e^{\epsilon_0 h}}{\lambda} := \mu.$$

This assumption requires the dissipativity constant λ to be large enough, the Lipschitz constant L to be moderate and the delay parameter h to be small.

Proposition 6.1. *Suppose that Assumptions 6.1–6.5 hold. Then every solution $\boldsymbol{u}_t$ of* (6.4) *with $\boldsymbol{u}_0 = \boldsymbol{\phi} \in \mathcal{C}([-h,0],\ell^2)$ satisfies*

$$\|\boldsymbol{u}_t\|^2_{\mathcal{C}([-h,0],\ell^2)} \le K_1\|\boldsymbol{\phi}\|^2_{\mathcal{C}([-h,0],\ell^2)}e^{-(\epsilon_0-\mu)t} + K_2,$$

where

$$K_1 = 2\left(1 + \frac{4L_f^2 e^{\epsilon_0 h}}{\lambda\epsilon_0}\left(1 - e^{-\epsilon_0 h}\right)\right), \quad K_2 = \frac{4}{\lambda\epsilon_0}\left(\frac{\mu}{\epsilon_0 - \mu} + 1\right)\|\boldsymbol{g}\|^2. \tag{6.5}$$

Proof. Multiply the ith component of (6.1) by $u_i(t)$ and summing over $i \in \mathbb{Z}$, and taking into account the negative definite property of Λ to obtain

$$\frac{\mathrm{d}}{\mathrm{d}t}\|\boldsymbol{u}(t)\|^2 \le -2\lambda\|\boldsymbol{u}(t)\|^2 + 2\sum_{i\in\mathbb{Z}} u_i(t) f_i(u_i(t-\theta_i)) + 2\sum_{i\in\mathbb{Z}} u_i(t) g_i. \tag{6.6}$$

The second and third term on the right hand of (6.6) are estimated respectively, by

$$2\sum_{i\in\mathbb{Z}} f_i(u_i(t-\theta_i))u_i(t) \le \frac{\lambda}{2}\|\boldsymbol{u}(t)\|^2 + \frac{4}{\lambda}\sum_{i\in\mathbb{Z}} |f_i(u_i(t-\theta_i))|^2,$$

$$2\sum_{i\in\mathbb{Z}} g_i u_i(t) \le \frac{\lambda}{2}\|\boldsymbol{u}(t)\|^2 + \frac{4}{\lambda}\|\boldsymbol{g}\|^2.$$

Combining these gives

$$\frac{\mathrm{d}}{\mathrm{d}t}\|\boldsymbol{u}(t)\|^2 \le -\lambda\,\|\boldsymbol{u}(t)\|^2 + \frac{4}{\lambda}\|\boldsymbol{g}\|^2 + \frac{4}{\lambda}\sum_{i\in\mathbb{Z}} |f_i(u_i(t-\theta_i))|^2. \tag{6.7}$$

The inequality (6.7) along with the Lipschitz property of f_i give

$$\begin{aligned}\frac{\mathrm{d}}{\mathrm{d}t}e^{\epsilon_0 t}\|\boldsymbol{u}(t)\|^2 &= e^{\epsilon_0 t}\frac{\mathrm{d}}{\mathrm{d}t}\|\boldsymbol{u}(t)\|^2 + \epsilon_0 e^{\epsilon_0 t}\|\boldsymbol{u}(t)\|^2 \\ &\le (\epsilon_0 - \lambda)\, e^{\epsilon_0 t}\|\boldsymbol{u}(t)\|^2 + \frac{4}{\lambda}\|\boldsymbol{g}\|^2 e^{\epsilon_0 t} + \frac{4}{\lambda}e^{\epsilon_0 t}\sum_{i\in\mathbb{Z}} |f_i(u_i(t-\theta_i))|^2 \\ &\le (\epsilon_0 - \lambda)\, e^{\epsilon_0 t}\|\boldsymbol{u}(t)\|^2 + \frac{4}{\lambda}\|\boldsymbol{g}\|^2 e^{\epsilon_0 t} + \frac{4L^2}{\lambda}e^{\epsilon_0 t}\sum_{i\in\mathbb{Z}} |u_i(t-\theta_i)|^2,\end{aligned}$$

which can then be integrated to obtain

$$\begin{aligned}e^{\epsilon_0 t}\|\boldsymbol{u}(t)\|^2 &\le \|\boldsymbol{u}(0)\|^2 + (\epsilon_0 - \lambda)\int_0^t e^{\epsilon_0 s}\|\boldsymbol{u}(s)\|^2\,\mathrm{d}s \\ &\quad + \frac{4}{\lambda\epsilon_0}\|\boldsymbol{g}\|^2(e^{\epsilon_0 t} - 1) + \frac{4L_f^2}{\lambda}\int_0^t e^{\epsilon_0 s}\sum_{i\in\mathbb{Z}} |u_i(s-\theta_i)|^2\,\mathrm{d}s.\end{aligned}$$

By Lemma 6.4 this becomes

$$\begin{aligned}e^{\epsilon_0 t}\|\boldsymbol{u}(t)\|^2 &\le \|\boldsymbol{u}(0)\|^2 + (\epsilon_0 - \lambda)\int_0^t e^{\epsilon_0 s}\|\boldsymbol{u}(s)\|^2\,ds + \frac{4}{\lambda\epsilon_0}\|\boldsymbol{g}\|^2(e^{\epsilon_0 t} - 1) \\ &\quad + \frac{4L_f^2 e^{\epsilon_0 h}}{\lambda\epsilon_0}\|\boldsymbol{\phi}\|^2_{\mathcal{C}([-h,0],\ell^2)}\left(1 - e^{-\epsilon_0 h}\right) + \frac{4L_f^2 e^{\epsilon_0 h}}{\lambda}\int_0^t e^{\epsilon_0 s}\|\boldsymbol{u}(s)\|^2_{\ell^2}\,\mathrm{d}s.\end{aligned}$$

With $\lambda > \epsilon_0$ this reduces to

$$e^{\epsilon_0 t}\|\boldsymbol{u}(t)\|^2 \le K_0 + \frac{4}{\lambda\epsilon_0}\|\boldsymbol{g}\|^2(e^{\epsilon_0 t} - 1) + \frac{4L_f^2 e^{\epsilon_0 h}}{\lambda}\int_0^t e^{\epsilon_0 s}\|\boldsymbol{u}_s\|^2_{\mathcal{C}([-h,0],\ell^2)}\,\mathrm{d}s, \tag{6.8}$$

where

$$K_0 = \left(1 + \frac{4L_f^2 e^{\epsilon_0 h}}{\lambda \epsilon_0}\left(1 - e^{-\epsilon_0 h}\right)\right)\|\phi\|^2_{C([-h,0],\ell^2)}.$$

Let $\theta \in [-h, 0]$. Replacing t by $t+\theta$ in (6.8) and using

$$\|u(t+\theta)\| = \|\phi(t+\theta)\| \le \|\phi\|_{C([-h,0],\ell^2)}, \qquad t+\theta < 0,$$

we obtain

$$\|u(t+\theta)\|^2 \le K_0 e^{-\epsilon_0(t+\theta)} + \frac{4}{\lambda\epsilon_0}\|g\|^2(1 - e^{-\epsilon_0(t+\theta)})$$
$$+ \frac{4L_f^2 e^{\epsilon_0 h}}{\lambda} e^{-\epsilon_0(t+\theta)} \int_0^{t+\theta} e^{\epsilon_0 r}\|u_s\|^2_{C([-h,0],\ell^2)}\, ds.$$

Using the fact that $\theta \in [-h, 0]$ and neglecting the negative terms then yields

$$e^{\epsilon_0 t}\|u_t\|^2_{C([-h,0],\ell^2)} \le K(t) + \mu \int_0^t e^{\epsilon_0 s}\|u_s\|^2_{C([-h,0],\ell^2)}\, ds,$$

where

$$K(t) = K_0 + \frac{4}{\lambda\epsilon_0}\|g\|^2 e^{\epsilon_0 t}, \qquad \mu = \frac{4L_f^2 e^{\epsilon_0 h}}{\lambda}.$$

Then by the Gronwall inequality

$$e^{\epsilon_0 t}\|u_t\|^2_{C([-h,0],\ell^2)} \le K(t) + \mu \int_0^t e^{\mu(t-s)} K(s)\, ds.$$

The integral here breaks down to two parts,

$$\mu e^{\mu t} K_0 \int_0^t e^{-\mu s}\, ds = e^{\mu t} K_0 (1 - e^{-\mu t}) = K_0(e^{\mu t} - 1)$$

and

$$\mu e^{\mu t}\frac{4}{\lambda\epsilon_0}\|g\|^2 \int_0^t e^{(\epsilon_0 - \mu)s}\, ds = \frac{4\mu}{\lambda\epsilon_0(\epsilon_0 - \mu)}\|g\|^2(e^{\epsilon_0 t} - e^{\mu t}),$$

where the assumption $\lambda > \epsilon_0 > \mu$ was used. Neglecting the negative terms then yields

$$\|u_t\|^2_{C([-h,0],\ell^2)} \le K(t)e^{-\epsilon_0 t} + \mu e^{-\epsilon_0 t}\int_0^t e^{\mu(t-s)} K(s)\, ds$$
$$\le K_0 e^{-\epsilon_0 t} + \frac{4}{\lambda\epsilon_0}\|g\|^2 + K_0 e^{-(\epsilon_0-\mu)t} + s\frac{4\mu}{\lambda\epsilon_0(\epsilon_0-\mu)}\|g\|^2 e^{-(\epsilon_0-\mu)t}$$
$$\le \left(2K_0 + \frac{4\mu}{\lambda\epsilon_0(\epsilon_0-\mu)}\|g\|^2\right) e^{-(\epsilon_0-\mu)t} + \frac{4}{\lambda\epsilon_0}\|g\|^2.$$

In conclusion

$$\|u_t\|^2_{C([-h,0],\ell^2)} \le K_1\|\phi_0\|^2_{C([-h,0],\ell^2)} e^{-(\epsilon_0-\mu)t} + K_2, \tag{6.9}$$

where K_1 and K_2 are as defined in (6.5). □

6.2.3 *Uniqueness of solutions*

Having the existence of global solutions of problem (6.4) we now establish their uniqueness.

Lemma 6.5. *Suppose that Assumptions 6.1–6.5 hold. Then the solution $\boldsymbol{u}(\cdot)$ of problem (6.4) is unique.*

Proof. By Lemma 6.2 the mapping $\mathcal{U} \mapsto \mathfrak{F}(\mathcal{U})$ maps $\mathcal{C}([-h,0],\ell^2)$ to ℓ^2 and is globally Lipschitz continuous with Lipschitz constant $4\nu + \lambda + L_f$.

Hence, suppose that we have two different solutions $\boldsymbol{u}$, $\boldsymbol{v}$ of problem (6.4) with the same initial condition $\boldsymbol{u}(s) = \boldsymbol{v}(s) = \boldsymbol{\phi}(s)$ for all $s \in [-h,0]$. Setting $\boldsymbol{w} = \boldsymbol{u} - \boldsymbol{v}$, we obtain

$$\frac{\mathrm{d}}{\mathrm{d}t}\|\boldsymbol{w}(t)\|^2 \leq 2(4\nu + \lambda + L_f)\|\boldsymbol{w}_t\|^2_{\mathcal{C}([-h,0],\ell^2)}.$$

Integrating from 0 to t then gives

$$\|\boldsymbol{w}(t)\|^2 \leq 2(4\nu + \lambda + L_f)\int_0^t \|\boldsymbol{w}_s\|^2_{\mathcal{C}([-h,0],\ell^2)}\mathrm{d}s + \|\boldsymbol{w}(0)\|^2.$$

Let $\theta \in [-h,0]$. Replacing t by $t+\theta$ in the inequality above and using $\|\boldsymbol{w}(t+\theta)\| = 0$ when $t+\theta < 0$. We obtain

$$\|\boldsymbol{w}(t+\theta)\|^2 \leq 2(4\nu + \lambda + L_f)\int_0^{t+\theta} \|\boldsymbol{w}_s\|^2_{\mathcal{C}([-h,0],\ell^2)}\mathrm{d}s + \|\boldsymbol{w}(0)\|^2.$$

Then take the supremum on $\theta \in [-h,0]$,

$$\|\boldsymbol{w}_t\|^2_{\mathcal{C}([-h,0],\ell^2)} \leq 2(4\nu + \lambda + L_f)\int_0^t \|\boldsymbol{w}_s\|^2_{\mathcal{C}([-h,0],\ell^2)}\mathrm{d}s + \|\boldsymbol{w}(0)\|^2.$$

By Gronwall's inequality, we have

$$\|\boldsymbol{w}_t\|^2_{\mathcal{C}([-h,0],\ell^2)} \leq 2(4\nu + \lambda + L_f)e^{2(4\nu+\lambda+L_f)t}\|\boldsymbol{w}(0)\|^2 + \|\boldsymbol{w}(0)\|^2. \tag{6.10}$$

Since $\boldsymbol{w}(0) = 0$, we obtain that $\boldsymbol{w} \equiv 0$. □

The proof of the next corollary follows easily using (6.10).

Corollary 6.1. *The map $(t, \boldsymbol{\phi}) \mapsto \boldsymbol{u}_t$ is continuous.*

6.3 Asymptotic behaviour

When Assumptions 6.1–6.5 hold, Theorem 6.1 and Lemma 6.5 ensure the local existence and uniqueness of solutions of the delayed lattice system (6.4), while Proposition 6.1 implies that every local solution of (6.1) can be extended globally. With the uniqueness of the solution, this will allow us to define a semi-group in terms of the solution mapping $\varphi : \mathbb{R}^+ \times \mathcal{C}([-h,0],\ell^2) \to \mathcal{C}([-h,0],\ell^2)$ by

$$\varphi(t, \boldsymbol{\phi}) = \boldsymbol{u}_t,$$

where $\boldsymbol{u}_t$ is the unique solution to (6.4) with $\boldsymbol{u}_0(s) = \boldsymbol{\phi}(s)$ for $s \in [-h, 0]$. The semi-group $\{\varphi(t)\}_{t\geq 0}$ is continuous in its variables by Corollary 6.1. It also follows from inequality (6.1) that the semi-group has a bounded absorbing set.

Corollary 6.2. *The bounded set defined by*

$$Q := \left\{\mathcal{U} \in \mathcal{C}([-h,0],\ell^2) : \|\mathcal{U}\|_{\mathcal{C}([-h,0],\ell^2)} \leq \sqrt{1+K_2}\right\},$$

where K_2 is defined as in (6.5), *is absorbing for the semi-group $\{\varphi(t)\}_{t\geq 0}$.*

Notice that the absorbing set Q does not need to be positively invariant under the semi-group $\{\varphi(t)\}_{t\geq 0}$, but it is contained in a larger absorbing set which is positively invariant, namely the set $\bigcup_{0\leq t\leq T(Q)} \varphi(t, Q)$, where $T(Q)$ is the time for Q to absorb itself under $\{\varphi(t)\}_{t\geq 0}$. We next show the existence of a global attractor for the semi-group $\{\varphi(t)\}_{t\geq 0}$ by using Theorem 2.3. To that end, it remains to prove the asymptotic compactness because of Corollary 6.2.

6.3.1 *Tails estimate*

To show the asymptotic compactness of the semi-group $\{\varphi(t)\}_{t\geq 0}$, we need to estimate the tails of solutions of (6.4).

Lemma 6.6. *Suppose that Assumptions 6.1–6.5 hold and let B be a bounded set of $\mathcal{C}([-h,0],\ell^2)$. Then, for any $\varepsilon > 0$ there exist $T(\varepsilon, B)$ and $I(\varepsilon, B)$ such that*

$$\max_{s\in[-h,0]} \sum_{|i|>2I(\varepsilon,B)} |u_i(t+s)|^2 < \varepsilon, \quad t \geq T(\varepsilon, B),$$

for any initial condition $\boldsymbol{\phi} \in B$ and the corresponding solution $\boldsymbol{u}(\cdot)$ of (6.4) *with $\boldsymbol{u}_o = \boldsymbol{\phi}$.*

Proof. Let $\xi : \mathbb{R} \to [0,1]$ be the smooth function defined in (3.8), with a constant C_0 such that $|\xi'(s)| \leq C_0$ for all $s \geq 0$. Let k be a fixed (and large) integer to be specified later, and set

$$v_i(t) = \xi_k(|i|)u_i(t) \text{ with } \xi_k(|i|) = \xi\Big(\frac{|i|}{k}\Big), \quad i \in \mathbb{Z}.$$

Multiply the ith component of (6.1) by v_i, and summing over $i \in \mathbb{Z}$ gives

$$\frac{1}{2}\frac{\mathrm{d}}{\mathrm{d}t}\sum_{i\in\mathbb{Z}} \xi_k(|i|)\,|u_i|^2 = \langle \nu\Lambda \boldsymbol{u}, \boldsymbol{v}\rangle - \lambda \sum_{i\in\mathbb{Z}} \xi_k(|i|)|u_i(t)|^2 + \sum_{i\in\mathbb{Z}} \xi_k(|i|)u_i(t)f_i(u_i(t-\theta_i)) + \sum_{i\in\mathbb{Z}} \xi_k(|i|)u_i(t)g_i. \quad (6.11)$$

By Young's inequality,

$$\xi_k(|i|)u_i(t)f_i(u_i) \leq \frac{\lambda}{4}\xi_k(|i|)|u_i(t)|^2 + \frac{4}{\lambda}\xi_k(|i|)|f_i(u_i(t-\theta_i))|^2.$$

Hence by the Lipschitz property of f_i

$$\xi_k(|i|)u_i(t)f_i(u_i) \le \frac{\lambda}{4}\xi_k(|i|)|u_i(t)|^2 + \frac{4L_f^2}{\lambda}\xi_k(|i|)|u_i(t-\theta_i)|^2. \tag{6.12}$$

Then using Young's inequality again,

$$\sum_{i\in\mathbb{Z}}\xi_k(|i|)g_iu_i(t) \le \frac{\lambda}{4}\sum_{i\in\mathbb{Z}}\xi_k(|i|)|u_i(t)|^2 + \frac{4}{\lambda}\sum_{i\in\mathbb{Z}}\xi_k(|i|)g_i^2. \tag{6.13}$$

Finally, following the arguments in Chapter 3 there exists another constant positive C (depending on the bounded subset B) such that

$$\langle \Lambda \boldsymbol{u}(t), \boldsymbol{v}(t)\rangle \ge -\frac{C}{k} \text{ for all } t \ge 0.$$

Inserting the estimations (6.12) and (6.13) into (6.11), we obtain

$$\begin{aligned}\frac{1}{2}\frac{\mathrm{d}}{\mathrm{d}t}\sum_{i\in\mathbb{Z}}\xi_k(|i|)u_i^2(t) \le \nu\frac{C}{k} - \frac{1}{2}\lambda\sum_{i\in\mathbb{Z}}\xi_k(|i|)|u_i(t)|^2 + \frac{4}{\lambda}\sum_{i\in\mathbb{Z}}\xi_k(|i|)g_i^2 \\ + \frac{4L_f^2}{\lambda}\sum_{i\in\mathbb{Z}}\xi_k(|i|)|u_i(t-\theta_i)|^2.\end{aligned}$$

Then, arguing as in the proof of Proposition 6.1, we have

$$\begin{aligned}\frac{\mathrm{d}}{\mathrm{d}t}\left(e^{\epsilon_0 t}\sum_{i\in\mathbb{Z}}\xi_k(|i|)\,|u_i(t)|^2\right) \le (\epsilon_0-\lambda)e^{\epsilon_0 t}\sum_{i\in\mathbb{Z}}\xi_k(|i|)\,\,|u_i(t)|^2 + \frac{2\nu C}{k}e^{\epsilon_0 t} \\ + \frac{4}{\lambda}\sum_{i\in\mathbb{Z}}\xi_k(|i|)g_i^2 + \frac{4L_f^2}{\lambda}\sum_{i\in\mathbb{Z}}\xi_k(|i|)\,|u_i\,(t-\theta_i)|^2.\end{aligned}$$

Integrating the above differential inequality over $(0,t)$ we get

$$\begin{aligned}e^{\epsilon_0 t}\sum_{i\in\mathbb{Z}}\xi_k(|i|)\,|u_i(t)|^2 \le \sum_{i\in\mathbb{Z}}\xi_k(|i|)\,\,|u_i\,(0)|^2 + (\epsilon_0-\lambda)\int_0^t e^{\epsilon_0 s}\sum_{i\in\mathbb{Z}}\xi_k(|i|)\,|u_i\,(s)|^2\,\mathrm{d}s \\ + \frac{2}{\epsilon_0}\left(e^{\epsilon_0 t}-1\right)\left(\frac{\nu C}{k} + \frac{2}{\lambda}\sum_{i\in\mathbb{Z}}\xi_k(|i|)g_i^2\right) \\ + \frac{4L_f^2}{\lambda}\int_0^t e^{\epsilon_0 s}\sum_{i\in\mathbb{Z}}\xi_k(|i|)\,|u_i\,(s-\theta_i)|^2\,\mathrm{d}s.\end{aligned}$$

By Lemma 6.4

$$\begin{aligned}\int_0^t e^{\epsilon_0 s}\sum_{i\in\mathbb{Z}}\xi_k(|i|)|u_i\,(s-\theta_i)\,|^2\mathrm{d}s \le \frac{e^{\epsilon_0 h}}{\epsilon_0}\sum_{i\in\mathbb{Z}}\xi_k(|i|)|\phi_i|^2_{\mathcal{C}([-h,0],\mathbb{R})}\left(1-e^{-\epsilon_0 h}\right) \\ + e^{\epsilon_0 h}\int_0^t e^{\epsilon_0 s}\sum_{i\in\mathbb{Z}}\xi_k(|i|)|u_i\,(s)\,|^2\mathrm{d}s.\end{aligned}$$

Collecting all these estimates results in

$$e^{\epsilon_0 t}\sum_{i\in\mathbb{Z}}\xi_k(|i|)\left|u_i(t)\right|^2 \le \left(1+\frac{2L_f^2 e^{\epsilon_0 h}}{\lambda\epsilon_0}\left(1-e^{-\epsilon_0 h}\right)\right)\sum_{i\in\mathbb{Z}}\xi_k(|i|)|\phi_i|^2_{C([-h,0],\mathbb{R})}$$
$$+\frac{2}{\epsilon_0}\left(e^{\epsilon_0 t}-1\right)\left(\frac{\nu C}{K}+\frac{2}{\lambda}\sum_{i\in\mathbb{Z}}\xi_k(|i|)g_i^2\right)$$
$$+\left(\epsilon_0-\lambda+\frac{4L_f^2 e^{\epsilon_0 h}}{\lambda}\right)\int_0^t e^{\epsilon_0 s}\sum_{i\in\mathbb{Z}}\xi_k(|i|)\left|u_i(s)\right|^2 \mathrm{d}s,$$

i.e.,

$$e^{\epsilon_0 t}\sum_{i\in\mathbb{Z}}\xi_k(|i|)\left|u_i(t)\right|^2 \le R_k(t)+\mu\int_0^t e^{\epsilon_0 s}\sum_{i\in\mathbb{Z}}\xi_k(|i|)\left|u_i(s)\right|^2 \mathrm{d}s,$$

where

$$R_k(t)=R_B+\frac{2}{\epsilon_0}\left(e^{\epsilon_0 t}-1\right)\left(\frac{\nu C}{k}+\frac{2}{\lambda}\sum_{i\in\mathbb{Z}}\xi_k(|i|)g_i^2\right),\qquad \mu=\frac{4L_f^2 e^{\epsilon_0 h}}{\lambda}$$

with

$$R_B=\left(1+\frac{2L_f^2 e^{\epsilon_0 h}}{\lambda\epsilon_0}\left(1-e^{-\epsilon_0 h}\right)\right)\|B\|^2$$

since $\lambda>\epsilon_0$ and $\phi\in B$. As in the proof of Proposition 6.1 we can rewrite this expression as

$$e^{\epsilon_0 t}\sum_{i\in\mathbb{Z}}\xi_k(|i|)\left|u_{t,i}\right|^2_{C([-h,0],\mathbb{R})} \le R_k(t)+\mu\int_0^t e^{\epsilon_0 s}\sum_{i\in\mathbb{Z}}\xi_k(|i|)\left|u_{s,i}\right|^2_{C([-h,0],\mathbb{R})}. \tag{6.14}$$

Applying Gronwall's inequality to (6.14) gives

$$e^{\epsilon_0 t}\sum_{i\in\mathbb{Z}}\xi_k(|i|)\left|u_{t,i}\right|^2_{C([-h,0],\mathbb{R})} \le R_k(t)+\mu\int_0^t e^{\mu(t-s)}R_k(s)\,\mathrm{d}s.$$

The integral term in the above inequality satisfies

$$\mu\int_0^t e^{\mu(t-s)}R_k(s)\,\mathrm{d}s \le R_B e^{\mu t}+\frac{2e^{\epsilon_0 t}}{\epsilon_0(\epsilon_0-\mu)}\left(\frac{\nu C}{k}+\frac{2}{\lambda}\sum_{i\in\mathbb{Z}}\xi_k(|i|)g_i^2\right).$$

Hence

$$\sum_{i\in\mathbb{Z}}\xi_k(|i|)\left|u_{t,i}\right|^2_{C([-h,0],\mathbb{R})} \le R_B e^{-(\epsilon_0-\mu)t}+\frac{2}{\epsilon_0(\epsilon_0-\mu)}\left(\frac{\nu C}{k}+\frac{2}{\lambda}\sum_{i\in\mathbb{Z}}\xi_k(|i|)g_i^2\right),$$

which implies that

$$\sum_{|i|\ge k}\left|u_{t,i}\right|^2_{C([-h,0],\mathbb{R})} \le R_B e^{-(\epsilon_0-\mu)t}+\frac{2}{\epsilon_0(\epsilon_0-\mu)}\left(\frac{\nu C}{k}+\frac{2}{\lambda}\sum_{|i|\ge k}g_i^2\right). \tag{6.15}$$

The first term in (6.15) will be less than $\varepsilon/3$ if t is large enough, while the other two terms will each be less than $\varepsilon/3$ if $k \geq I(\varepsilon, B)$ for appropriately chosen $I(\varepsilon, B)$ and $T(\varepsilon, B)$. This gives

$$\max_{s\in[-h,0]} \sum_{|i|>2I(\varepsilon,B)} |u_i(t+s)|^2 < \varepsilon, \quad t \geq T(\varepsilon, B),$$

as was to be proved. □

6.3.2 *Existence of the global attractor*

In order to establish the existence of a global attractor, we need to prove that the semi-group $\{\varphi(t)\}_{t\geq 0}$ generated by the delay lattice system (6.4) is asymptotically compact.

Lemma 6.7. *Suppose that Assumptions 6.1–6.5 hold. Then the semi-group $\{\varphi(t)\}_{t\geq 0}$ is asymptotically compact.*

Proof. Let $t_n \to \infty$ and $\{\boldsymbol{\phi}^{(n)}\}_{n\in\mathbb{N}} \subset Q$, where Q is the bounded absorbing set of $\{\varphi(t)\}_{t\geq 0}$ in $C([-h,0],\ell^2)$ defined in Corollary 6.2. Consider $\boldsymbol{u}_{t_n}^{(n)} = \varphi(t_n, \boldsymbol{\phi}^{(n)})$. From (6.9) there exists $C > 0$ such that

$$\|\boldsymbol{u}_{t_n}^{(n)}(s)\| \leq C, \quad \forall s \in [-h,0], \quad \forall n \in \mathbb{N}.$$

For fixed $s \in [-h,0]$ we can find a subsequence (still denoted by $\boldsymbol{u}^{(n)}$) such that

$$\boldsymbol{u}^{(n)}(t_n+s) \rightharpoonup \boldsymbol{u}^*(s) \ \text{ in } \ell^2.$$

The weak convergence here is indeed strong, which follows from Lemma 6.6. In fact, there exists $N_1 > 0$ such that $t_n > T$ (where T is the constant in Lemma 6.6) when $n \geq N_1$. Moreover, for any $\varepsilon > 0$ there exist $I(\varepsilon)$ and $N_2(\varepsilon)$ such that

$$\sum_{|i|>I} |u_i^{(n)}(t_n+s)|^2 < \varepsilon, \quad \sum_{|i|>I} |u_i^*(s)|^2 < \varepsilon, \quad \sum_{|i|\leq I} |u_i^{(n)}(t_n+s) - u_i^*(s)|^2 < \varepsilon$$

if $n \geq \max\{N_1, N_2(\varepsilon)\}$. Hence

$$\begin{aligned}
\|\boldsymbol{u}^{(n)}(t_n+s) - \boldsymbol{u}^*(s)\|^2 &\leq \sum_{|i|\leq I} |u_i^{(n)}(t_n+s) - u_i^*(s)|^2 + \sum_{|i|>I} |u_i^{(n)}(t_n+s) - u_i^*(s)|^2 \\
&\leq \sum_{|i|\leq I} |u_i^{(n)}(t_n+s) - u_i^*(s)|^2 + 2\sum_{|i|>I} |u_i^{(n)}(t_n+s)|^2 \\
&\quad + 2\sum_{|i|>I} |u_i^*(s)|^2 \\
&< 5\varepsilon.
\end{aligned}$$

Thus, $\{\boldsymbol{u}^{(n)}(t_n+s)\}$ is precompact in ℓ^2 for any $s \in [-h,0]$. Since $\mathfrak{F}$ is a bounded map, Lemma 6.2 and the integral representation of solutions imply that

$$\|\boldsymbol{u}^{(n)}(t_n+s) - \boldsymbol{u}^{(n)}(t_n+t)\| \leq \int_s^t \|\mathfrak{F}(\boldsymbol{u}_{t_n+\tau}^{(n)})\| d\tau \leq C(t-s) \text{ if } -h \leq s < t \leq 0$$

for some $C > 0$. Then, the Ascoli-Arzelà theorem implies that $\{\boldsymbol{u}^{(n)}(t_n)\}$ is relatively compact in $\mathcal{C}([-h,0],\ell^2)$. □

Hence, by Theorem 2.3, the semi-group $\{\varphi(t)\}_{t\geq 0}$ generated by the delay LDS system (6.1) on $\mathcal{C}([-h,0],\ell^2)$ has a global attractor $\mathcal{A}$ given by

$$\mathcal{A} = \bigcap_{t\geq 0} \overline{\bigcup_{s\geq t} \varphi(s,Q)},$$

where Q is the absorbing set defined in Corollary 6.2.

6.4 End notes

Some of the proofs in this chapter are based on related proofs in Caraballo, Morillas & Valero [Caraballo *et al.* (2014)]. See also [Wang and Bai (2015); Zhao and Zhou (2007)] and [Zhao, Zhou and Wang (2009)].

Exponential attractors for two-dimensional nonlocal diffusion systems with delays were investigated by Yang, Wang & Kloeden [Yang *et al.* (2012a, 2022b)].

Neural lattice systems with delays were investigated in [Kloeden and Villarragut (2020); Zhou *et al.* (2021)] and [Wang *et al.* (2020a,b)], while [Zhang (2022)] considers FitzHugh-Nagumo lattice models with double time-delays.

Stochastic LDS with delays are considered in [Yan *et al.* (2010)] and [Sui *et al.* (2020, 2021a,b)] consider random and stochastic recurrent neural networks with delays, both finite and infinite.

6.5 Problems

Problem 6.1. What happens to the attractors when the delays converge to zero? In what sense do they converge to the attractor of the corresponding system without delays? (Hint: see [Kloeden (2006)].)

Problem 6.2. How do the proofs need to be changed when a weighted norm space ℓ^2_ρ is used instead of ℓ^2?

Problem 6.3. What is the major difference in analysis if the delays are time-dependent, i.e., $\theta_i(t)$ is considered instead of θ_i in (6.1)?

Chapter 7

Set-valued lattice models

In this chapter we study a set-valued version of the Laplacian lattice model (3.1) in Chapter 3 with the constant external forcing term being replaced by a compact convex subset of ℓ^2 . The resulting system is a set-valued system of ODEs, which is often called a system of differential inclusions. This simple situation illustrates the ideas involved. More complicated set-valued lattice systems will be considered in later chapters.

Consider the autonomous set-valued lattice system

$$\frac{\mathrm{d}u_i}{\mathrm{d}t} \in \nu\left(u_{i-1} - 2u_i + u_{i+1}\right) + f(u_i) + G_i, \quad i \in \mathbb{Z}, \tag{7.1}$$

under the following Assumptions.

Assumption 7.1. $f : \mathbb{R} \to \mathbb{R}$ is a continuously differentiable function satisfying

$$f(0) = 0, \quad f(s)s \leq -\alpha s^2, \quad f'(s) \leq D_f \qquad \forall\, s \in \mathbb{R},$$

for positive constants α, $D_f > 0$.

Assumption 7.2. $G_i = g_i[-1,1]$ for $i \in \mathbb{Z}$, where $\boldsymbol{g} = (g_i)_{i\in\mathbb{Z}} \in \ell^2$.

Here ℓ^2 is the separable Hilbert space defined as in Chapter 3, and Λ is the discrete Laplace operator defined as in Chapter 3 by $(\Lambda\boldsymbol{u})_i = u_{i-1} - 2u_i + u_{i+1}$ for $i \in \mathbb{Z}$ and any $\boldsymbol{u} = (u_i)_{i\in\mathbb{Z}} \in \ell^2$.

7.1 Set-valued lattice system on ℓ^2

Under Assumptions 7.1 and 7.2 the LDS (7.1) can be written as the infinite dimensional set-valued ordinary differential inclusion in the sequence space ℓ^2

$$\frac{\mathrm{d}\boldsymbol{u}}{\mathrm{d}t} \in \nu\Lambda\boldsymbol{u} + F(\boldsymbol{u}) + \boldsymbol{G} =: \mathfrak{F}(\boldsymbol{u}), \tag{7.2}$$

where $\boldsymbol{G} = (G_i)_{i\in\mathbb{Z}}$ and the Nemytskii operator $F : \ell^2 \to \ell^2$ is defined component wise by

$$F(\boldsymbol{u}) = (F_i(\boldsymbol{u}))_{i\in\mathbb{Z}} \quad \text{with} \quad F_i(\boldsymbol{u}) = f(u_i), \quad i \in \mathbb{Z}.$$

The proofs below require the fact shown in Chapter 3 that the mapping F satisfies a one-sided Lipschitz condition in ℓ^2, i.e.,

$$\langle F(\boldsymbol{u}) - F(\boldsymbol{v}), \boldsymbol{u} - \boldsymbol{v}\rangle \leq D_f \|\boldsymbol{u} - \boldsymbol{v}\|, \quad \forall\, \boldsymbol{u} = (u_i)_{i\in\mathbb{Z}}, \boldsymbol{v} = (v_i)_{i\in\mathbb{Z}} \in \ell^2. \tag{7.3}$$

Since $f(0) = 0$ with $\boldsymbol{v} = \boldsymbol{0}$ inequality (7.3) reduces to

$$\langle F(\boldsymbol{u}), \boldsymbol{u}\rangle \leq D_f \|\boldsymbol{u}\|.$$

Assumption 7.1 also ensures that $F : \ell^2 \to \ell^2$ is locally Lipschitz continuous on bounded subsets B of ℓ^2 with some Lipschitz constant L_B, i.e.,

$$\|F(\boldsymbol{u}) - F(\boldsymbol{v})\| \leq L_B \|u - v\|, \qquad \boldsymbol{u}, \boldsymbol{v} \in B. \tag{7.4}$$

In particular,

$$\|F(\boldsymbol{u})\| \leq \|F(\boldsymbol{u}) - F(\boldsymbol{0})\| + \|F(\boldsymbol{0})\| \leq L_B \|\boldsymbol{u}\|,$$

which shows that $F(\boldsymbol{u}) \in \ell^2$ for all $\boldsymbol{u} \in B \subset \ell^2$.

Compactness of the set G in ℓ^2

Lemma 7.1. *The set* $\boldsymbol{G} = (G_i)_{i\in\mathbb{Z}}$*, where* $G_i = g_i[-1, 1]$ *for* $i \in \mathbb{Z}$ *with* $\boldsymbol{g} = (g_i)_{i\in\mathbb{Z}} \in \ell^2$*, is a compact convex subset of* ℓ^2*.*

Proof. Consider an arbitrary sequence $\boldsymbol{x}^{(n)} = (x_i^{(n)})_{i\in\mathbb{Z}} \in \boldsymbol{G}$. Then $x_i^{(n)} \in G_i = g_i[-1, 1]$ for each $i \in \mathbb{Z}$ and $n \in \mathbb{N}$. Since each set G_i is compact in $\mathbb{R}$, there is a convergent subsequence $x_i^{(n_k)} \to x_i^* \in G_i$ as $n_k \to \infty$.

A diagonal subsequence argument can then be used to obtain a subsequence of $\boldsymbol{x}^{(n)}$, for simplicity labelled by $\boldsymbol{x}^{(n_k)}$, such that $x_i^{(n_k)} \to x_i^*$ as $n_k \to \infty$ for each $i \in \mathbb{Z}$. Note that $|x_i^*| \leq |g_i|$, since $x_i^* \in G_i$ for each $i \in \mathbb{Z}$, so $\boldsymbol{x}^* = (x_i^*)_{i\in\mathbb{Z}} \in \ell^2$.

It remains to show that $\boldsymbol{x}^{(n_k)}$ converges to $\boldsymbol{x}^*$ in ℓ^2. Since the convergence holds component wise in $\mathbb{R}$, for every ε and positive integer $I(\varepsilon)$ there exists an integer $N(\varepsilon, I(\varepsilon))$ such that

$$\sum_{|i|\leq I(\varepsilon)} \left|x_i^{(n_k)} - x_i^*\right|^2 \leq \varepsilon, \quad n_k \geq N(\varepsilon, I(\varepsilon)). \tag{7.5}$$

In addition,

$$\left|x_i^{(n_k)} - x_i^*\right|^2 \leq 2\left|x_i^{(n_k)}\right|^2 + 2\left|x_i^*\right|^2 \leq 4\left|g_i\right|^2$$

for each $i \in \mathbb{Z}$, so

$$\sum_{|i|>I(\varepsilon)} \left|x_i^{(n_k)} - x_i^*\right|^2 \leq 4 \sum_{|i|>I(\varepsilon)} g_i^2.$$

Since $\boldsymbol{g} \in \ell^2$, for every ε there exists an integer $\tilde{I}(\varepsilon)$ such that

$$\sum_{|i|>\tilde{I}(\varepsilon)} \left|x_i^{(n_k)} - x_i^*\right|^2 \leq 4 \sum_{|i|>\tilde{I}(\varepsilon)} g_i^2 \leq \varepsilon. \tag{7.6}$$

Combining the estimates (7.5) and (7.6) with $n_k \geq N(\varepsilon, \tilde{I}(\varepsilon))$ gives

$$\sum_{i\in\mathbb{Z}} \left|x_i^{(n_k)} - x_i^*\right|^2 = \sum_{|i|\leq\tilde{I}(\varepsilon)} \left|x_i^{(n_k)} - x_i^*\right|^2 + \sum_{|i|>\tilde{I}(\varepsilon)} \left|x_i^{(n_k)} - x_i^*\right|^2 \leq 2\varepsilon.$$

The convexity of the set $\boldsymbol{G}$ follows from the fact that its component sets are intervals and hence convex sets. □

The set-valued mapping $\mathfrak{F}$ on ℓ^2

The proof of Lemma 7.2 below requires the following inequality [Diamond and Kloeden (1994), Proposition 2.4.1(ii)]: for any nonempty compact subsets A_1, A_2, B_1, B_2 of ℓ^2, it holds

$$\text{dist}_{\ell^2}(A_1 + B_1, A_2 + B_2) \le \text{dist}_{\ell^2}(A_1, A_2) + \text{dist}_{\ell^2}(B_1, B_2) \tag{7.7}$$

where $A + B := \{a + b : a \in A, b \in B\}$.

Lemma 7.2. *The set-valued mapping $\boldsymbol{u} \mapsto \mathfrak{F}(\boldsymbol{u}) := \nu\Lambda\boldsymbol{u} + F(\boldsymbol{u}) + \boldsymbol{G}$ maps ℓ^2 onto a compact convex subset of ℓ^2 and is locally Lipschitz continuous on bounded subsets of ℓ^2.*

Proof. $\Lambda\boldsymbol{u}$ and $F(\boldsymbol{u})$ take values in ℓ^2 for each $\boldsymbol{u} \in \ell^2$, so $\mathfrak{F}(\boldsymbol{u})$ take values in ℓ^2 for each $\boldsymbol{u} \in \ell^2$. It is a compact convex subset of ℓ^2 because $\boldsymbol{G}$ is such a set. Moreover, it follows from the inequality (7.7) that

$$\begin{aligned}
\text{dist}_{\ell^2}\left(\mathfrak{F}(\boldsymbol{u}), \mathfrak{F}(\boldsymbol{v})\right) &= \text{dist}_{\ell^2}\left(\nu\Lambda\boldsymbol{u} + F(\boldsymbol{u}) + \boldsymbol{G}, \nu\Lambda\boldsymbol{v} + F(\boldsymbol{v}) + \boldsymbol{G}\right) \\
&\le \text{dist}_{\ell^2}\left(\nu\Lambda\boldsymbol{u}, \nu\Lambda\boldsymbol{v}\right) + \text{dist}_{\ell^2}\left(F(\boldsymbol{u}), F(\boldsymbol{v})\right) + \text{dist}_{\ell^2}\left(\boldsymbol{G}, \boldsymbol{G}\right) \\
&\le \nu\left\|\Lambda(\boldsymbol{u} - \boldsymbol{v})\right\| + \left\|F(\boldsymbol{u}) - F(\boldsymbol{v})\right\| \\
&\le 4\nu\left\|\boldsymbol{u} - \boldsymbol{v}\right\| + \left\|F(\boldsymbol{u}) - F(\boldsymbol{v})\right\|
\end{aligned}$$

for all $\boldsymbol{u}$, $\boldsymbol{v} \in \ell^2$. Similarly, with $\boldsymbol{u}$ and $\boldsymbol{v}$ interchanged,

$$\text{dist}_{\ell^2}\left(\mathfrak{F}(\boldsymbol{v}), \mathfrak{F}(\boldsymbol{u})\right) \le 4\nu\left\|\boldsymbol{v} - \boldsymbol{u}\right\| + \left\|F(\boldsymbol{v}) - F(\boldsymbol{u})\right\|.$$

Hence, with the Hausdorff metric,

$$\mathcal{H}_{\ell^2}\left(\mathfrak{F}(\boldsymbol{u}), \mathfrak{F}(\boldsymbol{v})\right) \le 4\nu\left\|\boldsymbol{u} - \boldsymbol{v}\right\| + \left\|F(\boldsymbol{u}) - F(\boldsymbol{v})\right\|,$$

i.e., the set-valued mapping $\boldsymbol{u} \mapsto \mathfrak{F}(\boldsymbol{u})$ is continuous with respect to the Hausdorff metric $\mathcal{H}_{\ell^2}$.

Finally, since the mapping $\boldsymbol{u} \mapsto F(\boldsymbol{u})$ is locally Lipschitz on bounded subsets, from (7.4)

$$\mathcal{H}_{\ell^2}\left(\mathfrak{F}(\boldsymbol{u}), \mathfrak{F}(\boldsymbol{v})\right) \le 4\nu\left\|\boldsymbol{u} - \boldsymbol{v}\right\| + L_B\left\|\boldsymbol{u} - \boldsymbol{v}\right\| = (4\nu + L_B)\left\|\boldsymbol{u} - \boldsymbol{v}\right\|$$

for $\boldsymbol{u}$, $\boldsymbol{v} \in B$, a bounded subset of ℓ^2, so the set-valued mapping $\boldsymbol{u} \mapsto \mathfrak{F}(\boldsymbol{u})$ is locally Lipschitz on bounded subsets of ℓ^2. □

7.2 Existence of solutions

This section concerns the existence of solutions for the set-valued lattice model (7.2). First, the definition of a solution to the lattice inclusion (7.2) is given as follows.

Definition 7.1. A function $\boldsymbol{u}(\cdot) = (u_i(t\cdot))_{i\in\mathbb{Z}} : [t_0, t_0+T) \to \ell^2$ is called a solution to the differential inclusion (7.2) if it is an absolutely continuous function $\boldsymbol{u}(t) = (u_i(t))_{i\in\mathbb{Z}} : [t_0, t_0+T] \to \ell^2$ such that

$$\frac{\mathrm{d}u_i(t)}{\mathrm{d}t} \in \nu(\Lambda\boldsymbol{u})_i + f(u_i(t)) + G_i, \quad \text{for all } \ i \in \mathbb{Z}, \text{ a.e..}$$

In particular, this means [Aubin and Cellina (1984); Smirnov (2002)] that there is a measurable selection $\boldsymbol{g}(t) \in \boldsymbol{G}$ such that

$$\frac{\mathrm{d}\boldsymbol{u}(t)}{\mathrm{d}t} = \nu\Lambda\boldsymbol{u} + F(\boldsymbol{u}) + \boldsymbol{g}(t), \quad \text{a.e..} \tag{7.8}$$

The lattice ODE (7.8) has the same form as the lattice ODE (3.4) in Chapter 3 and the coefficient functions have the same properties, which were stated above. It thus follows that for a given initial data $\boldsymbol{u}(0) = \boldsymbol{u}_o \in \ell^2$, there exists a global solution $\boldsymbol{u}(\,\cdot\,;\boldsymbol{u}_o) \in \mathcal{C}([0;\infty), \ell^2)$ of the inclusion system (7.2). In fact for each measurable selection $\boldsymbol{g}(t) \in \boldsymbol{G}$ there exists a unique solution of (7.8) since the vector field is locally Lipschitz.

The totality of these solutions over all selectors defines the *attainability set*

$$\Phi(t, \boldsymbol{u}_o) := \Big\{\boldsymbol{v} \in \ell^2 : \text{there exists a solution } \boldsymbol{u}(\cdot;\boldsymbol{u}_o) \text{ of (7.2) with} \\ \boldsymbol{u}(0;\boldsymbol{u}_o) = \boldsymbol{u}_o \text{ such that } \boldsymbol{v} = \boldsymbol{u}(t;\boldsymbol{u}_o)\Big\}. \tag{7.9}$$

Theorem 7.1. *Under Assumptions 7.1 and 7.2, the attainability set $\Phi(t, \boldsymbol{u}_o)$ of the lattice inclusion system (7.2) is a compact subset of ℓ^2 for each $\boldsymbol{u}_o \in \ell^2$ and $t \geq 0$.*

Proof. The result is trivial for $t = 0$ since $\Phi(0, \boldsymbol{u}_o) = \{\boldsymbol{u}_o\}$. To show that $\Phi(T, \boldsymbol{u}_o)$ is a compact set for a given $T > 0$, let $\boldsymbol{v}_k \in \Phi(T, \boldsymbol{u}_o)$ for $k \in \mathbb{N}$. Then there exist solutions $\boldsymbol{u}^{(k)}(0;\boldsymbol{u}_o)$ of (7.2) with $\boldsymbol{u}^{(k)}(T;\boldsymbol{u}_o) = \boldsymbol{v}_k$ and by Theorem 7.2 (given below in the next subsection) there exists a convergent subsequence in $\mathcal{C}([0,T], \ell^2)$ converging to a solution $\boldsymbol{u}^*(t)$ of (7.2). Clearly $\boldsymbol{u}^*(0) = \boldsymbol{u}_o$, so $\boldsymbol{u}^*(T) \in \Phi(T, \boldsymbol{u}_o)$. □

A compactness theorem

A simplified form of the next theorem was used in the proof of Theorem 7.1 above where the initial conditions were the same. The more general case considered here will be needed later. This is essentially a counterpart of Barbashin's Theorem, Theorem 2.5, in the present context.

Theorem 7.2. *Let $T > 0$. Suppose that Assumptions 7.1 – 7.2 hold and $\boldsymbol{u}_o^{(m)} \to \boldsymbol{u}_o$ as $m \to \infty$. Then for any sequence $\boldsymbol{u}^{(m)}(\cdot) = \boldsymbol{u}(\cdot;\boldsymbol{u}_o^{(m)})$ of solutions to the lattice inclusion system (7.2) with initial value $\boldsymbol{u}_o^{(m)}$, there is a subsequence $\boldsymbol{u}^{(m_k)}(\cdot) \to \boldsymbol{u}^*(\cdot) \in \mathcal{C}([0,T], \ell^2)$ with $m_k \to \infty$, where $\boldsymbol{u}^*(\cdot)$ is a solution of (7.2) with the initial value $\boldsymbol{u}^*(0) = \boldsymbol{u}_o$.*

Proof. Notice that it can be assumed without loss of generality that $\|\boldsymbol{u}_o^{(m)}\| \leq \|\boldsymbol{u}_o\| + 1$. Also, the solution $\boldsymbol{u}^{(m)}(\cdot)$ satisfies

$$\frac{\mathrm{d}\boldsymbol{u}^{(m)}(t)}{\mathrm{d}t} = \nu\Lambda\boldsymbol{u}^{(m)}(t) + F(\boldsymbol{u}^{(m)}(t)) + \boldsymbol{g}^{(m)}(t), \quad \text{a.e.,} \tag{7.10}$$

for some measurable selection $\boldsymbol{g}^{(m)}(t) \in \boldsymbol{G}$.

1. Convergent subsequence

First multiply the equation (7.10) by $u_i^{(m)}$ and summing over $i \in \mathbb{Z}$ to obtain

$$\begin{aligned}\frac{1}{2}\frac{\mathrm{d}}{\mathrm{d}t}\|\boldsymbol{u}^{(m)}(t)\|^2 &\leq \nu\langle\Lambda\boldsymbol{u}^{(m)}(t), \boldsymbol{u}^{(m)}(t)\rangle - \frac{\alpha}{2}\|\boldsymbol{u}^{(m)}(t)\|^2 + \frac{2}{\alpha}\|\boldsymbol{g}^{(m)}(t)\|^2 \\ &\leq -\frac{\alpha}{2}\|\boldsymbol{u}^{(m)}(t)\|^2 + \frac{2}{\alpha}\|\boldsymbol{g}\|^2\end{aligned}$$

since $\|\boldsymbol{g}^{(m)}(t)\| \leq \|\boldsymbol{g}\|$. Thus

$$\frac{\mathrm{d}}{\mathrm{d}t}\|\boldsymbol{u}^{(m)}(t)\| \leq -\alpha\|\boldsymbol{u}^{(m)}(t)\| + \frac{4}{\alpha}\|\boldsymbol{g}\|^2, \tag{7.11}$$

which can be integrated to give

$$\|\boldsymbol{u}^{(m)}(t)\|^2 \leq \|\boldsymbol{u}_o\|^2 + 1 + \frac{4}{\alpha}\|\boldsymbol{g}\|^2 := R_1^2, \quad \forall\, t \in [0, T]. \tag{7.12}$$

In addition, squaring the equation (7.11) gives

$$\begin{aligned}\left\|\frac{\mathrm{d}}{\mathrm{d}t}\boldsymbol{u}^{(m)}(t)\right\|^2 &\leq 2\Big(\alpha^2\|\boldsymbol{u}^{(m)}(t)\|^2 + \frac{16}{\alpha^2}\|\boldsymbol{g}\|^4\Big) \\ &\leq 2\Big(\alpha^2 R_1^2 + \frac{16}{\alpha^2}\|\boldsymbol{g}\|^4\Big) := R_2^2.\end{aligned}$$

Hence the sequence $\{\boldsymbol{u}^{(m)}(t)\}_{m\in\mathbb{N}}$ is uniformly bounded and equi-Lipschitz continuous on $[0, T]$. Then by the Ascoli-Arzelà Theorem there is a $\boldsymbol{u}^*(\cdot) \in \mathcal{C}([0, T], \ell^2)$ and a convergent subsequence $\{\boldsymbol{u}^{(m_k)}(\cdot)\}_{k\in\mathbb{N}}$ such that

$$\begin{aligned}\boldsymbol{u}^{(m_k)}(\cdot) \to \boldsymbol{u}^*(\cdot) \ \text{ strongly in } \mathcal{C}([0, T], \ell^2) \quad \text{as} \quad m_k \to \infty \\ \frac{\mathrm{d}}{\mathrm{d}t}\boldsymbol{u}^{(m_k)}(\cdot) \to \frac{\mathrm{d}}{\mathrm{d}t}\boldsymbol{u}^*(\cdot) \text{ weakly in } \mathcal{L}^1([0, T], \ell^2) \quad \text{as} \quad m_k \to \infty.\end{aligned}$$

The above estimate (7.12) gives component wise

$$\left|u_i^{(m)}(t)\right| \leq R_1, \quad \forall\, t \in [0, T], \quad i \in \mathbb{Z}.$$

Moreover,

$$\begin{aligned}\left|\frac{\mathrm{d}}{\mathrm{d}t}u_i^{(m)}(t)\right| &\leq \nu\,|(\Lambda\boldsymbol{u}(t))_i| + |f(u_i(t))| + \left|g_i^{(m)}(t)\right| \\ &\leq \|\nu\Lambda\boldsymbol{u}(t)\| + L_f(R_1)\,|\boldsymbol{u}_i(t)| + |g_i| \\ &\leq 4\nu\,\|\boldsymbol{u}(t)\| + L_f(R_1)\,\|\boldsymbol{u}(t)\| + \|\boldsymbol{g}\| \\ &\leq (4\nu + L_f(R_1))L_f(R_1) + \|\boldsymbol{g}\| := R_3, \quad \text{for all } i \in \mathbb{Z}, a.e.,\end{aligned}$$

where $L_f(R_1)$ is the local Lipschitz constant of the mapping f in the interval $[-R_1, R_1]$, cf. (7.4).

The Ascoli-Arzelà Theorem can then be used component wise to give a common convergent subsequence for each component, which will be denoted by the original subsequence, such that

$$u_i^{(m_k)}(\cdot) \to u_i^*(\cdot) \text{ strongly in } \mathcal{C}([0,T],\mathbb{R}), \quad i \in \mathbb{Z} \quad \text{as} \quad m_k \to \infty,$$

$$\frac{\mathrm{d}}{\mathrm{d}t} u_i^{(m_k)}(\cdot) \to \frac{\mathrm{d}}{\mathrm{d}t} u_i^*(\cdot) \text{ weakly in } \mathcal{L}^1([0,T],\mathbb{R}), \quad i \in \mathbb{Z} \quad \text{as} \quad m_k \to \infty.$$

A diagonal sequence argument then gives a common convergent subsequence for each component.

Finally, since the components $u_i^*(\cdot)$ of the limit function share the equi-Lipschitz continuity of the subsequence $\{u_i^{(m_k)}(\cdot)\}_{k\in\mathbb{N}}$, they are also absolutely continuous on $[0,T]$.

2. The limit as a solution of the lattice inclusion

Rearranging the lattice system (7.10) component wise for the convergent subsequence $\{\boldsymbol{u}^{(m_k)}(\cdot)\}_{k\in\mathbb{N}}$ gives

$$g_i^{(m_k)}(t) = \frac{\mathrm{d}}{\mathrm{d}t} u_i^{(m_k)}(t) - \nu(\Lambda \boldsymbol{u}^{(m_k)}(t))_i - f(u_i^{(m_k)}(t)), \quad i \in \mathbb{Z}, \text{ a.e..} \tag{7.13}$$

With the limit $u_i^*(\cdot)$ constructed above define

$$g_i^*(t) := \frac{\mathrm{d}}{\mathrm{d}t} u_i^*(t) - \nu(\Lambda \boldsymbol{u}^*(t))_i - f(u_i^*(t)), \quad i \in \mathbb{Z}, \text{ a.e..} \tag{7.14}$$

For $N > 1$ sum both sides of the equation (7.13) from $k = 1$ to N to obtain

$$\frac{1}{N}\sum_{k=1}^{N} g_i^{(m_k)}(t) = \frac{1}{N}\sum_{k=1}^{N} \frac{\mathrm{d}}{\mathrm{d}t} u_i^{(m_k)}(t) - \frac{1}{N}\sum_{k=1}^{N} (\Lambda \boldsymbol{u}^{(m_k)}(t))_i - \frac{1}{N}\sum_{k=1}^{N} f(u_i^{(m_k)}(t)) \tag{7.15}$$

for a.e. $i \in \mathbb{Z}$. Then by the Banach-Saks Theorem, Theorem 2.8, the terms on the right side of (7.15) have subsequences (still denoted by the same label) that converge strongly to the terms on the right hand side of (7.14) in $\mathcal{L}^1([0,T],\mathbb{R})$ for each $i \in \mathbb{Z}$ as $N \to \infty$. Moreover, because $g_i^{(m_k)}(t)(\cdot) \to g_i^*(\cdot)$ weakly in $\mathcal{L}^1([0,T],\mathbb{R})$, the Banach-Saks Theorem gives

$$\frac{1}{N}\sum_{k=1}^{N} g_i^{(m_k)}(\cdot) \to g_i^*(\cdot) \text{ strongly in } \mathcal{L}^1([0,T],\mathbb{R}) \text{ as } N \to \infty, \quad \forall\, i \in \mathbb{Z}.$$

It is clear that $g_i^*(t) \in G_i$ for $i \in \mathbb{Z}$ and $t \in [0,T]$, so $\boldsymbol{g}^*$ is an admissible selector. Since it satisfies equation (7.14), the limit $\boldsymbol{u}^*(\cdot)$ is a solution of the differential inclusion (7.2). This completes the proof of Theorem 7.2. □

7.3 Set-valued semi-dynamical systems with compact values

There is a large literature for autonomous set-valued dynamical systems, which are often called set-valued semi-groups or *general dynamical systems*, see e.g., [Szegö and Treccani (1969)]. Such systems are often generated by differential inclusions or differential equations without uniqueness of solutions [Aubin and Cellina (1984); Smirnov (2002); Tolstonogov (2000)]. This theory was mainly developed on the locally compact state space $\mathbb{R}^d$, but much of it holds here in the Hilbert space ℓ^2, when the attainability sets take compact values.

The definition of set-valued dynamical system was given in Chapter 2. Recall that the attainability sets $\Phi(t, \boldsymbol{u}_o)$ of the set-valued lattice model (7.2) were define by (7.9). The special structure of the lattice system (7.2) ensures that the set-valued mapping $(t, x) \mapsto \Phi(t, \boldsymbol{u})$ is in fact continuous in $(t, \boldsymbol{u}) \in \mathbb{R}^+ \times \ell^2$ with respect to the Hausdorff metric, i.e.,

$$\mathcal{H}_{\ell^2}\left(\Phi(t, \boldsymbol{u}), \Phi(t_0, \boldsymbol{u}_o)\right) \to 0 \quad \text{as} \quad (t, \boldsymbol{u}) \to (t_0, \boldsymbol{u}_o) \quad \text{in } \mathbb{R}^+ \times \ell^2.$$

This means the set-valued mapping $(t, x) \to \Phi(t, \boldsymbol{u})$ is both upper semi-continuous and lower semi continuous in both variables.

Theorem 7.3. *Under Assumptions 7.1 and 7.2 the set-valued lattice system* (7.2) *generates a set-valued semi-dynamical system* $\{\Phi(t)\}_{t\geq 0}$ *in* ℓ^2 *with compact attainability sets, which is continuous in both variables with respect to the Hausdorff metric* $\mathcal{H}_{\ell^2}$.

Proof. By definition, $\Phi(0, \boldsymbol{u}_o) = \{\boldsymbol{u}_o\}$ for every $\boldsymbol{u}_o \in \ell^2$, while the compactness of $\Phi(t, \boldsymbol{u}_o)$ for $t > 0$ was shown in Theorem 7.1.

1. Semi-group property: Recall that $\boldsymbol{u}(t) \in \Phi(t, \boldsymbol{u}_o)$ is a unique solution to the lattice ODE

$$\frac{\mathrm{d}\boldsymbol{u}(t)}{\mathrm{d}t} = \nu\Lambda \boldsymbol{u}(t) + F(\boldsymbol{u}(t)) + \boldsymbol{g}(t), \quad \text{a.e.,} \tag{7.16}$$

in ℓ^2 for an appropriate selector $\boldsymbol{g}(t) \in \boldsymbol{G}$. For every $\boldsymbol{v} \in \Phi(s, \Phi(t, \boldsymbol{u}_o))$, there exists $\boldsymbol{w} \in \Phi(t, \boldsymbol{u}_o)$ and a solution $\boldsymbol{u}^{(1)}(\cdot, \boldsymbol{w})$ of (7.2) such that $\boldsymbol{v} = \boldsymbol{u}^{(1)}(t; \boldsymbol{w})$. At the same time since $\boldsymbol{w} \in \Phi(t, \boldsymbol{u}_o)$, there exists a solution $\boldsymbol{u}^{(2)}(\cdot; \boldsymbol{u}_o)$ of (7.2) such that $\boldsymbol{w} = \boldsymbol{u}^{(2)}(t; \boldsymbol{u}_o)$. Let $\boldsymbol{u}^*(\cdot)$ be the concatenation of $\boldsymbol{u}^{(2)}(\cdot)$ and $\boldsymbol{u}^{(1)}(\cdot)$ and their corresponding selectors $\boldsymbol{g}^{(2)}(\tau) \in \boldsymbol{G}$ on $[0, t]$ and $\boldsymbol{g}^{(1)}(\tau) \in \boldsymbol{G}$ on $[t, s + t]$. Then $\boldsymbol{u}^*(\cdot; \boldsymbol{u}_o)$ is the solution to (7.16) with the selector $\boldsymbol{g}^*$ given by the concatenation of $\boldsymbol{g}^{(2)}$ and $\boldsymbol{g}^{(1)}$. This implies that $\boldsymbol{v} = \boldsymbol{u}^*(s + t; \boldsymbol{u}_o) \in \Phi(s + t, \boldsymbol{u}_o)$ and hence

$$\Phi(s, \Phi(t, \boldsymbol{u}_o)) \subset \Phi(s + t, \boldsymbol{u}_o). \tag{7.17}$$

On the other hand, for every $\boldsymbol{v} \in \Phi(s + t, \boldsymbol{u}_o)$ there exists a solution $\boldsymbol{u}(\cdot, \boldsymbol{u}_o)$ of (7.2) such that $\boldsymbol{v} = \boldsymbol{u}(s + t; \boldsymbol{u}_o)$. There also exists a selector $\boldsymbol{g}(\tau) \in \boldsymbol{G}$ on $[0, s + t]$ such that $\boldsymbol{u}(\cdot)$ is a unique solution of the ODE (7.2). Define $\boldsymbol{g}^{(1)}(\tau) = \boldsymbol{g}(\tau)$, $\boldsymbol{u}^1(\tau) = \boldsymbol{u}(\tau)$ on $[0, t]$ and $\boldsymbol{g}^{(2)}(\tau) = \boldsymbol{g}(\tau)$, $\boldsymbol{u}^{(2)}(\tau) = \boldsymbol{u}(\tau)$ on $[t, s + t]$. Then

$\boldsymbol{g}^{(1)}(\tau) \in \boldsymbol{G}$ on $[0,t]$ and $\boldsymbol{g}^{(2)}(\tau) \in \boldsymbol{G}$) on $[t, s+t]$, and thus $\boldsymbol{u}^{(1)}(\cdot)$ and $\boldsymbol{u}^{(2)}(\cdot)$ are the corresponding solutions to the ODE (7.2). By the uniqueness of solutions to (7.16), $\boldsymbol{u}^{(1)}(0) = \boldsymbol{u}_o$, $\boldsymbol{u}^{(1)}(t) = \boldsymbol{u}^{(2)}(t) = \boldsymbol{w}$ and $\boldsymbol{u}^{(2)}(s+t) = \boldsymbol{v}$. This implies that $\boldsymbol{v} \in \Phi(s, \boldsymbol{u}(t, \boldsymbol{u}_o)) \subset \Phi(s, \Phi(t, \boldsymbol{u}_o))$. hence

$$\Phi(s+t, \boldsymbol{u}_o) \subset \Phi(s, \Phi(t, \boldsymbol{u}_o)). \tag{7.18}$$

Combining (7.17) and (7.18) gives $\Phi(s+t, \boldsymbol{u}_o) = \Phi(s, \Phi(t, \boldsymbol{u}_o))$.

2. Lower semi-continuity in both t and u_o: Suppose (for contradiction) that $\Phi(t, \boldsymbol{u})$ does not converge lower semi continuously to $\Phi(t_0, \boldsymbol{u}_o)$ in ℓ^2 as $(t, u) \to (t_0, \boldsymbol{u}_o)$. Then there exist $\varepsilon_0 > 0$, $t_n \to t_0$ and $\boldsymbol{u}_o^{(n)} \to \boldsymbol{u}_o$ in ℓ^2 such that

$$\operatorname{dist}_{\ell^2}\left(\Phi(t_0, \boldsymbol{u}_o),\ \Phi(t_n, \boldsymbol{u}_o^{(n)})\right) \geq 2\,\varepsilon_0 \quad \text{for all} \quad n \in \mathbb{N}.$$

For each $n \in \mathbb{N}$ there is some $a_n \in \Phi(t_0, \boldsymbol{u}_o)$ with

$$\operatorname{dist}_{\ell^2}\left(a_n,\ \Phi(t_n, \boldsymbol{u}_o^{(n)})\right) \geq \frac{1}{2} \cdot \operatorname{dist}_{\ell^2}\left(\Phi(t_0, \boldsymbol{u}_o),\ \Phi(t_n, \boldsymbol{u}_o^{(n)})\right) \geq \varepsilon_0. \tag{7.19}$$

First, the lattice ODE (7.16) has a solution $\boldsymbol{u}^{(n)}(\cdot)$ in $[0, t_0]$ with a measurable selector $\boldsymbol{g}^{(n)}(t) \in \boldsymbol{G}$ satisfying $\boldsymbol{u}^{(n)}(0) = \boldsymbol{u}_o$ and $\boldsymbol{u}^{(n)}(t_0) = a_n$. Let $\tilde{\boldsymbol{g}}^{(n)}$ and $\tilde{\boldsymbol{u}}^{(n)}$ be the trivial extensions of $\boldsymbol{g}^{(n)}$ and $\boldsymbol{u}^{(n)}$ from $[0, t_0]$ to $[0, T]$, respectively, i.e.,

$$\begin{cases} \tilde{\boldsymbol{g}}^{(n)}(t) = \boldsymbol{g}^{(n)}(t), \quad \tilde{\boldsymbol{u}}^{(n)}(t) = \boldsymbol{u}^{(n)}(t) & \text{for} \quad t \in [0, t_0], \\ \tilde{\boldsymbol{g}}^{(n)}(t) = \tilde{\boldsymbol{u}}^{(n)}(t) = 0 & \text{for} \quad t \in (t_0, T]. \end{cases}$$

Then $\tilde{\boldsymbol{u}}^{(n)} : [0, T] \to \ell^2$ is a solution of the following equation

$$\frac{\mathrm{d}}{\mathrm{d}t}\boldsymbol{u}^n(t) = \nu\Lambda \boldsymbol{u}^n(t) + F(\boldsymbol{u}^n(t)) + \tilde{\boldsymbol{g}}^n(t), \ \text{a.e. on } [0, T] \tag{7.20}$$

satisfying $\tilde{\boldsymbol{u}}^{(n)}(0) = \boldsymbol{u}_o$. On the other hand, for every $n \in \mathbb{N}$, let $\boldsymbol{v}^{(n)}$: $[0, T] \to \ell^2$ be the unique solution to the equation (7.20) satisfying $\boldsymbol{v}^{(n)}(0) = \boldsymbol{u}_o^{(n)}$.

A priori bounds (7.11), (7.12) provide a bounded subset B of ℓ^2 containing all values of $\tilde{\boldsymbol{u}}^{(n)}(\cdot)$, $\boldsymbol{v}^{(n)}(\cdot)$ and a constant $L_B \geq 0$ such that all $\boldsymbol{v}(\cdot)$ are Lipschitz continuous with Lipschitz constant no larger than L_B on B.

In particular, the Lipschitz conditions (7.3), (7.4) ensure for a.e. $t \in [0, T]$

$$\begin{aligned} \left\| \frac{\mathrm{d}\tilde{\boldsymbol{u}}^{(n)}}{\mathrm{d}t}(t) - \frac{\mathrm{d}\boldsymbol{v}^{(n)}}{\mathrm{d}t}(t) \right\| &\leq \nu \|\Lambda\left(\tilde{\boldsymbol{u}}^{(n)}(t) - \boldsymbol{v}^n(t)\right)\| + \|F(\tilde{\boldsymbol{u}}^{(n)}(t)) - F(\boldsymbol{v}^n(t))\| \\ &\leq (4\nu + L_B)\, \|\tilde{\boldsymbol{u}}^{(n)}(t) - \boldsymbol{v}^n(t)\|. \end{aligned}$$

Gronwall's inequality leads to

$$\|\tilde{\boldsymbol{u}}^{(n)}(t) - \boldsymbol{v}^n(t)\| \leq \|\boldsymbol{u}_o - \boldsymbol{u}_o^{(n)}\|\, e^{(4\nu + L_B)\,t} \quad \text{for all} \quad t \in [0, T],$$

and finally

$$\begin{aligned} \operatorname{dist}_{\ell^2}\left(a_n,\ \Phi(t_n, \boldsymbol{u}_o^{(n)})\right) &\leq \|\tilde{\boldsymbol{u}}^{(n)}(t_0) - \boldsymbol{v}^{(n)}(t_n)\| \\ &\leq \|\tilde{\boldsymbol{u}}^{(n)}(t_0) - \boldsymbol{v}^{(n)}(t_0)\| + \|\boldsymbol{v}^{(n)}(t_0) - \boldsymbol{v}^{(n)}(t_n)\| \\ &\leq \|\boldsymbol{u}_o - \boldsymbol{u}_o^{(n)}\|\, e^{(4\nu + L_B)\,t_0} + L_B\, |t_n - t_0| \end{aligned}$$

which has the limit 0 as $n \to \infty$ in contradiction to inequality (7.19).

3. Upper semi-continuity in both t and u_o: Suppose (for contradiction) that $\Phi(t, \boldsymbol{u})$ does not converge upper semi continuously to $\Phi(t_0, \boldsymbol{u}_o)$ in ℓ^2 as $(t, u) \to (t_0, \boldsymbol{u}_o)$. Then there exist $\varepsilon_0 > 0$, $t_n \to t_0$ and $\boldsymbol{u}_o^{(n)} \to \boldsymbol{u}_o$ in ℓ^2 such that

$$\operatorname{dist}_{\ell^2}\left(\Phi(t_n, \boldsymbol{u}_o^{(n)}),\ \Phi(t_0, \boldsymbol{u}_o)\right) \geq 2\,\varepsilon_0 \quad \text{for all} \quad n \in \mathbb{N}.$$

This implies that for each $n \in \mathbb{N}$ there is some $a_n \in \Phi(t_n, \boldsymbol{u}_o^{(n)})$ with

$$\operatorname{dist}_{\ell^2}\left(a_n, \Phi(t_0, \boldsymbol{u}_o)\right) \geq \frac{1}{2} \cdot \operatorname{dist}_{\ell^2}\left(\Phi(t_n, \boldsymbol{u}_o^{(n)}), \Phi(t_0, \boldsymbol{u}_o)\right) \geq \varepsilon_0. \tag{7.21}$$

Similar to Part *2*, the lattice ODE (7.16) has a solution $\boldsymbol{u}^{(n)}(\cdot)$ in $[0, t_n]$ with a measurable selector $\boldsymbol{g}^{(n)}(t) \in \boldsymbol{G}$ satisfying $\boldsymbol{u}^{(n)}(0) = \boldsymbol{u}_o^{(n)}$ and $\boldsymbol{u}^{(n)}(t_n) = a_n$. Let $\tilde{\boldsymbol{g}}^{(n)}$ and $\tilde{\boldsymbol{u}}^{(n)}$ be the trivial extensions of $\boldsymbol{g}^{(n)}$ and $\boldsymbol{u}^{(n)}$ from $[0, t_n]$ to $[0, T]$, respectively. Lemma 7.2 leads to subsequences such that $\tilde{\boldsymbol{g}}^{(n_k)}$ converges weakly to some $\boldsymbol{g}^*$ in $\mathcal{L}^1(0, T; \ell^2)$ and $\frac{\mathrm{d}}{\mathrm{d}t}\tilde{\boldsymbol{u}}^{(n_k)}$ is weakly convergent in $\mathcal{L}^1(0, T; \ell^2)$. Furthermore, Ülger's Lemma 2.4 implies that $\boldsymbol{g}^*$ also has its values in the bounded norm-closed convex (and thus weakly compact) subset $\boldsymbol{G} \subset \ell^2$.

For the rest of this proof, the weak topology of ℓ^2 is metrised by means of a countable dense subset $\{\mathfrak{s}_k\}_{k \in \mathbb{N}}$ of the unit sphere of ℓ^2 which contains all the canonical unit vectors $(\ldots 0, 0, 1, 0, 0 \ldots)$, e.g.,

$$\mathfrak{d}_{weak}(\boldsymbol{v}_1, \boldsymbol{v}_2) := \sum_{k=1}^{\infty} 2^{-k} \frac{|\langle \mathfrak{s}_k,\, \boldsymbol{v}_1 - \boldsymbol{v}_2 \rangle|}{1 + |\langle \mathfrak{s}_k,\, \boldsymbol{v}_1 - \boldsymbol{v}_2 \rangle|}, \qquad (\boldsymbol{v}_1, \boldsymbol{v}_2 \in \ell^2).$$

The Ascoli-Arzelà Theorem in the form of Corollary 2.2 leads to a subsequence (again denoted by) $\{\tilde{\boldsymbol{u}}^{(n_k)}\}_{k \in \mathbb{N}}$ converging uniformly to some $\boldsymbol{u}^* \in \mathcal{C}([0, T], (\ell^2, \mathfrak{d}_{weak}))$. Indeed, the a priori bounds (7.11), (7.12) imply that $\{\tilde{\boldsymbol{u}}^{(n_k)}\}_{k \in \mathbb{N}}$ is equi-Lipschitz continuous in $[0, T]$ and $\left\{\tilde{\boldsymbol{u}}^{(n_k)}(t) \;\middle|\; t \in [0, T],\ k \in \mathbb{N}\right\}$ is norm-bounded, thus relatively weakly compact in ℓ^2.

We next show that $\boldsymbol{u}^*$ satisfies both $\boldsymbol{u}^*(0) = \boldsymbol{u}_o$ and

$$\frac{\mathrm{d}}{\mathrm{d}t}\boldsymbol{u}^* \;=\; \nu \Lambda \boldsymbol{u}^* + F(\boldsymbol{u}^*) + \boldsymbol{g}^*, \quad a.e.\ t \in [0, T].$$

In fact, for each $i \in \mathbb{Z}$, the choice of $\mathfrak{d}_{weak} : \ell^2 \times \ell^2 \to \mathbb{R}_+$ guarantees

$$\tilde{u}_i^{(n_k)}(\cdot) \to u_i^*(\cdot) \;\text{ uniformly in } \mathcal{C}([0, T], \mathbb{R}).$$

Moreover,

$$\frac{\mathrm{d}}{\mathrm{d}t}\tilde{u}_i^{(n_k)}(\cdot) \to \frac{\mathrm{d}}{\mathrm{d}t}u_i^*(\cdot) \text{ weakly in } \mathcal{L}^1([0, T], \mathbb{R}).$$

Rearranging the lattice system (7.10) component wise for the convergent subsequence $\{\tilde{\boldsymbol{u}}^{(n_k)}(\cdot)\}_{k \in \mathbb{N}}$ gives

$$\tilde{g}_i^{(n_k)}(t) = \frac{\mathrm{d}}{\mathrm{d}t}\tilde{u}_i^{(n_k)}(t) - \nu(\Lambda \tilde{\boldsymbol{u}}^{(n_k)}(t))_i - f(\tilde{u}_i^{(n_k)}(t)), \quad \text{a.e.}. \tag{7.22}$$

and for $N > 1$ sum both sides of the equation (7.22) from $k = 1$ to N to obtain

$$\frac{1}{N}\sum_{k=1}^{N}\tilde{g}_i^{(n_k)}(t) = \frac{1}{N}\sum_{k=1}^{N}\frac{\mathrm{d}}{\mathrm{d}t}\tilde{u}_i^{(n_k)}(t) - \frac{1}{N}\sum_{k=1}^{N}(\Lambda\tilde{\boldsymbol{u}}^{(n_k)}(t))_i - \frac{1}{N}\sum_{k=1}^{N}f(\tilde{u}_i^{(n_k)}(t)), \quad \text{a.e..} \tag{7.23}$$

It then follows by the Banach-Saks Theorem, Theorem 2.8, the terms on the right-hand side of (7.23) have subsequences (still denoted by the same label) that converge strongly in $\mathcal{L}^1([0,T],\mathbb{R})$ to

$$\frac{\mathrm{d}}{\mathrm{d}t}u_i^*(t) - \nu(\Lambda\boldsymbol{u}^*(t))_i - f(u_i^*(t))$$

as $N \to \infty$. Moreover, because $\tilde{g}_i^{(n_k)}(\cdot) \to g_i^*(\cdot)$ weakly in $\mathcal{L}^1([0,T],\mathbb{R})$, the Banach-Saks Theorem gives

$$\frac{1}{N}\sum_{k=1}^{N}\tilde{g}_i^{(n_k)}(\cdot) \to g_i^*(\cdot) \text{ strongly in } \mathcal{L}^1([0,T],\mathbb{R}) \text{ as } N \to \infty$$

and so, $g_i^* = \frac{\mathrm{d}}{\mathrm{d}t}u_i^*(t) - \nu(\Lambda\boldsymbol{u}^*(t))_i - f(u_i^*(t))$ holds a.e. in $[0,T]$ for each $i \in \mathbb{Z}$.

Finally, for every $k \in \mathbb{N}$,

$$\begin{aligned}\operatorname{dist}_{\ell^2}(a_{n_k}, \Phi(t_0,\boldsymbol{u}_o)) &= \operatorname{dist}_{\ell^2}\left(\tilde{\boldsymbol{u}}^{(n_k)}(t_{n_k}), \Phi(t_0,\boldsymbol{u}_o)\right) \le \|\tilde{\boldsymbol{u}}^{(n_k)}(t_{n_k}) - \boldsymbol{u}^*(t_0)\|\\ &\le \|\tilde{\boldsymbol{u}}^{(n_k)}(t_{n_k}) - \boldsymbol{u}^*(t_{n_k})\| + \|\boldsymbol{u}^*(t_{n_k}) - \boldsymbol{u}^*(t_0)\|\\ &\to 0 \text{ as } k \to \infty,\end{aligned}$$

which contradicts inequality (7.21).

Combining both results gives the asserted continuous convergence and completes the proof of the theorem. □

7.4 Existence of a global attractor

Theorem 7.4. *Suppose that Assumptions 7.1 and 7.2 are satisfied. The autonomous set-valued semi-dynamical system Φ generated by the lattice inclusion (7.2) on ℓ^2 has a (strong) global attractor $\mathcal{A} \in \ell^2$.*

Proof. The proof has two parts.

1. Existence of an absorbing set: Let $\boldsymbol{u}(t)$ be an arbitrary solution of the lattice inclusion (7.2) with the initial condition $\boldsymbol{u}_o$. Then there is a selector $\boldsymbol{g}(t) \in \boldsymbol{G}$ such that $\boldsymbol{u}(t)$ is a solution of the lattice ODE (7.16) with this selector. Taking the inner product of equation (7.16) with $\boldsymbol{u}$ and using Assumption 7.1 gives

$$\begin{aligned}\frac{\mathrm{d}}{\mathrm{d}t}\|\boldsymbol{u}(t)\|^2 &= 2\nu\langle\Lambda\boldsymbol{u}(t),\boldsymbol{u}(t)\rangle + 2\langle F(\boldsymbol{u}(t)),\boldsymbol{u}(t)\rangle + 2\langle\boldsymbol{g}(t),\boldsymbol{u}\rangle\\ &\le 2\sum_{i\in\mathbb{Z}}u_i(t)f(u_i(t)) + 2\sum_{i\in\mathbb{Z}}g_i(t)u_i(t) \le -\alpha\|\boldsymbol{u}(t)\|^2 + \frac{\|\boldsymbol{g}(t)\|^2}{\alpha},\end{aligned}$$

where the last step follows from Young's inequality. Hence, Gronwall's lemma implies that

$$\|\boldsymbol{u}(t)\|^2 \le \|\boldsymbol{u}_o\|^2 e^{-\alpha t} + \frac{\|\boldsymbol{g}\|^2}{\alpha}\left(1 - e^{-\alpha t}\right), \qquad t \ge 0, \tag{7.24}$$

since $\|\boldsymbol{g}(t)\| \le \|\boldsymbol{g}\|$.

Define the closed and bounded ball in ℓ^2 by

$$Q := \left\{ \boldsymbol{u} \in \ell^2 : \|\boldsymbol{u}\|^2 \le R^2 := 1 + \frac{\|\boldsymbol{g}\|^2}{\alpha} \right\}.$$

Then Q is an absorbing set for the set-valued semi-dynamical system Φ, because the estimate (7.24) holds for all solutions of the lattice inclusion (7.2). In fact, for any $\boldsymbol{u}_o$ in any bounded set $B \subset \ell^2$ it is straightforward to check that

$$\Phi(t, B) \subset Q, \qquad \forall t \ge \frac{2}{\alpha} \ln \|B\|,$$

where $\|B\| := \sup_{\boldsymbol{u} \in B} \|\boldsymbol{u}\|$.

Moreover, for every $\boldsymbol{u}_o \in Q$ and any solution $\boldsymbol{u}(t; u_o)$ of (7.2), it follows from estimate (7.24) that

$$\|\boldsymbol{u}(t; u_o)\|^2 \le R^2 e^{-\alpha t} + R^2\left(1 - e^{-\alpha t}\right) = R^2.$$

Hence, $\|\Phi(t, u_o)\|^2 \le R^2$, i.e., Q is strongly positive invariant under Φ.

2. Asymptotic tails property: In order to establish the asymptotic compactness property 2.1 for the set-valued semi-group system Φ in ℓ^2, the asymptotic tails estimate needs to be shown to hold. This can be shown as in Subsection 3.2.2 of Chapter 3 for each solution of the lattice inclusion (7.2) since the corresponding selectors take values in the constant set $\boldsymbol{G}$, which is uniformly bounded in ℓ^2.

In particular, consider the smooth cut-off function $\xi : \mathbb{R} \to [0, 1]$ as defined in (3.8). Then define $\xi_k(s) = \xi(s/k)$ for all $s \in \mathbb{R}$ and a fixed $k \in \mathbb{N}$ (its value will be specified later). Given any solution $\boldsymbol{u} \in \ell^2$ of the lattice inclusion (7.2) corresponding to a selector $\boldsymbol{g}(t) \in \boldsymbol{G}$, define $\boldsymbol{v} \in \ell^2$ component wise as $v_i := \xi_k(|i|)u_i$ for $i \in \mathbb{Z}$.

Taking the inner product of the lattice ODE (7.8) with $\boldsymbol{v}(t)$ gives

$$\frac{\mathrm{d}}{\mathrm{d}t}\langle \boldsymbol{u}(t), \boldsymbol{v}\rangle + \nu\langle \mathrm{D}^+\boldsymbol{u}(t), \mathrm{D}^+\boldsymbol{v}(t)\rangle = \langle F(\boldsymbol{u}(t)), \boldsymbol{v}(t)\rangle + \langle \boldsymbol{g}(t), \boldsymbol{v}(t)\rangle,$$

that is

$$\frac{\mathrm{d}}{\mathrm{d}t}\sum_{i\in\mathbb{Z}} \xi_k(|i|)|u_i(t)|^2 = -2\nu\langle \mathrm{D}^+\boldsymbol{u}(t), \mathrm{D}^+\boldsymbol{v}(t)\rangle$$
$$+2\sum_{i\in\mathbb{Z}} \xi_k(|i|)u_i(t)f(u_i(t)) + 2\sum_{i\in\mathbb{Z}} \xi_k(|i|)g_i(t)u_i(t).$$

Proceeding as in Subsection 3.2.2 of Chapter 3 for this solution leads to

$$\sum_{i\in\mathbb{Z}} \xi_k(|i|)|u_i(t;\boldsymbol{u}_o)|^2 \leq e^{-\alpha t}\|Q\|^2 + \frac{\varepsilon}{\alpha}$$

for $\boldsymbol{u}_o \in Q$ and hence

$$\sum_{i\in\mathbb{Z}} \xi_k(|i|)|u_i(t;\boldsymbol{u}_o)|^2 \leq \frac{2\varepsilon}{\alpha}, \quad t \geq T(\varepsilon) := \frac{1}{\alpha}\ln\frac{\alpha\|Q\|^2}{\varepsilon}.$$

This bound is uniform in all such solutions for any initial value $\boldsymbol{u}_o \in Q$, since the corresponding selectors are uniformly bounded.

This is the desired asymptotic tails property. The asymptotic compactness in ℓ^2 of the set-valued semi-group Φ in the absorbing set Q then follows by Lemma 2.5. This completes the proof of Theorem 7.4. □

7.5 Endnotes

The literature for autonomous set-valued dynamical systems and set-valued differential equations or differential inclusions is vast, see e.g., [Szegö and Treccani (1969); Aubin and Cellina (1984); Smirnov (2002)]. This theory was mainly developed on the locally compact state space $\mathbb{R}^d$, but many results hold in the Hilbert space ℓ^2 when the attainability sets take compact values as in this chapter. See [Tolstonogov (2000)] for differential inclusions in Banach spaces and [Hu and Papageorgiou (2000)] for set-valued analysis in general.

This chapter is based on unpublished work, but closely follows similar results in the literature. The proof of lower semi-continuity in Theorem 7.3 is due to Thomas Lorenz (personal communication).

More complicated models of set-valued lattice systems will be investigated in later chapters. Other related papers dealing with set-valued lattice models include [Morillas and Valero (2009, 2012)].

7.6 Problems

Problem 7.1. How must the proofs be changed when the compact convex set $\boldsymbol{G}$ is replaced by a set-valued mapping $u \mapsto \boldsymbol{G}(u)$ which is Lipschitz continuous in the Hausdorff metric and has compact convex values? What if the mapping is only upper semi-continuous?

Problem 7.2. How do the proofs need to be changed when a weighted-norm space ℓ^2_ρ is used instead of ℓ^2?

Problem 7.3. Is there an alternative to using the Banach-Sacks Theorem to give the strong convergence of $g_i^{(m_k)}$ to g_i^* in the proofs of Theorem 7.2 and Theorem 7.3?

Chapter 8

Second order lattice dynamical systems

In this chapter we consider the second order lattice system

$$\frac{\mathrm{d}^2 u_i}{\mathrm{d}t^2} + \mu \frac{\mathrm{d}u_i}{\mathrm{d}t} + \nu(\mathrm{A}\boldsymbol{u})_i + \lambda_i u_i + f_i(u_i) = g_i, \quad i \in \mathbb{Z}, \tag{8.1}$$

with the initial conditions

$$u_i(0) = u_{o,i}, \quad \dot{u}_i(0) = v_{o,i}, \quad i \in \mathbb{Z}, \tag{8.2}$$

where μ is a positive constant, λ_i, g_i, $u_{o,i}$, $v_{o,i} \in \mathbb{R}$ for $i \in \mathbb{Z}$ and $\boldsymbol{u} = (u_i)_{i\in\mathbb{Z}}$.

Here A is a non-negative and self-adjoint linear operator on a sequence space, with the decomposition $\mathrm{A} = \mathrm{D}_p\bar{\mathrm{D}}_p = \bar{\mathrm{D}}_p\mathrm{D}_p$, where D_p is defined by

$$(\mathrm{D}_p\boldsymbol{u})_i = \sum_{j=-p}^{p} d_{i+j}u_{i+j}, \; \forall \boldsymbol{u} = (u_i)_{i\in\mathbb{Z}}, \tag{8.3}$$

and $\bar{\mathrm{D}}_p$ is the adjoint of D_p. Notice that $-\mathrm{A}$ can be considered as a generalised discrete Laplacian operator. A simple example is the discrete Laplacian operator Λ corresponding to $(\mathrm{A}\boldsymbol{u})_i = 2u_i - u_{i-1} - u_{i+1}$.

The coefficients in (8.3) are assumed to satisfy

Assumption 8.1. For some positive constant M_d

$$|d_{i+j}| \le M_d, \quad i \in \mathbb{Z}, \quad -p \le j \le p.$$

The lattice system (8.1)–(8.2) will be studied in the weighted sequence space ℓ^2_ρ defined in Section 1.3 and used in Section 5. For the reader's convenience, recall the separable Hilbert space

$$\ell^2_\rho := \Big\{\boldsymbol{u} = (u_i)_{i\in\mathbb{Z}} : \sum_{i\in\mathbb{Z}} \rho_i u_i^2 < \infty\Big\}$$

with the inner product

$$\langle \boldsymbol{u}, \boldsymbol{v}\rangle_\rho := \sum_{i\in\mathbb{Z}} \rho_i u_i v_i \quad \text{for } \boldsymbol{u} = (u_i)_{i\in\mathbb{Z}}, \boldsymbol{v} = (v_i)_{i\in\mathbb{Z}} \in \ell^2_\rho$$

and the norm

$$\|\boldsymbol{u}\|_\rho := \sqrt{\sum_{i\in\mathbb{Z}} \rho_i u_i^2} \quad \text{for } \boldsymbol{u} = (u_i)_{i\in\mathbb{Z}} \in \ell^2_\rho.$$

Here the weights are again assumed to satisfy the Assumption 1.2, which is restated below for the reader's convenience.

Assumption 8.2. There exist positive constants γ_0 and $\gamma_1 > 0$ such that

$$\rho_{i\pm1} \leq \gamma_0\rho_i, \quad |\rho_{i\pm1} - \rho_i| \leq \gamma_1\rho_i, \quad \forall i \in \mathbb{Z}.$$

Instead of Assumption 1.1, here it is assumed that

Assumption 8.3. There exists $M_\rho \in (0,\infty)$ such that $0 < \rho_i \leq M_\rho$ for all $i \in \mathbb{Z}$.

Example 8.1. Two examples of weights satisfying Assumptions 8.2 and 8.3 are

$$\rho_i = \frac{1}{(1+\epsilon^2 i^2)^q} \text{ with } q > 1/2, \quad \rho_i = e^{-\epsilon|i|}, \text{ with } \epsilon > 0.$$

In addition, the following assumptions are made on the parameters λ_i, nonlinear terms f_i and the forcing terms g_i for $i \in \mathbb{Z}$.

Assumption 8.4. There exist $L_f > 0$ and $\boldsymbol{K} = (K_i)_{i\in\mathbb{Z}} \in \ell^2_\rho$ such that

$$|f_i(0)| \leq K_i, \quad |f_i(s_1) - f_i(s_2)| \leq L_f|s_1 - s_2|, \quad \forall s_1, s_2 \in \mathbb{R}, i \in \mathbb{Z}.$$

Assumption 8.5. $\boldsymbol{g} = (g_i)_{i\in\mathbb{Z}} \in \ell^2_\rho$;

Assumption 8.6. There exist two positive constants $m_\lambda, M_\lambda > 0$ such that

$$0 < m_\lambda \leq \lambda_i \leq M_\lambda < +\infty, \quad \forall i \in \mathbb{Z}.$$

Finally, define the n-fold product of ℓ^2_ρ by

$$\mathfrak{E}_n := \ell^2_\rho \times \ell^2_\rho \times \cdots \times \ell^2_\rho,$$

with norm

$$\left\|(\boldsymbol{u}^{(1)},\cdots,\boldsymbol{u}^{(n)})\right\|_{\mathfrak{E}_n} = \left(\sum_{k=1}^n \left\|\boldsymbol{u}^{(k)}\right\|_\rho^2\right)^{1/2}, \quad (\boldsymbol{u}^{(1)},\cdots,\boldsymbol{u}^{(n)}) \in \mathfrak{E}_n.$$

Then $\mathfrak{E}_n$ is a separable Hilbert space.

8.1 Existence and uniqueness of solution

Set $\boldsymbol{u}_o := (u_{o,i})_{i\in\mathbb{Z}}$ and $\boldsymbol{v}_o := (v_{o,i})_{i\in\mathbb{Z}}$. Then the system (8.1)–(8.2) can be written as

$$\frac{\mathrm{d}^2\boldsymbol{u}}{\mathrm{d}t^2} + \mu\frac{\mathrm{d}\boldsymbol{u}}{\mathrm{d}t} + \nu \mathrm{A}\boldsymbol{u} + \boldsymbol{\lambda u} + F(\boldsymbol{u}) = \boldsymbol{g}, \quad u(0) = \boldsymbol{u}_o, \quad \dot{\boldsymbol{u}}(0) = \boldsymbol{v}_o, \tag{8.4}$$

where $\boldsymbol{u} = (u_i)_{i\in\mathbb{Z}}$, $\mathrm{A}\boldsymbol{u} = \big((\mathrm{A}\boldsymbol{u})_i\big)_{i\in\mathbb{Z}}$, $\boldsymbol{\lambda u} = (\lambda_i u_i)_{i\in\mathbb{Z}}$, $F(\boldsymbol{u}) = (f_i(u_i))_{i\in\mathbb{Z}}$, $\boldsymbol{g} = (g_i)_{i\in\mathbb{Z}}$. Notice that the system (8.4) is equivalent to

$$\begin{cases} \frac{\mathrm{d}\boldsymbol{u}}{\mathrm{d}t} = \boldsymbol{v}, & \boldsymbol{u}(0) = \boldsymbol{u}_o \\ \frac{\mathrm{d}\boldsymbol{v}}{\mathrm{d}t} = -\mu\boldsymbol{v} - \nu\mathrm{A}\boldsymbol{u} - \boldsymbol{\lambda u} - F(\boldsymbol{u})\boldsymbol{g}, & \boldsymbol{v}(0) = \boldsymbol{v}_o. \end{cases} \tag{8.5}$$

For any $\boldsymbol{u} = (u_i)_{i\in\mathbb{Z}}$, $\boldsymbol{v} = (v_i)_{i\in\mathbb{Z}} \in \ell^2_\rho$, define a bilinear form $\langle\cdot,\cdot\rangle_{\lambda,\rho}$ by

$$\langle \boldsymbol{u}, \boldsymbol{v}\rangle_{\lambda,\rho} = \langle \mathrm{D}_p\boldsymbol{u}, \mathrm{D}_p\boldsymbol{v}\rangle_\rho + \sum_{i\in\mathbb{Z}} \rho_i\lambda_i u_i v_i. \tag{8.6}$$

Then $\langle\cdot,\cdot\rangle_{\lambda,\rho}$ is an inner product on ℓ^2_ρ. Moreover, the norm $\|\cdot\|_{\lambda,\rho}$ induced by (8.6) is equivalent to the ℓ^2_ρ norm $\|\cdot\|_\rho$. In fact,

$$\begin{aligned} m_\lambda \|\boldsymbol{u}\|^2_\rho \le \|\boldsymbol{u}\|^2_{\lambda,\rho} &\le \left((2p+1)^2 M_d^2 \gamma_0^p + M_\lambda\right)\|\boldsymbol{u}\|^2_\rho \\ &\le \left((2p+1)^2 M_d^2\gamma_0^p + M_\lambda\right) M_\rho \|\boldsymbol{u}\|^2, \end{aligned}$$

where $\|\cdot\|$ is the norm in ℓ^2.

Denote by $\ell^2_{\lambda,\rho}$ the space ℓ^2_ρ equipped with the inner product $\langle\cdot,\cdot\rangle_{\lambda,\rho}$ and norm $\|\cdot\|_{\lambda,\rho}$ and define $\mathfrak{H} = \ell^2_{\lambda,\rho} \times \ell^2_\rho$. Then $\mathfrak{H}$ is a separable Hilbert space with the associated inner product and norm

$$\langle \boldsymbol{x}^{(1)}, \boldsymbol{x}^{(2)}\rangle_{\mathfrak{H}} = \langle \boldsymbol{u}^{(1)}, \boldsymbol{u}^{(2)}\rangle_{\lambda,\rho} + \langle \boldsymbol{v}^{(1)}, \boldsymbol{v}^{(2)}\rangle_\rho, \quad \|\boldsymbol{x}\|^2_{\mathfrak{H}} = \langle \boldsymbol{x}, \boldsymbol{x}\rangle_{\mathfrak{H}} = \|\boldsymbol{u}\|^2_{\lambda,\rho} + \|\boldsymbol{v}\|^2_\rho$$

for

$$\boldsymbol{x}^{(j)} = \begin{pmatrix} \boldsymbol{u}^{(j)} \\ \boldsymbol{v}^{(j)} \end{pmatrix} = \begin{pmatrix} u_i^{(j)} \\ v_i^{(j)} \end{pmatrix}_{i\in\mathbb{Z}} \in \mathfrak{H}, \quad j = 1, 2, \quad \boldsymbol{x} = \begin{pmatrix} \boldsymbol{u} \\ \boldsymbol{v} \end{pmatrix} \in \mathfrak{H}.$$

Now consider $\delta \in \mathbb{R}^+$ and define

$$\boldsymbol{w} = \boldsymbol{v} + \delta\boldsymbol{u}, \quad \boldsymbol{y} = \begin{pmatrix} \boldsymbol{u} \\ \boldsymbol{w} \end{pmatrix}$$

where $\boldsymbol{u}$, $\boldsymbol{v}$ satisfy the system (8.5). Then system (8.5) can be written as

$$\frac{\mathrm{d}}{\mathrm{d}t}\boldsymbol{y}(t) + \mathfrak{L}\boldsymbol{y}(t) = \mathfrak{F}(\boldsymbol{y}(t)), \quad \boldsymbol{y}_o = \begin{pmatrix} \boldsymbol{u}_o \\ \boldsymbol{v}_o + \delta\boldsymbol{u}_o \end{pmatrix}, \tag{8.7}$$

where

$$\mathfrak{L} = \begin{pmatrix} \delta & -1 \\ \boldsymbol{\lambda} + \nu\mathrm{A} - \delta(\mu - \delta) & \mu - \delta \end{pmatrix}, \quad \mathfrak{F}(\boldsymbol{y}) = \begin{pmatrix} 0 \\ -F(\boldsymbol{u}) + \boldsymbol{g} \end{pmatrix}.$$

Theorem 8.1. *Let $T > 0$ and assume that Assumptions 8.4–8.6 hold. Then given any initial data $\boldsymbol{y}_o \in \mathfrak{H}$, the system (8.7) admits a unique solution $\boldsymbol{y}(\cdot;\boldsymbol{y}_o) \in \mathcal{C}^1([0,T),\mathfrak{H})$. Moreover, the solution $\boldsymbol{y}(\cdot;\boldsymbol{y}_o)$ generates a continuous semi-group $\{\varphi(t)\}_{t\geq 0}$ on the state space $\mathfrak{H}$.*

Proof. For any $t \in \mathbb{R}$, $\boldsymbol{g} \in \ell^2_\rho$, and $\boldsymbol{y} \in \mathfrak{H}$, by Assumptions 8.4–8.6, the function $\mathfrak{F}$ in (8.7) is continuous in t and globally Lipschitz continuous in $\boldsymbol{y}$ from $\mathfrak{H}$ to $\mathfrak{H}$ with a uniformly bounded Lipschitz constant. Also, observe that the linear operator $\mathfrak{L}$ in (8.7) has $\mathfrak{H}$ as its domain and generates an analytic semi-group on $\mathfrak{H}$. Therefore, by Theorem 6.1.7 of [Pazy (2007)], the initial value problem (8.7) possesses a unique global classical solution $\boldsymbol{y}(\cdot;\boldsymbol{y}_o) \in \mathcal{C}^1([0,+\infty),\mathfrak{H})$, which is continuous in its initial value.

Define the mapping $\varphi(t) : \mathfrak{H} \to \mathfrak{H}$ by

$$\varphi(t,\boldsymbol{y}_o) = \boldsymbol{y}(t;\boldsymbol{y}_o), \quad t \geq 0.$$

Then $\{\varphi(t)\}_{t\geq 0}$ is a continuous semi-group, and will be referred to as the continuous semi-group generated by the system (8.7). □

Now for any $t \geq 0$ and $\boldsymbol{y}_o \in \mathfrak{H}$, set $\boldsymbol{x}(t) = \mathcal{T}(\delta)\boldsymbol{y}(t;\boldsymbol{y}_o)$, where $\boldsymbol{y}(t;\boldsymbol{y}_o)$ is the solution to the system (8.7), and $\mathcal{T}(\delta) = \begin{pmatrix} 1 & 0 \\ -\delta & 1 \end{pmatrix}$ is an isomorphism on $\mathfrak{H}$. Then $\boldsymbol{x}(t) = \begin{pmatrix} \boldsymbol{u}(t) \\ \boldsymbol{v}(t) \end{pmatrix}$ is the unique solution to the system (8.5), and moreover

$$\tilde{\varphi}(t,\boldsymbol{x}_o) := \boldsymbol{x}(t;\boldsymbol{x}_o) = \boldsymbol{y}(t;\boldsymbol{y}_o) + \begin{pmatrix} 0 \\ -\delta \boldsymbol{u}(t;\boldsymbol{x}_o) \end{pmatrix} \quad \text{for} \quad \boldsymbol{x}_o \in \mathfrak{H}, t \geq 0, \tag{8.8}$$

defines a continuous semi-group $\{\tilde{\varphi}(t)\}_{t\geq 0}$, referred to as the semi-group generated by the system (8.5).

8.2 Existence of a bounded absorbing set

In this subsection, we prove the existence of a closed and bounded absorbing set for the semi-group $\{\tilde{\varphi}(t)\}_{t\geq 0}$ in $\mathfrak{H}$ defined by (8.8). For technical purpose, the parameter δ above is chosen to be

$$\delta = \frac{\mu m_\lambda}{\mu^2 + 2m_\lambda}. \tag{8.9}$$

Lemma 8.1. *Suppose that the number γ_1 in Assumption 8.3 satisfies*

$$0 < \gamma_1 \leq \min\left\{ \frac{\delta}{\nu M_d \gamma_2}, \frac{\mu}{4\nu M_d \gamma_2^2 (2p+1)^2} \right\}, \tag{8.10}$$

where $\gamma_2 = \gamma_2(p) := \gamma_0^{p-1} + \gamma_0^{p-2} + \cdots + \gamma_0 + 1$. Then for any $\boldsymbol{y} = \begin{pmatrix} \boldsymbol{u} \\ \boldsymbol{w} \end{pmatrix} \in \mathfrak{H}$,

$$\langle \mathfrak{L}\boldsymbol{y}, \boldsymbol{y} \rangle_{\mathfrak{H}} \geq \frac{\delta}{2} \|\boldsymbol{y}\|^2_{\mathfrak{H}} + \frac{\mu}{4} \|\boldsymbol{w}\|^2_\rho = \frac{\delta}{2} \left(\|\boldsymbol{u}\|^2_{\rho,\lambda} + \|\boldsymbol{w}\|^2_\rho \right) + \frac{\mu}{4} \|\boldsymbol{w}\|^2_\rho .$$

Proof. For any $\boldsymbol{y} = \begin{pmatrix} \boldsymbol{u} \\ \boldsymbol{w} \end{pmatrix} \in \mathfrak{H}$, one has

$$\langle \mathfrak{L}\boldsymbol{y}, \boldsymbol{y}\rangle_{\mathfrak{H}} = \delta \|\boldsymbol{u}\|^2_{\rho,\lambda} + (\mu - \delta)\|\boldsymbol{w}\|^2_{\rho} - \delta(\mu-\delta)\langle \boldsymbol{u}, \boldsymbol{w}\rangle_{\rho}$$
$$-\nu\langle \mathrm{D}_p\boldsymbol{w}, \mathrm{D}_p\boldsymbol{u}\rangle_{\rho} + \nu\langle \mathrm{A}\boldsymbol{u}, \boldsymbol{w}\rangle_{\rho}. \tag{8.11}$$

Writing $\boldsymbol{\rho} \odot \boldsymbol{w} = (\rho_i w_i)_{i\in\mathbb{Z}}$, then

$$\begin{aligned}\langle \mathrm{A}\boldsymbol{u}, \boldsymbol{w}\rangle_{\rho} &= \sum_{i\in\mathbb{Z}} (\mathrm{D}_p\boldsymbol{u})_i(\mathrm{D}_p(\boldsymbol{\rho}\odot\boldsymbol{w}))_i \\ &= \langle \mathrm{D}_p\boldsymbol{u}, \mathrm{D}_p\boldsymbol{w}\rangle_{\rho} + \sum_{i\in\mathbb{Z}} (\mathrm{D}_p\boldsymbol{u})_i\big((\mathrm{D}_p(\boldsymbol{\rho}\odot\boldsymbol{w}))_i - \rho_i(\mathrm{D}_p\boldsymbol{w})_i\big),\end{aligned}$$

and therefore (8.11) becomes

$$\langle \mathfrak{L}\boldsymbol{y}, \boldsymbol{y}\rangle_{\mathfrak{H}} = \delta \|\boldsymbol{u}\|^2_{\rho,\lambda} + (\mu - \delta)\|\boldsymbol{w}\|^2_{\rho} - \delta(\mu-\delta)\langle \boldsymbol{u}, \boldsymbol{w}\rangle_{\rho}$$
$$+\nu \sum_{i\in\mathbb{Z}} (\mathrm{D}_p\boldsymbol{u})_i\big((\mathrm{D}_p(\boldsymbol{\rho}\odot\boldsymbol{w}))_i - \rho_i(\mathrm{D}_p\boldsymbol{w})_i\big). \tag{8.12}$$

Assumption 8.3 implies that $\rho_{i\pm j} \le \gamma_0^j \rho_i$. Then by definition of γ_2 we have

$$|\rho_{i+j} - \rho_i| \le \gamma_1(\rho_{i+j-1} + \cdots + \rho_{i+1} + \rho_i) \le \gamma_1\gamma_2(j)\rho_i, \quad -p \le j \le p,$$

and consequently

$$\big(\mathrm{D}_p(\boldsymbol{\rho}\odot\boldsymbol{w})\big)_i - \rho_i(\mathrm{D}_p\boldsymbol{w})_i = \sum_{j=-p}^{j=p} d_{i+j}(\rho_{i+j} - \rho_i)w_{i+j} \le \gamma_1\gamma_2(p)M_d \sum_{j=-p}^{j=p} \rho_i w_{i+j}.$$

It then follows that the last term of (8.12) satisfies

$$\begin{aligned}&\left|\sum_{i\in\mathbb{Z}} (\mathrm{D}_p\boldsymbol{u})_i\big((\mathrm{D}_p(\boldsymbol{\rho}\odot\boldsymbol{w}))_i - \rho_i(\mathrm{D}_p\boldsymbol{w})_i\big)\right| \\ &\le \gamma_1\gamma_2(p)M_d \sum_{i\in\mathbb{Z}} \rho_i|(\mathrm{D}_p\boldsymbol{u})_i| \cdot \sum_{j=-p}^{j=p} |w_{i+j}| \\ &\le \frac{1}{2}\gamma_1\gamma_2(p)M_d \sum_{i\in\mathbb{Z}} \left(\rho_i(\mathrm{D}_p\boldsymbol{u})_i^2 + \rho_i\left(\sum_{j=-p}^{j=p}|w_{i+j}|\right)^2\right) \\ &\le \frac{1}{2}\gamma_1\gamma_2(p)M_d\left(\|\mathrm{D}_p\boldsymbol{u}\|^2_{\rho} + 2(2p+1)\sum_{i\in\mathbb{Z}}\sum_{j=-p}^{j=p} M_d^{|j|}\rho_{i+j}w_{i+j}^2\right) \\ &\le \frac{1}{2}\gamma_1\gamma_2(p)M_d\|\mathrm{D}_p\boldsymbol{u}\|^2_{\rho} + \gamma_1\gamma_2^2(p)(2p+1)^2M_d\|\boldsymbol{w}\|^2_{\rho}.\end{aligned} \tag{8.13}$$

Inserting (8.13) into (8.12) results in

$$\begin{aligned}\langle \mathfrak{L}\boldsymbol{y},\boldsymbol{y}\rangle_{\mathfrak{H}} &\geq \delta \|\boldsymbol{u}\|_{\rho,\lambda}^2 + (\mu-\delta)\|\boldsymbol{w}\|_\rho^2 - \delta(\mu-\delta)\langle \boldsymbol{u},\boldsymbol{v}\rangle_\rho \\ &\quad - \frac{1}{2}\nu\gamma_1\gamma_2(p)M_d \|D\boldsymbol{u}\|_\rho^2 - \nu\gamma_1\gamma_2^2(p)(2p+1)^2 M_d \|\boldsymbol{w}\|_\rho^2 \\ &\geq \frac{\delta}{2}\|\boldsymbol{u}\|_{\rho,\lambda}^2 + \left(\frac{3}{4}\mu - \delta\right)\|\boldsymbol{w}\|_\rho^2 + \frac{1}{2}\left(\delta - \nu\gamma_1\gamma_2(p)M_d\right)\|\mathrm{D}_p\boldsymbol{u}\|_\rho^2 \\ &\quad - \delta(\mu-\delta)\langle \boldsymbol{u},\boldsymbol{w}\rangle_\rho + \left(\frac{\mu}{4} - \nu\gamma_1\gamma_2^2(p)(2p+1)^2 M_d\right)\|\boldsymbol{w}\|_\rho^2.\end{aligned}$$

It then follows from the condition (8.10) that

$$\langle \mathfrak{L}\boldsymbol{y},\boldsymbol{y}\rangle_{\mathfrak{H}} \geq \frac{\delta}{2}\|\boldsymbol{u}\|_{\rho,\lambda}^2 + \left(\frac{\mu}{2}-\delta\right)\|\boldsymbol{w}\|_\rho^2 - \frac{\delta\mu}{\sqrt{m_\lambda}}\|\boldsymbol{u}\|_{\rho,\lambda}\cdot\|\boldsymbol{w}\|_\rho + \frac{\mu}{4}\|\boldsymbol{w}\|_\rho^2. \tag{8.14}$$

Notice the condition (8.9) implies that $4\cdot\frac{\delta}{2}\cdot(\frac{\mu}{2}-\delta) = \frac{\delta^2\mu^2}{m_\lambda}$. The desired assertion then follows directly from (8.14). □

Lemma 8.2. *Suppose that*

$$\delta - \frac{8L_f^2}{\mu m_\lambda} > 0. \tag{8.15}$$

Then there exists a closed and bounded absorbing set Q in $\mathfrak{H}$ for the semi-group $\{\tilde{\varphi}(t)\}_{t\geq 0}$ defined by (8.8) such that for any bounded subset B of $\mathfrak{H}$ there exists $T_B > 0$ yielding $\tilde{\varphi}(t,B) \subset Q$ for all $t \geq T_B$. In particular, there exists $T_Q > 0$ so that $\tilde{\varphi}(t,Q) \subset Q$ for all $t \geq T_Q$.

Proof. Let $\boldsymbol{y}(t;\boldsymbol{y}_o) = \begin{pmatrix}\boldsymbol{u}(t)\\ \boldsymbol{w}(t)\end{pmatrix} \in \mathfrak{H}$ be the solution of (8.7) with initial condition $\boldsymbol{y}_o \in \mathfrak{H}$. Taking inner product $\langle\cdot,\cdot\rangle_{\mathfrak{H}}$ of (8.7) with $\boldsymbol{y}(t)$ gives

$$\frac{1}{2}\frac{\mathrm{d}}{\mathrm{d}t}\|\boldsymbol{y}(t;\boldsymbol{y}_o)\|_{\mathfrak{H}}^2 + \langle \mathfrak{L}\boldsymbol{y},\boldsymbol{y}\rangle_{\mathfrak{H}} = \langle \mathfrak{F}(\boldsymbol{y}),\boldsymbol{y}\rangle_{\mathfrak{H}} = -\langle F(\boldsymbol{u}),\boldsymbol{w}\rangle_\rho + \langle \boldsymbol{g},\boldsymbol{w}\rangle_\rho. \tag{8.16}$$

The terms on the right hand side of (8.16) satisfy, respectively,

$$\begin{aligned} -\langle F(\boldsymbol{u}),\boldsymbol{w}\rangle_\rho &\leq \sum_{i\in\mathbb{Z}} \rho_i\left(|f_i(u_i)-f_i(0)|\cdot|w_i| + |f_i(0)|\cdot|w_i|\right) \\ &\leq \sum_{i\in\mathbb{Z}} \rho_i\left(L_f|u_i|\cdot|w_i| + K_i|w_i|\right) \\ &\leq \sum_{i\in\mathbb{Z}} \rho_i\left(\frac{4L_f^2}{\mu m_\lambda}\lambda_i u_i^2 + \frac{4K_i^2}{\mu} + \frac{\mu w_i^2}{8}\right) \\ &\leq \frac{4L^2}{\mu m_\lambda}\|\boldsymbol{y}\|_{\mathfrak{H}}^2 + 4\mu\|\boldsymbol{K}\|_\rho^2 + \frac{\mu}{8}\|\boldsymbol{w}\|_\rho^2, \\ \langle \boldsymbol{g},\boldsymbol{w}\rangle_\rho &\leq \frac{4}{\mu}\|\boldsymbol{g}\|_\rho^2 + \frac{\mu}{8}\|\boldsymbol{w}\|_\rho^2.\end{aligned}$$

By Lemma 8.1 we deduce from (8.16) that

$$\frac{\mathrm{d}}{\mathrm{d}t}\|\boldsymbol{y}(t;\boldsymbol{y}_o)\|_{\mathfrak{H}}^2 \le \left(-\delta + \frac{8L_f^2}{\mu m_\lambda}\right)\|\boldsymbol{y}\|_{\mathfrak{H}}^2 + \frac{8}{\mu}\left(\|\boldsymbol{g}\|_\rho^2 + \|\boldsymbol{K}\|_\rho^2\right). \tag{8.17}$$

Then applying Gronwall's inequality to (8.17) gives for $t \ge 0$,

$$\|\boldsymbol{y}(t;\boldsymbol{y}_o)\|_{\mathfrak{H}}^2 \le e^{-\zeta t}\|\boldsymbol{y}_o\|_{\mathfrak{H}}^2 + \frac{8}{\mu\zeta}\left(\|\boldsymbol{g}\|_\rho^2 + \|\boldsymbol{K}\|_\rho^2\right), \tag{8.18}$$

where $\zeta := \delta - \frac{8L_f^2}{\mu m_\lambda} > 0$ by the condition (8.15).

Note that for $\boldsymbol{y}_o = \begin{pmatrix} \boldsymbol{u}_o \\ \boldsymbol{v}_o + \delta\boldsymbol{u}_o \end{pmatrix}$ and $\boldsymbol{x}_o = \begin{pmatrix} \boldsymbol{u}_o \\ \boldsymbol{v}_o \end{pmatrix}$, we have

$$\|\boldsymbol{y}_o\|_{\mathfrak{H}}^2 \le \|\boldsymbol{u}_o\|_{\rho,\lambda}^2 + \|\boldsymbol{v}_o\|_\rho^2 + \delta^2\|\boldsymbol{u}_o\|_\rho^2 \le \left(1 + \frac{\delta^2}{m_\lambda}\right)\|\boldsymbol{x}_o\|_{\mathfrak{H}}^2.$$

Hence it follows from (8.18) that

$$\|\boldsymbol{y}(t;\boldsymbol{y}_o)\|_{\mathfrak{H}}^2 \le m_\lambda\|\boldsymbol{x}_o\|_{\mathfrak{H}}^2 e^{-\zeta t} + \frac{8}{\mu\zeta}\left(\|\boldsymbol{g}\|_\rho^2 + \|\boldsymbol{K}\|_\rho^2\right), \tag{8.19}$$

which implies that for any bounded $B \subset \mathfrak{H}$, and $\boldsymbol{x}_o \in B$

$$\begin{aligned}\|\boldsymbol{x}(t;\boldsymbol{x}_o)\|_{\mathfrak{H}}^2 &= \left\|\boldsymbol{y}(t;\boldsymbol{y}_o) + \begin{pmatrix} 0 \\ -\delta\boldsymbol{u}(t)\end{pmatrix}\right\|_{\mathfrak{H}}^2 \le 2\|\boldsymbol{y}(t;\boldsymbol{y}_o)\|_{\mathfrak{H}}^2 + 2\delta^2\|\boldsymbol{u}(t)\|_\rho^2 \\ &\le 2m_\lambda\|\boldsymbol{y}(t;\boldsymbol{y}_o)\|_{\mathfrak{H}}^2 \le 2m_\lambda\|\boldsymbol{x}_o\|_{\mathfrak{H}}^2\cdot e^{-\zeta t} + \frac{16}{\mu\zeta}\left(\|\boldsymbol{g}\|_\rho^2 + \|\boldsymbol{K}\|_\rho^2\right).\end{aligned}$$

Hence there exists $T_B > 0$ such that

$$\|\boldsymbol{x}(t;\boldsymbol{x}_o)\|_{\mathfrak{H}}^2 \le 1 + \frac{16}{\mu\zeta}\left(\|\boldsymbol{g}\|_\rho^2 + \|\boldsymbol{K}\|_\rho^2\right) \text{ for all } t \ge T_B. \tag{8.20}$$

This concludes that the set

$$Q := \left\{\boldsymbol{x} \in \mathfrak{H} : \|\boldsymbol{x}\|_{\mathfrak{H}}^2 \le 1 + \frac{16}{\mu\zeta}\left(\|\boldsymbol{g}\|_\rho^2 + \|\boldsymbol{K}\|_\rho^2\right)\right\}$$

is an absorbing set for the semi-group $\{\tilde{\varphi}(t)\}_{t\ge0}$ generated by system (8.5). □

8.3 Existence of a global attractor

Lemma 8.3. *Let condition* (8.15) *hold. Then the semi-group* $\{\tilde{\varphi}(t)\}_{t\ge0}$ *generated by system* (8.5) *is asymptotically compact on* Q.

Proof. Let ξ be the smooth cut-off function defined in (3.8), and recall that there exists a positive constant C_0 such that $|\xi'(s)| \le C_0$ for $s \in \mathbb{R}^+$. Let k be a sufficient large integer (to be specified later) and set $\xi_k(|i|) := \xi\left(\frac{|i|}{k}\right)$.

Let $\boldsymbol{x}(t;\boldsymbol{x}_o) = \begin{pmatrix} u_i(t;\boldsymbol{x}_o) \\ v_i(t;\boldsymbol{x}_o)\end{pmatrix}_{i\in\mathbb{Z}}$ and $\boldsymbol{y}(t;\boldsymbol{y}_o) = \begin{pmatrix} u_i(t;\boldsymbol{y}_o) \\ w_i(t;\boldsymbol{y}_o)\end{pmatrix}_{i\in\mathbb{Z}}$ be the solutions to (8.5) and (8.7), respectively, and set

$$\tilde{\boldsymbol{u}} = (\tilde{u}_i)_{i\in\mathbb{Z}}, \quad \tilde{\boldsymbol{w}} = (\tilde{w}_i)_{i\in\mathbb{Z}} \quad \text{with} \quad \tilde{u}_i = \xi_k(|i|)\,u_i, \quad \tilde{w}_i = \xi_k(|i|)\,w_i, \quad \forall i \in \mathbb{Z}.$$

Denoting $\tilde{\boldsymbol{y}} = \begin{pmatrix} \tilde{\boldsymbol{u}} \\ \tilde{\boldsymbol{w}} \end{pmatrix} = \begin{pmatrix} \tilde{u}_i \\ \tilde{w}_i \end{pmatrix}_{i\in\mathbb{Z}}$, and taking inner product $\langle\cdot,\cdot\rangle_{\mathfrak{H}}$ of (8.7) with $\tilde{\boldsymbol{y}}$ gives

$$\left\langle \frac{\mathrm{d}}{\mathrm{d}t}\boldsymbol{y}, \tilde{\boldsymbol{y}} \right\rangle_{\mathfrak{H}} + \langle \mathfrak{L}\boldsymbol{y}, \tilde{\boldsymbol{y}}\rangle_{\mathfrak{H}} = \langle \mathfrak{F}(\boldsymbol{y}), \tilde{\boldsymbol{y}}\rangle_{\mathfrak{H}}. \tag{8.21}$$

We next estimate each term in (8.21). First,

$$\left\langle \frac{\mathrm{d}}{\mathrm{d}t}\boldsymbol{y}, \tilde{\boldsymbol{y}} \right\rangle_{\mathfrak{H}} = \sum_{i\in\mathbb{Z}} \rho_i (\mathrm{D}_p\dot{\boldsymbol{u}})_i (\mathrm{D}_p\tilde{\boldsymbol{u}})_i + \frac{1}{2}\frac{\mathrm{d}}{\mathrm{d}t}\sum_{i\in\mathbb{Z}} \rho_i \xi_k(|i|)\left(\lambda_i u_i^2 + w_i^2\right),$$

in which

$$\begin{aligned}
(\mathrm{D}_p\tilde{\boldsymbol{u}})_i &= \xi_k(|i|)(\mathrm{D}_p\boldsymbol{u})_i + (\mathrm{D}_p\tilde{\boldsymbol{u}})_i - \xi_k(|i|)(\mathrm{D}_p\boldsymbol{u})_i \\
&= \xi_k(|i|)(\mathrm{D}_p\boldsymbol{u})_i + \sum_{j=-p}^{p} d_{i+j}\left(\xi_k(|i+j|) - \xi_k(|i|)\right) u_{i+j} \\
&\le \xi_k(|i|)(\mathrm{D}_p\boldsymbol{u})_i + \frac{2pM_dC_0}{k}\sum_{j=-p}^{p} |u_{i+j}|, \\
(\mathrm{D}_p\dot{\boldsymbol{u}})_i &= (\mathrm{D}_p(\boldsymbol{w} - \delta\boldsymbol{u}))_i = \sum_{j=-p}^{p} (w_{i+j} - \delta u_{i+j}).
\end{aligned}$$

Therefore,

$$\begin{aligned}
\left\langle \frac{\mathrm{d}}{\mathrm{d}t}\boldsymbol{y}, \tilde{\boldsymbol{y}} \right\rangle_{\mathfrak{H}} &\ge \frac{1}{2}\sum_{i\in\mathbb{Z}} \rho_i\xi_k(|i|)\left((\mathrm{D}_p\boldsymbol{u})_i^2 + \lambda_i u_i^2 + w_i^2\right) \\
&\quad - \frac{2pC_0M_d^2}{k}\sum_{i\in\mathbb{Z}} \rho_i \left(\sum_{j=-p}^{p} |w_{i+j} - \delta u_{i+j}|\right)\cdot \sum_{k=-p}^{p} |u_{i+k}| \\
&\ge \frac{1}{2}\sum_{i\in\mathbb{Z}} \rho_i\xi_k(|i|)\,|y_i|_{\mathfrak{H}}^2 - \frac{C_1}{k}\|\boldsymbol{y}\|_{\mathfrak{H}}^2,
\end{aligned} \tag{8.22}$$

with

$$C_1 := 4pC_0M_d^2(2p+1)(m_\lambda + \delta)\gamma_2(p)/m_\lambda, \quad |y_i|_{\mathfrak{H}}^2 = (\mathrm{D}_p\boldsymbol{u})_i^2 + \lambda_i u_i^2 + w_i^2. \tag{8.23}$$

Next, the second term of (8.21) can be written as

$$\begin{aligned}
\langle \mathfrak{L}\boldsymbol{y}, \tilde{\boldsymbol{y}}\rangle_{\mathfrak{H}} &= \delta\langle \mathrm{D}_p\boldsymbol{u}, \mathrm{D}_p\tilde{\boldsymbol{u}}\rangle_\rho - \langle \mathrm{D}_p\boldsymbol{w}, \mathrm{D}_p\tilde{\boldsymbol{u}}\rangle_\rho + \delta\langle \boldsymbol{\lambda}\boldsymbol{u}, \tilde{\boldsymbol{u}}\rangle_\rho - \langle \boldsymbol{\lambda}\boldsymbol{w}, \tilde{\boldsymbol{u}}\rangle_\rho \\
&\quad + \langle \boldsymbol{\lambda}\boldsymbol{u}, \tilde{\boldsymbol{w}}\rangle_\rho + \nu\langle \mathrm{A}\boldsymbol{u}, \tilde{\boldsymbol{w}}\rangle_\rho - \delta(\mu-\delta)\langle \boldsymbol{u}, \tilde{\boldsymbol{w}}\rangle_\rho + (\mu-\delta)\langle \boldsymbol{w}, \tilde{\boldsymbol{w}}\rangle_\rho.
\end{aligned} \tag{8.24}$$

The first term on the right hand side of (8.24) satisfies

$$\begin{aligned}
\langle \mathrm{D}_p\boldsymbol{u}, \mathrm{D}_p\tilde{\boldsymbol{u}}\rangle_\rho &= \sum_{i\in\mathbb{Z}} \rho_i(\mathrm{D}_p\boldsymbol{u})_i\big(\xi_k(|i|)(\mathrm{D}_p\boldsymbol{u})_i + (\mathrm{D}_p\tilde{\boldsymbol{u}})_i - \xi_k(|i|)(\mathrm{D}_p\boldsymbol{u})_i\big) \\
&= \sum_{i\in\mathbb{Z}} \xi_k(|i|)\rho_i(\mathrm{D}_p\boldsymbol{u})_i^2 - \sum_{i\in\mathbb{Z}} \rho_i(\mathrm{D}_p\boldsymbol{u})_i\big((\mathrm{D}_p\tilde{\boldsymbol{u}})_i - \xi_k(|i|)(\mathrm{D}_p\boldsymbol{u})_i\big)
\end{aligned}$$

in which

$$\sum_{i\in\mathbb{Z}} \rho_i \left|(\mathrm{D}_p\boldsymbol{u})_i \left((\mathrm{D}_p\tilde{\boldsymbol{u}})_i - \xi_k(|i|)(\mathrm{D}_p\boldsymbol{u})_i\right)\right| \le \frac{2M_d^2 pC_0}{k}\sum_{i\in\mathbb{Z}}\rho_i \sum_{j=-p}^{p}|u_{i+j}|\sum_{j=-p}^{p}|u_{i+j}|$$
$$\le \frac{C_1}{k}\|\boldsymbol{y}\|_{\mathfrak{H}}^2. \tag{8.25}$$

Thus

$$\langle \mathrm{D}_p\boldsymbol{u}, \mathrm{D}_p\tilde{\boldsymbol{u}}\rangle_\rho \ge \sum_{i\in\mathbb{Z}}\xi_k(|i|)\,\rho_i(\mathrm{D}_p\boldsymbol{u})_i^2 - \frac{C_1}{k}\|\boldsymbol{y}\|_{\mathfrak{H}}^2. \tag{8.26}$$

Similar to (8.25) we have

$$\sum_{i\in\mathbb{Z}} \rho_i \left|(\mathrm{D}_p\boldsymbol{u})_i \left((\mathrm{D}_p\tilde{\boldsymbol{y}})_i - \xi_k(|i|)(\mathrm{D}_p\boldsymbol{w})_i\right)\right| \le \frac{C_1}{k}\|\boldsymbol{y}\|_{\mathfrak{H}}^2,$$
$$\sum_{i\in\mathbb{Z}} \rho_i \left|(\mathrm{D}_p\boldsymbol{w})_i \left((\mathrm{D}_p\tilde{\boldsymbol{u}})_i - \xi_k(|i|)(\mathrm{D}_p\boldsymbol{u})_i\right)\right| \le \frac{C_1}{k}\|\boldsymbol{y}\|_{\mathfrak{H}}^2,$$

and hence

$$\nu\langle \mathrm{A}\boldsymbol{u}, \boldsymbol{y}\rangle_\rho - (\mathrm{D}_p\boldsymbol{w}, \mathrm{D}_p\tilde{\boldsymbol{u}})_\rho = \nu\sum_{i\in\mathbb{Z}}\rho_i(\mathrm{D}_p\boldsymbol{u})_i\big((\mathrm{D}_p\tilde{\boldsymbol{y}})_i - \xi_k(|i|)(\mathrm{D}_p\boldsymbol{w})_i\big)$$
$$-\sum_{i\in\mathbb{Z}}\rho_i(\mathrm{D}_p\boldsymbol{w})_i\big((\mathrm{D}_p\tilde{\boldsymbol{u}})_i - \xi_k(|i|)(\mathrm{D}_p\boldsymbol{u})_i\big)$$
$$\ge -\frac{(\nu+1)C_1}{k}\|\boldsymbol{y}\|_{\mathfrak{H}}^2. \tag{8.27}$$

The remaining terms of (8.24) satisfy, respectively

$$\langle \boldsymbol{\lambda}\boldsymbol{u}, \tilde{\boldsymbol{u}}\rangle_\rho = \sum_{i\in\mathbb{Z}}\xi_k(|i|)\,\rho_i\lambda_i u_i^2, \qquad -\langle \boldsymbol{\lambda}\boldsymbol{w}, \tilde{\boldsymbol{u}}\rangle_\rho + \langle \boldsymbol{\lambda}\boldsymbol{u}, \tilde{\boldsymbol{w}}\rangle_\rho = 0, \tag{8.28}$$

$$\langle \boldsymbol{u}, \tilde{\boldsymbol{w}}\rangle_\rho = \sum_{i\in\mathbb{Z}}\xi_k(|i|)\,\rho_i u_i w_i, \qquad \langle \boldsymbol{w}, \tilde{\boldsymbol{w}}\rangle_\rho = \sum_{i\in\mathbb{Z}}\xi_k(|i|)\,\rho_i w_i^2. \tag{8.29}$$

Collecting (8.26)–(8.29) and inserting into (8.24) leads to

$$\langle \mathfrak{L}\boldsymbol{y}, \tilde{\boldsymbol{y}}\rangle_{\mathfrak{H}} \ge \sum_{i\in\mathbb{Z}}\xi_k(|i|)\left(\frac{\delta}{2}|y_i|_{\mathfrak{H}}^2 + \frac{\mu}{4}\rho_i w_i^2\right) - \frac{(\nu+2)C_1}{k}\|\boldsymbol{y}\|_{\mathfrak{H}}^2, \tag{8.30}$$

where C_1 and $|y_i|_{\mathfrak{H}}^2$ are defined as in (8.23).

The term $\langle \mathfrak{F}(\boldsymbol{y}), \tilde{\boldsymbol{y}}\rangle_{\mathfrak{H}}$ on the right hand side of (8.21) can be written as

$$-\langle F(\boldsymbol{u}), \tilde{\boldsymbol{w}}\rangle_\rho + \langle \boldsymbol{g}, \tilde{\boldsymbol{w}}\rangle_\rho,$$

in which

$$
\begin{aligned}
-\langle F(\boldsymbol{u}), \tilde{\boldsymbol{w}}\rangle_\rho &\le \sum_{i\in\mathbb{Z}} \xi_k(|i|)\,\rho_i\big(|f_i(u_i)-f_i(0)||w_i| + |f_i(0)||w_i|\big)\\
&\le \sum_{i\in\mathbb{Z}} \xi_k(|i|)\,\rho_i\,(L_f|u_i|\cdot|w_i| + K_i\cdot|w_i|)\\
&\le \sum_{i\in\mathbb{Z}} \xi_k(|i|)\,\rho_i\left(\frac{4L_f^2}{\mu m_\lambda}\lambda_i u_i^2 + \frac{\mu}{16}w_i^2 + \frac{4}{\mu}K_i^2 + \frac{\mu}{16}w_i^2\right)\\
&\le \sum_{i\in\mathbb{Z}} \xi_k(|i|)\,\frac{4L^2}{\mu m_\lambda}\,|y_i|_{\mathfrak{H}}^2 + \sum_{i\in\mathbb{Z}} \xi_k(|i|)\,\rho_i\left(\frac{\mu}{8}w_i^2 + \frac{4}{\mu}K_i^2\right),
\end{aligned}
$$

$$
\text{and}\quad \langle \boldsymbol{g}, \tilde{\boldsymbol{w}}\rangle_\rho \le \sum_{i\in\mathbb{Z}} \xi_k(|i|)\,\rho_i\left(\frac{4}{\mu}g_i^2 + \frac{\mu}{8}w_i^2\right),
$$

where $|y_i|_{\mathfrak{H}}^2$ is defined as in (8.23). Therefore

$$
\langle \mathfrak{F}(\boldsymbol{y}), \tilde{\boldsymbol{y}}\rangle_{\mathfrak{H}} \le \sum_{i\in\mathbb{Z}} \xi_k(|i|)\,\rho_i\left(\frac{\mu}{4}w_i^2 + \frac{4}{\mu}|g_i|^2 + \frac{4}{\mu}K_i^2\right). \tag{8.31}
$$

Now inserting inequalities (8.22), (8.30) and (8.31) into (8.21) we obtain

$$
\begin{aligned}
\frac{\mathrm{d}}{\mathrm{d}t}\sum_{i\in\mathbb{Z}} \xi_k(|i|)\,|y_i(t;\boldsymbol{y}_o)|_{\mathfrak{H}}^2 &\le \left(-\delta + \frac{8L^2}{\mu m_\lambda}\right)\sum_{i\in\mathbb{Z}} \xi_k(|i|)\,|y_i(t;\boldsymbol{y}_o)|_{\mathfrak{H}}^2 + \frac{C_2}{k}\,\|\boldsymbol{y}(t;\boldsymbol{y}_o)\|_{\mathfrak{H}}^2\\
&\quad + \frac{8}{\mu}\sum_{i\in\mathbb{Z}} \xi_k(|i|)\,\rho_i\,(g_i^2 + K_i^2),
\end{aligned} \tag{8.32}
$$

where $|y_i(t;\boldsymbol{y}_o)|_{\mathfrak{H}}^2$ is defined as (8.23) and $C_2 := 2C_1\,(2+\nu+m_\lambda+\delta)$.

Recalling that $\zeta = \delta - \frac{8L^2}{\mu m_\lambda}$ and applying Gronwall's inequality to (8.31) from T_Q to t, we obtain

$$
\begin{aligned}
\sum_{i\in\mathbb{Z}} \xi_k(|i|)\,|y_i(t;\boldsymbol{y}_o)|_{\mathfrak{H}}^2 &\le e^{-\zeta(t-T_Q)}\,\|\boldsymbol{y}(T_Q,\boldsymbol{y}_o)\|_E^2 + \frac{C_2}{k}\int_{T_Q}^t e^{-\eta(t-\tau)}\,\|\boldsymbol{y}(\tau,\boldsymbol{y}_o)\|_{\mathfrak{H}}^2\,\mathrm{d}\tau\\
&\quad + \frac{8}{\mu\zeta}\sum_{i\in\mathbb{Z}} \xi_k(|i|)\,\rho_i\,(g_i^2 + K_i^2).
\end{aligned}
$$

For the solution $\boldsymbol{x} = (x_i)_{i\in\mathbb{Z}}$ to (8.5), set $|x_i|_{\mathfrak{H}}^2 := (\mathrm{D}_p\boldsymbol{u})_i^2 + \lambda_i u_i^2 + v_i^2$. Clearly, $|x_i|_{\mathfrak{H}}^2 \le m_\lambda\,|y_i|_{\mathfrak{H}}^2$ and thus

$$
\begin{aligned}
\sum_{i\in\mathbb{Z}} \xi_k(|i|)\,\rho_i\,|x_i(t;\boldsymbol{x}_o)|_{\mathfrak{H}}^2 &\le m_\lambda e^{-\zeta(t-T_Q)}\,\|\boldsymbol{y}(T_Q;\boldsymbol{y}_o)\|_{\mathfrak{H}}^2\\
&\quad + \frac{m_\lambda C_2}{k}\int_{T_Q}^t e^{-\zeta(t-\tau)}\,\|\boldsymbol{y}(\tau;\boldsymbol{y}_o)\|_{\mathfrak{H}}^2\,\mathrm{d}\tau + \frac{8m_\lambda}{\mu\kappa}\sum_{i\in\mathbb{Z}} \xi_k(|i|)\,\rho_i\,(g_i^2 + K_i^2).
\end{aligned} \tag{8.33}
$$

It remains to estimate each term on the right hand side of (8.33) for $\boldsymbol{x}_o \in Q\cap\mathfrak{H}$. First, by (8.19)

$$
\begin{aligned}
e^{-\eta(t-T_Q)}\,\|\boldsymbol{y}(T_Q;\boldsymbol{y}_o)\|_{\mathfrak{H}}^2 \;&\le\; m_\lambda\,\|\boldsymbol{x}_o\|_{\mathfrak{H}}^2\,e^{-\zeta(t-T_Q)} + \frac{8}{\mu\eta}\left(\|\boldsymbol{g}\|_\rho^2 + \|\boldsymbol{K}\|_\rho^2\right)\\
&\overset{t\to\infty}{\longrightarrow} 0,
\end{aligned}
$$

which implies that for $\varepsilon > 0$, there exists $T_1(\varepsilon, Q) \geq T_Q$, such that for $t \geq T_1(\varepsilon, Q)$,

$$m_\lambda e^{-\zeta(t-T_Q)} \|\boldsymbol{y}(T_Q; \boldsymbol{y}_o)\|_{\mathfrak{H}}^2 \leq \frac{\varepsilon}{3}. \tag{8.34}$$

Then by (8.18), for $t \geq T_Q$,

$$\int_{T_Q}^{t} e^{-\zeta(t-\tau)} \|\boldsymbol{y}(\tau; \boldsymbol{y}_o)\|_{\mathfrak{H}}^2 \, \mathrm{d}\tau \leq \int_{T_Q}^{t} e^{-\zeta(t-\tau)} C_3 \mathrm{d}\tau \leq \frac{C_3}{\zeta},$$

where $C_3 := 1 + \frac{8}{\mu\eta}\left(\|\boldsymbol{g}\|_\rho^2 + \|\boldsymbol{K}\|_\rho^2\right)$. Hence choosing k large enough gives

$$\frac{m_\lambda C_2}{k} \int_{T_Q}^{t} e^{-\zeta(t-\tau)} \|\boldsymbol{y}(\tau; \boldsymbol{y}_o)\|_{\mathfrak{H}}^2 \, \mathrm{d}\tau \leq \frac{\varepsilon}{3}. \tag{8.35}$$

In addition, Assumption 8.5 ensures that for k large enough

$$\frac{8m_\lambda}{\mu\kappa} \sum_{i\in\mathbb{Z}} \xi_k(|i|)\, \rho_i \left(g_i^2 + K_i^2\right) = \frac{8m_\lambda}{\mu\kappa} \sum_{|i|\geq k} \rho_i \left(g_i^2 + K_i^2\right) \leq \frac{\varepsilon}{3}. \tag{8.36}$$

Combining (8.33)–(8.36) then results in

$$\sum_{|i|\geq k} |x_i(t; \boldsymbol{X}_o)|_{\mathfrak{H}}^2 \leq \sum_{i\in\mathbb{Z}} \xi_k(|i|)\, |x_i(t; \boldsymbol{x}_o)|_{\mathfrak{H}}^2 \leq \varepsilon$$

for t and k large enough, depending on ε. This proves the asymptotic tails property and completes the proof of Lemma 8.3. □

The proof of the next theorem then follows immediately from Lemma 8.2, Lemma 8.3 and Theorem 8.1.

Theorem 8.2. *Let condition* (8.15) *hold. Then the semi-group* $\{\tilde{\varphi}(t)\}_{t\geq 0}$ *generated by system* (8.5) *possesses a unique global attractor given by*

$$\mathcal{A} = \bigcap_{\tau\geq T_Q} \overline{\bigcup_{t\geq\tau} \tilde{\varphi}(t; Q)} \subset \mathfrak{H}.$$

8.4 End notes

Second order stochastic lattice systems were considered in [Han (2012, 2013)]. This chapter is based on material in [Zhou (2002a)] and [Han (2012)]. See also [Abdallah (2005, 2009)], [Fan and Wang (2007); Fan and Yang (2010)] and [Zhou (2002b)].

The discrete Schrödinger system was considered in [Karachalios and Yannacopoulos (2005)] and Klein-Gordon-Schrödinger lattice systems were considered in [Abdallah (2006); Huang *et al.* (2009); Yin *et al.* (12007)].

For second order stochastic LDS see [Han (2012, 2013)] and Wang, Li & Xu [Wang *et al.* (2010)].

8.5 Problems

Problem 8.1. Show that the attractors of finite dimensional approximations of the lattice system (8.1) converge upper semi-continuously to the attractor of (8.1).

Problem 8.2. Formulate the lattice system as the discretisation of the following PDE for damped coupled nonlinear oscillators:

$$u_{tt} - \mu u_{xxt} - \nu u_{xx} + \lambda u + f(u) + h(u_t) = g, \quad \mu, \nu > 0.$$

Problem 8.3. Establish assumptions on f, g and h such that the lattice system derived in Problem 8.2 has a global attractor.

Chapter 9

Discrete time lattice systems

In theoretical ecology models describing the spatial dispersal and the temporal evolution of species with non-overlapping generations are often based on integro-difference equations [Lütscher (2019)]. They essentially involve the iteration of continuous functions on an ambient domain. They can also arise from temporal discretisations of integro-differential equations or as time-1-maps of evolutionary differential equations.

Discrete time lattice models, i.e., lattice difference equations, occur when the spatial variable in integro-difference equations is discretised and also with the temporal discretisation of lattice differential equations, see Chapter 4.

9.1 Autonomous systems

First, we consider autonomous lattice difference equations of the form

$$u_{\mathfrak{i}}(t+1) = f_{\mathfrak{i}}(u_{\mathfrak{i}}(t)) + \sum_{\mathfrak{j}\in\mathbb{Z}^d} \kappa_{\mathfrak{i},\mathfrak{j}} g_{\mathfrak{j}}(u_{\mathfrak{j}}(t)), \qquad \mathfrak{i}\in\mathbb{Z}^d, \quad t\in\mathbb{Z}^+, \tag{9.1}$$

with connection weights $\kappa_{\mathfrak{i},\mathfrak{j}}$, where $\mathfrak{i}, \mathfrak{j} \in \mathbb{Z}^d$.

Remark 9.1. Representative functions $f_{\mathfrak{i}}(s)$ and $g_{\mathfrak{i}}(s)$ are

$$ase^{-b|s|}, \qquad \frac{as}{b+|s|}, \qquad s\in\mathbb{R},$$

which arise in the spatial Ricker and Beverton-Holt equations of theoretical ecology.

9.1.1 *Preliminaries*

Due to the global connection structure of $\sum_{\mathfrak{j}\in\mathbb{Z}^d} \kappa_{\mathfrak{i},\mathfrak{j}} g_{\mathfrak{j}}(u_{\mathfrak{j}}(t))$, the lattice system will be considered in a weighted space of bi-infinite sequences with vectorial indices $\mathfrak{i} = (i_1, \cdots, i_d) \in \mathbb{Z}^d$ with $|\mathfrak{i}| = \sum_{j=1}^d |i_j|$ defined in Section 1.3.

Throughout this chapter the weights $\rho_{\mathfrak{i}}$ are assumed to satisfy Assumption 1.1, which is repeated in Assumption 9.1 below for the reader's convenience.

Assumption 9.1. $\rho_{\mathfrak{i}} > 0$ for all $\mathfrak{i} \in \mathbb{Z}^d$ and $\rho_\Sigma := \sum_{\mathfrak{i}\in\mathbb{Z}^d} \rho_{\mathfrak{i}} < \infty$.

In addition, it will be assumed that

Assumption 9.2. The aggregate interconnection strength coefficients $\kappa_{\mathfrak{i},\mathfrak{j}}$ are reciprocal-weighted finite in the sense that

$$\sup_{\mathfrak{i}\in\mathbb{Z}^d}\sum_{\mathfrak{j}\in\mathbb{Z}^d}\frac{\kappa_{\mathfrak{i},\mathfrak{j}}^2}{\rho_{\mathfrak{j}}}\le\kappa\quad\text{for some }\kappa>0.$$

The nonlinear terms $f_{\mathfrak{i}}$ and $g_{\mathfrak{i}}$ are assumed to satisfy the Assumptions 9.3–9.5 below.

Assumption 9.3. $f_{\mathfrak{i}}(0)=g_{\mathfrak{i}}(0)=0$ for each $\mathfrak{i}\in\mathbb{Z}^d$.

Assumption 9.4. for each $\mathfrak{i}\in\mathbb{Z}^d$ the functions $f_{\mathfrak{i}}$, $g_{\mathfrak{i}}:\mathbb{R}\to\mathbb{R}$ are globally Lipschitz continuous with

$$|f_{\mathfrak{i}}(x)-f_{\mathfrak{i}}(y)|\le L_f|x-y|,\quad |g_{\mathfrak{i}}(x)-g_{\mathfrak{i}}(y)|\le L_{g,\mathfrak{i}}|x-y|,\qquad\forall\, x,y\in\mathbb{R},$$

where $L_f>0$ and $\sup_{\mathfrak{i}\in\mathbb{Z}^d}\sum_{\mathfrak{j}\in\mathbb{Z}^d}\frac{\kappa_{\mathfrak{i},\mathfrak{j}}^2L_{g,\mathfrak{j}}^2}{\rho_{\mathfrak{j}}}\le\hat{L}_g^2$ for some $\hat{L}_g$.

Assumption 9.5. for each $\mathfrak{i}\in\mathbb{Z}^d$ the functions $f_{\mathfrak{i}}$, $g_{\mathfrak{i}}:\mathbb{R}\to\mathbb{R}$ are uniformly bounded with

$$|f_{\mathfrak{i}}(x)|,|g_{\mathfrak{i}}(x)|\le M_{\mathfrak{i}}\qquad\forall\, x\in\mathbb{R},\quad\text{for some}\quad \boldsymbol{M}=(M_{\mathfrak{i}})_{\mathfrak{i}\in\mathbb{Z}}\in\ell_\rho^2.$$

Remark 9.2. For greater generality, the index set $\mathbb{Z}^d$ could be replaced by an index set $\mathbb{I}$ which is the union of an increasing sequence of finite subsets $\mathbb{I}_N$. In the set up here, an infinite subset of $\mathbb{Z}^d$ could be considered simply by setting the interconnection weights $\kappa_{\mathfrak{i},\mathfrak{j}}$ equal to zero for indices outside this set. An example with finitely many interconnections in section 9.3 represents the admissible indices by $\mathfrak{j}-\mathfrak{i}\in\mathbb{Z}_N^d$ for each $\mathfrak{i}\in\mathbb{Z}^d$, where $\mathbb{Z}_N^d:=\{\mathfrak{i}\in\mathbb{Z}^d:|\mathfrak{i}|\le N\}$.

For $\boldsymbol{u}\in\ell_\rho^2$, define the mapping $\mathfrak{F}(\boldsymbol{u})=(\mathfrak{F}_{\mathfrak{i}}(\boldsymbol{u}))_{\mathfrak{i}\in\mathbb{Z}^d}$ componentwise by

$$\mathfrak{F}_{\mathfrak{i}}(\boldsymbol{u})=f_{\mathfrak{i}}(u_{\mathfrak{i}})+\sum_{\mathfrak{j}\in\mathbb{Z}^d}\kappa_{\mathfrak{i},\mathfrak{j}}g_{\mathfrak{j}}(u_{\mathfrak{j}}),\qquad \mathfrak{i}\in\mathbb{Z}^d,\quad \boldsymbol{u}\in\ell_\rho^2.$$

Lemma 9.1. *$\mathfrak{F}$ maps ℓ_ρ^2 to ℓ_ρ^2 and is uniformly bounded with*

$$\|\mathfrak{F}(\boldsymbol{u})\|_\rho\le\|\boldsymbol{M}\|_\rho\sqrt{2(1+\kappa\rho_\Sigma)},\qquad\forall\boldsymbol{u}\in\ell_\rho^2.$$

Moreover $\mathfrak{F}(\mathbf{0})=\mathbf{0}$.

Proof. Let $\boldsymbol{u}\in\ell_\rho^2$. Then for each $\mathfrak{i}\in\mathbb{Z}^d$,

$$\begin{aligned}\sqrt{\rho_{\mathfrak{i}}}|\mathfrak{F}_{\mathfrak{i}}(\boldsymbol{u})|&\le\sqrt{\rho_{\mathfrak{i}}}|f_{\mathfrak{i}}(u_{\mathfrak{i}})|+\sqrt{\rho_{\mathfrak{i}}}\sum_{\mathfrak{j}\in\mathbb{Z}^d}\frac{1}{\sqrt{\rho_{\mathfrak{j}}}}|\kappa_{\mathfrak{i},\mathfrak{j}}|\sqrt{\rho_{\mathfrak{j}}}|g_{\mathfrak{j}}(u_{\mathfrak{j}})|\\&\le\sqrt{\rho_{\mathfrak{i}}}M_{\mathfrak{i}}+\sqrt{\rho_{\mathfrak{i}}}\sum_{\mathfrak{j}\in\mathbb{Z}^d}\frac{1}{\sqrt{\rho_{\mathfrak{j}}}}|\kappa_{\mathfrak{i},\mathfrak{j}}|\sqrt{\rho_{\mathfrak{j}}}M_{\mathfrak{j}}.\end{aligned}$$

Squaring both sides gives

$$
\begin{aligned}
\rho_{\mathrm{i}}|\mathfrak{F}_{\mathrm{i}}(\boldsymbol{u})|^2 &\leq 2\rho_{\mathrm{i}}M_{\mathrm{i}}^2 + 2\rho_{\mathrm{i}}\left(\sum_{\mathrm{j}\in\mathbb{Z}^d}\frac{1}{\sqrt{\rho_{\mathrm{j}}}}|\kappa_{\mathrm{i},\mathrm{j}}|\sqrt{\rho_{\mathrm{j}}}M_{\mathrm{j}}\right)^2 \\
&\leq 2\rho_{\mathrm{i}}M_{\mathrm{i}}^2 + 2\rho_{\mathrm{i}}\sum_{\mathrm{j}\in\mathbb{Z}^d}\frac{\kappa_{\mathrm{i},\mathrm{j}}^2}{\rho_{\mathrm{j}}}\sum_{\mathrm{j}\in\mathbb{Z}^d}\rho_{\mathrm{j}}M_{\mathrm{j}}^2 \\
&\leq 2\rho_{\mathrm{i}}M_{\mathrm{i}}^2 + 2\rho_{\mathrm{i}}\kappa\|\boldsymbol{M}\|_\rho^2,
\end{aligned}
$$

so summing over $\mathrm{i}\in\mathbb{Z}^d$ results in

$$
\|\mathfrak{F}(\boldsymbol{u})\|_\rho^2 \leq 2\sum_{\mathrm{i}\in\mathbb{Z}^d}\rho_{\mathrm{i}}M_{\mathrm{i}}^2 + 2\kappa\|\boldsymbol{M}\|_\rho^2\sum_{\mathrm{i}\in\mathbb{Z}^d}\rho_{\mathrm{i}} \leq 2\left(1+2\kappa\rho_\Sigma\right)\|\boldsymbol{M}\|_\rho^2.
$$

Hence $\mathfrak{F}$ maps ℓ_ρ^2 to ℓ_ρ^2 and is uniformly bounded.

The property $\mathfrak{F}(\mathbf{0}) = \mathbf{0}$ follows immediately from Assumption 9.3. □

Lemma 9.2. *The mapping $\mathfrak{F} : \ell_\rho^2 \to \ell_\rho^2$ is Lipschitz continuous with*

$$
\|\mathfrak{F}(\boldsymbol{u}) - \mathfrak{F}(\boldsymbol{v})\|_\rho \leq \sqrt{2(L_f^2 + \hat{L}_g^2\rho_\Sigma)}\|\boldsymbol{u}-\boldsymbol{v}\|_\rho, \qquad \forall \boldsymbol{u},\boldsymbol{v}\in\ell_\rho^2.
$$

Proof. Let $\boldsymbol{u}$, $\boldsymbol{v}\in\ell_\rho^2$. Then for each $\mathrm{i}\in\mathbb{Z}^d$,

$$
\begin{aligned}
\sqrt{\rho_{\mathrm{i}}}|\mathfrak{F}_{\mathrm{i}}(\boldsymbol{u}) - \mathfrak{F}_{\mathrm{i}}(\boldsymbol{v})| &\leq \sqrt{\rho_{\mathrm{i}}}|f_{\mathrm{i}}(u_{\mathrm{i}}) - f_{\mathrm{i}}(v_{\mathrm{i}})| + \sqrt{\rho_{\mathrm{i}}}\sum_{\mathrm{j}\in\mathbb{Z}^d}\frac{|\kappa_{\mathrm{i},\mathrm{j}}|}{\sqrt{\rho_{\mathrm{j}}}}\sqrt{\rho_{\mathrm{j}}}|g_{\mathrm{j}}(u_{\mathrm{j}}) - g_{\mathrm{i}}(v_{\mathrm{j}})| \\
&\leq \sqrt{\rho_{\mathrm{i}}}L_f|u_{\mathrm{i}} - v_{\mathrm{i}}| + \sqrt{\rho_{\mathrm{i}}}\sum_{\mathrm{j}\in\mathbb{Z}^d}\frac{|\kappa_{\mathrm{i},\mathrm{j}}|}{\sqrt{\rho_{\mathrm{j}}}}\sqrt{\rho_{\mathrm{j}}}L_{g,\mathrm{j}}|u_{\mathrm{j}} - v_j|.
\end{aligned}
$$

Squaring both sides gives

$$
\begin{aligned}
\rho_{\mathrm{i}}|\mathfrak{F}_{\mathrm{i}}(\boldsymbol{u}) - \mathfrak{F}_{\mathrm{i}}(\boldsymbol{v})|^2 &\leq 2\rho_{\mathrm{i}}L_f^2|u_{\mathrm{i}} - v_{\mathrm{i}}|^2 + 2\rho_{\mathrm{i}}\left(\sum_{\mathrm{j}\in\mathbb{Z}^d}\frac{|\kappa_{\mathrm{i},\mathrm{j}}|}{\sqrt{\rho_{\mathrm{j}}}}\sqrt{\rho_{\mathrm{j}}}L_{g,\mathrm{j}}|u_{\mathrm{j}} - v_j|\right)^2 \\
&\leq 2L_f^2\rho_{\mathrm{i}}|u_{\mathrm{i}} - v_{\mathrm{i}}|^2 + 2\rho_{\mathrm{i}}\sum_{\mathrm{j}\in\mathbb{Z}^d}\frac{\kappa_{\mathrm{i},\mathrm{j}}^2L_{g,\mathrm{j}}^2}{\rho_{\mathrm{j}}}\sum_{\mathrm{j}\in\mathbb{Z}^d}\rho_{\mathrm{j}}|u_{\mathrm{j}} - v_j|^2 \\
&\leq 2L_f^2\rho_{\mathrm{i}}|u_{\mathrm{i}} - v_{\mathrm{i}}|^2 + 2\rho_{\mathrm{i}}\hat{L}_g^2\|\boldsymbol{u}-\boldsymbol{v}\|_\rho^2.
\end{aligned}
$$

Then summing over $\mathrm{i}\in\mathbb{Z}^d$ results in

$$
\begin{aligned}
\|\mathfrak{F}(\boldsymbol{u}) - \mathfrak{F}(\boldsymbol{v})\|_\rho^2 &= \sum_{\mathrm{i}\in\mathbb{Z}^d}\rho_{\mathrm{i}}|\mathfrak{F}_{\mathrm{i}}(\boldsymbol{u}) - \mathfrak{F}_{\mathrm{i}}(\boldsymbol{v})|^2 \\
&\leq 2L_f^2\sum_{\mathrm{i}\in\mathbb{Z}^d}\rho_{\mathrm{i}}|u_{\mathrm{i}} - v_{\mathrm{i}}|^2 + 2\hat{L}_g^2\|\boldsymbol{u}-\boldsymbol{v}\|_\rho^2\sum_{\mathrm{i}\in\mathbb{Z}^d}\rho_{\mathrm{i}} \\
&\leq 2(L_f^2 + \hat{L}_g^2\rho_\Sigma)\|\boldsymbol{u}-\boldsymbol{v}\|_\rho^2,
\end{aligned}
$$

which implies that $\mathfrak{F}$ is Lipschitz continuous on ℓ_ρ^2. □

Compactness of the nonlinear operator $\mathfrak{F}$ on ℓ^2_ρ

We first show that the nonlinear operator $\mathfrak{F}$ satisfies a uniform tails condition using a method adapted from Bates, Lu & Wang [Bates *et al.* (2001)], see Chapter 3.

Lemma 9.3. *For every $\varepsilon > 0$ there exists an $I(\varepsilon) \in \mathbb{N}$ such that*

$$\sum_{|\mathrm{i}|>I(\varepsilon)} \rho_{\mathrm{i}} \left|\mathfrak{F}_{\mathrm{i}}(\boldsymbol{u})\right|^2 \leq \varepsilon$$

for all $\boldsymbol{u} \in \ell^2_\rho$.

Proof. The proof again requires the cut-off function ξ defined in (3.8), and for a large fixed k (to be determined later in the proof) set

$$v_{\mathrm{i}} = \xi_k(|\mathrm{i}|)u_{\mathrm{i}} \quad \text{with} \quad \xi_k(|\mathrm{i}|) = \xi\left(\frac{|\mathrm{i}|}{k}\right), \quad \mathrm{i} \in \mathbb{Z}^d.$$

As in the proofs of the Lemma 9.1 above we have

$$\rho_{\mathrm{i}}|\mathfrak{F}_{\mathrm{i}}(\boldsymbol{u})|^2 \leq 2\rho_{\mathrm{i}}M_{\mathrm{i}}^2 + 2\rho_{\mathrm{i}}\left(\sum_{\mathrm{j}\in\mathbb{Z}^d} \frac{1}{\sqrt{\rho_{\mathrm{j}}}}|\kappa_{\mathrm{i},\mathrm{j}}|\sqrt{\rho_{\mathrm{j}}}M_{\mathrm{j}}\right)^2.$$

Then multiplying by $\xi_k(|\mathrm{i}|)$ and summing over $\mathrm{i} \in \mathbb{Z}^d$,

$$\begin{aligned}
\sum_{\mathrm{i}\in\mathbb{Z}^d} \xi_k(|\mathrm{i}|)\rho_{\mathrm{i}}|\mathfrak{F}_{\mathrm{i}}(\boldsymbol{u})|^2 &\leq 2\sum_{\mathrm{i}\in\mathbb{Z}^d} \xi_k(|\mathrm{i}|)\rho_{\mathrm{i}}M_{\mathrm{i}}^2 + 2\sum_{\mathrm{i}\in\mathbb{Z}^d} \xi_k(|\mathrm{i}|)\rho_{\mathrm{i}} \sum_{\mathrm{j}\in\mathbb{Z}^d} \frac{\kappa_{\mathrm{i},\mathrm{j}}^2}{\rho_{\mathrm{j}}} \sum_{\mathrm{j}\in\mathbb{Z}^d} \rho_{\mathrm{j}}M_{\mathrm{j}}^2 \\
&\leq 2\sum_{|\mathrm{i}|>k} \rho_{\mathrm{i}}M_{\mathrm{i}}^2 + 2\kappa\|\boldsymbol{M}\|_\rho^2 \sum_{|\mathrm{i}|>k} \rho_{\mathrm{i}} \\
&\leq \varepsilon, \qquad \forall k \geq I(\varepsilon),
\end{aligned}$$

since both series on the right side converge. The desired assertion then follows directly. □

Theorem 9.1. *The nonlinear operator $\mathfrak{F}$ on ℓ^2_ρ is compact.*

Proof. Define the closed and bounded subset of ℓ^2_ρ

$$Q = \left\{\boldsymbol{u} \in \ell^2_\rho : \|\boldsymbol{u}\|_\rho \leq \|\boldsymbol{M}\|_\rho\sqrt{2(1+\kappa\rho_\Sigma)}\right\}.$$

Then by Lemma 9.1, $\mathfrak{F}(\ell^2_\rho) \subset Q$.

Now consider a sequence $\{\boldsymbol{u}^{(k)}\}_{k\in\mathbb{N}}$ in ℓ^2_ρ. Then $\mathfrak{F}(\boldsymbol{u}^{(k)}) \in Q$ for each $k \in \mathbb{N}$. Since Q is bounded there exists $\boldsymbol{v}^* = (v^*_{\mathrm{i}})_{\mathrm{i}\in\mathbb{Z}^d} \in \ell^2_\rho$ and a subsequence of $\{\mathfrak{F}(\boldsymbol{u}^{(k)})\}_{k\in\mathbb{N}}$, still denoted by $\{\mathfrak{F}(\boldsymbol{u}^{(k)})\}_{k\in\mathbb{N}}$, such that

$$\mathfrak{F}(\boldsymbol{u}^{(k)}) \rightharpoonup \boldsymbol{v}^* \quad \text{weakly in } \ell^2_\rho \text{ as } k \to \infty. \tag{9.2}$$

We next show that the convergence in (9.2) is, in fact, strong in ℓ^2_ρ. To that end, let $\varepsilon > 0$ and using Lemma 9.3 there exists $I_1(\varepsilon) \in \mathbb{N}$ such that

$$\sum_{|\mathrm{i}|>I_1(\varepsilon)} \rho_\mathrm{i}|\mathfrak{F}_\mathrm{i}(\boldsymbol{u}^{(k)})|^2 \leq \frac{\varepsilon^2}{8}, \qquad \forall k \in \mathbb{N}.$$

Moreover, since $\boldsymbol{v}^* \in \ell^2_\rho$, there exists $I_2(\varepsilon) \in \mathbb{N}$ such that

$$\sum_{|\mathrm{i}|>I_2(\varepsilon)} \rho_\mathrm{i}(v^*_\mathrm{i})^2 < \frac{\varepsilon^2}{8}.$$

On the other hand, due to the weak convergence (9.2), it follows that

$$\mathfrak{F}_\mathrm{i}(\boldsymbol{u}^{(k)}) \to v^*_\mathrm{i} \qquad \text{as } k \to \infty \text{ for each } \mathrm{i} \in \mathbb{Z}.$$

In particular, for each $|\mathrm{i}| \leq I(\varepsilon)$, where $I(\varepsilon) = \max\{I_1(\varepsilon), I_2(\varepsilon)\}$, there exists $N(\varepsilon) \in \mathbb{N}$ such that

$$\sum_{|\mathrm{i}|\leq I(\varepsilon)} \rho_\mathrm{i}\left|\mathfrak{F}_\mathrm{i}(\boldsymbol{u}^{(k)}) - v^*_\mathrm{i}\right|^2 < \frac{\varepsilon^2}{2} \quad \text{for all } k \geq N(\varepsilon).$$

Finally, for $k \geq N(\varepsilon)$ it follows that

$$\begin{aligned}\|\mathfrak{F}(\boldsymbol{u}^{(k)}) - \boldsymbol{v}^*\|^2_\rho &= \sum_{|i|\leq I(\varepsilon)} \rho_\mathrm{i}\left|(\mathfrak{F}_\mathrm{i}(\boldsymbol{u}^{(k)}) - v^*_\mathrm{i}\right|^2 + \sum_{|i|>I(\varepsilon)} \rho_\mathrm{i}\left|(\mathfrak{F}_\mathrm{i}(\boldsymbol{u}^{(k)}) - v^*_\mathrm{i}\right|^2 \\ &< \frac{\varepsilon^2}{2} + 2\sum_{|i|>I(\varepsilon)} \left(\rho_\mathrm{i}\left|\mathfrak{F}_\mathrm{i}(\boldsymbol{u}^{(k)})\right|^2 + |v^*_\mathrm{i}|^2\right) < \varepsilon^2,\end{aligned}$$

i.e., $\left\{\mathfrak{F}(\boldsymbol{u}^{(k)})\right\}_{k\in\mathbb{N}}$ converges strongly to $\boldsymbol{v}^*$ in ℓ^2_ρ as $k \to \infty$. □

9.1.2 *Existence of a global attractor*

The lattice difference equation (9.1) can be written as a first order difference equation on ℓ^2_ρ

$$\boldsymbol{u}(t+1) = \mathfrak{F}(\boldsymbol{u}(t)), \quad t \in \mathbb{Z}^+.$$

It generates a discrete time semi-dynamical system $\{\varphi(t)\}_{t\in\mathbb{Z}^+}$ by iterating the mapping $\mathfrak{F}$, i.e.,

$$\varphi(t, \boldsymbol{u}) = \underbrace{\mathfrak{F} \circ \cdots \circ \mathfrak{F}}_{t \text{ times}}(\boldsymbol{u}), \ t \geq 1, \text{ with } \varphi(0, \boldsymbol{u}) = \boldsymbol{u} \quad \text{for all } \boldsymbol{u} \in \ell^2_\rho.$$

The mapping $\boldsymbol{u} \mapsto \varphi(t, \boldsymbol{u})$ is continuous on ℓ^2_ρ.

Recall that $\mathfrak{F}(\ell^2_\rho) \subset Q$ due to Lemma 9.1, and thus $\varphi(t, Q) \subset Q$ for all $t \geq 1$, i.e., the set Q is positive invariant under φ. Moreover, Q is an absorbing set for φ and, by Theorem 9.1, $\varphi(t, Q)$ is a compact subset of ℓ^2_ρ for each $t \geq 1$ and these sets form a nested sequence of compact subsets. It follows immediately that φ has a global attractor as stated in the following theorem.

Theorem 9.2. *The discrete time semi-dynamical system $\{\varphi(t)\}_{t\in\mathbb{Z}^+}$ on ℓ^2_ρ has a global attractor in ℓ^2_ρ given by*

$$\mathcal{A} = \bigcap_{t\geq 1} \varphi(t, Q).$$

Note that $\mathbf{0} \in \mathcal{A}$, since it is a constant, hence entire, solution of the system.

Remark 9.3. The above results also apply to lattice versions of integro-difference equations based on the Ricker and Beverton-Holt population models. Since the solutions are non-negative one restricts attention to the non-negative cone $\ell^{2,+}_\rho$ in ℓ^2_ρ, which is positive invariant under $\mathfrak{F}$, i.e., $\mathfrak{F}(\ell^{2,+}_\rho) \subset \ell^{2,+}_\rho$. In this case the absorbing set is $Q^+ = Q \cap \ell^{2,+}_\rho$.

9.1.3 *Finite dimensional approximations of the global attractor*

Numerical computations require a finite number of dimensions. Define

$$\mathbb{Z}^d_N = \left\{ \mathrm{i} \in \ell^2_\rho \,:\, |\mathrm{i}| \leq N \right\}.$$

Note that the cardinality of $\mathbb{Z}^d_N$ is $(2N+1)^d$, and consider the $(2N+1)^d$-dimensional difference equation

$$u_\mathrm{i}(t+1) = f_\mathrm{i}(u_\mathrm{i}(t)) + \sum_{\mathrm{j}\in\mathbb{Z}^d_N} \kappa_{\mathrm{i},\mathrm{j}} g_\mathrm{j}(u_\mathrm{j}(t)), \qquad \mathrm{i} \in \mathbb{Z}^d_N, \quad t \in \mathbb{Z}^+. \tag{9.3}$$

For $\mathrm{u} = (u_\mathrm{i})_{\mathrm{i}\in\mathbb{Z}^d_N} \in \mathbb{R}^{(2N+1)^d}$ define $\mathfrak{F}^{(N)}(\mathrm{u}) = \left(\mathfrak{F}^{(N)}_\mathrm{i}(\mathrm{u})\right)_{\mathrm{i}\in\mathbb{Z}^d_N}$ componentwise by

$$\mathfrak{F}^{(N)}_\mathrm{i}(\mathrm{u}) = f_\mathrm{i}(u_\mathrm{i}) + \sum_{\mathrm{j}\in\mathbb{Z}^d_N} \kappa_{\mathrm{i},\mathrm{j}} g_\mathrm{j}(u_\mathrm{j}), \qquad \mathrm{i} \in \mathbb{Z}^d_N, \quad \mathrm{u} \in \mathbb{R}^{(2N+1)^d}.$$

A finite dimensional discrete time semi-dynamical system $\{\varphi^{(N)}(t)\}_{t\in\mathbb{Z}^+}$ can be formed by iterating the mapping $\mathfrak{F}^{(N)}$. Moreover, counterparts of the estimates in Lemmas 9.1–9.3 for the mapping $\mathfrak{F}^{(N)}$ in the space $\mathbb{R}^{(2N+1)^d}$ hold uniformly in N.

The mapping $\mathfrak{F}^{(N)}$ maps $\mathbb{R}^{(2N+1)^d}$ into the compact subset

$$Q^{(N)} = \left\{ \mathrm{u} \in \mathbb{R}^{(2N+1)^d} \,:\, |\mathrm{u}|_\rho \leq \sqrt{2\beta(1+\kappa\rho_\Sigma)} \right\},$$

where $|\cdot|_\rho$ is the weighted norm of $\mathbb{R}^{(2N+1)^d}$ with the same weights as in the norm on ℓ^2_ρ, that is equivalent to the Euclidean norm. Hence, the compactness of the mapping $\mathfrak{F}^{(N)}$ holds trivially in $\mathbb{R}^{(2N+1)^d}$. Finally, $Q^{(N)}$ is positively invariant under $\varphi^{(N)}$. Hence the following theorem holds.

Theorem 9.3. *The finite dimensional discrete time semi-dynamical system $\{\varphi^{(N)}(t)\}_{t\in\mathbb{Z}^+}$ on $\mathbb{R}^{(2N+1)^d}$ has a global attractor in $\mathbb{R}^{(2N+1)^d}$ given by*

$$\mathcal{A}^{(N)} = \bigcap_{t\geq 1} \varphi^{(N)}(t, Q^{(N)}).$$

Upper semi-continuous convergence

As in Chapter 4 (with $d = 1$), the space $\mathbb{R}^{(2N+1)^d}$ can be embedded into ℓ^2_ρ by setting the missing components with indices $|\mathrm{i}| > N$ to 0. Similarly, $\varphi^{(N)}$ and $\mathfrak{F}^{(N)}$ can be extended to mappings from ℓ^2_ρ into ℓ^2_ρ. Considering the sets $\mathcal{A}^{(N)}$ as subsets of ℓ^2_ρ, it can be shown that they converge upper semi-continuously to $\mathcal{A}$ in ℓ^2_ρ.

Theorem 9.4. *The finite dimensional attractors $\mathcal{A}^{(N)}$ converge upper semi-continuously to the attractor in ℓ^2_ρ, i.e.,*

$$\lim_{N\to\infty} \mathrm{dist}_{\ell^2_\rho}\left(\mathcal{A}^{(N)}, \mathcal{A}\right) = 0.$$

The proof is adapted from a related result in Chapter 4. It requires the following lemmas, the first of which is a uniform tails result inside the finite dimensional attractors. The proof is essentially that of Lemma 9.3 with the points of the finite dimensional attractors considered as elements of ℓ^2_ρ.

Lemma 9.4. *For every $\varepsilon > 0$ there exists an integer $I(\varepsilon)$ such that*

$$\sum_{I(\varepsilon)\le|\mathrm{i}|} \rho_\mathrm{i}|a_\mathrm{i}|^2 \le \varepsilon \quad \text{for all } \boldsymbol{a} = (a_\mathrm{i})_{\mathrm{i}\in\mathbb{Z}^d_N} \in \mathcal{A}^{(N)}, N \gg 1.$$

Proof. Since the global attractor $\mathcal{A}^{(N)}$ is invariant under $\varphi^{(N)}$ and hence $\mathfrak{F}^{(N)}$, for any $\boldsymbol{a} = (a_\mathrm{i})_{\mathrm{i}\in\mathbb{Z}^d} \in \mathcal{A}^{(N)}$ there exists $\boldsymbol{u}^{(N)} \in \mathcal{A}^{(N)}$ such that $\boldsymbol{a} = \mathfrak{F}^{(N)}(\boldsymbol{u}^{(N)})$. From the proof of Lemma 9.3 for every $\varepsilon > 0$ there exists $I_1(\varepsilon)$ such that

$$\begin{aligned}\sum_{\mathrm{i}\in\mathbb{Z}^d} \xi_k(|\mathrm{i}|)\rho_\mathrm{i}|\mathfrak{F}^{(N)}_\mathrm{i}(\boldsymbol{u}^{(N)})|^2 &\le 2\sum_{\mathrm{i}\in\mathbb{Z}^d} \xi_k(|\mathrm{i}|)\rho_\mathrm{i}M_\mathrm{i}^2 + 2\sum_{\mathrm{i}\in\mathbb{Z}^d} \xi_k(|\mathrm{i}|)\rho_\mathrm{i}\sum_{\mathrm{j}\in\mathbb{Z}^d}\frac{\kappa^2_{\mathrm{i},\mathrm{j}}}{\rho_\mathrm{j}}\sum_{\mathrm{j}\in\mathbb{Z}^d}\rho_\mathrm{j}M_\mathrm{j}^2\\ &\le 2\sum_{|\mathrm{i}|>k}\rho_\mathrm{i}B_\mathrm{i}^2 + 2\kappa\|\boldsymbol{M}\|^2_\rho\sum_{|\mathrm{i}|>k}\rho_\mathrm{i}\\ &\le \varepsilon \quad \text{for all } k \ge I_1(\varepsilon), \quad N\in\mathbb{N},\end{aligned}$$

since both series on the right side converge, i.e.,

$$\sum_{\mathrm{i}\in\mathbb{Z}^d_N} \xi_k(|\mathrm{i}|)\rho_\mathrm{i}|a_\mathrm{i}|^2 \le \varepsilon \quad \text{for all } k \ge I_1(\varepsilon), \quad N\in\mathbb{N}. \tag{9.4}$$

The left hand side of (9.4) vanishes trivially for $k > N$, but for $N \gg 1$ there exists an $I(\varepsilon) \ge I_1(\varepsilon)$ with $I(\varepsilon) < N$, such that

$$\sum_{I(\varepsilon)\le|\mathrm{i}|<N} \rho_\mathrm{i}|a_\mathrm{i}|^2 = \sum_{\mathrm{i}\in\mathbb{Z}^d_N} \xi_k(|\mathrm{i}|)\rho_\mathrm{i}|a_\mathrm{i}|^2 \le \varepsilon,$$

where the sum on the left hand side needs not be zero. Moreover, this holds uniformly in $\boldsymbol{a} \in \mathcal{A}^{(N)}$. □

Lemma 9.5. *Let $\boldsymbol{a}^{(N)} \in \mathcal{A}^{(N)}$ for $N \in \mathbb{N}$. Then, there exists a subsequence $\{\boldsymbol{a}^{(N_j)}\}_{j\in\mathbb{N}}$ and an element $\boldsymbol{a}^* \in Q$ such that $\{\boldsymbol{a}^{(N_j)}\}$ converge to $\boldsymbol{a}^*$ in ℓ^2_ρ as $j \to \infty$.*

Proof. Consider $\boldsymbol{a}^{(N)}$ as elements of ℓ^2_ρ. In particular, $\boldsymbol{a}^{(N)} \in Q$ for all $N \in \mathbb{N}$. Since Q is a closed and bounded convex subset of ℓ^2_ρ it is the weakly compact. Hence there exists $\boldsymbol{a}^* = (a^*_{\mathbf{i}})_{\mathbf{i}\in\mathbb{Z}^d} \in \ell^2_\rho$ and a subsequence $\{\boldsymbol{a}^{(N_j)}\}_{j\in\mathbb{N}}$ converging weakly to $\boldsymbol{a}^*$ in ℓ^2_ρ, i.e.,

$$\boldsymbol{a}^{(N_j)} \rightharpoonup \boldsymbol{a}^* \qquad \text{as } j \to \infty.$$

Similarly, as in the proof of Theorem 9.1, but using Lemma 9.4 instead of Lemma 9.3 it can be shown that the weak convergence is, in fact, strong convergence, i.e.,

$$\|\boldsymbol{a}^{(N_j)} - \boldsymbol{a}^*\|_\rho \to 0 \quad \text{as } j \to \infty,$$

and consequently, $\boldsymbol{a}^* \in Q$. □

Lemma 9.6. *The limit point $\boldsymbol{a}^*$ in Lemma 9.5 belongs to the global attractor $\mathcal{A}$.*

Proof. Let $\boldsymbol{a}^{(N)} \in \mathcal{A}^{(N)}$. Then the dynamical system $\varphi^{(N)}$ has an entire trajectory $\boldsymbol{u}^{(N)}$ in $\mathcal{A}^{(N)}$ through $\boldsymbol{a}^{(N)}$, i.e., $\boldsymbol{u}^{(N)}(t+1) = \varphi^{(N)}(1, \boldsymbol{u}^{(N)}(t))$ for all $t \in \mathbb{Z}$ with $\boldsymbol{u}^{(N)}(0) = \boldsymbol{a}^{(N)}$. Repeating the above convergence argument for each $t \in \mathbb{Z}$ with the sequence $\{\boldsymbol{u}^{(N)}(t)\}_{N\in\mathbb{N}}$ there is a $\boldsymbol{u}^*(t) \in Q$ and a subsequence $\{\boldsymbol{u}^{(N_j)}(t)\}_{N\in\mathbb{N}}$ which converges strongly to $\boldsymbol{u}^*(t)$ in ℓ^2_ρ. This holds for each $t \in \mathbb{Z}$. A diagonal argument can be used to construct a commonly indexed subsequence for each $t \in \mathbb{Z}$. In view of the trivial extension of finite dimensional elements to ℓ^2_ρ it follows that

$$\mathfrak{F}(\boldsymbol{u}^{(N_j)}(t)) = \mathfrak{F}^{(N_j)}(\boldsymbol{u}^{(N_j)}(t)) = \boldsymbol{u}^{(N_j)}(t+1), \qquad j \in \mathbb{N}, t \in \mathbb{Z}.$$

Then by the continuity of the mapping $\mathfrak{F}$ and the convergence of the subsequences in ℓ^2_ρ it follows that $\mathfrak{F}(\boldsymbol{u}^*(t)) = \boldsymbol{u}^*(t+1)$, i.e., $\boldsymbol{u}^*(t+1) = \varphi(1, \boldsymbol{u}^*(t))$, for each $t \in \mathbb{Z}$. Thus the points $(\boldsymbol{u}^*(t))_{t\in\mathbb{N}}$ forms a bounded entire trajectory of the semi-dynamical system φ. By a characterisation of attractors, Corollary 2.1, this bounded entire trajectory is contained in the global attractor $\mathcal{A}$, and in particular, $\boldsymbol{a}^* \in \mathcal{A}$. □

Now we are ready to finish the proof of Theorem 9.4. To that end, suppose (for contradiction) that the assertion of Theorem 9.4 is false. Then, there exist $\varepsilon_0 > 0$ and a sequence $\{\boldsymbol{a}^{(N_j)}\}_{j\in\mathbb{N}}$ with $\boldsymbol{a}^{(N_j)} \in \mathcal{A}^{(N_j)}$ for $j \in \mathbb{N}$ such that

$$\operatorname{dist}_{\ell^2_\rho}\left(\boldsymbol{a}^{(N_j)}, \mathcal{A}\right) \geq \varepsilon_0 \quad \text{for all } j \in \mathbb{N}. \tag{9.5}$$

However, by Lemma 9.5, there exists a subsequence $\{\boldsymbol{a}^{(N_{j_k})}\}_{k\in\mathbb{N}}$ of $\{\boldsymbol{a}^{(N_j)}\}_{j\in\mathbb{N}}$ such that

$$\operatorname{dist}_{\ell^2_\rho}\left(\boldsymbol{a}^{(N_{j_k})}, \mathcal{A}\right) \to 0 \quad \text{as} \quad N_{j_k} \to \infty.$$

This contradicts with (9.5) and completes the proof. □

9.2 Convergent sequences of interconnection weights

We now consider autonomous lattice difference equations of the form (9.1) for a sequence of interconnection weights $\kappa_{\mathrm{i,j}}^{(\iota)}$, $\iota \in \mathbb{N}$,

$$u_{\mathrm{i}}(t+1) = f_{\mathrm{i}}(u_{\mathrm{i}}(t)) + \sum_{\mathrm{j}\in\mathbb{Z}^d} \kappa_{\mathrm{i,j}}^{(\iota)} g_{\mathrm{j}}(u_{\mathrm{j}}(t)), \qquad \mathrm{i} \in \mathbb{Z}^d, \quad t \in \mathbb{Z}^+. \tag{9.6}$$

with weights satisfy the same properties as those in Assumptions 9.1 and 9.2 uniformly in $\iota \in \mathbb{N}$ as well as the convergence property:

Assumption 9.6. $\kappa_{\mathrm{i,j}}^{(\iota)} \to \kappa_{\mathrm{i,j}}$ as $\iota \to \infty$ in the sense that for every $\varepsilon > 0$ there exists $N(\varepsilon) \in \mathbb{N}$ such that

$$\sum_{\mathrm{j}\in\mathbb{Z}^d} \frac{(\kappa_{\mathrm{i,j}}^{(\iota)} - \kappa_{\mathrm{i,j}})^2}{\rho_j} \le \varepsilon^2 \quad \text{for all } \iota \ge N(\varepsilon), \qquad \mathrm{i} \in \mathbb{Z}^d.$$

It follows from Assumptions 9.1 to 9.3 that the lattice system (9.6) generates a discrete time semi-dynamical system $\varphi^{(\iota)}$ on ℓ_ρ^2 for each $\iota \in \mathbb{N}$ and that these systems have the same closed and bounded positive invariant absorbing set

$$Q = \left\{ \boldsymbol{u} \in \ell_\rho^2;\ \|\boldsymbol{u}\|_\rho \le \|\boldsymbol{M}\|_\rho \sqrt{2(1+\kappa\rho_\Sigma)} \right\} \subset \ell_\rho^2,$$

as the discrete time semi-dynamical system φ generated by the lattice system (9.1). Moreover, the corresponding mappings $\mathfrak{F}^{(\iota)} : \ell_\rho^2 \to \ell_\rho^2$ defined componentwise by

$$\mathfrak{F}_{\mathrm{i}}^{(\iota)}(\boldsymbol{u}) = f_{\mathrm{i}}(u_{\mathrm{i}}) + \sum_{\mathrm{j}\in\mathbb{Z}^d} \kappa_{\mathrm{i,j}}^{(\iota)} g_{\mathrm{j}}(u_{\mathrm{j}}), \qquad \mathrm{i} \in \mathbb{Z}^d, \quad \boldsymbol{u} \in \ell_\rho^2,$$

are compact. Hence each $\varphi^{(\iota)}$ has a global attractor $\mathcal{A}^{(\iota)}$ in ℓ_ρ^2, which is a subset of Q. The aim is to show that these attractors $\mathcal{A}^{(\iota)}$ converge upper semi continuously in ℓ_ρ^2 to the global attractor $\mathcal{A}$ of the discrete time semi-dynamical system φ generated by the lattice system (9.1) under the additional Assumption 9.6.

The following estimates are needed for the proof of upper semi-continuous convergence later.

Lemma 9.7. *For any $\varepsilon > 0$ there exists $N(\varepsilon) \in \mathbb{N}$ such that*

$$\|\mathfrak{F}^{(\iota)}(\boldsymbol{u}) - \mathfrak{F}(\boldsymbol{u})\|_\rho \le \varepsilon \sqrt{\rho_\Sigma} \|\boldsymbol{M}\|_\rho, \quad \forall \iota \ge N(\varepsilon), \boldsymbol{u} \in \ell_\rho^2.$$

Proof. Let $\boldsymbol{u} \in \ell_\rho^2$. Then for each $\mathrm{i} \in \mathbb{Z}^d$ and $\iota \in \mathbb{N}$,

$$\begin{aligned}\sqrt{\rho_{\mathrm{i}}} \left| \mathfrak{F}_{\mathrm{i}}^{(\iota)}(\boldsymbol{u}) - \mathfrak{F}_{\mathrm{i}}(\boldsymbol{u}) \right| &\le \sqrt{\rho_{\mathrm{i}}} \sum_{\mathrm{j}\in\mathbb{Z}^d} \left| \kappa_{\mathrm{i,j}}^{(\iota)} - \kappa_{\mathrm{i,j}} \right| |g_{\mathrm{j}}(u_{\mathrm{j}})| \\ &\le \sqrt{\rho_{\mathrm{i}}} \sum_{\mathrm{j}\in\mathbb{Z}^d} \frac{1}{\sqrt{\rho_{\mathrm{j}}}} \left| \kappa_{\mathrm{i,j}}^{(\iota)} - \kappa_{\mathrm{i,j}} \right| \sqrt{\rho_{\mathrm{j}}} M_{\mathrm{j}}.\end{aligned}$$

Squaring both sides and using Assumption 9.6, there exists $N(\varepsilon) > 0$ such that

$$\begin{aligned}\rho_{\mathrm{i}}\left|\mathfrak{F}_{\mathrm{i}}^{(\iota)}(\boldsymbol{u})-\mathfrak{F}_{\mathrm{i}}(\boldsymbol{u})\right|^2 &\leq \rho_{\mathrm{i}}\left(\sum_{\mathrm{j}\in\mathbb{Z}^d}\frac{\left|\kappa_{\mathrm{i,j}}^{(\iota)}-\kappa_{\mathrm{i,j}}\right|}{\sqrt{\rho_{\mathrm{j}}}}\sqrt{\rho_{\mathrm{j}}}M_{\mathrm{j}}\right)^2\\ &\leq \rho_{\mathrm{i}}\sum_{\mathrm{j}\in\mathbb{Z}^d}\frac{\left|\kappa_{\mathrm{i,j}}^{(\iota)}-\kappa_{\mathrm{i,j}}\right|^2}{\rho_{\mathrm{j}}}\sum_{\mathrm{j}\in\mathbb{Z}^d}\rho_{\mathrm{j}}M_{\mathrm{j}}^2\\ &\leq \rho_{\mathrm{i}}\varepsilon^2\|\boldsymbol{M}\|_\rho^2 \quad \text{for } \iota \geq N(\varepsilon).\end{aligned}$$

Finally, summing the above inequality over $\mathrm{i} \in \mathbb{Z}^d$ gives

$$\|\mathfrak{F}^{(\iota)}(\boldsymbol{u})-\mathfrak{F}(\boldsymbol{u})\|_\rho^2 \leq \varepsilon^2\|\boldsymbol{M}\|_\rho^2\sum_{\mathrm{i}\in\mathbb{Z}^d}\rho_{\mathrm{i}} = \varepsilon^2\|\boldsymbol{M}\|_\rho^2\rho_\Sigma$$

for $\iota \geq N(\varepsilon)$ and all $\boldsymbol{u} \in \ell_\rho^2$. □

Then, by Lemmas 9.2 and 9.5, we have that for any $\boldsymbol{u}$, $\boldsymbol{v} \in \ell_\rho^2$

$$\begin{aligned}\|\mathfrak{F}^{(\iota)}(\boldsymbol{u})-\mathfrak{F}(\boldsymbol{v})\|_\rho &\leq \|\mathfrak{F}^{(\iota)}(\boldsymbol{u})-\mathfrak{F}(\boldsymbol{u})\|_\rho + \|\mathfrak{F}(\boldsymbol{u})-\mathfrak{F}(\boldsymbol{v})\|_\rho\\ &\leq \varepsilon\rho_\Sigma\|\boldsymbol{M}\|_\rho + \sqrt{2(L_f^2+\hat{L}_g^2\rho_\Sigma)}\|\boldsymbol{u}-\boldsymbol{v}\|_\rho \quad \text{for } \iota \geq N(\varepsilon).\end{aligned}$$

Upper semi-continuous convergence of attractors

Theorem 9.5. *The global attractors $\mathcal{A}^{(\iota)}$ of the lattice systems (9.6) converge upper semi-continuously to the global attractor $\mathcal{A}$ of the lattice system (9.1) in ℓ_ρ^2, i.e.,*

$$\lim_{\iota\to\infty}\operatorname{dist}_{\ell_\rho^2}\left(\mathcal{A}^{(l)},\mathcal{A}\right) = 0.$$

The proof is by contradiction and is essentially the same as that of Theorem 9.4, so will not be repeated here. However, it requires the following lemmas instead of Lemmas 9.5 and 9.6.

Lemma 9.8. *Let $\boldsymbol{a}^{(\iota)} \in \mathcal{A}^{(\iota)}$ for $\iota \in \mathbb{N}$. Then, there exists a subsequence $\{\boldsymbol{a}^{(\iota_j)}\}_{j\in\mathbb{N}}$ and an element $\boldsymbol{a}^* \in Q$ such that $\boldsymbol{a}^{(\iota_j)}$ converges to $\boldsymbol{a}^*$ in ℓ_ρ^2 as $j \to \infty$.*

Proof. Consider each $\boldsymbol{a}^{(\iota)}$ as an element of ℓ_ρ^2. In particular, $\boldsymbol{a}^{(\iota)} \in Q$ for all $\iota \in \mathbb{N}$. Since Q is a closed and bounded subset of ℓ_ρ^2, by a diagonal argument there exist $\boldsymbol{a}^* = (a_{\mathrm{i}}^*)_{\mathrm{i}\in\mathbb{Z}^d} \in \ell_\rho^2$ and a subsequence $\{\boldsymbol{a}^{(\iota_j)}\}_{j\in\mathbb{N}}$ converging weakly to $\boldsymbol{a}^*$ in ℓ_ρ^2, i.e.,

$$\boldsymbol{a}^{(\iota_j)} \rightharpoonup \boldsymbol{a}^* \quad \text{as } j \to \infty.$$

Now note that the tails estimate in Lemma 9.3 is uniform for all mappings $\mathfrak{F}^{(\iota)}$ since they all satisfy the same bounds as the mapping $\mathfrak{F}$. First, one uses the weak compactness of the set Q to obtain a weakly convergent subsequence. Then, similar to the proof of Theorem 9.1, one uses the uniform tails estimate to show that this convergence is, in fact, strong convergence. □

Lemma 9.9. *The limit point $\boldsymbol{a}^*$ in Lemma 9.8 is contained in the global attractor $\mathcal{A}$.*

Proof. Let $\boldsymbol{a}^{(\iota)} \in \mathcal{A}^{(\iota)}$. Then the dynamical system generated by the mapping $\mathfrak{F}^{(\iota)}$ has an entire trajectory $\boldsymbol{u}^{(\iota)}$ in $\mathcal{A}^{(\iota)}$ through $\boldsymbol{a}^{(\iota)}$, i.e.,

$$\boldsymbol{u}^{(\iota)}(t+1) = \mathfrak{F}^{(\iota)}(\boldsymbol{u}^{(\iota)}(t)) \quad \text{for all} \quad t \in \mathbb{Z} \quad \text{with} \quad \boldsymbol{u}^{(\iota)}(0) = \boldsymbol{a}^{(\iota)}.$$

Repeating the above convergence argument for each $t \in \mathbb{Z}$ with the sequence $\{\boldsymbol{u}^{(\iota_j)}(t)\}_{j\in\mathbb{N}}$ there is a $\boldsymbol{u}^*(t) \in Q$ and a subsequence $\{\boldsymbol{u}^{(\iota_j)}(t)\}_{j\in\mathbb{N}}$ which converge strongly to $\boldsymbol{u}^*(t)$ in ℓ^2_ρ. This holds for each $t \in \mathbb{Z}$ and a diagonal argument can be used to construct commonly indexed subsequences for each $t \in \mathbb{Z}$. Then

$$\begin{aligned}\|\boldsymbol{u}^*(t+1) - \mathfrak{F}(\boldsymbol{u}^*(t))\|_\rho &\le \|\boldsymbol{a}^*(t+1) - \boldsymbol{u}^{(\iota_j)}(t+1)\|_\rho + \|\boldsymbol{u}^{(\iota_j)}(t+1) - \mathfrak{F}(\boldsymbol{u}^*(t))\|_\rho \\ &\le \|\boldsymbol{u}^*(t+1) - \boldsymbol{u}^{(\iota_j)}(t+1)\|_\rho + \|\mathfrak{F}^{(\iota_j)}(\boldsymbol{u}^{(\iota_j)}(t)) - \mathfrak{F}(\boldsymbol{u}^*(t))\|_\rho\end{aligned}$$

since $\boldsymbol{u}^{(\iota_j)}(t+1) = \mathfrak{F}^{(\iota_j)}(\boldsymbol{u}^{(\iota_j)}(t))$.

Notice that due to Lemma 9.5, for every $\varepsilon > 0$ there exists $N(\varepsilon) > 0$ such that

$$\|\mathfrak{F}^{(\iota_j)}(\boldsymbol{u}^{(\iota_j)}(t)) - \mathfrak{F}(\boldsymbol{u}^*(t))\|_\rho \le \varepsilon\|\boldsymbol{M}\|_\rho\sqrt{\rho_\Sigma} + \sqrt{2(L_f^2 + \hat{L}_g^2\rho_\Sigma)}\|\boldsymbol{u}^{(\iota_j)}(t) - \boldsymbol{u}^*(t)\|_\rho$$

for $\iota_j \ge N(\varepsilon)$. By strong convergence there is an $N^*(\varepsilon, t)$ such that

$$\|\boldsymbol{u}^{(\iota_j)}(t) - \boldsymbol{u}^*(t)\|_\rho < \varepsilon, \quad \|\boldsymbol{u}^*(t+1) - \boldsymbol{u}^{(\iota_j)}(t+1)\|_\rho < \varepsilon$$

for $\iota_j \ge N^*(\varepsilon, t)$. It follows that

$$\|\boldsymbol{u}^*(t+1) - \mathfrak{F}(\boldsymbol{u}^*(t))\|_\rho < \left(1 + \sqrt{2(L_f^2 + \hat{L}_g^2\rho_\Sigma)}\right)\varepsilon$$

for all $\varepsilon < 0$. Hence $\|\boldsymbol{u}^*(t+1) - \mathfrak{F}(\boldsymbol{u}^*(t))\|_\rho = 0$, which yields that $\boldsymbol{u}^*(t+1) = \mathfrak{F}(\boldsymbol{u}^*(t))$.

The argument above holds for all $t \in \mathbb{Z}$, so the points $\{\boldsymbol{u}^*(t)\}_{t\in\mathbb{N}}$ form a bounded entire trajectory of semi-dynamical system φ generated by the mapping $\mathfrak{F}$. By a characterisation of attractors, Corollary 2.1, this bounded entire trajectory is contained in the global attractor $\mathcal{A}$. Hence, in particular, $\boldsymbol{a}^* \in \mathcal{A}$. □

9.3 Lattice systems with finitely many interconnections

Another kind of approximation of the lattice differential equation (9.1) is when each node interacts with only finitely many other nodes (see Chapter 16 for Hopfield lattice models) as in the following lattice difference equation,

$$u_{\mathfrak{i}}(t+1) = f_{\mathfrak{i}}(u_{\mathfrak{i}}(t)) + \sum_{\mathfrak{j}-\mathfrak{i}\in\mathbb{Z}^d_N} \kappa_{\mathfrak{i},\mathfrak{j}}g_{\mathfrak{j}}(u_{\mathfrak{j}}(t)), \qquad \mathfrak{i} \in \mathbb{Z}^d, \quad t \in \mathbb{Z}^+.$$

The essential difference with the finite dimensional system (9.3) is that the system (9.3) is defined for all indices $\mathfrak{i} \in \mathbb{Z}^d$ rather than just for those in $\mathfrak{i} \in \mathbb{Z}^d_N$ and the finite sum is centered on the component index. It can be reformulated as

$$u_{\mathfrak{i}}(t+1) = f_{\mathfrak{i}}(u_{\mathfrak{i}}(t)) + \sum_{\mathfrak{j}\in\mathbb{Z}^d} \kappa_{\mathfrak{i},\mathfrak{j}}^{(N)} g_{\mathfrak{j}}(u_{\mathfrak{j}}(t)), \qquad \mathfrak{i}\in\mathbb{Z}^d, \quad t\in\mathbb{Z}^+,$$

with interconnection weights defined by

$$\kappa_{\mathfrak{i},\mathfrak{j}}^{(N)} := \begin{cases} \kappa_{\mathfrak{i},\mathfrak{j}} & \text{if } |\mathfrak{j}-\mathfrak{i}| \le N \\ 0 & \text{otherwise} \end{cases}.$$

Clearly these $\kappa_{\mathfrak{i},\mathfrak{j}}^{(N)}$ satisfy the same properties as $\kappa_{\mathfrak{i},\mathfrak{j}}$.

For $\boldsymbol{u} = (u_{\mathfrak{i}})_{\mathfrak{i}\in\mathbb{Z}^d} \in \ell^2_\rho$, define $\widehat{\mathfrak{F}}^{(N)}(\boldsymbol{u}) = \left(\widehat{\mathfrak{F}}_{\mathfrak{i}}^{(N)}(\boldsymbol{u})\right)_{\mathfrak{i}\in\mathbb{Z}^d}$ componentwise by

$$\widehat{\mathfrak{F}}_{\mathfrak{i}}^{(N)}(\boldsymbol{u}) = f_{\mathfrak{i}}(u_{\mathfrak{i}}) + \sum_{\mathfrak{j}-\mathfrak{i}\in\mathbb{Z}^d_N} \kappa_{\mathfrak{i},\mathfrak{j}} g_{\mathfrak{j}}(u_{\mathfrak{j}}), \qquad \mathfrak{i}\in\mathbb{Z}^d, \quad \boldsymbol{u}\in\ell^2_\rho.$$

All of the estimates and results above for the mapping $\mathfrak{F}^{(N)}$ hold for the mapping $\widehat{\mathfrak{F}}^{(N)}$ uniformly in $N\in\mathbb{N}$. In particular, $\widehat{\mathfrak{F}}^{(N)}$ is a compact operator and generates discrete time semi-dynamical system $\widehat{\varphi}^{(N)}$ on ℓ^2_ρ with a common absorbing set Q for all $N\in\mathbb{N}$.

The proof of the following theorem is essentially the same as the proofs above, and hence omitted.

Theorem 9.6. *The discrete time semi-dynamical system $\widehat{\varphi}^{(N)}$ on ℓ^2_ρ has a global attractor $\widehat{\mathcal{A}}^{(N)}$ given by*

$$\widehat{\mathcal{A}}^{(N)} = \bigcap_{t\ge 1} \widehat{\varphi}^{(N)}(t, Q).$$

Moreover, these attractors converge upper semi-continuously to the attractor $\mathcal{A}$, i.e.,

$$\lim_{N\to\infty} \operatorname{dist}_{\ell^2_\rho}\left(\widehat{\mathcal{A}}^{(N)}, \mathcal{A}\right) = 0.$$

9.4 Nonautonomous systems

The nonautonomous version of the lattice difference equations (9.1) has the form

$$u_{\mathfrak{i}}(t+1) = f_{\mathfrak{i}}(t, u_{\mathfrak{i}}(t)) + \sum_{\mathfrak{j}\in\mathbb{Z}^d} \kappa_{\mathfrak{i},\mathfrak{j}} g_{\mathfrak{j}}(t, u_{\mathfrak{j}}(t)), \qquad \mathfrak{i}\in\mathbb{Z}^d, \quad t\in\mathbb{Z}, \tag{9.7}$$

with weights $\kappa_{\mathfrak{i},\mathfrak{j}}$, where $\mathfrak{i}, \mathfrak{j}\in\mathbb{Z}^d$.

As earlier, that the weights $\rho_{\mathfrak{i}}$ and aggregate interconnection coefficients are assumed to satisfy Assumption 9.1 and Assumption 9.2. In addition, the coefficient functions in (9.7) satisfy the following nonautonomous versions of Assumption 9.2 to Assumption 9.5.

Assumption 9.7. $f_{\mathfrak{i}}(t,0) = g_{\mathfrak{i}}(t,0) = 0$ for each $\mathfrak{i}\in\mathbb{Z}^d$, $t\in\mathbb{Z}$.

Assumption 9.8. For each $\mathfrak{i} \in \mathbb{Z}^d$ the functions $f_\mathfrak{i}$, $g_\mathfrak{i} : \mathbb{Z} \times \mathbb{R} \to \mathbb{R}$ are globally Lipschitz continuous with

$$|f_\mathfrak{i}(t,x)-f_\mathfrak{i}(t,y)| \le L_f(t)|x-y|, \ |g_\mathfrak{i}(t,x)-g_\mathfrak{i}(t,y)| \le L_{g,\mathfrak{i}}(t)|x-y|, \ \forall\, t \in \mathbb{Z},\ x,y \in \mathbb{R},$$

in which $L_f(t) \in \mathbb{R}_+$ for all $t \in \mathbb{Z}$ and $L_{g,\mathfrak{i}}$ satisfies

$$\sup_{\mathfrak{i}\in\mathbb{Z}^d} \sum_{\mathfrak{j}\in\mathbb{Z}^d} \frac{\kappa_{\mathfrak{i},\mathfrak{j}}^2 L_{g,\mathfrak{j}}^2(t)}{\rho_\mathfrak{j}} = \hat{L}_g^2(t), \quad \text{for all } t \in \mathbb{Z}.$$

Assumption 9.9. For each $\mathfrak{i} \in \mathbb{Z}^d$ the functions $f_\mathfrak{i}$, $g_\mathfrak{i} : \mathbb{Z} \times \mathbb{R} \to \mathbb{R}$ are bounded, i.e., there exists some $(M_i(t))_{i\in\mathbb{Z}} \in \ell_\rho^2$ for each $t \in \mathbb{Z}$ such that

$$|f_\mathfrak{i}(t,x)|, |g_\mathfrak{i}(t,x)| \le M_\mathfrak{i}(t) \qquad \forall\, t \in \mathbb{Z},\ x \in \mathbb{R}.$$

Now define the operator $\mathfrak{F}(t,\boldsymbol{u}) = (\mathfrak{F}_\mathfrak{i}(t,\boldsymbol{u}))_{\mathfrak{i}\in\mathbb{Z}^d}$ for all $t \in \mathbb{Z}$, $\boldsymbol{u} \in \ell_\rho^2$ componentwise by

$$\mathfrak{F}_\mathfrak{i}(t,\boldsymbol{u}) = f_\mathfrak{i}(t,u_\mathfrak{i}) + \sum_{\mathfrak{j}\in\mathbb{Z}^d} \kappa_{\mathfrak{i},\mathfrak{j}} g_\mathfrak{j}(t,u_\mathfrak{j}), \qquad \mathfrak{i} \in \mathbb{Z}^d,\ \ \boldsymbol{u} \in \ell_\rho^2,\ \ t \in \mathbb{Z}.$$

These mappings map ℓ_ρ^2 to ℓ_ρ^2 and satisfy similar properties as their autonomous counterparts. The proofs are essentially the same, namely the uniform bounds

$$\|\mathfrak{F}(t,\boldsymbol{u})\|_\rho \le \|\boldsymbol{M}(t)\|_\rho \sqrt{2(1+\kappa\rho_\Sigma)}, \quad \|\boldsymbol{M}(t)\|_\rho = \sqrt{\sum_{\mathfrak{i}\in\mathbb{Z}^d} \rho_\mathfrak{i} M_i^2(t)}, \quad \forall \boldsymbol{u} \in \ell_\rho^2,\ \ t \in \mathbb{Z},$$

and the global Lipschitz condition

$$\|\mathfrak{F}(t,\boldsymbol{u}) - \mathfrak{F}(t,\boldsymbol{v})\|_\rho \le \sqrt{2(L_f^2(t) + \rho_\Sigma \hat{L}_g^2(t))}\|\boldsymbol{u}-\boldsymbol{v}\|_\rho, \quad \forall \boldsymbol{u},\boldsymbol{v} \in \ell_\rho^2,\ \ t \in \mathbb{Z}.$$

In addition, $\mathfrak{F}(t,\boldsymbol{0}) = \boldsymbol{0}$ for all $t \in \mathbb{Z}$.

The uniform tails condition in Lemma 9.3 also holds for each $t \in \mathbb{Z}$. Consequently, the following theorem holds.

Theorem 9.7. *The nonlinear operator $\mathfrak{F}(t,\cdot)$ on ℓ_ρ^2 is compact for each $t \in \mathbb{Z}$.*

9.4.1 *Existence of a pullback attractor*

The non-autonomous lattice difference equation (9.3) can be written as a non-autonomous difference equation on ℓ_ρ^2,

$$\boldsymbol{u}(t+1) = \mathfrak{F}(t,\boldsymbol{u}(t)), \quad t \in \mathbb{Z}. \tag{9.8}$$

It generates a discrete time two-parameter semi-group or process $\{\psi(t,t_0)\}_{(t,t_0)\in\mathbb{R}_\ge^2}$ through composition of the mapping $\mathfrak{F}(t,\cdot)$, i.e.,

$$\psi(t,t_0,\boldsymbol{u}) := \mathfrak{F}(t-1,\mathfrak{F}(t-2,\mathfrak{F}(\cdots,\mathfrak{F}(t_0,\boldsymbol{u})))), \quad t \ge t_0,\ \ t_0 \in \mathbb{Z},$$

with $\psi(t_0,t_0,\boldsymbol{u}) = \boldsymbol{u}$ for all $\boldsymbol{u} \in \ell_\rho^2$ and $t_0 \in \mathbb{Z}$. Moreover, the mapping $\boldsymbol{u} \mapsto \psi(t,t_0,\boldsymbol{u})$ is Lipschitz continuous on ℓ_ρ^2.

For each $t \in \mathbb{Z}$ define the closed and bounded subset of ℓ^2_ρ

$$Q_t = \left\{ \boldsymbol{u} \in \ell^2_\rho : \|\boldsymbol{u}\|_\rho \leq \|\boldsymbol{M}(t-1)\|_\rho \sqrt{2(1+\kappa\rho_\Sigma)} \right\}, \quad t \in \mathbb{Z},$$

and set

$$\mathcal{Q} = \{Q_t : t \in \mathbb{Z}\}.$$

Since $\mathfrak{F}(t, \ell^2_\rho) \subset Q_{t+1}$ for each $t \in \mathbb{Z}$, it follows that $\mathfrak{F}(t, Q_t) \subset Q_{t+1}$ for each $t \in \mathbb{Z}$. Thus $\mathcal{Q}$ is positive invariant under ψ. Moreover, it is an absorbing set in both forward and pullback senses for the non-autonomous process ψ. By Theorem 9.7, $\psi(t, t_0, Q_{t_0})$ is a compact subset of ℓ^2_ρ for each $t \geq t_0 + 1$ and all $t_0 \in \mathbb{Z}$ and these sets form a nested sequence of compact subsets. The following theorem then follows immediately.

Theorem 9.8. *The discrete time process* $\{\psi(t, t_0)\}_{(t,t_0)\in\mathbb{R}^2_\geq}$ *generated by the system* (9.8) *on* ℓ^2_ρ *has a global pullback attractor* $\mathcal{A} = (A_t)_{t\in\mathbb{Z}}$ *in* ℓ^2_ρ *with nonempty compact component sets given by*

$$A_t = \bigcap_{t_0 \leq t} \psi(t, t_0, Q_{t_0}), \quad t \in \mathbb{Z}.$$

Note, in particular, that $\mathbf{0} \in A_t$, since it is a constant, hence an entire solution of the system and thus belongs to the pullback attractor.

9.4.2 *Existence of a forward ω-limit sets*

The family pullback attractor $\mathcal{A} = (A_t)_{t\in\mathbb{Z}}$ given by Theorem 9.8 needs not be forward attracting, i.e., satisfy

$$\lim_{t\to\infty} \operatorname{dist}_{\ell^2_\rho} (\psi(t, t_0, B_{t_0}), A_t) = 0, \qquad \text{fixed } t_0,$$

for all families $\mathcal{B} = (B_t)_{t\in\mathbb{Z}}$ of bounded component subsets B_t of ℓ^2_ρ.

The problem is that there may be other forward ω-limit points starting in $\mathcal{Q}$ than those starting within $\mathcal{A}$. Recall that forward ω-limit set starting at time t_0 in the component subset Q_{t_0} of $\mathcal{Q}$ is defined by

$$\omega_{Q_{t_0}, t_0} := \bigcap_{t \geq t_0} \psi(t, t_0, Q_{t_0}),$$

since the family $\mathcal{Q}$ is positive invariant. It is a nonempty compact subset of ℓ^2_ρ under the above assumptions. Note that

$$\lim_{t\to\infty} \operatorname{dist}_{\ell^2_\rho} \left(\psi(t, t_0, B_{t_0}), \omega_{B_{t_0}, t_0}\right) = 0.$$

Since $\omega_{Q_{t_0}, t_0} \subset \omega_{Q_{\tau_0}, \tau_0}$ for $t_0 \leq \tau_0$, the set

$$\omega^\infty_{\mathcal{Q}} := \overline{\bigcup_{t_0 \in \mathbb{Z}} \omega_{Q_{t_0}, t_0}},$$

is nonempty and closed in ℓ^2_ρ.

Lemma 9.10. *Suppose there is a nonempty compact subset $\mathcal{K}$ of ℓ_ρ^2 such that*

$$\lim_{t\to\infty} \operatorname{dist}_{\ell_\rho^2}\left(\psi\left(t, t_0, Q_{t_0}\right), \mathcal{K}\right) = 0, \quad \textit{for all } t_0 \in \mathbb{Z}.$$

Then $\boldsymbol{\omega}_Q^\infty$ is a compact subset of ℓ_ρ^2.

The proof for Lemma 9.10 follows by noting that $\boldsymbol{\omega}_{Q_{t_0},t_0} \subset \mathcal{K}$ for all $t_0 \in \mathbb{Z}$. Notice that the condition in Lemma 9.10 is a uniform asymptotic compactness condition. The compactness of the process mapping ψ for each initial time t_0 does not suffice to ensure the compactness of $\boldsymbol{\omega}_Q^\infty$.

Note that $\boldsymbol{\omega}_Q^\infty = \overline{\bigcup_{t_0 \geq T^*} \boldsymbol{\omega}_{Q_{t_0},t_0}}$ for any finite $T^* \in \mathbb{Z}$. This is quite reasonable since the future asymptotic behaviour should not depend on that of the system in the distant past. In fact, the system itself needs not be defined for all $t \in \mathbb{Z}$, but just for $t \geq T^*$ for some $T^* \in \mathbb{Z}$. This is more realistic in many biological and physical problems.

The set $\boldsymbol{\omega}_Q^\infty$ characterises the forward asymptotic behaviour of the nonautonomous system ψ. It was called the *forward attracting set* of the nonautonomous system in [Kloeden and Yang (2021)] and is closely related to the Haraux-Vishik *uniform attractor*, but it may be smaller and does not require the system to be defined for all time or the attraction to be uniform in the initial time.

9.5 Endnotes

This chapter is based on the papers [Kloeden (2022a,b)]. In addition, random attractors are considered in [Kloeden (2022a)] and skew products in [Kloeden (2022b)].

Related problems for integro-difference equations are investigated in Huynh, Kloeden & Pötzsche [Hyunh *et al.* (2020)]. See [Kloeden and Yang (2021)] for more on the Haraux-Vishik uniform attractor.

9.6 Problems

Problem 9.1. Determine the steady state solutions of the discrete lattice model (9.1) when $f_i(x) = -x$ and the $g_i(x)$ are non-negative monotone increasing functions. When are there at most countably many steady states?

Problem 9.2. Write out the proof of Lemma 9.10.

Problem 9.3. What are the relationship and difference between the forward omega-limit set established in Section 9.4.2 and a forward attractor, if exists?

Chapter 10

Three topics in brief

There are a number of papers in the literature that consider the finite dimensionality of attractors, and the existence of exponential attractors in various kinds of dissipative lattice systems. In addition, the existence of traveling waves in lattice systems has attracted some attention. These topics are briefly formulated here and some typical results are presented without detailed proofs. The purpose here is to draw their attention to interested readers, who can then study them in more detail from the cited papers.

10.1 Finite dimension of lattice attractors

Lattice dynamical systems are infinite dimensional systems, but their attractors are often finite dimensional. It is interesting to obtain an estimate of the dimension of such an attractor. This section is based on [Zhou and Shi (2006)]. The following definitions are taken from [Chueshov (2015)].

Definition 10.1. Let $\mathcal{K}$ be a compact set in a metric space $\mathfrak{X}$. The *fractal dimension* $\dim_f(\mathcal{K})$ is defined by

$$\dim_f(\mathcal{K}) := \limsup_{\epsilon \to 0} \frac{\ln N(\mathcal{K}, \epsilon)}{1/\epsilon}$$

where $N(\mathcal{K}, \epsilon)$ is the minimal number of closed sets of diameter 2ϵ covering $\mathcal{K}$.

Recall that a real-valued function $\varrho(x)$ on $\mathfrak{X}$ is called a *semi-norm* on $\mathfrak{X}$ if

$$\varrho(x + y) \leq \varrho(x) + \varrho(y) \quad \text{and } \varrho(\lambda x) = |\lambda| \varrho(x) \quad \text{for all } x, y \in \mathfrak{X}, \lambda \in \mathbb{R}.$$

A semi-norm $\varrho(x)$ on $\mathfrak{X}$ is said to be *compact* if and only if $\varrho(x_m) \to 0$ for any sequence $\{x_m\} \subset \mathfrak{X}$ such that $x_m \rightharpoonup 0$ (weak convergence) in $\mathfrak{X}$.

Proposition 10.1. *Let $\mathfrak{H}$ be a separable Hilbert space and $\mathcal{K}$ be a bounded closed set in $\mathfrak{H}$. Assume that there exists a mapping $V : \mathcal{K} \to \mathfrak{X}$ such that*

(i) $\mathcal{K} \subseteq V(\mathcal{K})$,

(ii) V is Lipschitz on $\mathcal{K}$, i.e., there exists $L_V > 0$ such that

$$\|V(x) - V(y)\|_{\mathfrak{H}} \leq L_V \|x - y\|_{\mathfrak{H}},$$

(iii) there exist compact seminorms $\varrho_1(x)$ and $\varrho_2(x)$ on $\mathfrak{H}$ such that

$$\|V(x) - V(y)\|_{\mathfrak{H}} \leq \zeta \|x - y\|_{\mathfrak{H}} + C\left(\varrho_1(x - y) + \varrho_2(V(x) - V(y))\right)$$

for x, $y \in \mathcal{K}$, where $0 < \zeta < 1$ and $C > 0$ are constants.

Then $\mathcal{K}$ is a compact set in $\mathfrak{X}$ of finite fractal dimension.

As an illustrative example, consider the first order lattice system

$$\dot{u}_i - \nu(\Lambda \boldsymbol{u})_i + \lambda_i u_i + f_i(u_i) = g_i, \quad i \in \mathbb{Z}, \quad t > 0, \tag{10.1}$$

with initial data $u_i(0) = u_{o,i} \in \mathbb{R}$ for $i \in \mathbb{Z}$, where Λ is the discrete Laplacian operator defined as in Section 3.1, $\boldsymbol{u} = (u_i)_{i\in\mathbb{Z}}$, $g_i \in \mathbb{R}$, and $\lambda_i \in \mathbb{R}_+$ with

$$m_\lambda := \inf_{i\in\mathbb{Z}} \lambda_i > 0.$$

Throughout this section the functions f_i are assumed to satisfy

Assumption 10.1.

$$x f_i(x) \geq 0, \text{ for } x \in \mathbb{R}, i \in \mathbb{Z}, \quad \text{and} \quad \sup_{i\in\mathbb{Z}} \max_{x\in[-r,r]} |f_i(x)| \leq \Gamma_f(r), \quad r \in \mathbb{R}^+$$

where $\Gamma_f : \mathbb{R}^+ \to \mathbb{R}^+$ is a positive valued continuous function.

Assumption 10.2. $\boldsymbol{g} = (g_i)_{i\in\mathbb{Z}} \in \ell^2$.

Theorem 10.1. *Under Assumptions 10.1 and 10.2, the lattice system* (10.1) *has a global attractor $\mathcal{A}$ in ℓ^2, which has finite fractal dimension.*

Proof. The existence of the attractor $\mathcal{A}$ in ℓ^2 follows as for example in Chapter 3. To show that $\mathcal{A}$ has finite fractal dimension in ℓ^2 we need to apply Proposition 10.1. The details are given in [Zhou and Shi (2006)] Theorem 4. Here we only brief the main idea.

The essential step is to show that the difference between any two solutions of the lattice system (10.1), $\boldsymbol{u}(t; \boldsymbol{u}_o)$ and $\boldsymbol{v}(t; \boldsymbol{v}_o)$ starting at points $\boldsymbol{u}_o, \boldsymbol{v}_o \in \mathcal{A}$ respectively satisfy the estimate

$$\|\boldsymbol{u}(T) - \boldsymbol{v}(T)\| \leq e^{-m_\lambda T/2} \|\boldsymbol{u}_o - \boldsymbol{v}_o\| + \sqrt{\frac{2\Gamma_f(R)}{m_\lambda}} \sum_{|i|\leq I(\epsilon_0)} \max_{s\in[0,T]} |u_i(s) - v_i(s)|,$$

where $R = \frac{2\|\boldsymbol{g}\|^2}{m_\lambda}$ is the radius of the absorbing set for the lattice system (10.1), ϵ_0 is a positive constant such that $f_i'(x) \geq -m_\lambda/2$ for all $i \in \mathbb{Z}$ and $|x| \leq \epsilon_0$, and $I(\epsilon_0)$ is a positive integer such that $\sum_{|i|\geq I(\epsilon_0)} u_i^2 \leq \epsilon_0^2$ for every $\boldsymbol{u} = (u_i)_{i\in\mathbb{Z}}$ in the attractor $\mathcal{A}$.

Set

$$\varrho(\boldsymbol{u}_o - \boldsymbol{v}_o) = \sum_{|i| \le I(\varepsilon_0)} \max_{s \in [0,T]} |u_i(s) - v_i(s)| .$$

Then ϱ is a compact semi-norm with respect to $\boldsymbol{u}_o - \boldsymbol{v}_o$ in ℓ^2. In fact, if $\boldsymbol{u}_o - \boldsymbol{v}_o \rightharpoonup 0$, i.e., weakly, in ℓ^2, then $\max_{s\in[0,T]} |u_i(s) - v_i(s)| \to 0$ for all $|i| \le I(\epsilon_0)$, due to the uniform continuity of the solutions $\boldsymbol{u}(s)$, $\boldsymbol{v}(s)$ on $[0, T]$ and the estimate

$$\|\boldsymbol{u}(t) - \boldsymbol{v}(t)\| \le e^{(-m_\lambda + \Gamma_f(R))t} \|\boldsymbol{u}_o - \boldsymbol{v}_o\|_{\ell^2}, \quad t \ge 0.$$

It then follows by Proposition 10.1 that $\mathcal{A}$ has finite fractal dimension. □

In [Zhou and Shi (2006)] it is also shown that global attractors for certain partly dissipative lattice systems and second order lattice systems also have finite fractal dimensions. These systems will be considered in the next section in the context of exponential attractors.

10.2 Exponential attractors

Global attractor sometimes attracts orbits at a relatively slow speed and it might take an unexpected long time to be reached. Exponential attractors are an alternative tool for the study of long-term behaviour of solutions which attract all bounded sets exponentially fast. When they exist they contain the global attractor.

We start with some concepts from Eden, Foias & Kalantarov [Eden *et al.* (1998)] about exponential attractors for a map defined on a separable Hilbert space. Let $\mathfrak{H}$ be a separable Hilbert space, and $\mathcal{B}$ be the collection of all bounded subset of $\mathfrak{H}$.

Definition 10.2. A compact set $\mathcal{M}$ is called an exponential attractor for a semigroup $\{\varphi(t)\}_{t\ge 0}$ if

(i) $\mathcal{M} \subseteq \mathcal{B}$,
(ii) $\mathcal{M}$ is positive invariant under φ, i.e. $\varphi(\mathcal{M}) \subseteq \mathcal{M}$,
(iii) $\mathcal{M}$ has finite fractal dimension $\mathrm{Dim}_f(\mathcal{M})$,
(iv) there exist positive constants C_1 and C_2 such that for every $\boldsymbol{u} \in B \in \mathcal{B}$ and positive integer k,

$$\mathrm{dist}(\varphi^k(\boldsymbol{u}), \mathcal{M}) \le C_1 e^{-C_2 k},$$

where φ^k represents the k-times composition of φ.

Recall that a map $\mathcal{S}$ is called a κ-contraction on $B \in \mathcal{B}$ if it is a contraction with respect the Kuratowski measure of noncompactness of B, i.e., there is a positive number $q < 1$ such that

$$\kappa(\mathcal{S}(B)) < q\kappa(B)$$

for every $B \in \mathcal{B}$. The Kuratowski measure of noncompactness of $B \in \mathcal{B}$ is defined by

$$\kappa(B) = \inf\{d > 0 : \text{there exists an open cover of } B \text{ with sets of diameter} \le d\}.$$

Its properties were given in Section 2.3.1.

The following definitions are taken from [Eden *et al.* (1998)].

Definition 10.3. A mapping $\mathcal{S} : B \to B$ is said to satisfy the discrete squeezing property if there exists an orthogonal projection P_N of rank N such that for every $\boldsymbol{u}$ and $\boldsymbol{v}$ in B,

$$\|\mathrm{P}_N(\mathcal{S}(\boldsymbol{u}) - \mathcal{S}(\boldsymbol{v}))\| \leq \|(\mathrm{I} - \mathrm{P}_N)(\mathcal{S}(\boldsymbol{u}) - \mathcal{S}(\boldsymbol{v}))\| \Rightarrow \|\mathcal{S}(\boldsymbol{u}) - \mathcal{S}(\boldsymbol{v})\| \leq \frac{1}{8}\|\boldsymbol{u} - \boldsymbol{v}\|.$$

The proof of the following fundamental lemma can be found in [Dung and Nicolaenko (2001); Eden *et al.* (1994, 1998)]. In Lemma 10.1 and Theorem 10.2 below, let $\mathfrak{H}$ be a separable Hilbert space, $\{\varphi(t)\}_{t\geq 0}$ be a continuous semi-group on $\mathfrak{H}$, and let Q be a compact absorbing set of φ that absorbs all bounded subsets of $\mathfrak{H}$ in finite time.

Lemma 10.1. *Suppose there exists a time $T^* > 0$ such that the mapping $\varphi^* := \varphi(T^*)$ is a Lipschitz κ-contraction map on Q with Lipschitz constant L_* and satisfies the discrete squeezing property on Q with projections P_N. Then φ^* has an exponential attractor $\mathfrak{M}^*$ on Q with $\dim_f(\mathfrak{M}^*) \leq CN \ln L_*$, where C is a postive constant.*

Theorem 10.2. *Under the same assumptions as in Lemma 10.1, and further assume that the mapping $\varphi(t, \boldsymbol{u})$ is Lipschitz from $[0, T] \times Q$ into Q for every $T > 0$. Then*

(i) $\mathfrak{M} = \bigcup_{0\leq t\leq T^*} \varphi(t, \mathfrak{M}^*)$ is an exponential attractor for the semi-group $\{\varphi(t)\}_{t\geq 0}$ on Q with

$$\dim_f(\mathfrak{M}) \leq \dim_f(\mathfrak{M}^*) + 1.$$

(ii) $\mathfrak{M}$ attracts all points of $B \in \mathcal{B}$ at a uniform exponential rate, i.e., there exist positive constants C_1, C_2 such that

$$\mathrm{dist}(\varphi(t, \boldsymbol{u}), \mathfrak{M}) \leq C_1 e^{-C_2 t} \quad \textit{for all } \boldsymbol{u} \in B \in \mathcal{B}, t \geq 0.$$

Remark 10.1. The existence of an exponential attractor implies that of a global attractor $\mathcal{A}$ with

$$\mathcal{A} = \bigcap_{t\geq 0} \varphi(t, \mathfrak{M}) \subseteq \mathfrak{M}.$$

A global attractor may, however, still exist under much weaker conditions without ensuring the existence of an exponential attractor.

10.2.1 *Application to general lattice systems*

The results of Theorem 10.2 were extended to lattice dynamical systems on the weighted space ℓ^2_ρ of infinite real-valued sequences in [Han (2011b)].

The weights $(\rho_i)_{i\in\mathbb{Z}}$ are assumed to satisfy the same conditions as in Chapter 8. For the reader's convenience, they are repeated below.

Assumption 10.3. There exists a positive constant M_ρ such that $0 \le \rho_i \le M_\rho < \infty$ for all $i \in \mathbb{Z}$.

Assumption 10.4. There exist positive constants γ_0, $\gamma_1 > 0$, which are independent of $i \in \mathbb{Z}$, such that

$$\rho_{i\pm1} \le \gamma_0\rho_i, \quad |\rho_{i\pm1} - \rho_i| \le \gamma_1\rho_i, \quad \forall i \in \mathbb{Z}.$$

The results presented below apply to various kinds of lattice models. They can be found in [Han (2011b)] as well as in [Zhao and Zhou (2009)] and are given without proof here. The following general abstract set up is required.

Consider the following locally coupled lattice dynamical system

$$\frac{\mathrm{d}}{\mathrm{d}t}u_i(t) = F_i(u_{j|j\in J_{ip}}(t)), \quad i \in \mathbb{Z} \tag{10.2}$$

with initial condition

$$u_i(0) = u_{o,i}, \quad i \in \mathbb{Z}, \quad t > 0,$$

where $\boldsymbol{u} = (u_i)_{i\in\mathbb{Z}}$, $J_{ip} = \{j \in \mathbb{Z} : |i - j| \le p\}$ and p is a fixed positive integer.

[Zhao and Zhou (2009)] established conditions for the existence of an exponential attractor for the system (10.2) in non-weighted space ℓ^2 with $\rho_i \equiv 1$ for all $i \in \mathbb{Z}$ when the solution operator associated to (10.2) generates a continuous semi-group $\{\varphi(t)\}_{t\ge0}$ on ℓ^2 and the mappings F_i satisfy a uniform growth property, which was later relaxed in [Han (2011b)].

Definition 10.4. For $T > 0$, a function $\boldsymbol{u} = (u_i)_{i\in\mathbb{Z}} : [0, T) \to \ell^2_\rho$ is called a solution of (10.2) if $\boldsymbol{u} \in \mathcal{C}([0, T), \ell^2_\rho)$, and for each $i \in \mathbb{Z}$ the component $u_i \in \mathcal{C}^1((0, T), \mathbb{R})$ satisfies (10.2).

Notice that if $\boldsymbol{u} : [0, T) \to \ell^2_\rho$ is a solution of (10.2), then

$$u_i(t) = u_i(0) + \int_0^t F_i(u_j(s)|_{j\in J_{ip}})\mathrm{d}s, \quad i \in \mathbb{Z}, \ t \in [0, T).$$

Finally, for a positive number N, define orthogonal projection $\mathrm{P}_N : \ell^2_\rho \to \ell^2_\rho$ by

$$(\mathrm{P}_N\boldsymbol{u})_i = \begin{cases} u_i, & |i| \le N, \\ 0, & |i| > N. \end{cases}$$

and set $\mathrm{P}^c_N = \mathrm{I} - \mathrm{P}_N$, where I is the identity operator of ℓ^2_ρ.

The following result was established in [Han (2011b)], in which the function $F_i \in \mathcal{C}(\mathbb{R}^{2q+1}; \mathbb{R})$ is assumed to satisfy appropriate conditions under which the solution $\boldsymbol{u}(t) = (u_i(t))_{i\in\mathbb{Z}}$ of (10.2) exists globally on ℓ^2_ρ in $\mathbb{R}_+$ for any initial data $\boldsymbol{u}_o = (u_{o,i})_{i\in\mathbb{Z}} \in \ell^2_\rho$.

Theorem 10.3. *Assume that the solution $\boldsymbol{u}(t; \boldsymbol{u}_o)$ of the system* (10.2) *generates a continuous semi-group $\{\varphi(t)\}_{t\ge0}$ on ℓ^2_ρ. Moreover, if $\{\varphi(t)\}_{t\ge0}$ satisfies conditions*

(a) $\{\varphi(t)\}_{t\geq 0}$ *has a closed bounded absorbing set* $Q \subset \ell^2_\rho$ *such that* $\varphi(t,Q) \subseteq Q$ *for* $t \geq T_Q$ *with some* $T_Q > 0$ *and* $\{\varphi(t)\}_{t\geq 0}$ *is Lipschitz continuous from* $[0,T] \times Q$ *into* Q *for each* $T > 0$,

(b) *there exist time* $T^* > T_Q$, *constants* $\zeta \in (0,1)$, $C > 0$ *and positive integer* N *such that for any* $\boldsymbol{u}_o, \boldsymbol{v}_o \in Q$,

$$\|\varphi(T^*,\boldsymbol{u}_o)-\varphi(T^*,\boldsymbol{v}_o)\|_\rho \leq \zeta\|\boldsymbol{u}_o-\boldsymbol{v}_o\|_\rho + C \max_{t\in[t_o,T^*]} \|\mathrm{P}_N(\varphi(t,\boldsymbol{u}_o)-\varphi(t,\boldsymbol{v}_o))\|_\rho, \tag{10.3}$$

$$\sum_{|i|>N} \rho_i(\varphi(T^*)\boldsymbol{u}_o - \varphi(T^*)\boldsymbol{v}_o)_i^2 \leq \frac{1}{128}\|\boldsymbol{u}_o - \boldsymbol{v}_o\|_\rho^2, \quad \forall \boldsymbol{u}_o, \boldsymbol{v}_o \in Q. \tag{10.4}$$

Then $\varphi(T^*) := \varphi^*$ *has an exponential attractor* $\mathcal{M}^* \subset \ell^2_\rho$ *on*

$$\mathcal{Q} := \overline{\bigcup_{t\geq T_Q} \varphi(t,Q)}.$$

Moreover, $\mathcal{M} = \bigcup_{0\leq t\leq T^*} \varphi(t,\mathcal{M}) \subset \ell^2_\rho$ *is an exponential attractor for* $\{\varphi(t)\}_{t\geq 0}$ *on* $\mathcal{Q}$ *with* $\dim_f(\mathcal{M}) \leq \dim_f(\mathcal{M}^*) + 1$.

10.2.2 *First order lattice systems*

Consider the first order lattice system

$$\frac{\mathrm{d}u_i}{\mathrm{d}t} + \nu(\mathrm{A}\boldsymbol{u})_i + \lambda_i u_i + f_i(u_i) = g_i, \quad i \in \mathbb{Z}, \quad t > 0, \tag{10.5}$$

with initial data

$$u_i(0) = u_{o,i}, \quad i \in \mathbb{Z},$$

where $\boldsymbol{u} = (u_i)_{i\in\mathbb{Z}}$, $g_i \in \mathbb{R}$, $\lambda_i \in \mathbb{R}_+$, $u_{o,i} \in \mathbb{R}$ for $i \in \mathbb{Z}$, and A is a non-negative and self-adjoint linear operator on the space of infinite sequences with the decomposition $\mathrm{A} = \bar{\mathrm{D}}_p\mathrm{D}_p = \mathrm{D}_p\bar{\mathrm{D}}_p$, where D_p is defined by

$$(\mathrm{D}_p\boldsymbol{u})_i = \sum_{k=-p}^{p} d_k u_{i+k}, \; \forall \boldsymbol{u} = (u_i)_{i\in\mathbb{Z}}, \quad |d_k| \leq M_d \text{ (constant)}, \quad -p \leq k \leq p,$$

and $\bar{\mathrm{D}}_p$ is the adjoint of D_p. We make the following assumptions on functions f_i, λ_i, g_i, $i \in \mathbb{Z}$:

Assumption 10.5. $f_i \in \mathcal{C}^1(\mathbb{R},\mathbb{R})$ with $f_i(0) = 0$, and there exist a non-negative number D_f and a positive-valued continuous function $\Gamma_f : \mathbb{R}_+ \to \mathbb{R}_+$ such that

$$f_i'(s) \geq -D_f, \quad \sup_{i\in\mathbb{Z}} \max_{s\in[-r,r]} |f_i'(s)| \leq \Gamma_f(r), \; \forall r \in \mathbb{R}_+, \quad s \in \mathbb{R}, \; i \in \mathbb{Z}.$$

Assumption 10.6. There exist $\beta_i \in \mathbb{R}$ such that

$$sf_i(s) \geq -\beta_i^2, \quad \forall s \in \mathbb{R}, \quad \boldsymbol{\beta} = (\beta_i)_{i\in\mathbb{Z}} \in \ell^2_\rho.$$

Assumption 10.7. There exist two positive constants m_λ and M_λ such that

$$0 < m_\lambda \le \lambda_i \le M_\lambda < +\infty, \quad \forall i \in \mathbb{Z}.$$

Assumption 10.8. $\boldsymbol{g} = (g_i)_{i\in\mathbb{Z}} \in \ell^2_\rho$.

Assumption 10.9. There exists a positive number $\epsilon > 0$ such that

$$\inf_{(i,s)\in\mathcal{I}_\epsilon} f_i'(s) \ge -\frac{m_\lambda}{4} \quad \text{with } \mathcal{I}_\epsilon := \{(i,s) \in \mathbb{Z}\times\mathbb{R} : \sqrt{\rho_i}|s| \le \epsilon\}.$$

The system (10.5) can be written as

$$\frac{\mathrm{d}\boldsymbol{u}}{\mathrm{d}t} + \nu \mathrm{A}\boldsymbol{u} + \boldsymbol{\lambda u} + F(\boldsymbol{u}) = \boldsymbol{g}, \quad \boldsymbol{u}(0) = \boldsymbol{u}_o := (u_{o,i})_{i\in\mathbb{Z}}, \quad t > 0, \tag{10.6}$$

where $\boldsymbol{u} = (u_i)_{i\in\mathbb{Z}}$, $\boldsymbol{\lambda u} = (\lambda_i u_i)_{i\in\mathbb{Z}}$, $F(\boldsymbol{u}) = (f_i(u_i))_{i\in\mathbb{Z}}$ and $\boldsymbol{g} = (g_i)_{i\in\mathbb{Z}}$.

It can be shown as in [Zhao and Zhou (2009)] that, under appropriate assumptions, the lattice system (10.6) generates a continuous semi-group $\{\varphi(t)\}_{t\ge0}$ on ℓ^2_ρ, which has a bounded and closed absorbing ball Q. Moreover, the set

$$\mathcal{Q} := \overline{\bigcup_{t\ge T_Q} \varphi(t,Q)} \subseteq Q \in \ell^2_\rho$$

is a bounded, closed and positive invariant set for $\{\varphi(t)\}_{t\ge0}$, where T_Q is the time for Q to be absorbed by itself under the semi-group φ.

Theorem 10.4. *Let Assumptions 10.5–10.9 hold and γ_1 be small enough. Then*

(i) the semi-group $\{\varphi(t)\}_{t\ge0}$ is Lipschitz continuous from $[0,T]\times\mathcal{Q}$ into $\mathcal{Q}$ for each $T > 0$,
(ii) there exist a time T^* and a number $N \in \mathbb{N}$ such that the operator $\varphi(T^*) := \varphi^*$ satisfies conditions (10.3) and (10.4) of Theorem 10.3 on $\mathcal{Q}$,
(iii) φ^* has an exponential attractor $\mathfrak{M}^*$ on $\mathcal{Q}$,
(iv) $\mathfrak{M} = \bigcup_{0\le t\le T^*} \varphi(t,\mathfrak{M}^*)$ is an exponential attractor for $\{\varphi(t)\}_{t\ge0}$ on $\mathcal{Q}$ with $\dim_f(\mathfrak{M}) \le \dim_f(\mathfrak{M}^*) + 1$.

10.2.3 *Partly dissipative lattice systems*

Consider the initial value problem for the partly dissipative lattice system:

$$\begin{cases} \dot{u}_i + \nu(\mathrm{A}\boldsymbol{u})_i + \lambda_i u_i + f_i(u_i) + \alpha v_i = h_i, \\ \dot{v}_i + \sigma v_i - \mu u_i = g_i, \quad i \in \mathbb{Z}, \quad t > 0, \\ u_i(0) = u_{o,i}, \quad v_i(0) = v_{o,i}, \quad i \in \mathbb{Z}, \end{cases}$$

or equivalently,

$$\begin{cases} \dot{\boldsymbol{u}} + \nu\mathrm{A}\boldsymbol{u} + \boldsymbol{\lambda u} + F(\boldsymbol{u}) + \alpha\boldsymbol{v} = \boldsymbol{h}, \\ \dot{\boldsymbol{v}} + \sigma\boldsymbol{v} - \mu\boldsymbol{u} = \boldsymbol{g}, \quad t > 0, \\ \boldsymbol{u}(0) = \boldsymbol{u}_o := (u_{o,i})_{i\in\mathbb{Z}}, \quad \boldsymbol{v}(0) = \boldsymbol{v}_o := (v_{o,i})_{i\in\mathbb{Z}}, \end{cases} \tag{10.7}$$

where $\boldsymbol{u} = (u_i)_{i\in\mathbb{Z}}$, $\boldsymbol{v} = (v_i)_{i\in\mathbb{Z}}$, $\boldsymbol{\lambda u} = (\lambda_i u_i)_{i\in\mathbb{Z}}$, $F(\boldsymbol{u}) = (f_i(u_i))_{i\in\mathbb{Z}}$, $\boldsymbol{h} = (h_i)_{i\in\mathbb{Z}}$, $\boldsymbol{g} = (g_i)_{i\in\mathbb{Z}}$ with $f_i \in C^1(\mathbb{R},\mathbb{R})$. Here ν, α, σ, $\mu > 0$ are positive constants, while λ_i for $i \in \mathbb{Z}$ and A are defined as in the previous subsection 10.2.2. A representative model of this kind is the FitzHugh-Nagumo model, which will be investigated in detail in Chapter 18.

It can be shown as in [Zhao and Zhou (2009)] that the solution map of (10.7) generates a continuous semi-group φ on $\ell^2_\rho \times \ell^2_\rho$, which has a bounded and closed absorbing set Q. The existence of exponential attractors for (10.7) in the weighted space $\ell^2_\rho \times \ell^2_\rho$ by applying Theorem 10.1 under Assumptions 10.5–10.9 was investigated in [Han (2011b)]. The following result come from [Han (2011b)], where proofs can be found.

Theorem 10.5. *Suppose that Assumptions 10.5–10.9 hold, that γ_1 is small enough and that $\boldsymbol{h} \in \ell^2_\rho$. Then*

(i) the semi-group $\{\varphi(t)\}_{t\geq 0}$ is Lipschitz continuous from $[0,T] \times Q$ into Q for each T;

(ii) there exists a time T^* and a number $N \in \mathbb{N}$ such that the operator $\varphi^* := \varphi(T^*)$ satisfies conditions (10.3) *and* (10.4) *of Theorem 10.3 on Q;*

(iii) φ^* has an exponential attractor $\mathfrak{M}^*$ on

$$\mathcal{Q} := \overline{\bigcup_{t\geq T^*} \varphi(t,Q)} \subseteq Q \in \ell^2_\rho \times \ell^2_\rho;$$

(iv) $\mathfrak{M} := \bigcup_{0\leq t\leq T^*} \varphi(t,\mathfrak{M}^*)$ is an exponential attractor for $\{\varphi(t)\}_{t\geq 0}$ on $\mathcal{Q}$ and

$$\dim_f(\mathfrak{M}) \leq \dim_f(\mathfrak{M}^*) + 1.$$

10.2.4 *Second order lattice systems*

Consider the initial value problem for the second order lattice systems:

$$\begin{cases} \ddot{u}_i + \alpha\dot{u}_i + \nu(\mathrm{A}\boldsymbol{u})_i + \lambda_i u_i + f_i(u_i) = g_i, \ i \in \mathbb{Z}, \ t > 0, \\ u_i(0) = u_{o,i}, \quad \dot{u}_i(0) = v_{o,i}, \qquad i \in \mathbb{Z}, \end{cases} \tag{10.8}$$

where $\boldsymbol{u} = (u_i)_{i\in\mathbb{Z}}$, α, $\lambda_i > 0$, $g_i \in \mathbb{R}$, $f_i \in C^1(\mathbb{R},\mathbb{R})$, and A is defined as in the previous subsection 10.2.2. See also Chapter 8.

The system (10.8) can be written as

$$\begin{cases} \ddot{\boldsymbol{u}} + \alpha\dot{\boldsymbol{u}} + \nu\mathrm{A}\boldsymbol{u} + \lambda\boldsymbol{u} + F(\boldsymbol{u}) = \boldsymbol{g}, \quad t > 0, \\ \boldsymbol{u}(0) = \boldsymbol{u}_o := (u_{o,i})_{i\in\mathbb{Z}}, \quad \dot{\boldsymbol{u}}(0) = \boldsymbol{v}_o := (v_{o,i})_{i\in\mathbb{Z}}, \end{cases} \tag{10.9}$$

where $\boldsymbol{u} = (u_i)_{i\in\mathbb{Z}}$, $\mathrm{A}\boldsymbol{u} = ((\mathrm{A}\boldsymbol{u})_i)_{i\in\mathbb{Z}}$, $\lambda u = (\lambda_i u_i)_{i\in\mathbb{Z}}$, $F(\boldsymbol{u}) = (f_i(u_i))_{i\in\mathbb{Z}}$, $\boldsymbol{g} = (g_i)_{i\in\mathbb{Z}}$. Here the functions f_i satisfy the following assumptions.

Assumption 10.10. There exist positive real numbers K_i, $i \in \mathbb{Z}$, and L_f such that

$$|f_i(s)| \leq K_i, \quad |f_i(s_1) - f_i(s_2)| \leq L_f|s_1 - s_2|, \quad \forall s_1, s_2 \in \mathbb{R}, i \in \mathbb{Z}.$$

Assumption 10.11. There exist a function $\Gamma_f \in \mathcal{C}(\mathbb{R}_+;\mathbb{R}_+)$ and $I_0 \in \mathbb{N}$ such that

$$\Gamma_f(0) = 0, \quad \max_{|i|>I_0} \sup_{s\in[-r,r]} |f_i'(s)| \leq \Gamma_f(r), \quad \forall r \in \mathbb{R}_+.$$

The existence of an exponential attractor will be established in the weighted space $\ell_\rho^2 \times \ell_\rho^2$ for $(\boldsymbol{u}, \boldsymbol{v})$ where $\boldsymbol{v} := \dot{\boldsymbol{u}}$. The set up and notation from Chapter 8 will be used. For $\boldsymbol{u} = (u_i)_{i\in\mathbb{Z}}$, $\boldsymbol{v} = (v_i)_{i\in\mathbb{Z}} \in \ell_\rho^2$ and define a bilinear form by

$$\langle \boldsymbol{u}, \boldsymbol{v}\rangle_{\lambda,\rho} = \langle \mathrm{D}_p\boldsymbol{u}, \mathrm{D}_p\boldsymbol{v}\rangle_\rho + \sum_{i\in\mathbb{Z}} \rho_i\lambda_i u_i v_i.,$$

This bi-linear form $\langle\cdot,\cdot\rangle_{\lambda,\rho}$ is an inner product on ℓ_ρ^2 and its induced norm $\|\cdot\|_{\lambda,\rho}$ is equivalent to $\|\cdot\|_\rho$. Let $\ell_{\lambda,\rho}^2 := (\ell_\rho^2, (\cdot,\cdot)_{\lambda,\rho}, \|\cdot\|_{\lambda,\rho})$ and $\mathfrak{H} := \ell_{\lambda,\rho}^2 \times l_\rho^2$. Then $\mathfrak{H}$ is a separable Hilbert space.

Let $\zeta = \frac{\alpha\lambda}{2\alpha^2+4\lambda}$, and define $\boldsymbol{w} := \boldsymbol{v} + \zeta\boldsymbol{u}$. Set $\boldsymbol{y} = \begin{pmatrix} \boldsymbol{u} \\ \boldsymbol{w} \end{pmatrix}$ and

$$\mathfrak{L} = \begin{pmatrix} \varepsilon & -1 \\ \lambda + \mathrm{A} - \varepsilon(\alpha-\varepsilon) & \alpha - \varepsilon \end{pmatrix} I, \qquad \mathfrak{F}(\boldsymbol{y}) = \begin{pmatrix} 0 \\ -F(\boldsymbol{u}) + \boldsymbol{g} \end{pmatrix}.$$

Then system (10.9) is equivalent to the evolution equation

$$\frac{\mathrm{d}\boldsymbol{y}(t)}{\mathrm{d}t} + \mathfrak{L}\boldsymbol{y} = \mathfrak{F}(\boldsymbol{y}) \tag{10.10}$$

in $\mathfrak{H}$ with initial condition $\boldsymbol{y}(0) = (\boldsymbol{u}_o, \boldsymbol{v}_o + \zeta\boldsymbol{u}_o)$.

It was shown in [Zhao and Zhou (2009)], under appropriate assumptions, that the second order lattice system (10.10) generates a continuous semi-group $\{\varphi(t)\}_{t\geq 0}$ on $\mathfrak{H}$, which has a closed and bounded absorbing set Q. The following result on the existence of an exponential attractor comes from [Han (2011b)], where the proof can be found.

Theorem 10.6. *Suppose that Assumptions 10.7, 10.8, 10.10 and 10.11 hold and that γ_1 is small enough. Then there exists a time $T^* > 0$ such that $\varphi(T^*) \doteq \varphi^*$ has an exponential attractor $\mathcal{M}^*$ on*

$$\mathcal{Q} := \overline{\bigcup_{t\geq T^*} \varphi_3(t, Q)} \subseteq Q \subset \mathfrak{H}.$$

Moroever, $\mathcal{M} = \bigcup_{0\leq t\leq T^*} \varphi(t, \mathcal{M}^*)$ is an exponential attractor for $\{\varphi(t)\}_{t\geq 0}$ on $\mathcal{Q}$ with

$$\dim_f(\mathcal{M}) \leq \dim_f(\mathcal{M}^*) + 1.$$

10.3 Traveling waves for lattice neural field equations

Traveling waves or fronts are of particular interest in the neural sciences. Here, following [Faye (2018)], they will be briefly described in the context of a neural field lattice system. The proofs are very different from those for dissipative lattice systems considered elsewhere in this book and thus will be only very briefly sketched. Details can be found in the paper [Faye (2018)] and the references cited in it.

Consider the lattice differential equation

$$\frac{\mathrm{d}u_i(t)}{\mathrm{d}t} = -u_i(t) + \sum_{j\in\mathbb{Z}} \kappa_j f(u_{i-j}(t)), \quad i \in \mathbb{Z}, \quad t > 0, \tag{10.11}$$

where $u_i(t)$ denotes the membrane potential of neuron labeled i at time t and κ_j represents the strength of interactions associated to the neural network at position j on the lattice, while the firing rate of neurons f is a nonlinear function. Examples are Hopfield neural network models with infinite range interactions and a discretized Amari neural field equation, see Chapters 16 and 19.

Here we are interested in special kinds of entire solutions of the lattice system (10.11). First suppose that the lattice system (10.11) has two steady states $(u_i(t))_{i\in\mathbb{Z}}$ with $u_i(t) \equiv 0$ and $u_i(t) \equiv 1$.

Traveling waves are particular kinds of entire solutions of (10.11) of the form $u_i(t) = u(i - ct)$ for some $c \in \mathbb{R}$, where the mapping $u : \mathbb{R} \to \mathbb{R}$ satisfies

$$-c\frac{\mathrm{d}u(x)}{\mathrm{d}x} = -u(x) + \sum_{j\in\mathbb{Z}} \kappa_j f(u(x-j)), \quad x \in \mathbb{R}, \tag{10.12}$$

$$\lim_{x\to-\infty} u(x) = 1, \quad \lim_{x\to\infty} u(x) = 0, \tag{10.13}$$

where we set $x = i - ct$.

When $c = 0$, a *stationary wave solution* of the lattice system (10.11) is understood as a sequence $(u_i(t))_{i\in\mathbb{Z}} = \{\tilde{u}_i\}_{i\in\mathbb{Z}}$, independent of time, which verifies

$$\tilde{u}_i = \sum_{j\in\mathbb{Z}} \kappa_j f(\tilde{u}_{n-j}), \quad i \in \mathbb{Z},$$

$$\lim_{i\to-\infty} \tilde{u}_i = 1, \quad \lim_{i\to\infty} \tilde{u}_i = 0.$$

The mathematical study of traveling waves in neural networks goes back to the pioneering work of [Ermentrout and McLeod (1993)] for continuous neural field equations when the kinetics of the system is of bistable type. The existence and uniqueness (up to translation) of monotone traveling front solutions are established in [Ermentrout and McLeod (1993)]. If the support of the interactions is finite, the theory developed by [Mallet-Paret (1999)] can be used to study of traveling front solutions in general lattice differential equations. In [Bates *et al.* (2003)] traveling waves were studied in infinite range lattice differential equations with bistable dynamics. The main difference between these works and the setting here is that the nonlinearity appears within the infinite sum in (10.11), which means the the results in [Bates *et al.* (2003)] are not directly applicable. Nevertheless, some of the ideas developed in [Bates *et al.* (2003)] can be adapted to the present situation.

Existence of monotone travelling waves

The interaction weights $(\kappa_j)_{j\in\mathbb{Z}}$ in (10.11) are assumed to satisfy the normalization condition

Assumption 10.12. $\sum_{j\in\mathbb{Z}} \kappa_j = 1$,

and symmetric and convergent conditions

Assumption 10.13.

(i) $\kappa_j = \kappa_{-j} \geq 0$ for all $j \in \mathbb{Z}$ and $j_{\pm 1} > 0$;
(ii) $\sum_{j\in\mathbb{Z}} |j|\kappa_j < \infty$.

The first condition is a natural biological assumption and expresses the symmetric and excitatory nature of the considered neural network, while the second condition is a technical assumption that is necessary in the process of proving the existence and uniqueness of traveling front solutions.

From Assumption 10.12 the steady state solutions of the lattice system (10.11) with $u_i(t) \equiv u$ for all $i \in \mathbb{Z}$ for some $u \in \mathbb{R}$ satisfy the equation

$$-u + f(u) := F(u). \tag{10.14}$$

The nonlinear function f is assumed to have bistable nonlinearity.

Assumption 10.14.

(i) $f \in C_b^r(\mathbb{R})$ for $r \geq 2$ with $f(0) = 0$ and $f(1) = 1$ together with $f'(0) < 1$ and $f'(1) < 1$;
(ii) there exists a unique $\zeta \in (0, 1)$ such that $f(\zeta) = \zeta$ with $f'(\zeta) > 1$;
(iii) $u \mapsto f(u)$ is strictly nondecreasing on $[0, 1]$ and there exist $M_f > 1 > m_f > 0$ such that $m_f < f'(u) \leq M_f$ for all $u \in [0, 1]$.

Conditions (i)–(ii) ensure that $u_i(t) \equiv u$ for all $i \in \mathbb{Z}$ with $u \in \{0, \zeta, 1\}$ are steady state solutions of the lattice system (10.11) and that the function F in (10.14) is of bistable type. The third condition (iii) ensures that f is an increasing function, which is natural for a firing rate function. The assumed regularity $f \in C_b^r(\mathbb{R})$ is needed for the proof of the uniqueness result below.

The first result in [Faye (2018)] is about the existence of monotone traveling front solutions of (10.11), which means the existence of the solution profile u_* and speed c_*.

Theorem 10.7. *Suppose that Assumptions 10.12–10.14 hold. Then there exists a traveling wave solution $u_i(t) = u_*(n - c_*t)$ of the lattice system* (10.11) *such that the profile u_* satisfies* (10.12) *when $c_* \neq 0$, or* (10.12) *if $c_* = 0$. In the later case, the stationary wave solution will be denoted by $(u_i^*)_{i\in\mathbb{Z}}$. Moreover,*

(i) $\operatorname{sgn}(c_*) = \operatorname{sgn} \int_0^1 F(u)\, du$ *if $c_* \neq 0$;*

(ii) *if* $\operatorname{sgn}\int_0^1 F(u)\,du = 0$, *then* $c_* = 0$;
(iii) *if* $c_* \neq 0$, *then* $u_* \in \mathcal{C}^{r+1}(\mathbb{R})$ *and* $u'_* < 0$ *on* $\mathbb{R}$;
(iv) *if* $c_* = 0$, *then* $(u_i^*)_{i\in\mathbb{Z}}$ *is a strictly decreasing sequence.*

The proof of Theorem 10.7 relies on a strategy in Bates, Chen & Chmaij [Bates *et al.* (2003)] to regularise the traveling wave problem (10.12)–(10.13). This leads to considering a sequence of traveling waves problems for continuous neural field equations and applying the results of [Ermentrout and McLeod (1993)]. Specifically, one introduces

$$\mathcal{K}_\delta(x) := \sum_{j\in\mathbb{Z}} \kappa_j \delta(x-j)$$

where $\delta(x-j)$ stands for the delta Dirac mass at $x = j$. With this notation, we can write

$$\sum_{j\in\mathbb{Z}} \kappa_j f(\tilde{u}_{n-j})\mathcal{K}_\delta(x) = \mathcal{K}_\delta * f(u)[x],$$

where $*$ denotes the convolution on the real line. As a consequence, the traveling wave problem (10.12)–(10.13) can be written as

$$-c\frac{\mathrm{d}u(x)}{\mathrm{d}x} = -u(x) + \mathcal{K}_\delta * f(u)[x] \quad x \in \mathbb{R},,$$
$$\lim_{x\to-\infty} u(x) = 1, \quad \lim_{x\to\infty} u(x) = 0.$$

The kernel $\mathcal{K}_\delta$ is then approximated by a kernel $\mathcal{K}_m$ with compact support such that $\mathcal{K}_m * \phi \to \mathcal{K}_\delta * \phi$ as $m \to \infty$ uniformly on compact sets for all $\phi \in \mathcal{C}_c^\infty(\mathbb{R})$. This leads to a sequence of regular traveling wave problems

$$-c_m\frac{\mathrm{d}u_m(x)}{\mathrm{d}x} = -u_m(x) + \mathcal{K}_m * f(u_m)[x] \quad x \in \mathbb{R},$$
$$\lim_{x\to-\infty} u_m(x) = 1, \quad \lim_{x\to\infty} u_m(x) = 0,$$

for which weak solutions (c_m, u_m) can be found as in [Ermentrout and McLeod (1993)]. The final step is to pass to the limit and verify that the limiting front profiles satisfy all the properties stated in Theorem 10.7.

Another result in [Faye (2018)] is about the uniqueness of traveling front solutions with nonzero wave speed.

Theorem 10.8. *Let* (u_*, c_*) *be a solution to* (10.12) *as given in Theorem 10.7 such that* $c_* \neq 0$. *Let* $(\hat{u}, \hat{c})$ *be another solution to* (10.12). *Then* $c_* = \hat{c}$ *and, up to a translation,* $u_* = \hat{u}$.

Other results investigate the spectral stability of these traveling waves.

10.4 End notes

[Cholewa and Czaja (2020)] considers the finite dimensionality of both global and exponential attractors in autonomous Laplacian lattice systems. See also [Fan and Yang (2010); Zhou and Shi (2006); Zhou *et al.* (2008)].

The second section on exponential attractors is based on the paper [Han (2011b)], which makes use of results in [Zhao and Zhou (2009)], where exponential attractors were established in the space ℓ^2 instead of the space ℓ^2_ρ. See also [Abdallah (2008); Cholewa and Czaja (2020); Fan (2008); Fan and Yang (2010); Li *et al.* (2011)]. Czaja also considered pullback exponential attractors in [Czaja (2022)]. [Zhou and Han (2012, 2013)] considered pullback and uniform exponential attractors for non-autonomous LDS. Exponential attractors for two-dimensional nonlocal diffusion systems with delay were investigated by Yang, Wang & Kloeden [Yang *et al.* (2012a, 2022b)].

[Li and Lv (2009); Li and Wang (2007); Li and Zhong (2005)] considered attractors in partly dissipative LDS, both autonomous and nonautonomous. See [Han (2011c)] for the stochastic case.

The third section on traveling waves is based on the paper [Faye (2018)], which applies techniques introduced in Bates, Chen & Chmaj [Bates *et al.* (2003)] for traveling waves in other kinds of lattice systems. See also [Afraimovich and Nekorkin (1994); Chow (2003); Chow and Mallet-Paret (1995); Chow *et al.* (1998); Ernaux and Nicolis (1993)]. [Mallet-Paret (1999)] studies traveling front solutions in general lattice differential equations when the support of the interactions is finite.

10.5 Problems

Problem 10.1. Does a sequence of parameterised exponential attractors converge continuously (in the Hausdorff metric) rather than just upper semi-continuously to another exponential attractor, as the sequence of parameters approaching a constant?

Problem 10.2. Determine appropriate sequence spaces for travelling wave solutions.

PART 4
Stochastic and Random LDS

Chapter 11

Random dynamical systems

In this chapter we introduce basic concepts and theory of random dynamical systems. Throughout the chapter let $(\Omega, \mathcal{F}, \mathbb{P})$ be a probability space, where $\mathcal{F}$ is a σ-algebra on Ω and $\mathbb{P}$ is a probability measure.

11.1 Random ordinary differential equations

Let $\eta\colon [0,T]\times\Omega \to \mathbb{R}^m$ be an $\mathbb{R}^m$-valued stochastic process with continuous sample paths. In addition, let $g : \mathbb{R}^m \times \mathbb{R}^d \to \mathbb{R}^d$ be a continuous function. A random ordinary differential equation (RODE) in $\mathbb{R}^d$,

$$\frac{\mathrm{d}x}{\mathrm{d}t} = g(x, \eta_t(\omega)), \qquad x \in \mathbb{R}^d,$$

is a nonautonomous ordinary differential equation (ODE)

$$\frac{\mathrm{d}x}{\mathrm{d}t} = G_\omega(t,x) := g(x, \eta_t(\omega)) \tag{11.1}$$

for almost every realisation $\omega \in \Omega$.

A simple example of a scalar RODE is

$$\frac{\mathrm{d}x}{\mathrm{d}t} = -x + \sin W_t(\omega),$$

where W_t is a scalar Wiener process. Here $d = m = 1$, and $g(x,z) = -x + \sin z$. RODEs with other kinds of noise such as fractional Brownian motion have also been used.

For convenience, it will be assumed that the RODE (11.1) holds for all $\omega \in \Omega$, by restricting Ω to a subset of full probability if necessary, and that g is infinitely often continuously differentiable in its variables, although k-times continuously differentiable with k sufficiently large would suffice. In particular, g is then locally Lipschitz in x, so the initial value problem

$$\frac{\mathrm{d}x}{\mathrm{d}t} = g(x(t,\omega), \eta_t(\omega)), \qquad x(0,\omega) = x_0(\omega), \tag{11.2}$$

where the initial value x_0 is an $\mathbb{R}^d$-valued random variable, has a unique pathwise solution $x(t,\omega)$ for every $\omega \in \Omega$, which will be assumed to exist on the finite time interval $[0,T]$ under consideration.

The solution of the RODE (11.2) is a stochastic process $\{X_t\}_{t\in[0,T]}$. Its sample paths $t \mapsto X_t(\omega)$ are continuously differentiable, but need not be further differentiable, since the vector field $G_\omega(t,x)$ of the nonautonomous ODE (11.1) is usually only at most continuous, but not differentiable in t, no matter how smooth the function g is in its variables.

11.1.1 *RODEs with canonical noise*

RODEs typically involve given stochastic processes in their vector fields which can differ from example to example. The theory of random dynamical systems, in contrast, is formulated abstractly in terms of a canonical noise process. This allows greatly generality and is, in particular, independent of the dimension of the driving noise process. The canonical noise process is represented by a measure theoretical autonomous dynamical system ϑ on the sample space Ω of some probability space $(\Omega, \mathcal{F}, \mathbb{P})$. Specifically, it is a group under composition of measure preserving transformations $\vartheta_t : \Omega \to \Omega$, $t \in \mathbb{R}$, i.e., satisfying

(i) $\vartheta_0(\omega) = \omega$ for all $\omega \in \Omega$;
(ii) $\vartheta_{s+t}(\omega) = \vartheta_s \circ \vartheta_t(\omega)$ for all $\omega \in \Omega$ and any $s, t \in \mathbb{R}$;
(iii) $(t,\omega) \mapsto \vartheta_t(\omega)$ is measurable for all $\omega \in \Omega$ and any $s, t \in \mathbb{R}$;
(iv) $\vartheta_t(\mathbb{P}) = \mathbb{P}$ for every $t \in \mathbb{R}$.

The notation $\vartheta_t(\mathbb{P}) = \mathbb{P}$ for the measure preserving property of ϑ_t with respect to $\mathbb{P}$ is just a compact way of writing $\mathbb{P}(\vartheta_t(A)) = \mathbb{P}(A)$ for all $A \in \mathcal{F}$ and $t \in \mathbb{R}$. The space $(\Omega, \mathcal{F}, \mathbb{P}, \{\vartheta_t\}_{t\in\mathbb{R}})$ is called a *metric dynamical system*.

In this context RODEs have the *canonical* form

$$\frac{\mathrm{d}x}{\mathrm{d}t} = g(x, \vartheta_t(\omega)), \tag{11.3}$$

where the vector field function $g : \mathbb{R}^d \times \Omega \to \mathbb{R}^d$ is assumed to be suitably smooth in its first variable and measurable in the second.

11.1.2 *Existence und uniqueness results for RODEs*

If the vector field function g in the RODE (11.2) is continuous in both of its variables and the sample paths of noise process η_t are also continuous, then classical existence and uniqueness theorems for ODEs still hold pathwise. On the other hand, if the sample paths of the noise process η_t are only measurable in t, then the existence and uniqueness of solutions are to be understood in the sense of Carathéodory.

The classical existence and uniqueness result due to Picard and Lindelöf holds under a local Lipschitz assumption. It can be proved using a convergence sequence of successive approximation.

Theorem 11.1. (The Picard-Lindelöf Theorem) *Let $G : [t_0, T] \times \mathbb{R}^d \to \mathbb{R}^d$ be continuous on a parallelepiped $R := \{(t,x) : t_0 \le t \le t_0 + a, |x - x_0| \le b\}$ and* *uniformly Lipschitz continuous in x and continuous in t. In addition, let M be a bound for $|G(t,x)|$ on R and denote by $T := \min\{a, b/M\}$. Then the initial value problem* (11.2) *has a unique solution $x = x(t; x_0)$ on $[t_0, t_0 + T]$.*

The next theorem drops the Lipschitz assumption and sacrifices the uniqueness of solutions.

Theorem 11.2. (Peano's Existence Theorem) *Let $G : [t_0, T] \times \mathbb{R}^d \to \mathbb{R}^d$ be continuous on a parallelepiped $R := \{(t,x) : t_0 \le t \le t_0 + a, |x - x_0| \le b\}$. In addition, let M be an upper bound for $|G(t,x)|$ on R and denote by $T := \min\{a, b/M\}$. Then the IVP* (11.2) *has at least one solution $x = x(t; x_0)$ on $[t_0, t_0 + T]$.*

Similarly, results for ODEs in Hilbert spaces, see e.g., [Deimling (1977)], can be adapted pathwise to RODEs in the spaces.

11.2 Random dynamical systems

Random ODEs formulated with a noise process often generate random dynamical systems (RDSs). These RDSs and their attractors, called random attractors, are briefly introduced here for a separable Banach state space $(\mathfrak{E}, \|\cdot\|_{\mathfrak{E}})$ [Arnold (1974)]. In this theory the RODE is written in the canonical form (11.3) with the vector field function $g : \mathfrak{E} \times \Omega \to \mathfrak{E}$ and a noise represented by a measure-preserving metric dynamical system $\vartheta = \{\vartheta_t\}_{t \in \mathbb{R}}$ acting on a probability space $(\Omega, \mathcal{F}, \mathbb{P})$ rather than as a specific noise process η_t as in (11.1).

Definition 11.1. $(\Omega, \mathcal{F}, \mathbb{P}, \{\vartheta_t\}_{t \in \mathbb{R}})$ is called a metric dynamical system, if

(i) $\vartheta : \mathbb{R} \times \Omega \to \Omega$ is $(\mathcal{B}(\mathbb{R}) \times \mathcal{F}, \mathcal{F})$-measurable,
(ii) ϑ_0 is the identity on Ω,
(iii) $\vartheta_{s+t} = \vartheta_t \circ \vartheta_s$ for all $s, t \in \mathbb{R}$ and $\vartheta_t \mathbb{P} = \mathbb{P}$ for all $t \in \mathbb{R}$.

Definition 11.2. A stochastic process $\{\varphi(t,\omega)\}_{t \ge 0, \omega \in \Omega}$ is said to be a continuous random dynamical system (RDS) over $(\Omega, \mathcal{F}, \mathbb{P}, \{\vartheta_t\}_{t \in \mathbb{R}})$ with state space $\mathfrak{E}$ if

(i) (initial condition) $\varphi(0, \omega, \boldsymbol{x}) = \boldsymbol{x}$ for all $\omega \in \Omega$ and $\boldsymbol{x} \in \mathfrak{E}$,
(ii) (cocycle property) $\varphi(t + s, \omega, \boldsymbol{x}) = \varphi(t, \vartheta_s(\omega), \varphi(s, \omega, \boldsymbol{x}))$ for all $s, t \ge 0$, $\omega \in \Omega$ and $\boldsymbol{x} \in \mathfrak{E}$,
(iii) (continuity) the mapping $\varphi(t, \omega) : \mathfrak{E} \to \mathfrak{E}$, $\boldsymbol{x} \mapsto \varphi(t, \omega, \boldsymbol{x})$ is continuous for every $t \ge 0$ and $\omega \in \Omega$,
(iv) (measurability) the mapping $\varphi : [0, +\infty) \times \Omega \times \mathfrak{E} \to \mathfrak{E}$, $(t, \omega, \boldsymbol{x}) \mapsto \varphi(t, \omega, \boldsymbol{x})$ is $(\mathcal{B}[0, +\infty) \times \mathcal{F} \times \mathcal{B}(\mathfrak{E}), \mathcal{B}(\mathfrak{E}))$-measurable.

Notice that as stated in Remark 2.6, the definition of RDS is essentially a skew product flow with an ergodic driving dynamical system $(\Omega, \mathcal{F}, \mathbb{P}, \{\vartheta_t\}_{t \in \mathbb{R}})$ and a measurable cocycle mapping $\varphi : (t, \omega, \boldsymbol{x}) \mapsto \varphi(t, \omega, \boldsymbol{x})$.

Random attractors and their deterministic non-autonomous attractors consist of families of sets that are mapped onto each other by the cocycle mapping φ. For random attractors the component sets are labeled by $\omega \in \Omega$. Another way of expressing this is in terms of random sets defined as follows.

Definition 11.3. (random, bounded, compact, tempered set)

(i) A set-valued mapping $\omega \mapsto D(\omega) \subset \mathfrak{E}$ (which is often written as $D(\omega)$ for short) is said to be a random set if the mapping $\omega \mapsto \text{dist}_{\mathfrak{E}}(\boldsymbol{x}, D(\omega))$ is measurable for any $\boldsymbol{x} \in \mathfrak{E}$.

(ii) A random set $D(\omega)$ is said to be bounded if there exist $\boldsymbol{x}^* \in \mathfrak{E}$ and a random variable $r(\omega) > 0$ such that

$$D(\omega) \subset \{\boldsymbol{x} \in \mathfrak{E} : \|\boldsymbol{x} - \boldsymbol{x}^*\|_{\mathfrak{E}} \leq r(\omega), \boldsymbol{x}^* \in \mathfrak{E}\} \quad \forall\, \omega \in \Omega.$$

(iii) A random set $D(\omega)$ is called a compact random set if $D(\omega)$ is compact for all $\omega \in \Omega$.

(iv) A bounded random set $D(\omega)$ is said to be tempered with respect to $(\vartheta_t)_{t\in\mathbb{R}}$ if

$$\lim_{t\to+\infty} e^{-\epsilon t} \sup_{\boldsymbol{x}\in D(\vartheta_{-t}\omega)} \|\boldsymbol{x}\|_{\mathfrak{E}} = 0, \quad a.e.\ \omega \in \Omega, \quad \forall\, \epsilon > 0.$$

(v) A random variable $\omega \mapsto r(\omega) \in \mathbb{R}$ is said to be tempered with respect to $(\vartheta_t)_{t\in\mathbb{R}}$ if

$$\lim_{t\to+\infty} e^{-\epsilon t} \sup_{t\in\mathbb{R}} |r(\vartheta_{-t}\omega)| = 0, \quad a.e.\ \omega \in \Omega, \quad \forall\, \epsilon > 0.$$

Random sets are often denoted by $\mathcal{D} = \{D(\omega)\}_{\omega\in\Omega}$.

In the above definition and later, a property holds for a.e. $\omega \in \Omega$ means that there is $\widetilde{\Omega} \subset \Omega$ with $\mathbb{P}(\widetilde{\Omega}) = 1$ and $\vartheta_t(\widetilde{\Omega}) = \widetilde{\Omega}$ for all $t \in \mathbb{R}$ such that the property holds for all $\omega \in \widetilde{\Omega}$.

Throughout this chapter denote by $\mathscr{D}(\mathfrak{E})$ the collection of all tempered random sets in $\mathfrak{E}$. A random attractor of a random dynamical system is a random set which is a pullback attractor in the pathwise sense with respect to the attracting basin of tempered random subsets of $\mathfrak{E}$.

Definition 11.4. A random set $\mathscr{A} = \{A(\omega)\}_{\omega\in\Omega}$ from $\mathscr{D}(\mathfrak{E})$ is called a random $\mathscr{D}$ attractor (pullback $\mathscr{D}$ attractor or random attractor) for a random dynamical system $\{\varphi(t,\omega)\}_{t\geq 0,\omega\in\Omega}$ on $\mathfrak{E}$ if

(i) $\mathscr{A}$ is a random compact set,

(ii) $\mathscr{A}$ is φ-invariant, i.e., $\varphi(t,\omega,A(\omega)) = A(\vartheta_t(\omega))$ for a.e. $\omega \in \Omega$ and all $t \geq 0$;

(iii) $\mathscr{A}$ is pathwise pullback attracting in $\mathscr{D}(\mathfrak{E})$, i.e.,

$$\lim_{t\to\infty} \text{dist}_{\mathfrak{E}} \left(\varphi\left(t, \vartheta_{-t}(\omega), D(\vartheta_{-t}(\omega))\right), A(\omega)\right) = 0$$

for a.e. $\omega \in \Omega$ and all $\mathcal{D} = \{D(\omega)\}_{\omega\in\Omega} \in \mathscr{D}(\mathfrak{E})$.

Remark 11.1. The collection $\mathcal{D}$ of all tempered random subsets of $\mathfrak{E}$ is called the domain of attraction of $\mathcal{A}$. When $\mathcal{D}$ is chosen to be the collection of all finite (compact, resp.) deterministic subsets of $\mathfrak{E}$, the random $\mathcal{D}$ attractor is called a point (set, resp.) attractor.

The existence of a random attractor is ensured by the existence of a pullback absorbing set, defined as follows.

Definition 11.5. A random set $\mathcal{Q} = \{Q(\omega)\}_{\omega\in\Omega}$ is called a pullback absorbing set in $\mathcal{D}$, if for a.e. $\omega \in \Omega$ and all $\mathcal{D} = \{D(\omega)\}_{\omega\in\Omega} \in \mathcal{D}(\mathfrak{E})$, there exists $T_D(\omega) > 0$ such that

$$\varphi(t, \vartheta_{-t}(\omega), D(\vartheta_{-t}(\omega))) \subset Q(\omega), \quad \text{for all } t \geq T_D(\omega).$$

The following theorem is proved in Bates, Lisei and Lu [Bates *et al.* (2006)].

Theorem 11.3. *Let $\{\varphi(t,\omega)\}_{t\geq 0,\omega\in\Omega}$ be a continuous random dynamical system on $\mathfrak{E}$. If there exists a random closed absorbing set $\mathcal{Q} = \{Q(\omega)\}_{\omega\in\Omega}$ for $\{\varphi(t,\omega)\}_{t\geq 0,\omega\in\Omega}$ and satisfy that for a.e. $\omega \in \Omega$ each sequence $\boldsymbol{x}^{(n)} \in \varphi(t_n, \vartheta_{-t_n}\omega, Q(\vartheta_{-t_n}\omega))$ with $t_n \to \infty$ has a convergent subsequence in $\mathfrak{E}$. Then the RDS $\{\varphi(t,\omega)\}_{t\geq 0,\omega\in\Omega}$ has a unique global random $\mathcal{D}$ attractor $\mathcal{A} = \{A(\omega)\}_{\omega\in\Omega}$ with component sets defined by*

$$A(\omega) = \bigcap_{\tau \geq T_{Q(\omega)}} \overline{\bigcup_{t\geq\tau} \varphi(t_n, \vartheta_{-t}\omega, Q(\vartheta_{-t}\omega)}.$$

Remark 11.2. In [Bates *et al.* (2006)], the definition of continuous RDS requires the continuity of φ with respect to t, but only the continuity of φ in $\boldsymbol{x}$ is required for the proof of Theorem 11.3. Thus Theorem 11.3 also holds without the continuity of φ in t. In particular, it holds for discrete time random dynamical systems such as those resulting from the numerical discretisation of a RODE.

Notice that the random attractor in Theorem 11.3 is in pathwise pullback sense, and there is no analogous result for the existence of a random attractor with pathwise forward convergence. However, the ϑ_t-invariance of the probability measure $\mathbb{P}$ implies that

$$\begin{aligned}
&\mathbb{P}\{\omega \in \Omega : \text{dist}\,(\varphi(t, \vartheta_{-t}(\omega), K(\vartheta_{-t}(\omega)), A(\omega)) \geq \varepsilon\} \\
&= \mathbb{P}\{\omega \in \Omega : \text{dist}\,(\varphi(t, \omega, K(\omega)), A(\vartheta_t(\omega)) \geq \varepsilon\}
\end{aligned}$$

for any $\varepsilon > 0$. Since $\mathbb{P}$-almost sure convergence implies convergence in probability, a random pullback attractor also converges in the forwards sense, but only in the weaker sense of convergence in probability. This allows individual sample path to have large deviations from the attractor, but all converge in this probabilistic sense.

11.3 Random attractors for general RDS in weighted spaces

Here we provide an important result on existence of random attractors for general RDS in weighted space of infinite sequences established in Han, Shen and Zhou [Han *et al.* (2011)]. In particular, consider the weighted space of sequence ℓ^p_ρ defined as in Section 1.3 with the weights assumed to satisfy Assumption 1.2. For the reader's convenience, the assumption is restated as follows.

Assumption 11.1. There exist some positive constants γ_0 and γ_1 such that $0 < \rho_i \le \gamma_0$ and $\rho_i \le \gamma_1 \rho_{i\pm 1}$ for all $i \in \mathbb{Z}$.

Recall Definition 11.3(v) and denote by $\mathfrak{D}(\ell^p_\rho)$ the set of all tempered random sets of ℓ^p_ρ with respect to $\{\vartheta_t\}_{t\in\mathbb{R}}$. Similar to the existence of attractors for deterministic lattice systems, the existence of random attractors for an RDS usually requires a "light-tail" property for the RDS, defined as follows. In what follows $\varphi_i(t,\omega,\cdot)$ denotes the ith component of $\varphi(t,\omega,\cdot)$.

Definition 11.6. A random dynamical system $\{\varphi(t,\omega)\}_{t\ge 0,\omega\in\Omega}$ is said to be random asymptotically null in $\mathcal{D} = \{D(\omega)\}_{\omega\in\Omega} \in \mathfrak{D}(\ell^p_\rho)$ if for a.e. $\omega \in \Omega$ and every $\varepsilon > 0$, there exist $T(\varepsilon,\omega,D(\omega)) > 0$ and $I(\varepsilon,\omega,D(\omega)) \in \mathbb{N}$ such that

$$\Bigg(\sum_{|i|>I(\varepsilon,\omega,D(\omega))} \rho_i \left|\varphi_i(t,\vartheta_{-t}\omega,\boldsymbol{u}(\vartheta_{-t}\omega))\right|^p \Bigg)^{1/p} \le \varepsilon$$

for all $t \ge T(\varepsilon,\omega,D(\omega))$ and $\boldsymbol{u}(\omega) \in D(\omega)$.

Existence of random attractors

The proof of the following theorem extends the results of Bates, Lisei and Lu [Bates *et al.* (2006)] to random lattice systems on ℓ^p_ρ.

Theorem 11.4. *Let Assumption 11.1 hold and let $\{\varphi(t,\omega)\}_{t\ge 0,\omega\in\Omega}$ be a continuous random dynamical system on ℓ^p_ρ. Suppose that*

(i) there exists a bounded closed random absorbing set $\mathcal{Q} = \{Q(\omega)\}_{\omega\in\Omega} \in \mathfrak{D}(\ell^p_\rho)$ such that for a.e. $\omega \in \Omega$ and any $\mathcal{D} = \{D(\omega)\}_{\omega\in\Omega} \in \mathfrak{D}(\ell^p_\rho)$, there exists $T_D(\omega) > 0$ for which $\varphi(t,\vartheta_{-t}\omega, D(\vartheta_{-t}\omega)) \subset Q(\omega)$ for all $t \ge T_D(\omega)$,

(ii) the RDS $\{\varphi(t,\omega)\}_{t\ge 0,\omega\in\Omega}$ is random asymptotically null on $\mathcal{Q} = \{Q(\omega)\}_{\omega\in\Omega}$.

Then the RDS $\{\varphi(t,\omega)\}_{t\ge 0,\omega\in\Omega}$ possesses a unique global random $\mathfrak{D}$ attractor $\mathcal{A} = \{A(\omega)\}_{\omega\in\Omega}$ with component sets given by

$$A(\omega) = \bigcap_{\tau \ge T_Q(\omega)} \overline{\bigcup_{t\ge\tau} \varphi(t,\vartheta_{-t}\omega, Q(\vartheta_{-t}\omega))}. \tag{11.4}$$

Proof. The proof is based on Theorem 11.3. Consider a time sequence $t_n \to \infty$ as $n \to \infty$. For a.e. $\omega \in \Omega$, set

$$\boldsymbol{u}_o^{(n)}(\omega) \in Q(\vartheta_{-t_n}\omega), \quad \boldsymbol{u}^{(n)}(\omega) = \varphi(t_n,\vartheta_{-t_n}\omega,\boldsymbol{u}_o^{(n)}(\omega)), \quad n = 1,2,\cdots,$$

where $u_i^{(n)}(\omega) = \varphi_i(t_n, \vartheta_{-t_n}\omega, \boldsymbol{u}_o^{(n)}(\omega))$ for $i \in \mathbb{Z}$. By Assumption (i), there exists $N_1(\omega, Q) \in \mathbb{N}$ such that $t_n \geq T_Q(\omega)$ if $n \geq N_1(\omega, Q)$. Hence

$$\boldsymbol{u}^{(n)}(\omega) = \varphi(t_n, \vartheta_{-t_n}\omega, \boldsymbol{u}_o^{(n)}(\omega)) \in Q(\omega), \quad \forall n \geq N_1(\omega, Q).$$

We next show that the set $\tilde{Q} = \{u^{(n)}(\omega) = \varphi(t_n, \vartheta_{-t_n}\omega, \boldsymbol{u}_o^{(n)}(\omega))\}_{n \geq N_1(\omega,Q)}$ is pre-compact. This will be proved directly by showing there is a finite covering by open balls for each arbitrarily small radius. Given any given $\varepsilon > 0$, by Assumption (ii), there exists $T_1(\varepsilon, \omega, Q) > 0$, $I_0(\varepsilon, \omega, Q)$, and $N_2(\varepsilon, \omega, Q) \in \mathbb{N}$ such that $t_n \geq T_1(\varepsilon, \omega, Q)$ for all $n \geq N_2(\varepsilon, \omega, Q)$, and

$$\sup_{n \geq N_1(\omega,Q)} \Big(\sum_{|i| > I_0(\varepsilon,\omega,Q)} \rho_i \left| \varphi_i(t_n, \vartheta_{-t_n}\omega, \boldsymbol{u}_o^{(n)}(\omega)) \right|^p \Big)^{1/p} \leq \frac{\varepsilon}{2}.$$

Let $N_3(\varepsilon, \omega, Q) = \max\{N_1(\omega, Q), N_2(\varepsilon, \omega, Q)\}$. Then for any $n \geq N_3(\varepsilon, \omega, Q)$, $\boldsymbol{u}^{(n)}(\omega) = (u_i^{(n)}(\omega))_{i \in \mathbb{Z}}$ can be decomposed into

$$\boldsymbol{u}^{(n)}(\omega) = (v_i^{(n)}(\omega))_{i \in \mathbb{Z}} + (w_i^{(n)}(\omega))_{i \in \mathbb{Z}} = \boldsymbol{v}^{(n)}(\omega) + \boldsymbol{w}^{(n)}(\omega), \tag{11.5}$$

where

$$v_i^{(n)}(\omega) = \begin{cases} u_i^{(n)}(\omega), & |i| \leq I_0(\varepsilon, \omega, Q) \\ 0, & |i| > I_0(\varepsilon, \omega, Q) \end{cases}, \quad w_i^{(n)}(\omega) = \begin{cases} 0, & |i| \leq I_0(\varepsilon, \omega, Q) \\ u_i^{(n)}(\omega), & |i| > I_0(\varepsilon, \omega, Q) \end{cases}.$$

For $\omega \in \Omega$, let

$$r_0(\omega) = \sup_{\boldsymbol{u}(\omega) \in Q} \|\boldsymbol{u}(\omega)\|_{\rho,p}, \quad \rho^-(\omega) = \min_{|i| \leq I_0(\varepsilon,\omega,Q)} \rho_i^{1/p}, \quad \rho^+(\omega) = \max_{|i| \leq I_0(\varepsilon,\omega,Q)} \rho_i^{1/p}. \tag{11.6}$$

Then for $n \geq N_3(\varepsilon, \omega, Q)$, $\rho_i^{\frac{1}{p}} |v_i^{(n)}(\omega)| \leq r_0(\omega)$ and $|v_i^{(n)}(\omega)| \leq \frac{r_0(\omega)}{\rho^-(\omega)}$ for $|i| \leq I_0(\epsilon, \omega, Q)$, and moreover

$$\left\| v^{(n)}(\omega) \right\|_{\rho,p} = \Big(\sum_{|i| \leq I_0(\varepsilon,\omega,Q)} \rho_i |u_i^{(n)}(\omega)|^p \Big)^{\frac{1}{p}} \leq \left\| u^{(n)}(\omega) \right\|_{\rho,p} \leq r_0(\omega),$$

$$\left\| w^{(n)}(\omega) \right\|_{\rho,p} = \Big(\sum_{|i| > I_0(\varepsilon,\omega,Q)} \rho_i \left| u_i^{(n)}(\omega) \right|^p \Big)^{\frac{1}{p}} \leq \frac{\varepsilon}{2}.$$

Now define a random set

$$B(\omega) = \left\{ v = (v_i)_{|i| \leq I_0(\varepsilon,\omega,Q)} \in \mathbb{R}^{2I_0(\varepsilon,\omega,Q)+1} : v_i \in \mathbb{R}, \quad |v_i| \leq \frac{r_0(\omega)}{\rho^-(\omega)} \right\},$$

and a random number

$$n_{\varepsilon,\omega}(B(\omega)) := \left(\left\lfloor \frac{2r_0(\omega)\rho^+(\omega)\sqrt{2I_0(\varepsilon,\omega,Q)+1}}{\varepsilon\rho^-(\omega)} \right\rfloor + 1 \right)^{2I_0(\varepsilon,\omega,Q)+1},$$

where $\lfloor m \rfloor$ denotes the greatest integer which is less than or equal to the positive number m, and $r_0(\omega)$, $\rho^+(\omega)$ and $\rho^-(\omega)$ are defined by (11.6). Then

$B(\omega)$ is a $(2I_0(\varepsilon,\omega,Q)+1)$-dimensional regular polyhedron which can be covered by $n_{\varepsilon,\omega}(B(\omega))$ open balls of radius $\varepsilon/(2\sqrt{\rho^+(\omega)})$ centered at $\boldsymbol{u}_l^* = (u_{l,i}^*)_{|i|\le I_0(\varepsilon,\omega,Q)}$, $u_{l,i}^*\in\mathbb{R}$, $1\le l\le n_{\varepsilon,\omega}(B(\omega))$, in the norm of $\mathbb{R}^{2I_0(\varepsilon,\omega,Q)+1}$.

For each $1\le l\le n_{\varepsilon,\omega}(B(\omega))$, we set $\boldsymbol{v}_l=(v_{l,i})_{i\in\mathbb{Z}}\in\ell_\rho^p$ such that

$$v_{l,i}=\begin{cases} u_{l,i}^*, & |i|\le I_0(\varepsilon,\omega,Q),\\ 0, & |i|>I_0(\varepsilon,\omega,Q).\end{cases}$$

Then for $\boldsymbol{v}^{(n)}(\omega)=(v_i^{(n)}(\omega))_{i\in\mathbb{Z}}$ with $n\ge N_3(\varepsilon,\omega,Q)$ in the decomposition (11.5), there exists $l_0\in\{1,2,...,n_{\varepsilon,\omega}(B(\omega))\}$ such that

$$\begin{aligned}\left\|\boldsymbol{v}^{(n)}(\omega)-\boldsymbol{v}_{l_0}\right\|_{\rho,p} &= \Big(\sum_{|i|\le I_0(\varepsilon,\omega,Q)}\rho_i\left|u_i^{(n)}(\omega)-u_{l_0,i}\right|^p\Big)^{\frac1p}\\ &\le \rho^+(\omega)\Big(\sum_{|i|\le I_0(\varepsilon,\omega,Q)}\left|u_i^{(n)}(\omega)-u_{l_0,i}\right|^p\Big)^{\frac1p}\le\frac{\varepsilon}{2},\end{aligned}$$

and hence, we have

$$\begin{aligned}\left\|\boldsymbol{u}^{(n)}(\omega)-\boldsymbol{v}_{l_0}\right\|_{\rho,p} &= \left\|\boldsymbol{v}^{(n)}(\omega)-\boldsymbol{v}_{l0}+\boldsymbol{w}^{(n)}(\omega)\right\|_{\rho,p}\\ &\le \left\|\boldsymbol{v}^{(n)}(\omega)-\boldsymbol{v}_{l_0}\right\|_{\rho,p}+\left\|\boldsymbol{w}^{(n)}(\omega)\right\|_{\rho,p}\le\varepsilon.\end{aligned}$$

Therefore, $\{\boldsymbol{u}^{(n)}(\omega)=\varphi(t_n,\vartheta_{-t_n}\omega,\boldsymbol{u}_o^{(n)}(\omega))\}_{n\ge N_3(\varepsilon,\omega,Q)}$ can be covered by $n_{\varepsilon,\omega}(B(\omega))$ open balls in ℓ_ρ^p of radius ε centered at $\boldsymbol{v}_l=(v_{l,i})_{i\in\mathbb{Z}}$, $1\le l\le n_{\varepsilon,\omega}(B(\omega))$. By Theorem 11.3, the RDS $\{\varphi(t,\omega)\}_{t\ge0,\omega\in\Omega}$ possesses a unique global random $\mathscr{D}$ attractor $\mathscr{A}$ with component sets $A(\omega)$ given by (11.4). □

For the special case $\rho\equiv1$ and $p=2$, $\{\varphi(t,\omega)\}_{t\ge0,\omega\in\Omega}$ is a continuous RDS on the separable Hilbert space ℓ^2. In this case [Zhao and Zhou (2009)] constructed some sufficient conditions for the existence of a global random attractor. Theorem 11.4 above generalises the results in [Zhao and Zhou (2009)] to the weighted state space ℓ_ρ^p.

11.4 Stochastic differential equations as RODEs

Any finite dimensional Itô stochastic differential equation (SDE) with regular coefficients can be transformed to a RODE. This is easily illustrated for a scalar SDE with additive noise. For example, consider the SDE

$$\mathrm{d}X_t=f(X_t)\,\mathrm{d}t+\mathrm{d}W_t, \tag{11.7}$$

where W_t is a scalar Wiener process. Let $\mathcal{O}_t$ be the stochastic stationary Ornstein–Uhlenbeck process satisfying the linear SDE

$$\mathrm{d}\mathcal{O}_t=-\mathcal{O}_t\,\mathrm{d}t+\mathrm{d}W_t, \tag{11.8}$$

and set $z(t):=X_t-\mathcal{O}_t$.

Subtracting integral versions of the SDEs (11.7) and (11.8) gives

$$z(t) = z(0) + \int_0^t \left(f(z(s) + \mathcal{O}_s) + \mathcal{O}_s\right) \mathrm{d}s.$$

It then follows by continuity and the fundamental theorem of integral and differential calculus that z is pathwise differentiable and satisfies the RODE

$$\frac{\mathrm{d}z}{\mathrm{d}t} = f(z + \mathcal{O}_t) + \mathcal{O}_t.$$

Similarly, for a scalar Itô SDE with linear multiplicative noise

$$\mathrm{d}X_t = f(t, X_t)\,\mathrm{d}t + g(t)X_t \mathrm{d}W_t$$

the random transformation

$$z(t) = \mathcal{T}(t)X_t, \quad \text{with} \quad \mathcal{T}(t) := \exp\left(\frac{1}{2}\int_0^t g^2(s)\mathrm{d}s - \int_0^t g(s)\mathrm{d}W_s\right)$$

leads to the RODE

$$\frac{\mathrm{d}z}{\mathrm{d}t} = \mathcal{T}(t) f\left(t, \mathcal{T}^{-1}(t)z(t)\right),$$

or, more specifically,

$$\frac{\mathrm{d}}{\mathrm{d}t}z(t) = f\left(t, e^{-\frac{1}{2}\int_0^t g(s)^2\mathrm{d}s + \int_0^t g(s)\mathrm{d}W_s} z(t)\right) e^{\frac{1}{2}\int_0^t g(s)^2\mathrm{d}s - \int_0^t g(s)\mathrm{d}W_s}.$$

Such transformations often allow one to formulate random dynamical systems generated by SDEs via an associated RODE. In particular, solutions to finite dimensional stochastic differential equations, if they exist, generate random dynamical systems [Arnold (1974)].

They can sometimes be used for stochastic partial differential equations (SPDEs) and infinite dimensional SDEs modeling lattice systems with special type of noise due to the following lemma from [Caraballo, Kloeden and Schmalfuß (2004)].

Proposition 11.1. *Let $\{\varphi(t,\omega)\}_{t\geq 0,\omega\in\Omega}$ be an RDS. Suppose that there exists a mapping $\mathcal{T} : \Omega \times \mathfrak{E} \to \mathfrak{E}$ such that $\mathcal{T}(\omega, \cdot)$ is a homeomorphism on $\mathfrak{E}$ for every $\omega \in \Omega$, and the mappings $\mathcal{T}(\cdot, \boldsymbol{x})$ and $\mathcal{T}^{-1}(\cdot, \boldsymbol{x})$ are measurable. Then the mapping*

$$(t, \omega, \boldsymbol{x}) \mapsto \tilde{\varphi}(t, \omega, \boldsymbol{x}) := \mathcal{T}^{-1}(\vartheta_t\omega, \varphi(t, \omega, \mathcal{T}(\omega, \boldsymbol{x})))$$

is also a random dynamical system.

The RDS $\{\tilde{\varphi}(t,\omega)\}_{t\geq 0,\omega\in\Omega}$ is called a *conjugated* RDS for $\{\varphi(t,\omega)\}_{t\geq 0,\omega\in\Omega}$.

11.5 End notes

Random ordinary differential equations (RODEs) are investigated in detail in [Han and Kloeden (2017a)]. See also [Bunke (1972)] and [Neckel and Rupp (2014)]. RODEs are used extensively in the context of random dynamical systems in [Arnold (1974)]. For SDEs see [Evans (2013); Kloeden and Platen (1992)].

The transformation of SDEs to RODEs was first established independently by [Doss (1977)] and [Sussmann (1977)] in the case of commutative noise and later generalised to all SDEs by [Imkeller and Schmalfuß (2001)].

See [Crauel and Kloeden (2015)] for an overview article on both non-autonomous and random attractors. See [Cui and Kloeden (2018)] for forward attractors of non-autonomous random dynamical systems. Existence of random attractors and invariant Markov measures supported by the random attractor for random dynamical systems were investigated in [Crauel and Flandoli (1994)].

This chapter uses material from the papers Bates, Lisei and Lu [Bates *et al.* (2006)] and Han, Shen and Zhou [Han *et al.* (2011)].

11.6 Problems

Problem 11.1. Formulate the RODE (11.2) with a stationary Ornstein-Uhlenbeck process as its driving noise as a random dynamical system. *(Hint: The stationary Ornstein-Uhlenbeck process satisfies an Itô SDE.)*

Problem 11.2. Let $\vartheta : \Omega \to \Omega$ be a measure preserving transformation, and let $\mathcal{S} : \mathfrak{X} \times \Omega \to \mathfrak{X}$ be a measurable map. Set $\mathcal{S}_n := \mathcal{S} \circ \vartheta^{n-1}$, and define

$$\varphi(n, \omega) = \begin{cases} \mathcal{S}_n(\omega) \circ \mathcal{S}_{n-1}(\omega) \circ \cdots \circ \mathcal{S}_1(\omega), & n > 0 \\ \mathrm{Id}, & n = 0 \\ \mathcal{S}_{n+1}^{-1}(\omega) \circ \mathcal{S}_{n+2}^{-1}(\omega) \circ \cdots \circ \mathcal{S}_0^{-1}, & n < 0. \end{cases}$$

Show that $\varphi(n, \omega)$ defines a random dynamical system.

Chapter 12

Stochastic LDS with additive noise

The study on existence of global random attractors for first order stochastic LDS with additive white noise was initiated by Bates, Lisei & Lu [Bates *et al.* (2006)].

In this chapter, a stochastic lattice system with additive white noise will be investigated using weighted sequence spaces. In particular, we consider the following first order stochastic lattice differential equations

$$\mathrm{d}u_i = \big(-\lambda u_i - f_i(u_i) + g_i + (\mathrm{A}\boldsymbol{u})_i \big)\,\mathrm{d}t + \mathfrak{a}_i\,\mathrm{d}W_i(t), \tag{12.1}$$

where λ is a positive constant, u_i, $g_i \in \mathbb{R}$, $\mathfrak{a}_i \in \mathbb{R}_+$, $f_i \in \mathcal{C}^1(\mathbb{R})$ satisfies appropriate dissipative conditions for $i \in \mathbb{Z}$, and $\{W_i(t)\}_{i\in\mathbb{Z}}$ are mutually independent two-sided Wiener processes. Same as in Chapter 8, here A is a non-negative and self-adjoint linear operator on a sequence space with the decomposition

$$\mathrm{A} = \mathrm{D}_p\bar{\mathrm{D}}_p = \bar{\mathrm{D}}_p\mathrm{D}_p, \quad \text{with} \quad (\mathrm{D}_p\boldsymbol{u})_i = \sum_{j=-p}^{p} d_{i+j}u_{i+j}, \quad \boldsymbol{u} = (u_i)_{i\in\mathbb{Z}}, \tag{12.2}$$

and $\bar{\mathrm{D}}_p$ is the adjoint of D_p. Similar to Section 10.2.2, here it is assumed that

$$|d_i| \le M_d, \quad \mathrm{i} \in \mathbb{Z}.$$

The aim here is to show that the stochastic LDS (12.1) generates a random dynamical system that possesses a global random attractor on the weighted space ℓ^2_ρ defined as in Section 1.3. The weights ρ_i are assumed to satisfy the following assumption.

Assumption 12.1. There exist positive constants M_ρ and γ such that

$$0 < \rho_i \le M_\rho, \qquad \rho_i \le \gamma\rho_{i\pm1} \quad \text{for all } i \in \mathbb{Z}.$$

12.1 Random dynamical systems generated by stochastic LDS

Here we show that stochastic lattice system (12.1) generates a random dynamical system in the weighted space ℓ^2_ρ. To that end, the stochastic lattice system (12.1) is first transformed to a random lattice system and then the existence and uniqueness of solutions of the resulting random lattice system in ℓ^2_ρ is established.

Let Ω be defined by

$$\Omega = \{\omega \in C(\mathbb{R}, \ell^2) : \omega(0) = 0\}, \tag{12.3}$$

let $\mathcal{F}$ be the Borel σ-algebra on Ω generated by the compact open topology, and let $\mathbb{P}$ be the corresponding Wiener measure on $\mathcal{F}$. Given $\mathfrak{a} = (\mathfrak{a}_i)_{i\in\mathbb{Z}}$, define

$$\boldsymbol{W}(t) = \boldsymbol{W}(t,\omega) = \sum_{i\in\mathbb{Z}} \mathfrak{a}_i W_i(t) \boldsymbol{e}^i, \tag{12.4}$$

where $\boldsymbol{e}^i$ represents the bi-infinite sequence with the element having 1 at position i and 0 for all other components.

Lemma 12.1. *Assume that* $(\mathfrak{a}_i)_{i\in\mathbb{Z}} \in \ell^2$. *Then* $\boldsymbol{W}(t)$ *defined by* (12.4) *is a two-sided Brownian motion or Wiener process on* $(\Omega, \mathcal{F}, \mathbb{P})$ *with values in* ℓ^2. *Moreover,* $\boldsymbol{W}(t,\omega)$ *is adapted to* $\mathcal{F}_s^t$.

Proof. Let $\mathfrak{Q} \in \mathfrak{L}(\ell^2)$ be the diagonal operator with diagonal elements $\mathfrak{a}_i^2$. Clearly $\mathfrak{Q}$ is a trace class symmetric nonnegative operator on ℓ^2, and $\mathfrak{a}_i^2$ are eigenvalues for $\mathfrak{Q}$. Noticing that $\{W_i\}_{i\in\mathbb{Z}}$ is a cylindrical Brownian motion (see, e.g., [Da Prato and Zabczyk (2014)]), and $\{\boldsymbol{e}^i\}_{i\in\mathbb{Z}}$ is an orthonormal basis of ℓ^2, it then follows by Proposition 2.1.10 in [Prévôt and Röckner (2007)] that $\boldsymbol{W}(t) = \sum_{i\in\mathbb{Z}} \mathfrak{a}_i W_i(t)\boldsymbol{e}^i$ is a $\mathfrak{Q}$-Brownian motion in ℓ^2.

Observe that $(\Omega, \mathcal{F}, \mathbb{P}, (\vartheta_t)_{t\in\mathbb{R}})$ is a metric dynamical system. Let

$$\mathcal{F}_s^t = \sigma\{\boldsymbol{W}(t_2) - \boldsymbol{W}(t_1) : s \le t_1 \le t_2 \le t\}$$

be a filtration on Ω, where $\sigma\{\boldsymbol{W}(t_2) - \boldsymbol{W}(t_1) : s \le t_1 \le t_2 \le t\}$ is the smallest sigma algebra generated by the random variable $\boldsymbol{W}(t_2) - \boldsymbol{W}(t_1)$ for all t_1, t_2 such that $s \le t_1 \le t_2 \le t$. Note that here

$$\vartheta_\tau^{-1}\mathcal{F}_s^t = \mathcal{F}_{s+\tau}^{t+\tau}.$$

Thus $\boldsymbol{W}(t,\omega)$ in (12.4) is adapted to $\mathcal{F}_s^t$. □

Throughout the rest of this chapter, we consider the completion of the probability space, still denoted by $(\Omega, \mathcal{F}, \mathbb{P})$.

12.1.1 *Ornstein-Uhlenbeck process*

The stochastic lattice equation (12.5) can be transformed to a random ordinary differential equation without the white noise, by utilizing an Ornstein-Uhlenbeck process, defined as follows.

Definition 12.1. A real-valued Ornstein-Uhlenbeck process $\mathcal{O}(t)$ is defined as the solution to the linear stochastic differential equation

$$\mathrm{d}\mathcal{O}(t) + \lambda\mathcal{O}(t) = \sigma \mathrm{d}W(t),$$

where $\lambda > 0$, $\sigma > 0$ and $W(t)$ is the real-valued two-sided Brownian motion.

The following properties of the Ornstein-Uhlenbeck process defined above are taken from [Arnold (1998); Caraballo and Lu (2008); Caraballo, Kloeden and Schmalfuß (2004); Duan *et al.* (2003)].

Lemma 12.2. *Let* $\Omega_0 = \{\omega \in \mathcal{C}(\mathbb{R}, \mathbb{R}) : \omega(0) = 0\} := \mathcal{C}_0(\mathbb{R}, \mathbb{R})$. *There exists a* ϑ_t*-invariant set* $\widetilde{\Omega}_0 \in \mathcal{F}$ *of* Ω_0 *of full* $\mathbb{P}$ *measure such that*

(i)

$$\lim_{t\to\pm\infty} \frac{|\omega(t)|}{t} = 0, \quad \textit{for} \quad \omega \in \widetilde{\Omega}_0,$$

and for such ω, *the random variable given by*

$$\mathcal{O}(\omega) := -\lambda \int_{-\infty}^{0} e^{\lambda s}\omega(s)\mathrm{d}s$$

is well-defined;

(ii) the random variable $|\mathcal{O}(\omega)|$ *is tempered, i.e., for* $\omega \in \widetilde{\Omega}_0$ *and all* $\epsilon > 0$,

$$\lim_{t\to+\infty} e^{-\epsilon t} \sup_{t\in\mathbb{R}} |\mathcal{O}(\vartheta_{-t}\omega)| = 0;$$

(iii) the mapping

$$(t, \omega) \mapsto \mathcal{O}(\vartheta_t\omega) = -\lambda \int_{-\infty}^{0} e^{\lambda s}\omega(t+s)\mathrm{d}s + \sigma\omega(t)$$

is a stationary solution of Ornstein-Uhlenbeck equation (13.4) *with continuous sample paths, where* $W(t)\omega = W(t, \omega) = \omega(t)$ *for* $\omega \in \widetilde{\Omega}_0$ *and* $t \in \mathbb{R}$;

(iv)

$$\lim_{t\to\pm\infty} \frac{|\mathcal{O}(\vartheta_t\omega)|}{t} = 0, \quad \lim_{t\to\pm\infty} \frac{1}{t}\int_0^t \mathcal{O}(\vartheta_s\omega)\,\mathrm{d}s = 0.$$

12.1.2 *Transformation to a random ordinary differential equation*

Set $F(\boldsymbol{u}) = (f_i(u_i))_{i\in\mathbb{Z}}$ and $\boldsymbol{g} = (g_i)_{i\in\mathbb{Z}}$. Then the system (12.1) can be written as

$$\mathrm{d}\boldsymbol{u}(t) = (-\lambda\boldsymbol{u} - F(\boldsymbol{u}) + \boldsymbol{g} + \mathrm{A}\boldsymbol{u})\,\mathrm{d}t + \mathrm{d}\boldsymbol{W}(t), \quad \boldsymbol{u} = (u_i)_{i\in\mathbb{Z}}. \tag{12.5}$$

To transform the stochastic ordinary differential equation (12.5) into a random ordinary differential equation, let $\mathcal{O}_i$ be the real-valued Ornstein-Uhlenbeck process satisfying the linear stochastic differential equation

$$\mathrm{d}\mathcal{O}_i(t) + \lambda\mathcal{O}_i(t) = \mathfrak{a}_i\mathrm{d}W_i(t) \quad i \in \mathbb{Z}.$$

Set $\boldsymbol{\mathcal{O}}(t) := (\mathcal{O}_i(t))_{i\in\mathbb{Z}}$, then due to Lemma 12.1, the process $\boldsymbol{\mathcal{O}}(\vartheta_t\omega)$ solves the following equation on ℓ^2 provided $(\mathfrak{a}_i)_{i\in\mathbb{Z}} \in \ell^2$:

$$\mathrm{d}\boldsymbol{\mathcal{O}}(\vartheta_t\omega) + \lambda\boldsymbol{\mathcal{O}}(\vartheta_t\omega)\mathrm{d}t = \mathrm{d}\boldsymbol{W}(t), \quad t \geq 0.$$

Notice that the process $\boldsymbol{\mathcal{O}}$ with values in ℓ^2 is infinite dimensional, and is different from the scalar Ornstein-Uhlenbeck process $\mathcal{O}$ define in Section 12.1.1. In fact, each component of $\boldsymbol{\mathcal{O}}$ is a scalar Ornstein-Uhlenbeck process that satisfies all properties in Lemma 12.2. Define $\|\boldsymbol{\mathcal{O}}\|^2 = \sum_{i\in\mathbb{Z}} \mathcal{O}_i^2$. Then the following corollary on properties of $\boldsymbol{\mathcal{O}}(\vartheta_t\omega)$ follow directly from Lemma 12.2.

Corollary 12.1. *Let Ω be define as in (12.3). There exists a ϑ_t-invariant set $\widetilde{\Omega} \in \mathcal{F}$ of Ω of full $\mathbb{P}$ measure such that*

(i)

$$\lim_{t\to\pm\infty} \frac{\|\omega(t)\|}{t} = 0, \quad \text{for} \quad \omega \in \widetilde{\Omega},$$

and for such ω, the random variable given by

$$\boldsymbol{\mathcal{O}}(\omega) := -\lambda \int_{-\infty}^{0} e^{\lambda s}\omega(s)\mathrm{d}s$$

is well-defined;

(ii) the random variable $\|\boldsymbol{\mathcal{O}}(\omega)\|$ is tempered, i.e., for $\omega \in \widetilde{\Omega}$ and all $\epsilon > 0$,

$$\lim_{t\to+\infty} e^{-\epsilon t} \sup_{t\in\mathbb{R}} \|\boldsymbol{\mathcal{O}}(\vartheta_{-t}\omega)\| = 0;$$

(iii) the mapping

$$(t,\omega) \mapsto \boldsymbol{\mathcal{O}}(\vartheta_t\omega) = -\lambda \int_{-\infty}^{0} e^{\lambda s}\omega(t+s)\mathrm{d}s$$

is a stationary solution of Ornstein-Uhlenbeck equation (13.4) with continuous sample paths, where $\boldsymbol{W}(t)\omega = \boldsymbol{W}(t,\omega) = \omega(t)$ for $\omega \in \widetilde{\Omega}$ and $t \in \mathbb{R}$;

(iv)

$$\lim_{t\to\pm\infty} \frac{\|\boldsymbol{\mathcal{O}}(\vartheta_t\omega)\|}{t} = 0, \quad \lim_{t\to\pm\infty} \frac{1}{t}\int_0^t \boldsymbol{\mathcal{O}}(\vartheta_s\omega)\,\mathrm{d}s = \mathbf{0}.$$

In the end, let

$$\boldsymbol{v}(t,\omega) = \boldsymbol{u}(t,\omega) - \boldsymbol{\mathcal{O}}(\vartheta_t\omega), \quad \omega \in \Omega, \quad t \in \mathbb{R}.$$

The stochastic differential equation (12.5) then becomes the following random differential equation with random coefficients, but without white noise:

$$\frac{\mathrm{d}\boldsymbol{v}(t)}{\mathrm{d}t} = -\lambda\boldsymbol{v} + \mathrm{A}\boldsymbol{v} - F(\boldsymbol{v} + \boldsymbol{\mathcal{O}}(\vartheta_t\omega)) + \mathrm{A}\boldsymbol{\mathcal{O}}(\vartheta_t\omega) + \boldsymbol{g}. \tag{12.6}$$

12.1.3 *Existence and uniqueness of solutions*

In this subsection we establish the existence and uniqueness of solutions to the RODE (12.6). In addition to the Assumption 12.1, the following assumptions are made on the nonlinear terms f_i and g_i.

Assumption 12.2. There exists a function $\Gamma_f(r) \in \mathcal{C}(\mathbb{R}^+, \mathbb{R}^+)$ such that

$$\sup_{i\in\mathbb{Z}} \max_{s\in[-r,r]} |f_i'(s)| \le \Gamma_f(r), \ \forall r \in \mathbb{R}^+;$$

Assumption 12.3. $\boldsymbol{g} = (g_i)_{i\in\mathbb{Z}} \in \ell^2_\rho$.

Assumption 12.4. $f_i \in \mathcal{C}^1(\mathbb{R}, \mathbb{R})$ satisfies

$$sf_i(s) \ge \alpha s^{2(q+1)} - \beta_i^2, \quad |f_i(s)| \le M_f|s|(|s|^{2q} + 1)$$

for all $s \in \mathbb{R}$, $i \in \mathbb{Z}$, where α, β_i, M_f are positive constants, $q \in \mathbb{N}$, and $\boldsymbol{\beta} := (\beta_i)_{i\in\mathbb{Z}} \in \ell^2$.

We first show that for any $\boldsymbol{v}_o \in \ell^2$ and $\boldsymbol{g} \in \ell^2$, the problem (12.6) is well-posed in ℓ^2, presented in the following lemma.

Lemma 12.3. *Let Assumptions 12.1–12.4 hold. Then for* $T > 0$, $\omega \in \Omega$ *and any initial data* $\boldsymbol{v}_o \in \ell^2$, *the problem* (12.6) *has a unique solution* $\boldsymbol{v}(t; \boldsymbol{v}_o, \omega, \boldsymbol{g})$ *with* $\boldsymbol{v}(0; \boldsymbol{v}_o, \omega, \boldsymbol{g}) = \boldsymbol{v}_o$.

Proof. For $\boldsymbol{v} \in \ell^2$, $t \in \mathbb{R}$, $\omega \in \Omega$ and $\boldsymbol{g} \in \ell^2$, let

$$\mathfrak{F}(\boldsymbol{v}, \omega) = -\lambda \boldsymbol{v} + \mathrm{A}\boldsymbol{v} - F(\boldsymbol{v} + \boldsymbol{\mathcal{O}}(\omega)) + \mathrm{A}\boldsymbol{\mathcal{O}}(\omega) + \boldsymbol{g}. \tag{12.7}$$

Then by Assumption 12.4 and Corollary 12.1, $\mathfrak{F}$ is measurable in both variables, locally bounded and locally Lipschitz in $\boldsymbol{v}$. Moreover, the mapping $(\boldsymbol{v}, t) \mapsto \mathfrak{F}(\boldsymbol{v}, \vartheta_t\omega)$ is continuous. By [Arnold (1998); Chueshov (2002); Pazy (2007)], given any $\boldsymbol{v}_o$, $\boldsymbol{g} \in \ell^2$, the problem (12.6) possesses a unique local solution $\boldsymbol{v}(\cdot; \omega, \boldsymbol{v}_o, \boldsymbol{g}) \in \mathcal{C}([0, T_{\max}), \ell^2)$ satisfying the integral equation

$$\boldsymbol{v}(t) = \boldsymbol{v}_o + \int_0^t \big((-\lambda \boldsymbol{v}(s) + \mathrm{A}\boldsymbol{v}(s) - F(\boldsymbol{v}(s) + \boldsymbol{\mathcal{O}}(\vartheta_s\omega)) + \mathrm{A}\boldsymbol{\mathcal{O}}(\vartheta_s\omega) + \boldsymbol{g}\big)\mathrm{d}s \tag{12.8}$$

for $t \in [0, T_{\max})$, where $0 < T_{\max} \le T$.

We next show that $T_{max} = T$. In fact, it follows from (12.8) that

$$\|\boldsymbol{v}(t)\|^2 = \|\boldsymbol{v}_o\|^2 + 2\int_0^t \Big(- \lambda\|\boldsymbol{v}(s)\|^2 + \langle \mathrm{A}\boldsymbol{v}(s) + \mathrm{A}\boldsymbol{\mathcal{O}}(\vartheta_s\omega), \boldsymbol{v}(s)\rangle + \langle \boldsymbol{g}, \boldsymbol{v}(s)\rangle - \sum_{i\in\mathbb{Z}} v_i \cdot f_i\big(v_i + \mathcal{O}_i(\vartheta_s\omega)\big)\Big)\mathrm{d}s. \tag{12.9}$$

Observe that

$$\langle \mathrm{A}\boldsymbol{v}, \boldsymbol{v}\rangle \le (2p+1)M_d\|\boldsymbol{v}\|^2,$$

$$\langle \mathrm{A}\boldsymbol{\mathcal{O}}(\vartheta_s\omega), \boldsymbol{v}\rangle \le \frac{2}{\lambda}(2p+1)^2M_d^2\|\boldsymbol{\mathcal{O}}(\vartheta_s\omega)\|^2 + \frac{\lambda}{8}\|\boldsymbol{v}\|^2,$$

$$\langle \boldsymbol{g}, \boldsymbol{v}\rangle \le \frac{1}{\lambda}\|\boldsymbol{g}\|^2 + \frac{\lambda}{4}\|\boldsymbol{v}\|^2.$$

In addition, by Young's inequality, the last term in (12.9) satisfies

$$\begin{aligned}
&-\sum_{i\in\mathbb{Z}} v_i \cdot f_i(v_i + \mathcal{O}_i(\vartheta_s\omega))\\
&= -\sum_{i\in\mathbb{Z}} (v_i + \mathcal{O}_i(\vartheta_s\omega)) \cdot f_i(v_i + \mathcal{O}_i(\vartheta_s\omega)) + \sum_{i\in\mathbb{Z}} \mathcal{O}_i(\vartheta_s\omega) \cdot f_i(v_i + \mathcal{O}_i(\vartheta_s\omega))\\
&\le -\alpha \sum_{i\in\mathbb{Z}} (v_i + \mathcal{O}_i(\vartheta_s\omega))^{2q+2} + \sum_{i\in\mathbb{Z}} \beta_i^2 + M_f(1 + \frac{2}{\lambda}M_f)\|\boldsymbol{\mathcal{O}}(\vartheta_s\omega)\|^2 + \frac{\lambda}{8}\|\boldsymbol{v}\|^2\\
&\quad + \frac{1}{(2q+2)\left(\frac{2q+2}{2q+1}\alpha\right)^{2q+1}} \sum_{i\in\mathbb{Z}} (M_f|\mathcal{O}_i(\vartheta_s\omega)|)^{2q+2} + \alpha \sum_{i\in\mathbb{Z}} |v_i + \mathcal{O}_i(\vartheta_s\omega)|^{2q+2}\\
&\le C_1 \cdot (\|\boldsymbol{\mathcal{O}}(\vartheta_s\omega)\|^2 + \|\boldsymbol{\mathcal{O}}(\vartheta_s\omega)\|^{2q+2} + 1) + \frac{\lambda}{8}\|\boldsymbol{v}\|^2,
\end{aligned}$$

where C_1 is a positive constant depending on λ, α, M_f, $\|\boldsymbol{\beta}\|$, and q.

Inserting the above inequalities into (12.9), we obtain

$$\begin{aligned}
\|\boldsymbol{v}(t)\|^2 \le \|\boldsymbol{v}_o\|^2 + \int_0^t \Big((-\lambda + 2(2p+1)M_d)\,\|\boldsymbol{v}(s)\|^2 + C_2\Big((M_d^2+1)\|\boldsymbol{\mathcal{O}}(\vartheta_s\omega)\|^2\\
+ \|\boldsymbol{\mathcal{O}}(\vartheta_s\omega)\|^{2q+2} + \|\boldsymbol{g}\|^2 + 1\Big)\mathrm{d}s, \qquad (12.10)
\end{aligned}$$

where C_2 is a positive constant depending on λ, α, L_f, $\|\boldsymbol{\beta}\|$, p and q. Applying Gronwall's inequality to (12.10) then results in

$$\begin{aligned}
\|\boldsymbol{v}(t)\|^2 \le C_2 e^{(-\lambda+2(2p+1)M_d)t}\Bigg(\|\boldsymbol{v}_o\|^2\\
+ \int_0^t \Big((M_d^2+1)\|\boldsymbol{\mathcal{O}}(\vartheta_s\omega)\|^2 + \|\boldsymbol{\mathcal{O}}(\vartheta_s\omega)\|^{2q+2} + \|\boldsymbol{g}\|^2 + 1\Big)\mathrm{d}s\Bigg).
\end{aligned}$$

It then follows directly from Corollary 12.1 that the existence interval of solution $\boldsymbol{v}(\cdot;\omega,\boldsymbol{v}_o,\boldsymbol{g})$ is $[0,T)$, i.e., $T_{max} = T$. □

Notice that due to Lemma 12.3 above, given any $\omega \in \Omega$ and $\boldsymbol{g} \in \ell^2$, the solution of (12.6) with initial data $\boldsymbol{v}_o \in \ell^2$ satisfies $\boldsymbol{v}(\cdot;\omega,\boldsymbol{v}_o,\boldsymbol{g}) \in \ell^2 \subset \ell^2_\rho$ for all $t \ge 0$. We next show the Lipschitz continuity of solutions to (12.6) in bounded set of ℓ^2_ρ on $[0,T]$ for all $T > 0$.

Lemma 12.4. *Let Assumptions 12.1–12.4 hold, and let $\boldsymbol{v}(t;\omega,\boldsymbol{x}_o,\boldsymbol{g}),\boldsymbol{v}(t;\omega,\boldsymbol{y}_o,\boldsymbol{h})$ be two solutions of* (12.6) *with initial data $\boldsymbol{x}_o\in\ell^2$, $\boldsymbol{y}_o\in\ell^2$ and $\boldsymbol{g}$ being replaced by $\boldsymbol{g}\in\ell^2$, $\boldsymbol{h}\in\ell^2$, respectively. Then there exists $C>0$ such that*

$$\|\boldsymbol{v}(t;\omega,\boldsymbol{x}_o,\boldsymbol{g})-\boldsymbol{v}(t;\omega,\boldsymbol{y}_o,\boldsymbol{h})\|_\rho\le C\left(\|\boldsymbol{x}_o-\boldsymbol{y}_o\|_\rho+\|\boldsymbol{g}-\boldsymbol{h}\|_\rho\right).$$

Proof. For simplicity of notations, set $\boldsymbol{x}(t,\omega)=\boldsymbol{v}(t;\omega,\boldsymbol{x}_o,\boldsymbol{g})$, $\boldsymbol{y}(t,\omega)=\boldsymbol{v}(t;\omega,\boldsymbol{y}_o,\boldsymbol{h})$, and let $\boldsymbol{z}(t,\omega)=\boldsymbol{x}(t,\omega)-\boldsymbol{y}(t,\omega)$. Then $\boldsymbol{z}(t,\omega)\in\ell^2\subset\ell^2_\rho$ for $\omega\in\Omega$ satisfies the random ordinary differential equation (RODE)

$$\frac{\mathrm{d}\boldsymbol{z}}{\mathrm{d}t}=-\lambda\boldsymbol{z}+\mathrm{A}\boldsymbol{z}-F(\boldsymbol{x}+\boldsymbol{\mathcal{O}}(\vartheta_t\omega))+F(\boldsymbol{y}+\boldsymbol{\mathcal{O}}(\vartheta_t\omega))+\boldsymbol{g}-\boldsymbol{h}.$$

Taking inner product $\langle\cdot,\cdot\rangle_\rho$ of the above RODE with $\boldsymbol{z}$ on ℓ^2_ρ gives

$$\begin{aligned}\frac{1}{2}\frac{\mathrm{d}\|\boldsymbol{z}\|_\rho^2}{\mathrm{d}t}&=-\lambda\|\boldsymbol{z}\|_\rho^2+\langle\mathrm{A}\boldsymbol{z},\boldsymbol{z}\rangle_\rho+\langle\boldsymbol{g}-\boldsymbol{h},\boldsymbol{z}\rangle_\rho\\&\quad+\langle-F(\boldsymbol{x}+\boldsymbol{\mathcal{O}}(\vartheta_t\omega))+F(\boldsymbol{y}+\boldsymbol{\mathcal{O}}(\vartheta_t\omega)),\boldsymbol{z}\rangle_\rho,\end{aligned}$$

in which

$$\begin{aligned}|\langle\mathrm{A}\boldsymbol{z},\boldsymbol{z}\rangle_\rho|&\le C_3\|\boldsymbol{z}\|_\rho^2,\\|\langle-F(\boldsymbol{x}+\boldsymbol{\mathcal{O}}(\vartheta_t\omega))+F(\boldsymbol{y}+\boldsymbol{\mathcal{O}}(\vartheta_t\omega)),\boldsymbol{z}\rangle_\rho|&\le C_4(T,\boldsymbol{x}_o,\boldsymbol{y}_o,\|\boldsymbol{\mathcal{O}}(\vartheta_t\omega)\|)\\\langle\boldsymbol{g}-\boldsymbol{h},\boldsymbol{z}\rangle_\rho&\le\|\boldsymbol{g}-\boldsymbol{h}\|_\rho^2\|\boldsymbol{z}\|_\rho^2.\end{aligned}$$

Therefore

$$\frac{\mathrm{d}\|\boldsymbol{z}\|_\rho^2}{\mathrm{d}t}\le(-\lambda+2\tilde{\gamma}+2C_5)\|\boldsymbol{z}\|_\rho^2+\frac{1}{\lambda}\|\boldsymbol{g}-\boldsymbol{h}\|_\rho^2,\tag{12.11}$$

where

$$\tilde{\gamma}:=M_d(p+1+\gamma+\gamma^2+\cdots+\gamma^p)\tag{12.12}$$

for some constant C_5 depending on T, $\boldsymbol{x}_o$, $\boldsymbol{y}_o$, and $\|\boldsymbol{\mathcal{O}}(\vartheta_t\omega)\|$ for $t\in[0,T]$.

Applying Gronwall's inequality to (12.11) on $[0,T)$ implies the desired assertion for some constant C depending on T, M_d, p, γ_1, T, $\boldsymbol{x}_o$, $\boldsymbol{y}_o$, and $\sup_{0\le t\le T}\|\boldsymbol{\mathcal{O}}(\vartheta_t\omega)\|$. □

With Lemmas 12.3 and 12.4, we are now ready to extend the existence result on ℓ^2 for $\boldsymbol{v}_o\in\ell^2$ and $\boldsymbol{g}\in\ell^2$ to the existence on ℓ^2_ρ for $\boldsymbol{v}_o\in\ell^2_\rho$ and $\boldsymbol{g}\in\ell^2_\rho$.

Theorem 12.1. *Let $T>0$ and Assumptions 12.1–12.4 hold. Then for every $\omega\in\Omega$ and any initial data $\boldsymbol{v}_o\in\ell^2_\rho$, the problem* (12.6) *admits a unique solution $\boldsymbol{v}(\cdot;\omega,\boldsymbol{v}_o,\boldsymbol{g})\in\mathcal{C}([0,T),\ell^2_\rho)$ with $\boldsymbol{v}(0;\omega,\boldsymbol{v}_o,\boldsymbol{g})=\boldsymbol{v}_o$ and $\boldsymbol{v}(t;\omega,\boldsymbol{v}_o,\boldsymbol{g})$ continuous in $\boldsymbol{v}_o$, $\boldsymbol{g}\in\ell^2_\rho$.*

Proof. Let $\tilde{\ell}^2$ be the space ℓ^2 equipped with the weighted norm $\|\cdot\|_\rho$. Thanks to Lemma 12.4, we see that for given $T > 0$, there exists a mapping $\mathcal{S} \in \mathcal{C}(\tilde{\ell}^2 \times \tilde{\ell}^2, \mathcal{C}([0,T), \ell_\rho^2))$ such that $\mathcal{S}(\boldsymbol{v}_o, \boldsymbol{g}) = v(t;\omega, \boldsymbol{v}_o, \boldsymbol{g})$, where $\boldsymbol{v}(t;\omega,\boldsymbol{v}_o,\boldsymbol{g})$ is the solution of (12.6) on $[0,T)$ with $\boldsymbol{v}(0;\omega,\boldsymbol{v}_o,\boldsymbol{g}) = \boldsymbol{v}_o$. Since $\tilde{\ell}^2 \times \tilde{\ell}^2$ is dense in $\ell_\rho^2 \times \ell_\rho^2$, the mapping $\mathcal{S}$ can be extended uniquely to a continuous mapping $\widetilde{\mathcal{S}} : \ell_\rho^2 \times \ell_\rho^2 \to \mathcal{C}([0,T], \ell_\rho^2)$.

For given $\boldsymbol{v}_o \in \ell_\rho^2$ and $\boldsymbol{g} \in \ell_\rho^2$, we have $\boldsymbol{v}(\cdot;\omega,\boldsymbol{v}_o,\boldsymbol{g}) := \widetilde{\mathcal{S}}(\boldsymbol{v}_o,\boldsymbol{g}) \in \mathcal{C}([0,T),\ell_\rho^2)$ for $T > 0$. It remains to show that $\boldsymbol{v}(t;\omega,\boldsymbol{v}_o,\boldsymbol{g})$ satisfies the integral equation (12.8) for $t \in [0,T)$. In fact, since $\tilde{\ell}^2 \times \tilde{\ell}^2$ is dense in $\ell_\rho^2 \times \ell_\rho^2$, there exist two sequences $\{\boldsymbol{v}_o^{(n)}\} \subset \tilde{\ell}^2$, $\{\boldsymbol{g}^{(n)}\} \subset \tilde{\ell}^2$ such that

$$\|\boldsymbol{v}_o^{(n)} - \boldsymbol{v}_o\|_\rho \to 0 \quad \text{and} \quad \|\boldsymbol{g}^{(n)} - \boldsymbol{g}\|_\rho \to 0 \quad \text{as} \quad n \to \infty.$$

Let $\boldsymbol{v}^{(n)} = (v_i^{(n)})_{i\in\mathbb{Z}} = \boldsymbol{v}(t;\omega,\boldsymbol{v}_o^{(n)},\boldsymbol{g}^{(n)}) \in \tilde{\ell}^2$ be the solution to the equation (12.6) with initial condition $\boldsymbol{v}(0) = \boldsymbol{v}_o^{(n)}$ and $\boldsymbol{g}$ replaced by $\boldsymbol{g}^{(n)}$. It then satisfies the integral equation

$$\boldsymbol{v}^{(n)}(t) = \boldsymbol{v}_o^{(n)} + \int_0^t \big((-\lambda \boldsymbol{v}^{(n)}(s) + \mathrm{A}\boldsymbol{v}^{(n)}(s) - F(\boldsymbol{v}^{(n)}(s) + \boldsymbol{\mathcal{O}}(\vartheta_s\omega)) \\ + \mathrm{A}\boldsymbol{\mathcal{O}}(\vartheta_s\omega) + \boldsymbol{g}^{(n)}\big)\mathrm{d}s. \tag{12.13}$$

Multiplying the ith component of (12.13) by $\rho_i v_i^{(n)}(t)$ and summing over $i \in \mathbb{Z}$ gives

$$\|\boldsymbol{v}^{(n)}(t)\|_\rho^2 = \|\boldsymbol{v}_o^{(n)}\|_\rho^2 + 2\int_0^t \Big(-\lambda\|\boldsymbol{v}^{(n)}(s)\|_\rho^2 + \sum_{i\in\mathbb{Z}} \rho_i v_i^{(n)}(\mathrm{A}\boldsymbol{v}^{(n)})_i + \sum_{i\in\mathbb{Z}} \rho_i v_i^{(n)}(\mathrm{A}\boldsymbol{\mathcal{O}})_i \\ + \sum_{i\in\mathbb{Z}} \rho_i g_i^{(n)} v_i^{(n)} - \sum_{i\in\mathbb{Z}} \rho_i v_i^{(n)} f_i(v_i^{(n)} + \mathcal{O}_i(\vartheta_s\omega))\Big)\mathrm{d}s,$$

in which

$$\Big|\sum_{i\in\mathbb{Z}} \rho_i v_i^{(n)}(\mathrm{A}\boldsymbol{v}^{(n)})_i\Big| \le M_d \sum_{i\in\mathbb{Z}} \sqrt{\rho_i}|v_i^{(n)}| \sum_{j=i-p}^{i+p} \sqrt{\rho_i}|v_j^{(n)}| \le \tilde{\gamma}\|\boldsymbol{v}^{(n)}\|_\rho^2,$$

where $\tilde{\gamma}$ is the same as in (12.12),

$$\Big|\sum_{i\in\mathbb{Z}} \rho_i v_i^{(n)}(\mathrm{A}\boldsymbol{\mathcal{O}}(\vartheta_s\omega))_i\Big| \le \frac{M_d}{2}(2p+1)\left(\|\boldsymbol{v}^{(n)}\|_\rho^2 + M_\rho\|\boldsymbol{\mathcal{O}}(\vartheta_s\omega)\|^2\right),$$

$$\Big|\sum_{i\in\mathbb{Z}} \rho_i g_i^{(n)} v_i^{(n)}\Big| \le \frac{1}{\lambda}\|\boldsymbol{g}^{(n)}\|_\rho^2 + \frac{\lambda}{4}\|\boldsymbol{v}^{(n)}\|_\rho^2,$$

and again by Young's inequality

$$
\begin{aligned}
&-\sum_{i\in\mathbb{Z}} \rho_i v_i^{(n)} \cdot f_i(v_i^{(n)} + \mathcal{O}_i(\vartheta_s\omega)) \\
&= -\sum_{i\in\mathbb{Z}} \rho_i \big(v_i^{(n)} + \mathcal{O}_i(\vartheta_s\omega)\big) \cdot f_i(v_i^{(n)} + \mathcal{O}_i(\vartheta_s\omega)) + \sum_{i\in\mathbb{Z}} \rho_i \mathcal{O}_i(\vartheta_s\omega) \cdot f_i(v_i^{(n)} + \mathcal{O}_i(\vartheta_s\omega)) \\
&\le -\alpha \sum_{i\in\mathbb{Z}} \rho_i \big(v_i^{(n)} + \mathcal{O}_i(\vartheta_s\omega)\big)^{2q+2} + \|\boldsymbol{\beta}\|_\rho^2 + M_f\Big(1 + \frac{2}{\lambda} M_f\Big) M_\rho \|\boldsymbol{\mathcal{O}}(\vartheta_s\omega)\|^2 + \frac{\lambda}{8}\|\boldsymbol{v}\|_\rho^2 \\
&\quad + \frac{M_f^{2q+2} M_\rho}{(2q+2)\left(\frac{2q+2}{2q+1}\alpha\right)^{2q+1}} \sum_{i\in\mathbb{Z}} (|\mathcal{O}_i(\vartheta_s\omega)|)^{2q+2} + \alpha \sum_{i\in\mathbb{Z}} \rho_i |v_i^{(n)} + \mathcal{O}_i(\vartheta_s\omega)|^{2q+2} \\
&\le C \cdot (\|\boldsymbol{\mathcal{O}}(\vartheta_s\omega)\|^2 + \|\boldsymbol{\mathcal{O}}(\vartheta_s\omega)\|^{2q+2} + 1) + \frac{\lambda}{8}\|\boldsymbol{v}\|_\rho^2,
\end{aligned}
$$

where C depends on T, α, $\|\boldsymbol{\beta}\|_\rho^2$, M_f, M_ρ, λ, q.

Summarizing the above, we have

$$
\begin{aligned}
\|\boldsymbol{v}^{(n)}(t)\|_\rho^2 \le \|\boldsymbol{v}_o^{(n)}\|_\rho^2 + \int_0^t \Big(&(-\lambda + M_d(2p+1))\|\boldsymbol{v}^{(n)}(s)\|^2 \\
&+ \tilde{C}\big(\|\boldsymbol{\mathcal{O}}(\vartheta_s\omega)\|^2 + \|\boldsymbol{\mathcal{O}}(\vartheta_s\omega)\|^{2q+2} + 1 + \|\boldsymbol{g}^{(n)}\|_\rho^2\big)\Big) \mathrm{d}s, \quad (12.14)
\end{aligned}
$$

where $\tilde{C}$ is a positive constant depending on T, α, $\|\boldsymbol{\beta}\|_\rho^2$, M_d, M_f, M_ρ, λ, p, q, but does not depend on n. Therefore, $\boldsymbol{v}(t;\omega,\boldsymbol{v}_o,\boldsymbol{g}) \in \ell_\rho^2$ is the limit function of a subsequence of $\{\boldsymbol{v}^{(n)} = \boldsymbol{v}(t;\omega,\boldsymbol{v}_o^{(n)},\boldsymbol{g}^{(n)})\}$ in ℓ_ρ^2 and $\boldsymbol{v}(t;\omega,\boldsymbol{v}_o,\boldsymbol{g})$ satisfies the equation (12.8) for all $t \in [0,T)$. □

12.1.4 *Random dynamical systems generated by random LDS*

It will be shown here that the solution to (12.6) generates a random dynamical system with state space ℓ_ρ^2. Throughout the rest of this chapter, when the context is clear we may write $\boldsymbol{v}(t;\omega,\boldsymbol{v}_o,\boldsymbol{g})$ as $\boldsymbol{v}(t;\omega,\boldsymbol{v}_o)$ when $\boldsymbol{v}(t;\omega,\boldsymbol{v}_o,\boldsymbol{g})$ is shown to exist in Theorem 12.1.

Theorem 12.2. *Let $\omega \in \Omega$ and Assumptions 12.1–12.4 hold. Then the solution to the RODE (12.6) generates a continuous random dynamical system $\{\varphi(t,\omega)\}_{t\ge0,\omega\in\Omega}$ over $(\Omega,\mathcal{F},\mathbb{P},(\vartheta_t)_{t\in\mathbb{R}})$ with state space ℓ_ρ^2 defined by*

$$
\varphi(t,\omega,\boldsymbol{v}_o) := \boldsymbol{v}(t;\omega,\boldsymbol{v}_o) \quad \text{for} \quad \boldsymbol{v}_o \in \ell_\rho^2,\ t\ge 0,\ \omega\in\Omega.
$$

Moreover, the mapping

$$
\tilde{\varphi}(t,\omega,\boldsymbol{u}_o) := \varphi(t,\omega)(\boldsymbol{u}_o - \boldsymbol{\mathcal{O}}(\omega)) + \boldsymbol{\mathcal{O}}(\vartheta_t\omega) \quad \text{for} \quad \boldsymbol{u}_o \in \ell_\rho^2, t\ge 0, \omega\in\Omega, \quad (12.15)
$$

defines a continuous RDS $\{\tilde{\varphi}(t,\omega)\}_{t\ge0,\omega\in\Omega}$ over $(\Omega,\mathcal{F},\mathbb{P},(\vartheta_t)_{t\in\mathbb{R}})$ associated with (12.5).

Proof. By Theorem 12.1, for any $T > 0$ and $\boldsymbol{v}_o \in \ell^2_\rho$, the equation (12.6) has a solution $\boldsymbol{v}(t;\omega,\boldsymbol{v}_o)$ with $\boldsymbol{v}(0;\omega,\boldsymbol{v}_o) = \boldsymbol{v}_o$ that exists on $[0,\infty)$. Since $\boldsymbol{v}(\cdot;\omega,\boldsymbol{v}_o)$ depends continuously on the initial data $\boldsymbol{v}_o$ and $\boldsymbol{g}$ for all $\omega \in \Omega$, there exists a continuous modification of $\{\varphi(t)\}_{t\geq 0}$ such that $\varphi(\cdot,\omega,\cdot) : [0,\infty) \times \ell^2_\rho \to \ell^2_\rho$ is continuous for all ω and satisfies

$$\varphi(t,\omega,\boldsymbol{v}_o) = \boldsymbol{v}_o + \int_0^t \mathfrak{F}(\varphi(s,\omega,\boldsymbol{v}_o),\vartheta_s\omega)\mathrm{d}s, \qquad \omega \in \Omega, \tag{12.16}$$

where $\mathfrak{F}$ is defined by (12.7).

The measurability of $\tilde{\varphi}$ is inherited from that of $\mathcal{O}(\omega)$. Thus it only remains to verify the cocycle property. To that end, consider $t, \tau \geq 0$, and set

$$\mathcal{S}(s,\omega,\boldsymbol{v}_o) = \begin{cases} \varphi(s,\omega,\boldsymbol{v}_o), & 0 \leq s \leq \tau, \\ \varphi(s-\tau,\vartheta_\tau\omega,\varphi(\tau,\omega,\boldsymbol{v}_o)), & \tau < s \leq \tau + t. \end{cases}$$

Then for $s = \tau + t$ we have

$$\mathcal{S}(\tau+t,\omega,\boldsymbol{v}_o) = \varphi\big(t,\vartheta_\tau\omega,\varphi(\tau,\omega,\boldsymbol{v}_o)\big), \quad \tau, t \geq 0. \tag{12.17}$$

On the other hand, by (12.16) we have

$$\begin{aligned}\varphi\big(t,\vartheta_\tau\omega,\varphi(\tau,\omega,\boldsymbol{v}_o)\big) &= \boldsymbol{v}_o + \int_0^\tau \mathfrak{F}\big(\varphi(s,\omega,\boldsymbol{v}_o),\mathcal{O}(\vartheta_s\omega)\big)\mathrm{d}s \\ &\quad + \int_\tau^{\tau+t} \mathfrak{F}\big(\varphi(s-\tau,\vartheta_\tau\omega,\varphi(\tau,\omega,\boldsymbol{v}_o)),\mathcal{O}(\vartheta_\tau\omega)\big)\mathrm{d}s \\ &= \boldsymbol{v}_o + \int_0^{\tau+t} \mathfrak{F}\big(\mathcal{S}(s,\omega,\boldsymbol{v}_o),\vartheta_s\omega\big)\mathrm{d}s. \end{aligned} \tag{12.18}$$

The equations (12.17) and (12.18) together imply that

$$\mathcal{S}(\tau+t,\omega,\boldsymbol{v}_o) = \boldsymbol{v}_o + \int_0^{\tau+t} \mathfrak{F}\big(\mathcal{S}(s,\omega,\boldsymbol{v}_o),\vartheta_s\omega\big)\mathrm{d}s,$$

i.e., $\mathcal{S}(t,\omega,\boldsymbol{v}_o)$ is also a solution to (12.16). Thus, due to uniqueness of solutions, $\mathcal{S}(\tau+t,\omega,\boldsymbol{v}_o) = \varphi(\tau+t,\omega,\boldsymbol{v}_o)$, which, together with (12.17) give the cocycle property of φ.

It follows directly from Proposition 11.1 that the mapping $\tilde{\varphi}$ defined in (12.15) is also an RDS. □

From now on, the RDS $\{\tilde{\varphi}(t,\omega)\}_{t\geq 0,\omega\in\Omega}$ defined by (12.15) will be referred to as the RDS associated with the system (12.5).

12.2 Existence of global random attractors in weighted space

In this section, we investigate the existence of a global random attractor for the random dynamical system $\{\tilde{\varphi}(t,\omega)\}_{t\geq 0,\omega\in\Omega}$ associated with (12.5) defined by (12.15) in Theorem 12.2 in weighted space ℓ^2_ρ.

To that end, we first prove the existence of tempered random bounded absorbing sets and then verify the asymptotic nullness property in Definition 11.6 for the RDS $\{\tilde{\varphi}(t,\omega)\}_{t\geq 0,\omega\in\Omega}$ in ℓ^2_ρ. As in Chapter 13, the domain of attraction considered here is again the set of all tempered random sets of ℓ^2_ρ, denoted by $\mathscr{D}(\ell^2_\rho)$.

12.2.1 *Existence of tempered random bounded absorbing sets*

Lemma 12.5. *Let Assumptions 12.1–12.4 hold. In addition, assume that*

$$\lambda > \tilde{\gamma} = M_d(p+1+\gamma+\cdots+\gamma^p). \tag{12.19}$$

Then the random dynamical system $\{\tilde{\varphi}(t,\omega)\}_{t\geq 0,\omega\in\Omega}$ possesses a tempered bounded closed random absorbing set $\mathcal{Q} = \{Q(\omega)\}_{\omega\in\Omega} \in \mathscr{D}(\ell^2_\rho)$ such that for any $\mathcal{D} = \{D(\omega)\}_{\omega\in\Omega} \in \mathscr{D}(\ell^2_\rho)$ and each $D(\omega) \in \mathcal{D}$, there exists $T_{\mathcal{D}}(\omega) > 0$ such that $\tilde{\varphi}(t,\vartheta_{-t}\omega, D(\vartheta_{-t}\omega)) \subset Q(\omega)$ for all $t \geq T_{\mathcal{D}}(\omega)$. In particular, there exists $T_{\mathcal{Q}}(\omega) > 0$ such that $\tilde{\varphi}(t,\vartheta_{-t}\omega, Q(\vartheta_{-t}\omega)) \subset \mathcal{Q}$ for all $t \geq T_{\mathcal{Q}}(\omega)$.

Proof. Given $\boldsymbol{u}_o$ and $\boldsymbol{g}$ in ℓ^2 and $\omega\in\Omega$, let $\boldsymbol{v}(t,\omega) = \boldsymbol{v}(t;\omega,\boldsymbol{v}_o(\omega),\boldsymbol{g})$ be a solution of equation (12.6) with $\boldsymbol{v}_o(\omega) = \boldsymbol{u}_o - \boldsymbol{\mathcal{O}}(\omega) \in \ell^2$, then $\boldsymbol{v}(t,\omega) \in \ell^2$ for all $t \geq 0$. Moreover, $\|\boldsymbol{v}\|_\rho$ satisfies

$$\begin{aligned}\frac{1}{2}\frac{\mathrm{d}\,\|\boldsymbol{v}(t)\|_\rho^2}{\mathrm{d}t} &= -\lambda\|\boldsymbol{v}\|_\rho^2 + \langle \mathrm{A}\boldsymbol{v},\boldsymbol{v}\rangle_\rho + \langle \mathrm{A}\boldsymbol{\mathcal{O}}(\vartheta_t\omega),\boldsymbol{v}\rangle_\rho \\ &\quad + \langle \boldsymbol{g},\boldsymbol{v}\rangle_\rho - \sum_{i\in\mathbb{Z}} \rho_i v_i \cdot f_i\big(v_i + \mathcal{O}_i(\vartheta_t\omega)\big).\end{aligned} \tag{12.20}$$

First, notice that under the condition (12.19) there exists $\tilde{\lambda} > 0$ such that

$$0 < \tilde{\lambda} < \lambda - \tilde{\gamma}.$$

Then following similar estimates to those in the proof of Theorem 12.1, we obtain

$$|\langle \mathrm{A}\boldsymbol{\mathcal{O}}(\vartheta_t\omega),\boldsymbol{v}\rangle_\rho| \leq \frac{\tilde{\lambda}}{4}\|\boldsymbol{v}\|_\rho^2 + \frac{M_\rho}{\tilde{\lambda}}(2p+1)\|\boldsymbol{\mathcal{O}}(\vartheta_t\omega)\|^2,$$

$$-\sum_{i\in\mathbb{Z}} \rho_i v_i \cdot f_i\big(v_i + \mathcal{O}_i(\vartheta_t\omega)\big) \leq C(\|\boldsymbol{\mathcal{O}}(\vartheta_t\omega)\|^2 + \|\boldsymbol{\mathcal{O}}(\vartheta_t\omega)\|^{2q+2} + 1) + \frac{\tilde{\lambda}}{4}\|\boldsymbol{v}\|_\rho^2,$$

$$|\langle \boldsymbol{g},\boldsymbol{v}\rangle_\rho| \leq \frac{1}{\tilde{\lambda}}\|\boldsymbol{g}\|_\rho^2 + \frac{\tilde{\lambda}}{4}\|\boldsymbol{v}\|_\rho^2,$$

which can be inserted into (12.20) to give

$$\frac{\mathrm{d}\,\|\boldsymbol{v}\|_\rho^2}{\mathrm{d}t} \leq -\frac{\tilde{\lambda}}{2}\|\boldsymbol{v}\|_\rho^2 + C\left(1 + \|\boldsymbol{\mathcal{O}}(\vartheta_t\omega)\|^2 + \|\boldsymbol{\mathcal{O}}(\vartheta_t\omega)\|^{2q+2} + \|\boldsymbol{g}\|_\rho^2\right), \tag{12.21}$$

where C is a positive constant dependent of $\tilde{\lambda}, \alpha, M_f, M_d, \|\boldsymbol{\beta}\|_\rho, p$, and γ.

Applying Gronwall's inequality to (12.21), we obtain that for $t > 0$

$$\|\boldsymbol{v}(t,\omega)\|_\rho^2 \leq e^{-\frac{1}{2}\tilde{\lambda}t}\|\boldsymbol{v}_o\|_\rho^2 + \frac{2}{\tilde{\lambda}}\|\boldsymbol{g}\|_\rho^2\left(1 - e^{-\frac{\tilde{\lambda}}{2}t}\right) + C\mathcal{J}(t,\omega) \tag{12.22}$$

where

$$\mathcal{J}(t,\omega) = \int_0^t e^{-\frac{1}{2}\tilde{\lambda}(t-s)} \left(1 + \|\mathcal{O}(\vartheta_s\omega)\|^2 + \|\mathcal{O}(\vartheta_s\omega)\|^{2q+2}\right) \mathrm{d}s.$$

Replacing ω by $\vartheta_{-t}\omega$ in $\mathcal{J}(t,\omega)$ gives

$$\begin{aligned}\mathcal{J}(t,\vartheta_{-t}\omega) &= \int_0^t e^{-\frac{1}{2}\tilde{\lambda}(t-s)} \left(1 + \|\mathcal{O}(\vartheta_{s-t}\omega)\|^2 + \|\mathcal{O}(\vartheta_{s-t}\omega)\|^{2q+2}\right) \mathrm{d}s \\ &= \int_0^t e^{-\frac{1}{2}\tilde{\lambda}\tau} \left(1 + \|\mathcal{O}(\vartheta_{-\tau}\omega)\|^2 + \|\mathcal{O}(\vartheta_{-\tau}\omega)\|^{2q+2}\right) \mathrm{d}\tau.\end{aligned}$$

By Corollary 12.1, it follows that $\|\mathcal{O}(\omega)\|^2 + \|\mathcal{O}(\omega)\|^{2q+2}$ is tempered, and there exists a $0 \le T_0 < \infty$ such that

$$\|\mathcal{O}(\vartheta_{-\tau}\omega)\|^2 + \|\mathcal{O}(\vartheta_{-\tau}\omega)\|^{2q+2} \le e^{\frac{\tilde{\lambda}}{4}\tau}, \quad \forall \tau \ge T_0, \tag{12.23}$$

and consequently

$$\mathcal{J}(t,\vartheta_{-t}\omega) \le R(\omega)$$

where

$$R(\omega) := \int_0^{T_0} e^{-\frac{1}{2}\tilde{\lambda}\tau} \left(1 + \|\mathcal{O}(\vartheta_{-\tau}\omega)\|^2 + \|\mathcal{O}(\vartheta_{-\tau}\omega)\|^{2q+2}\right) \mathrm{d}\tau + \frac{4}{\tilde{\lambda}} e^{-\frac{\tilde{\lambda}}{4}T_0}.$$

It is straightforward to check that $R(\omega)$ is tempered. In fact, for any $c > 0$,

$$\begin{aligned}e^{-ct}R(\vartheta_{-t}\omega) &= e^{-ct}\frac{4}{\tilde{\lambda}} + e^{-ct}\int_0^{T_0} e^{-\frac{1}{2}\tilde{\lambda}\tau} \left(1 + \|\mathcal{O}(\vartheta_{-\tau}\omega)\|^2 + \|\mathcal{O}(\vartheta_{-\tau}\omega)\|^{2q+2}\right) \mathrm{d}\tau \\ &\to 0 \quad \text{as} \quad t \to +\infty. \end{aligned} \tag{12.24}$$

Summarizing (12.22) through (12.24) results in

$$\|\boldsymbol{v}(t;\vartheta_{-t}\omega, \boldsymbol{v}_o(\vartheta_{-t}\omega), \boldsymbol{g})\|_\rho^2 \le e^{-\frac{1}{2}\tilde{\lambda}t}\|\boldsymbol{v}_o\|_\rho^2 + \frac{2}{\tilde{\lambda}}\|\boldsymbol{g}\|_\rho^2 + CR(\omega).$$

This implies that for any $\mathcal{D} = \{D(\omega)\}_{\omega\in\Omega} \in \mathscr{D}(\ell_\rho^2)$ and $\boldsymbol{u}_o \in D(\vartheta_{-t}\omega)$

$$\begin{aligned}\|\tilde{\varphi}(t,\vartheta_{-t}\omega,\boldsymbol{u}_o)\|_\rho^2 &\le 4e^{-\frac{1}{2}\tilde{\lambda}t}\left(M_\rho\|\mathcal{O}(\vartheta_{-t}\omega)\|^2 + \sup_{\boldsymbol{u}_o\in D(\vartheta_{-t}\omega)} \|\boldsymbol{u}_o\|_\rho^2\right) \\ &\quad + 2M_\rho\|\mathcal{O}(\omega)\|^2 + \frac{2}{\tilde{\lambda}}\|\boldsymbol{g}\|_\rho^2 + CR(\omega).\end{aligned}$$

where

$$\lim_{t\to+\infty} e^{-\frac{1}{2}\tilde{\lambda}t}\left(M_\rho\|\mathcal{O}(\vartheta_{-t}\omega)\|^2 + \sup_{\boldsymbol{u}_o\in D(\vartheta_{-t}\omega)} \|\boldsymbol{u}_o\|_\rho^2\right) = 0.$$

Finally, for each $\omega \in \Omega$, define

$$\tilde{R}(\omega) = \left(2M_\rho\|\mathcal{O}(\omega)\|^2 + \frac{2}{\tilde{\lambda}}\|\boldsymbol{g}\|_\rho^2 + CR(\omega)\right)^{\frac{1}{2}}.$$

It then follows that $\mathcal{Q} = \{Q(\omega)\}_{\omega\in\Omega}$ with $Q(\omega)$ defined by

$$Q(\omega) = \overline{\left\{ \boldsymbol{u} \in \ell^2_\rho : \|\boldsymbol{u}\|_\rho \le \tilde{R}(\omega) \right\}} \subset \ell^2_\rho \tag{12.25}$$

is a closed tempered random absorbing set for the RDS $\{\tilde{\varphi}(t,\omega)\}_{t\ge 0,\omega\in\Omega}$. □

12.2.2 *Existence of global random attractors*

In this subsection, we show that the RDS $\{\tilde{\varphi}(t,\omega)\}_{t\ge 0,\omega\in\Omega}$ has a global random attractor in ℓ^2_ρ. This will be done by showing that the RDS $\{\tilde{\varphi}(t,\omega)\}_{t\ge 0,\omega\in\Omega}$ is aymptotically null on $\mathcal{Q} = \{Q(\omega)\}_{\omega\in\Omega}$ where $Q(\omega)$ is defined as in (12.25).

Theorem 12.3. *Let Assumptions 12.1–12.4 and the condition* (12.19) *hold. Then the RDS* $\{\tilde{\varphi}(t,\omega)\}_{t\ge 0,\omega\in\Omega}$ *associated with the system* (12.5) *possesses a unique global random* $\mathfrak{D}$ *attractor* $\mathcal{A} = \{A(\omega)\}_{\omega\in\Omega}$ *with* $A(\omega)$ *given by*

$$A(\omega) = \bigcap_{\tau\ge T_Q(\omega)} \overline{\bigcup_{t\ge\tau} \tilde{\varphi}(t,\vartheta_{-t}\omega, Q(\vartheta_{-t}\omega))} \in \ell^2_\rho.$$

Proof. Let $\boldsymbol{u}(t;\omega,\boldsymbol{u}_o,\boldsymbol{g}) = (u_i(t;\,\omega,\boldsymbol{u}_o,\boldsymbol{g}))_{i\in\mathbb{Z}}$ be a solution of (12.5). Then

$$\boldsymbol{v}(t;\omega,\boldsymbol{v}_o(\omega),\boldsymbol{g}) = (v_i(t;\omega,\boldsymbol{v}_o(\omega),\boldsymbol{g}))_{i\in\mathbb{Z}} = \boldsymbol{u}(t;\omega,\boldsymbol{u}_o,\boldsymbol{g}) - \mathcal{O}(\vartheta_t\omega) \in \ell^2_\rho$$

is a solution of equation (12.6) with initial data $\boldsymbol{v}_o(\omega) = \boldsymbol{u}_o - \mathcal{O}(\omega) \in \ell^2_\rho$.

Given any $n > 0$, define a truncation operator $\Xi_n : \ell^2_\rho \to \ell^2$, $\boldsymbol{v} = (v_i)_{i\in\mathbb{Z}} \mapsto \Xi_n\boldsymbol{v} = ((\Xi_n\boldsymbol{v})_i)_{i\in\mathbb{Z}}$ by

$$(\Xi_n\boldsymbol{v})_j = \begin{cases} v_j & \text{if} \quad |j|\le n \\ 0 & \text{if} \quad |j|>n \end{cases}. \tag{12.26}$$

Then Ξ_n is continuous and for any $\boldsymbol{v}_o \in \ell^2_\rho$ and $\boldsymbol{g} \in \ell^2_\rho$

$$\boldsymbol{v}(t;\omega,\boldsymbol{v}_o,\boldsymbol{g}) = \lim_{n\to\infty} \boldsymbol{v}_n \quad \text{with} \quad \boldsymbol{v}_n = \boldsymbol{v}(t;\omega,\Xi_n\boldsymbol{v}_o,\Xi_n\boldsymbol{g}). \tag{12.27}$$

Set $\boldsymbol{u}_o^{(n)} = \Xi_n\boldsymbol{u}_o$ and $\boldsymbol{g}^{(n)} = \Xi_n\boldsymbol{g}$, where Ξ_n is the truncation operator defined in (12.26). Then $\boldsymbol{u}_o^{(n)}$, $\boldsymbol{g}^{(n)} \in \ell^2$ and $\boldsymbol{u}(t;\omega,u_o^{(n)},\boldsymbol{g}^{(n)}) \to \boldsymbol{u}(t;\omega,\boldsymbol{u}_o,\boldsymbol{g})$ in ℓ^2_ρ when $n\to\infty$. Let $\boldsymbol{v}_n(t) = \boldsymbol{v}(t;\omega,\boldsymbol{u}_o^{(n)}(\omega),\boldsymbol{g}^{(n)})$ be the solution of (12.6) with $\boldsymbol{g} = \boldsymbol{g}^{(n)}$ satisfying $\boldsymbol{v}_n(0) = \boldsymbol{v}_o^{(n)}(\omega) = \boldsymbol{u}_o^{(n)} - \mathcal{O}(\omega)$.

Let $\xi \in C^1(\mathbb{R}^+,\mathbb{R})$ be the cut-off function defined as in (3.8), and recall that there exists $C_0 > 0$ such that $|\xi'(s)| \le C_0$ for $s\in\mathbb{R}^+$. Let k be a suitable large integer (to be specified later) and define $\xi_k(s) := \xi(s/k)$ for $s\in\mathbb{R}$. For each $n\in\mathbb{N}$, set $\tilde{v}_i^{(n)} = \xi_k(|i|)v_i^{(n)}$ for all $i\in\mathbb{Z}$, and write $\tilde{\boldsymbol{v}}^{(n)} = (\tilde{v}_i^{(n)})_{i\in\mathbb{Z}}$.

Taking the inner product $\langle\cdot,\cdot\rangle_\rho$ of (12.6) with $\tilde{\boldsymbol{v}}^{(n)}$ gives

$$\begin{aligned}\frac{\mathrm{d}}{\mathrm{d}t}\sum_{i\in\mathbb{Z}} \xi_k(|i|)\rho_i(v_i^{(n)})^2 &\le (-\lambda+\tilde{\gamma})\sum_{i\in\mathbb{Z}} \xi_k(|i|)\rho_i(v_i^{(n)})^2 \\ &\quad + \sum_{i\in\mathbb{Z}} \rho_i\xi_k(|i|)\Big(v_i^{(n)}(\mathrm{A}\mathcal{O})_i + v_i^{(n)}g_i^{(n)} - f_i(v_i^{(n)}+\mathcal{O}_i)v_i^{(n)}\Big),\end{aligned}$$

and following similar estimations to those in Lemma 12.5 we obtain

$$\begin{aligned}\frac{\mathrm{d}}{\mathrm{d}t}\sum_{i\in\mathbb{Z}}\xi_k(|i|)\rho_i(v_i^{(n)})^2 \le& -\frac{\tilde{\lambda}}{2}\sum_{i\in\mathbb{Z}}\xi_k(|i|)\rho_i(v_i^{(n)})^2+\frac{C_1M_d}{k}\|\boldsymbol{v}^{(n)}\|_\rho^2\\ &+\sum_{i\in\mathbb{Z}}\xi_k(|i|)\rho_i(\beta_i^2+\frac{2}{\tilde{\lambda}}g_i^2)\\ &+C_2\sum_{i\in\mathbb{Z}}\xi_k(|i|)\big(|\mathcal{O}_i(\vartheta_t\omega)|^2+|\mathcal{O}_i(\vartheta_t\omega)|^2+|\mathcal{O}_i(\vartheta_t\omega)|^{2q+2}\big),\end{aligned}$$

where $C_1>0$ and $C_2>0$ are constants dependent on $\lambda,\alpha,M_f,M_\rho$.

Applying Gronwall's inequality to the above differential inequality from T_Q to t $(t\ge T_Q)$ with ω being replaced by $\vartheta_{-t}\omega$ gives

$$\sum_{i\in\mathbb{Z}}\xi_k(|i|)\rho_i|v_i^{(n)}(t;\vartheta_{-t}\omega,\boldsymbol{v}_o^{(n)}(\vartheta_{-t}\omega),\boldsymbol{g}^{(n)})|^2\le\mathcal{J}_1(t)+\mathcal{J}_2(t)+\int_0^{t-T_Q}\mathcal{J}_3(s)\mathrm{d}s,\tag{12.28}$$

where

$$\mathcal{J}_1(t)=e^{-\frac{\tilde{\lambda}}{2}(t-T_Q)}\|\boldsymbol{v}^{(n)}(T_Q;\vartheta_{-t}\omega,\boldsymbol{v}_o^{(n)}(\vartheta_{-t}\omega),\boldsymbol{g}^{(n)})\|_\rho^2$$

$$\begin{aligned}\mathcal{J}_2(t)=&\frac{2}{\tilde{\lambda}}\left(1-e^{-\frac{\tilde{\lambda}}{2}t}\right)\Big(\sum_{|i|\ge k}\rho_i(\beta_i^2+\frac{2}{\tilde{\lambda}}(g_i^{(n)})^2)\\ &+\frac{C_1M_d}{k}\|\boldsymbol{v}^{(n)}(T_Q,\vartheta_{-t}\omega,\boldsymbol{v}_o^{(n)}(\vartheta_{-t}\omega),\boldsymbol{g}^{(n)})\|_\rho^2\Big),\end{aligned}$$

$$\mathcal{J}_3(s)=e^{-\frac{\tilde{\lambda}}{2}s}\sum_{|i|\ge k}\big(|\mathcal{O}_i(\vartheta_s\omega)|^2+|\mathcal{O}_i(\vartheta_s\omega)|^2+|\mathcal{O}_i(\vartheta_s\omega)|^{2q+2}\big).$$

We next estimate each of the terms on the right hand side of (12.28). For any $\varepsilon>0$, $\boldsymbol{u}_o^{(n)}\in Q(\vartheta_{-t}\omega)\cap\ell^2$, there exists $T_1(\varepsilon,\omega,Q)\ge T_Q$ such that

$$\mathcal{J}_1(t)\le\frac{\varepsilon}{8}\quad\text{for}\quad t\ge T_1.\tag{12.29}$$

Then by the fact that $\boldsymbol{\beta}\in\ell_\rho^2$ and $\boldsymbol{g}^{(n)}\in\ell_\rho^2$, we have that there exists $K(\varepsilon,\omega)>0$ such that

$$\mathcal{J}_2(t)\le\frac{\varepsilon}{4}\quad\text{for}\quad k>K.\tag{12.30}$$

Recall that due to (12.23) there exists $T_0<\infty$ such that $\|\mathcal{O}(\vartheta_{-s}\omega)\|^2+\|\mathcal{O}(\vartheta_{-s}\omega)\|^{2q+2}\le e^{\frac{\tilde{\lambda}}{4}s}$ for all $s\ge T_0$. Thus there exists $T_2=T_2(\varepsilon,\omega,Q)$ such that

$$\int_{T_2}^{t-T_Q}\mathcal{J}_3(s)\mathrm{d}s\le\frac{\varepsilon}{16}\quad\text{for}\quad t>T_Q+T_2.\tag{12.31}$$

In addition, since $\mathcal{O}(\vartheta_{-\tau}\omega)\in\ell^2$, there exists $K_2(\varepsilon,\omega)\in\mathbb{N}$ such that

$$\int_0^{T_2}\mathcal{J}_3(s)\mathrm{d}s\le\frac{\varepsilon}{16},\quad\text{for}\quad k>K_2.\tag{12.32}$$

It follows immediately from (12.31) and (12.32) that

$$\int_0^{t-T_Q} \mathcal{J}_3(s)\mathrm{d}s \leq \frac{\varepsilon}{8} \quad k > K_2, \quad t > T_Q + T_2. \tag{12.33}$$

Setting

$$T(\varepsilon,\omega,Q) = \max\{T_1(\varepsilon,\omega,Q), T_2(\varepsilon,\omega,Q)\}, \quad K(\varepsilon,\omega) = \max\{K_1(\varepsilon,\omega), K_2(\varepsilon,\omega)\},$$

collecting (12.29), (12.30) and (12.33) and inserting into (12.28) results in

$$\sum_{i\in\mathbb{Z}} \xi_k(|i|)\rho_i|v_{n,i}^{(m)}(t;\vartheta_{-t}\omega, \boldsymbol{v}_o^{(n)}(\vartheta_{-t}\omega), \boldsymbol{g}^{(n)})|^2 \leq \frac{\varepsilon}{2}, \tag{12.34}$$

for $t \geq T(\varepsilon,\omega,B_Q)$, and $k \geq K(\varepsilon,\omega)$.

On the other hand since $\mathcal{O}(\omega) \in \ell^2$, there exists $K_3(\varepsilon,\omega) \in \mathbb{N}$ such that

$$2\gamma \sum_{i\in\mathbb{Z}} \xi_k(|i|)|\mathcal{O}_i(\omega)|^2 \leq \frac{\varepsilon}{2} \quad \text{for} \quad k > K_3(\varepsilon,\omega). \tag{12.35}$$

Now setting $\tilde{K}(\varepsilon,\omega) = \max\{K(\varepsilon,\omega), K_3(\varepsilon,\omega)\}$ and combining (12.34)–(12.35) results in

$$\begin{aligned} &\sum_{|i|\geq 2k} \rho_i|u_i(t;\vartheta_{-t}\omega, \boldsymbol{u}_o^{(n)}(\vartheta_{-t}\omega), \boldsymbol{g}^{(n)})|^2 \\ &\leq 2\sum_{i\in\mathbb{Z}} \xi_k(|i|)\rho_i|v_{n,i}(t;\vartheta_{-t}\omega, \boldsymbol{v}_o^{(n)}(\vartheta_{-t}\omega), \boldsymbol{g}^{(n)})|^2 + 2\gamma\sum_{i\in\mathbb{Z}} \xi_k(|i|)|\mathcal{O}_i(\omega)|^2 \\ &\leq \varepsilon \quad \text{for} \quad k > \tilde{K}(\varepsilon,\omega), \quad t \geq \tilde{T}(\varepsilon,\omega,Q), \quad \forall\, n \geq 1. \end{aligned} \tag{12.36}$$

The asymptotic nullness of $\{\tilde{\varphi}(t,\omega)\}_{\omega\in\Omega}$ on $\mathcal{Q}$ follows immediately by letting $n \to \infty$ in (12.36). The existence of global random attractors then follows from Lemma 12.5 and Theorem 11.4. □

12.3 End notes

[Lv and Sun (2006)] extended the result of Bates, Lisei and Lu [Bates *et al.* (2006)] to stochastic LDS with additive white noise on the lattice $\mathbb{Z}^k, k \in \mathbb{N}$. [Han *et al.* (2011)] studied stochastics LDS with random connection weights and additive noise in weighted space. [Huang (2007)] studied the existence of a global random attractor for stochastic lattice FitzHugh-Nagumo equations with additive white noise. Random attractors for the Ladyzhenskaya model with additive noise were studied in [Zhao and Duan (2010)], and random attractors for a stochastic coupled fractional Ginzburg-Landau equation with additive noise were studied in [Shu *et al.* (2015)]

Wang, Li & Xu [Wang *et al.* (2010)] investigated the existence of global random attractors for second-order stochastic LDS with additive white noise, while [Han (2013)] studied second-order stochastic LDS with additive white noise on $\mathbb{Z}^k$ in weighted sequence space. [Zhao and Zhou (2009)] gave a sufficient condition for the existence of global random attractors for general stochastic LDS in the non-weighted space of infinite sequences and provided an application to damped sine-Gordon lattice systems with additive noise. In [Caraballo *et al.* (2011)], the existence of a random compact global attractor for a first-order stochastic LDS with additive noise and non-Lipschitz nonlinearity was investigated. Pullback and forward attractors for dissipative LDS with additive noise were studied in [Ban *et al.* (2009)]. Random attractors for stochastic lattice systems with delays and additive noise were studied in [Zhou and Wang (2013)].

A review chapter on stochastic LDS can be found in [Han (2015)]. For background material on stochastic differential equations and stochastic calculus, in particular Stratonovich integrals, see [Evans (2013); Kloeden and Platen (1992)]. For RODEs see [Han and Kloeden (2017a)].

12.4 Problems

Problem 12.1. Show that the random attractors of finite dimensional approximations of the lattice system (12.1) converge upper semi-continuously to the random attractor of (12.1) .

Problem 12.2. Verify each property in Corollary 12.1.

Problem 12.3. What are the technical difficulties to extend the existence result in Theorem 12.1 to the weighted space ℓ^p_ρ for $p > 2$?

Chapter 13

Stochastic LDS with multiplicative noise

Stochastic LDS with multiplicative white noise was first investigated by [Caraballo and Lu (2008)], who studied the existence of global random attractors for a first-order lattice systems with the discrete Laplacian operator and finite number of multiplicative white noise.

In this chapter, a stochastic lattice system with single multiplicative noise will be studied in the weighted space of infinite sequences. In addition to the difference in noise compared to Chapter 12, a more complex leading operator with randomly coupled nodes is considered here, i.e., the constant coefficients $d_{i,j}$ in the operators A and D_p defined in (12.2) are now stochastic processes. More precisely, we will study the following lattice system of first order stochastic differential equations

$$\mathrm{d}u_i = \Big(-\lambda u_i - f_i(u_i) + g_i + \sum_{j=-p}^{p} \eta_{i,j}(\vartheta_t\omega)u_{i+j}\Big)\,\mathrm{d}t + u_i \circ \mathrm{d}W(t), \quad i \in \mathbb{Z}, \tag{13.1}$$

where λ is a positive constant, u_i, $g_i \in \mathbb{R}$ and $f_i \in C^1(\mathbb{R})$ satisfy appropriate dissipative conditions for $i \in \mathbb{Z}$, $W(t)$ is a scalar two-sided Wiener process, and $\circ$ in the stochastic integral associated with $\mathrm{d}W(t)$ means the Stratonovich sense.

For $p \in \mathbb{N}$ the coupling weights $\eta_{i,-p}(\vartheta_t\omega), \ldots, \eta_{i,0}(\vartheta_t\omega), \ldots, \eta_{i,+p}(\vartheta_t\omega)$ in the nodes are stochastic processes, in which $(\vartheta_t)_{t\in\mathbb{R}}$ represents a metric dynamical system as defined in Section 11.2, on a suitable probability space $(\Omega, \mathcal{F}, \mathbb{P})$. In this chapter we use the probability space Ω_0 defined by

$$\Omega_0 = \{\omega \in C(\mathbb{R}, \mathbb{R}) : \omega(0) = 0\} = C_0(\mathbb{R}, \mathbb{R})$$

with $\mathcal{F}$ denoting the Borel σ-algebra on Ω_0 generated by the compact open topology, and $\mathbb{P}$ denoting the corresponding Wiener measure on $\mathcal{F}$. See, e.g., [Arnold (1998)].

13.1 Random dynamical systems generated by stochastic LDS

The aim of this section is to show that the stochastic LDS (13.1) generates a random dynamical system that possesses a global random attractor on the weighted space ℓ^2_ρ defined as in Section 1.3. The weights ρ_i are again assumed to satisfy the Assumption 12.1 as in Chapter 12.

Notice that $\eta_{i,-q}(\omega), \ldots, \eta_{i,0}(\omega), \ldots, \eta_{i,+q}(\omega)$ are random variables on the probability space $(\Omega_0, \mathcal{F}, \mathbb{P})$ and define the random linear operator $\mathrm{A}(\omega)(\cdot)$ componentwise by

$$\left(\mathrm{A}(\omega)\boldsymbol{u}\right)_i = \sum_{j=-q}^{q} \eta_{i,j}(\omega) u_{i+j}, \quad \boldsymbol{u} = (u_i)_{i\in\mathbb{Z}}, \quad \omega \in \Omega_0. \tag{13.2}$$

Then, the system (13.1) can be written as

$$\mathrm{d}\boldsymbol{u}(t) = (-\lambda \boldsymbol{u} - F(\boldsymbol{u}) + \boldsymbol{g} + \mathrm{A}(\vartheta_t\omega)\boldsymbol{u})\ \mathrm{d}t + \boldsymbol{u} \circ \mathrm{d}W(t), \tag{13.3}$$

where $\boldsymbol{u} = (u_i)_{i\in\mathbb{Z}}$, $F(\boldsymbol{u}) = (f_i(u_i))_{i\in\mathbb{Z}}$, and $\boldsymbol{g} = (g_i)_{i\in\mathbb{Z}}$.

13.1.1 *Transformation to a random LDS*

Similar to Chapter 12, the first step is to transform the stochastic equation (13.3) into a random ordinary differential equation. To that end, consider the scalar Ornstein-Uhlenbeck process on $(\Omega_0, \mathcal{F}, \mathbb{P}, (\vartheta_t)_{t\in\mathbb{R}})$ defined by

$$\mathcal{O}(\vartheta_t\omega) = -\int_{-\infty}^{0} e^s \vartheta_t\omega(s)\mathrm{d}s, \quad t \in \mathbb{R}, \quad \omega \in \Omega.$$

Then $\mathcal{O}(\vartheta_t\omega)$ solves the scalar Ornstein-Uhlenbeck equation

$$\mathrm{d}\mathcal{O} + \mathcal{O}\mathrm{d}t = \mathrm{d}W(t), \quad t \geq 0, \tag{13.4}$$

where $W(t)(\omega) = W(t, \omega) = \omega(t)$ for $\omega \in \Omega$ and $t \in \mathbb{R}$. Moreover, it possesses all the properties stated in Lemma 12.2 of Chapter 12.

Define the mapping $\mathcal{T}(\vartheta_t\omega) := e^{\mathcal{O}(\vartheta_t\omega)}\mathrm{Id}_{\ell^2_\rho}$. It is a homeomorphism in ℓ^2_ρ with the inverse operator defined as $\mathcal{T}^{-1}(\vartheta_t\omega) := e^{-\mathcal{O}(\vartheta_t\omega)}\mathrm{Id}_{\ell^2_\rho}$. By Lemma 12.2 and similar arguments to those in [Caraballo and Lu (2008)], both $\|\mathcal{T}(\vartheta_t\omega)\|$ and $\left\|\mathcal{T}^{-1}(\vartheta_t\omega)\right\|$ have sub-exponential growth as $t \to \pm\infty$ for $\omega \in \widetilde{\Omega}_0$, and therefore $\|\mathcal{T}\|$ and $\left\|\mathcal{T}^{-1}\right\|$ are both tempered.

The mapping of ϑ on $\widetilde{\Omega}_0$ possesses the same properties as the original one if we choose the trace σ-algebra with respect to $\widetilde{\Omega}_0$ to be denoted also by $\mathcal{F}$. Then we can change our metric dynamical system with respect to $\widetilde{\Omega}_0$, still denoted by $(\Omega_0, \mathcal{F}, \mathbb{P}, (\vartheta_t)_{t\in\mathbb{R}})$ for simplicity of exposition.

Now, for any $\omega \in \Omega_0$ and $t \in \mathbb{R}$, set

$$\boldsymbol{v}(t, \omega) = \mathcal{T}^{-1}(\vartheta_t\omega)\boldsymbol{u}(t, \omega) = e^{-\mathcal{O}(\vartheta_t\omega)}\boldsymbol{u}(t, \omega), \quad \omega \in \Omega_0, \quad t \in \mathbb{R},$$

where $\boldsymbol{u}(t, \omega)$ satisfies the equation (13.3). Then the system (13.3) can be written as the following random ordinary differential equation with random coefficients, but without white noise:

$$\frac{\mathrm{d}\boldsymbol{v}(t,\omega)}{\mathrm{d}t} = -\lambda v + \mathrm{A}(\vartheta_t\omega)\boldsymbol{v} + \mathcal{O}(\vartheta_t\omega)\boldsymbol{v} - e^{-\mathcal{O}(\vartheta_t\omega)}F(e^{\mathcal{O}(\vartheta_t\omega)}\boldsymbol{v}) + e^{-\mathcal{O}(\vartheta_t\omega)}\boldsymbol{g}. \tag{13.5}$$

13.1.2 *Existence and uniqueness of solutions to the random LDS*

The solutions to the RODE (13.5) are defined in the sense of following definition.

Definition 13.1. A function $\boldsymbol{u} : [0,T) \to \ell_\rho^2$ is called a solution of the following random differential equation

$$\frac{\mathrm{d}\boldsymbol{u}(t)}{\mathrm{d}t} = \mathfrak{F}(\boldsymbol{u}, \vartheta_t\omega), \quad a.e. \quad \boldsymbol{u} = (u_i)_{i\in\mathbb{Z}}, \quad \mathfrak{F} = (\mathfrak{F}_i)_{i\in\mathbb{Z}},$$

for $\omega \in \Omega$, if $\boldsymbol{u} \in \mathcal{C}([0,T), \ell_\rho^2)$ and

$$u_i(t) = u_i(0) + \int_0^t \mathfrak{F}_i(\boldsymbol{u}(s), \vartheta_s\omega)\mathrm{d}s \quad \text{for } i \in \mathbb{Z} \text{ and } t \in [0,T).$$

Remark 13.1. The solution defined in Definition 13.1 is in the Carathéodory sense, since $t \mapsto \mathrm{A}(\vartheta_t\omega)$ may only be measurable (see [Han and Kloeden (2017a)]), i.e., the differential equation holds only almost everywhere in time.

Here the Assumptions 12.1–12.3 are still assumed to hold, and Assumption 12.4 is replaced by Assumption 13.1 below.

Assumption 13.1. $f_i \in \mathcal{C}^1(\mathbb{R},\mathbb{R})$ $f_i(0) = 0$, and there exist constants $\alpha \geq 0$ and $\boldsymbol{\beta} = (\beta_i)_{i\in\mathbb{Z}} \in \ell_\rho^2$ such that

$$f_i'(s) \geq -\alpha, \quad sf_i(s) \geq -\beta_i^2, \quad \forall s \in \mathbb{R}, i \in \mathbb{Z}.$$

In addition, the random coupling coefficients are assumed to satisfy the Assumption 13.2 below.

Assumption 13.2. For $p \in \mathbb{N}$, $\eta(\omega)$ defined by

$$\eta(\omega) := \sup\{|\eta_{i,-p}(\omega)|, \ldots, |\eta_{i,0}(\omega)|, \ldots, |\eta_{i,+p}(\omega)| \,:\, i \in \mathbb{Z}\} \geq 0.$$

is tempered and $\eta(\vartheta_t\omega)$ $(< \infty)$ belongs to $\mathcal{L}^1_{\mathrm{loc}}(\mathbb{R})$ with respect to $t \in \mathbb{R}$ for each $\omega \in \Omega$ with

$$\mathbb{E}(\eta) = \lim_{t\to\pm\infty} \frac{1}{t}\int_0^t \eta(\vartheta_t\omega)\mathrm{d}s < \infty.$$

We first show that for $\boldsymbol{v}_o \in \ell^2$ and $\boldsymbol{g} \in \ell^2$, the problem (13.5) is well-posed in ℓ^2. To that end, we first show that the right-hand side of (13.5) has Lipschitz properties. More precisely, given $\boldsymbol{v} \in \ell^2$, $\omega \in \Omega$ and $\boldsymbol{g} \in \ell^2$, let

$$\mathfrak{F}(\boldsymbol{v},\omega) = -\lambda\boldsymbol{v} + \mathrm{A}(\omega)\boldsymbol{v} + \mathcal{O}(\omega)\boldsymbol{v} - e^{-\mathcal{O}(\omega)}F(e^{\mathcal{O}(\omega)}\boldsymbol{v}) + e^{-\mathcal{O}(\omega)}\boldsymbol{g}.$$

Then the following lemma hold.

Lemma 13.1. *Let Assumptions 12.1–12.3 and 13.1–13.2 hold. Then for any* $\boldsymbol{v}$, $\boldsymbol{v}^{(1)}$, $\boldsymbol{v}^{(2)}$ *in a bounded set* $B \subset \ell^2$ *and* $\omega \in \Omega$, *there exists a random variable* $\zeta_B(\omega)$ *such that*

$$\|\mathfrak{F}(\boldsymbol{v},\omega)\| \leq \zeta_B(\omega), \quad \left\|\mathfrak{F}(\boldsymbol{v}^{(1)},\omega) - \mathfrak{F}(\boldsymbol{v}^{(2)},\omega)\right\| \leq \zeta_B(\omega)\|\boldsymbol{v}^{(1)} - \boldsymbol{v}^{(2)}\|.$$

Proof. First, notice that $\mathfrak{F}(\boldsymbol{v},\omega)$ is continuous in $\boldsymbol{v}$ and measurable in ω from $\ell^2\times\Omega$ into ℓ^2. Then by (13.2) and Assumption 13.2,

$$\|\mathrm{A}(\omega)\boldsymbol{v}\| = \Big(\sum_{i\in\mathbb{Z}}\Big(\sum_{j=-p}^{p}\eta_{i,j}(\omega)v_{i+j}\Big)^2\Big)^{1/2} \le (2p+1)\eta(\omega)\cdot\|\boldsymbol{v}\|.$$

Moreover, by Assumption 12.2,

$$\Big\|F(e^{\mathcal{O}(\omega)}\boldsymbol{v})\Big\| \le \Gamma_f(e^{\mathcal{O}(\omega)}\|\boldsymbol{v}\|)\cdot e^{\mathcal{O}(\omega)}\|\boldsymbol{v}\|,$$

and thus,

$$\|\mathfrak{F}(\boldsymbol{v},\omega)\| \le \Big(\lambda+(2p+1)\eta(\omega)+|\mathcal{O}(\omega)|+\Gamma_f(e^{\mathcal{O}(\omega)}\|\boldsymbol{v}\|)\Big)\|\boldsymbol{v}\| + |e^{-\mathcal{O}(\omega)}|\cdot\|\boldsymbol{g}\|.$$

Furthermore, for any $\boldsymbol{v}^{(1)}=(v_i^{(1)})_{i\in\mathbb{Z}}$ and $\boldsymbol{v}^{(2)}=(v_i^{(2)})_{i\in\mathbb{Z}}\in\ell^2$ there hold

$$\Big\|F(e^{\mathcal{O}(\omega)}\boldsymbol{v}^{(1)})-F(e^{\mathcal{O}(\omega)}\boldsymbol{v}^{(2)})\Big\| \le \Gamma_f\Big(e^{\mathcal{O}(\omega)}\Big(\|\boldsymbol{v}^{(1)}\|+\|\boldsymbol{v}^{(2)}\|\Big)\Big)\cdot e^{\mathcal{O}(\omega)}\|\boldsymbol{v}^{(1)}-\boldsymbol{v}^{(2)}\|,$$

$$\begin{aligned}\Big\|\mathrm{A}(\omega)\boldsymbol{v}^{(1)}-\mathrm{A}(\omega)\boldsymbol{v}^{(2)}\Big\| &= \sum_{i\in\mathbb{Z}}\Big(\sum_{j=-p}^{p}\eta_{i,j}(\omega)\Big(v_{i+j}^{(1)}-v_{i+j}^{(2)}\Big)\Big)^{1/2}\\ &\le (2p+1)\eta(\omega)\cdot\|\boldsymbol{v}^{(1)}-\boldsymbol{v}^{(2)}\|.\end{aligned}$$

It then follows that

$$\begin{aligned}\Big\|\mathfrak{F}(\boldsymbol{v}^{(1)},\omega)-\mathfrak{F}(\boldsymbol{v}^{(2)},\omega)\Big\| \le \Big(&\lambda+(2p+1)\eta(\omega)+|\mathcal{O}(\omega)|\\ &+\Gamma_f\Big(e^{\mathcal{O}(\omega)}\Big(\|\boldsymbol{v}^{(1)}\|+\|\boldsymbol{v}^{(2)}\|\Big)\Big)\Big)\|\boldsymbol{v}^{(1)}-\boldsymbol{v}^{(2)}\|.\end{aligned}$$

Therefore, for any bounded set $B\subset\ell^2$ with $\sup_{\boldsymbol{u}\in B}\|\boldsymbol{u}\|\le r$, define a random variable $\zeta_B(\omega)$ by

$$\zeta_B(\omega)=\Big(\lambda+(2p+1)\eta(\omega)+|\mathcal{O}(\omega)|+\Gamma_f(e^{\mathcal{O}(\omega)}r)\Big)r+|e^{-\mathcal{O}(\omega)}|\cdot\|\boldsymbol{g}\|.$$

Then $\zeta_B(\omega)>0$ for every $\omega\in\Omega$, and satisfies

$$\begin{aligned}\int_t^{t+1}\zeta_B(\vartheta_t\omega)\mathrm{d}t = \int_t^{t+1}\Big(&(\lambda+(2p+1)\eta(\vartheta_s\omega)+|\mathcal{O}(\vartheta_s\omega)|+\Gamma_f(e^{\mathcal{O}(\omega)}r))r\\ &+e^{-\mathcal{O}(\vartheta_s\omega)}\|\boldsymbol{g}\|\Big)\mathrm{d}s<\infty,\quad \forall\, t\in\mathbb{R}.\end{aligned}$$

The desired assertions then follow directly. □

A major difficulty in studying the LDS (13.1) is due to that the coupling coefficients $\eta_{i,j}(\vartheta_t\omega)$ are only measurable. To handle this, approximate $\eta_{i,j}(\vartheta_t\omega)$ by a sequence of continuous functions $\eta_{i,j}^{(m)}(t,\omega)$ in t for each $\omega\in\Omega$. In fact, for each $\omega\in\Omega$, there exist a sequence $\eta_{i,j}^{(m)}(t,\omega)$ of continuous functions in $t\in\mathbb{R}$ (see [Adams and Fournier (2003); Robinson (2001)]) such that

$$\lim_{m\to\infty}\int_0^t|\eta_{i,j}^{(m)}(s,\omega)-\eta_{i,j}(\vartheta_s\omega)|\mathrm{d}s=0,\quad\forall\, t>0, \tag{13.6}$$

and

$$|\eta_{i,j}^{(m)}(t,\omega)| \le |\eta_{i,j}(\vartheta_t\omega)| \le |\eta(\vartheta_t\omega)|, \quad \forall t \in \mathbb{R}. \tag{13.7}$$

For $m \in \mathbb{N}$, consider the following random ordinary differential equations now in the classical rather than Carathéodory sense, see [Han and Kloeden (2017a)])

$$\begin{aligned}\frac{\mathrm{d}\boldsymbol{v}^{(m)}}{\mathrm{d}t} &= -\lambda \boldsymbol{v}^{(m)} + \mathrm{A}_m(t,\omega)\boldsymbol{v}^{(m)} + \mathcal{O}(\vartheta_t\omega)\boldsymbol{v}^{(m)} \\ &\quad - e^{-\mathcal{O}(\vartheta_t\omega)}F(e^{\mathcal{O}(\vartheta_t\omega)}\boldsymbol{v}^{(m)}) + e^{-\mathcal{O}(\vartheta_t\omega)}\boldsymbol{g}\end{aligned} \tag{13.8}$$

where the operator A_m is defined component wise by

$$(\mathrm{A}_m(t,\omega)\boldsymbol{v}^{(m)})_i = \sum_{j=-p}^{p} \eta_{i,j}^{(m)}(t,\omega)v_{i+j}^{(m)}.$$

Lemma 13.2. *Let Assumptions 12.1–12.3 and 13.1–13.2 hold, and $T > 0$. Then for every $\omega \in \Omega_0$ and any initial data $\boldsymbol{v}^{(m)}(0) = \boldsymbol{v}_o \in \ell^2$, the problem* (13.8) *has a solution $\boldsymbol{v}^{(m)}(t;\boldsymbol{v}_o)$ with $\boldsymbol{v}^{(m)}(0) = \boldsymbol{v}_o$.*

Proof. First, due to Lemma 13.1, the equation (13.8) has a unique solution $\boldsymbol{v}^{(m)}(\cdot;\omega,\boldsymbol{v}_o,\boldsymbol{g}) \in \mathcal{C}([0,T_{\max}^{(m)}),\ell^2)$ in the sense of Definition 13.1, that is,

$$\begin{aligned}v_i^{(m)}(t) = v_{o,i} + \int_0^t \Big(&-\lambda v_i^{(m)} + (\mathrm{A}_m(s,\omega)v^{(m)})_i + \mathcal{O}(\vartheta_s\omega)v_i^{(m)} \\ &- e^{-\mathcal{O}(\vartheta_s\omega)}f_i(e^{\mathcal{O}(\vartheta_s\omega)}v_i^{(m)}) + e^{-\mathcal{O}(\vartheta_s\omega)}g_i\Big)\mathrm{d}s.\end{aligned}$$

Then by the continuity of $\mathrm{A}_m(s,\omega)$ in s, $\boldsymbol{v}^{(m)}(\cdot;\omega,\boldsymbol{v}_o,\boldsymbol{g}) = (v_i^{(m)})_{i\in\mathbb{Z}}$ also satisfies the differential equation

$$\begin{aligned}\frac{\mathrm{d}v_i^{(m)}}{\mathrm{d}t} &= -\lambda v_i^{(m)} + (\mathrm{A}_m(t,\omega)v^{(m)})_i + \mathcal{O}(\vartheta_t\omega)v_i^{(m)} \\ &\quad - e^{-\mathcal{O}(\vartheta_t\omega)}f_i(e^{\mathcal{O}(\vartheta_s\omega)}v_i^{(m)}) + e^{-\mathcal{O}(\vartheta_t\omega)}g_i,\end{aligned} \tag{13.9}$$

for $t \in [0,T_{\max}^{(m)})$.

By Assumptions 12.2, 13.1–13.2, for every $t \in [0,T_{\max}^{(m)})$ there hold

$$(\mathrm{A}_m(t,\omega)v^{(m)})_i \cdot v_i^{(m)} \le \eta(\vartheta_t\omega)\cdot \left| v_i^{(m)}v_{i-q}^{(m)} + \cdots + v_i^{(m)}v_{i+q}^{(m)}\right|, \tag{13.10}$$

$$\begin{aligned}-\alpha e^{2\mathcal{O}(\vartheta_t\omega)}\cdot (v_i^{(m)})^2 &\le (e^{\mathcal{O}(\vartheta_t\omega)}v_i^{(m)})\cdot f_i(e^{\mathcal{O}(\vartheta_t\omega)}v_i^{(m)}) \\ &\le \Gamma_f(e^{\mathcal{O}(\vartheta_t\omega)}\|\boldsymbol{v}^{(m)}\|)\cdot e^{2\mathcal{O}(\vartheta_t\omega)}(v_i^{(m)})^2.\end{aligned} \tag{13.11}$$

Multiplying (13.9) by $v_i^{(m)}(t)$, summing it over $i \in \mathbb{Z}$, and using (13.10)–(13.11), we obtain that

$$\begin{aligned}\frac{\mathrm{d}\|\boldsymbol{v}^{(m)}(t)\|^2}{\mathrm{d}t} &\le \big(-\lambda + 2\alpha + 2\mathcal{O}(\vartheta_t\omega) + 2(2p+1)\eta(\vartheta_t\omega)\big)\cdot \|\boldsymbol{v}^{(m)}(t)\|^2 \\ &\quad + \frac{1}{\lambda}e^{-2\mathcal{O}(\vartheta_t\omega)}\|\boldsymbol{g}\|^2 \quad \forall\, t \in [0,T_{\max}^{(m)}).\end{aligned} \tag{13.12}$$

Applying Gronwall's inequality to (13.12) results in

$$\begin{aligned}\|\boldsymbol{v}^{(m)}(t)\|^2 \le\ & e^{2\alpha t+\int_0^t(|2\mathcal{O}(\vartheta_s\omega)|+2(2p+1)\eta(\vartheta_s\omega))\mathrm{d}s}\|\boldsymbol{v}_o\|^2\\ &+\frac{\|\boldsymbol{g}\|^2}{\lambda}\left(e^{(2\alpha-\lambda)t+\int_0^t(2\mathcal{O}(\vartheta_s\omega)+2(2q+1)\eta(\vartheta_s\omega))\mathrm{d}s}\right)\\ &\cdot\left(\int_0^t e^{(\lambda-2\alpha)s-2\mathcal{O}(\vartheta_s\omega)+\int_0^s(2\mathcal{O}(\vartheta_r\omega)+2(2p+1)\eta(\vartheta_r\omega))\mathrm{d}r}\mathrm{d}s\right)\\ \doteq\ & \mathcal{V}^2(t),\quad t\in[0,T_{\max}^{(m)}),\end{aligned} \tag{13.13}$$

where $|\mathcal{V}(t)|\in\mathcal{C}([0,T),\mathbb{R})$ is independent of m. This implies the interval of existence of the solution $\boldsymbol{v}^{(m)}(t)$ for the equation (13.8) is $[0,T)$. The proof is complete. □

We are now ready to show the existence and uniqueness of solutions to the problem (13.5), presented in the following lemma.

Lemma 13.3. *Let Assumptions 12.1–12.3 and 13.1–13.2 hold, and $T>0$. Then for every $\omega\in\Omega_0$ and any initial data $\boldsymbol{v}_o\in\ell^2$, the problem* (13.5) *has a solution $\boldsymbol{v}(t;\boldsymbol{v}_o,\omega,\boldsymbol{g})$ with $\boldsymbol{v}(0;\boldsymbol{v}_o,\omega,\boldsymbol{g})=\boldsymbol{v}_o$.*

Proof. By Lemma 13.1 and [Arnold (1998); Chueshov (2002); Pazy (2007)], the problem (13.5) possesses a unique local solution $\boldsymbol{v}(\cdot;\omega,\boldsymbol{v}_o,g)\in\mathcal{C}([0,T_{\max}),\ell^2)$ $(0<T_{\max}\le T)$ that satisfies the integral equation

$$\begin{aligned}\boldsymbol{v}(t)=\boldsymbol{v}_o+\int_0^t\big(&-\lambda\boldsymbol{v}+\mathrm{A}(\vartheta_s\omega)\boldsymbol{v}+\mathcal{O}(\vartheta_s\omega)\boldsymbol{v}\\ &-e^{-\mathcal{O}(\vartheta_s\omega)}F(e^{\mathcal{O}(\vartheta_s\omega)}\boldsymbol{v})+e^{-\mathcal{O}(\vartheta_s\omega)}\boldsymbol{g}\big)\mathrm{d}s,\quad t\in\mathbb{R}.\end{aligned}\tag{13.14}$$

It remains to show that the existence interval of the solution $\boldsymbol{v}(t)$ for the equation (13.5) is also $[0,T)$. In fact, by (13.13) in Lemma 13.2,

$$|v_i^{(m)}(t)|\le|\mathcal{V}(t)|,\quad \text{for all } m\in\mathbb{N},\quad t\in[0,T).\tag{13.15}$$

In addition, there exists $K(T,\omega)>0$ such that the right hand side of (13.9), denoted as $\mathfrak{F}_i^{(m)}(v^{(m)})$, satisfies

$$\begin{aligned}\left|\mathfrak{F}_i^{(m)}(v^{(m)}(t),\vartheta_t\omega)\right|=\Big|&-\lambda v_i^{(m)}+(\mathrm{A}_m(\vartheta_t\omega)v^{(m)})_i+\mathcal{O}(\vartheta_t\omega)v_i^{(m)}\\ &-e^{-\mathcal{O}(\vartheta_t\omega)}f_i(e^{\mathcal{O}(\vartheta_t\omega)}v_i^{(m)})+e^{-\mathcal{O}(\vartheta_t\omega)}g_i\Big|\\ \le\ & K(T,\omega),\quad t\in[0,T).\end{aligned}$$

Thus, for any $t,\ \tau\in[0,T)$, $m\in\mathbb{N}$,

$$|v_i^{(m)}(t)-v_i^{(m)}(\tau)|=\int_\tau^t|\mathfrak{F}_i^{(m)}(v^{(m)}(s),\vartheta_t\omega)|\mathrm{d}s\le K(T,\omega)\cdot|t-\tau|.$$

Then by the Ascoli-Arzelà Theorem, there exists a convergent subsequence $\{v_i^{(m_k)}(t) : t \in [0,T)\}$ of $\{v_i^{(m)}(t) : t \in [0,T)\}$ such that

$$v_i^{(m_k)}(t) \to v_i^*(t) \quad \text{as } k \to \infty \quad \text{for} \quad t \in [0,T),\ i \in \mathbb{Z}$$

and $v_i^*(t)$ is continuous in $t \in [0,T)$. Moreover, $|v_i^*(t)| \leq |\mathcal{V}(t)|$ for $t \in [0,T)$. It then follows from (13.6), (13.7), (13.15), Assumption 13.2, and the Lebesgue Dominated Convergence Theorem that

$$\lim_{k\to\infty} \int_0^t \left(\eta_{i,j}^{(m_k)}(s,\omega) - \eta_{i,j}(\vartheta_s\omega)\right) \mathrm{d}s = \int_0^t \lim_{k\to\infty} \left(\eta_{i,j}^{(m_k)}(s,\omega) - \eta_{i,j}(\vartheta_s\omega)\right) \mathrm{d}s = 0,$$

$$\lim_{k\to\infty} \left(\eta_{i,j}^{(m_k)}(s,\omega) - \eta_{i,j}(\vartheta_s\omega)\right) = 0 \quad \text{a.e. } s \in [0,T],$$

and

$$\begin{aligned}&\lim_{k\to\infty} \int_0^t \left(\eta_{i,j}^{(m_k)}(s,\omega) v_i^{(m_k)}(s) - \eta_{i,j}(\vartheta_s\omega) v_i^*(s)\right) \mathrm{d}s \\ &= \int_0^t \lim_{k\to\infty} \left(\eta_{i,j}^{(m_k)}(s,\omega) v_i^{(m_k)}(s) - \eta_{i,j}(\vartheta_s\omega) v_i^*(s)\right) \mathrm{d}s = 0.\end{aligned}$$

Thus, by replacing m by m_k in (13.9), and lettting $k \to \infty$, we have

$$\begin{aligned}v_i^*(t) = v_{o,i} + \int_0^t \Big(&- \lambda v_i^* + (\mathrm{A}(\vartheta_s\omega)v^*)_i + \mathcal{O}(\vartheta_s\omega)v_i^* - e^{-\mathcal{O}(\vartheta_s\omega)} \\ & f_i(e^{\mathcal{O}(\vartheta_s\omega)}v_i^*) + e^{-\mathcal{O}(\vartheta_s\omega)}g_i\Big)\mathrm{d}s\end{aligned}$$

for $t \in [0, T_{\max})$. By the uniqueness of solutions of (13.5), we have $v_i^*(t) = v_i(t)$ for $t \in [0, T_{\max})$. By (13.13), $\|\boldsymbol{v}(t)\| \leq |\mathcal{V}(t)|$ for $t \in [0, T_{\max})$, which implies that $T_{\max} = T$ for the solution of (13.5). The proof is complete. □

Next following the idea of [Wang (2006), Lemma 4.2] we prove that for any $\boldsymbol{v}_o \in \ell_\rho^2$ and $\boldsymbol{g} \in \ell_\rho^2$, the equation (13.5) has a unique solution $\boldsymbol{v}(t;\omega,\boldsymbol{v}_o,\boldsymbol{g})$ on $[0,T)$ in the sense of Definition 13.1 with $\boldsymbol{v}(0;\omega,\boldsymbol{v}_o,g) = \boldsymbol{v}_o$.

Theorem 13.1. *Let Assumptions 12.1–12.3 and 13.1–13.2 hold, and let $T > 0$. Then for every $\omega \in \Omega_0$ and any initial data $\boldsymbol{v}_o \in \ell_\rho^2$, the problem (13.5) admits a unique solution $\boldsymbol{v}(\cdot;\omega,\boldsymbol{v}_o,\boldsymbol{g}) \in \mathcal{C}([0,T),\ell_\rho^2)$ with $\boldsymbol{v}(0;\omega,\boldsymbol{v}_o,\boldsymbol{g}) = \boldsymbol{v}_o$. Moreover, the solution $\boldsymbol{v}(t;\omega,\boldsymbol{v}_o,\boldsymbol{g})$ is continuous in $\boldsymbol{v}_o$, $\boldsymbol{g} \in \ell_\rho^2$.*

Proof. For $\boldsymbol{x}_o = (x_{o,i})_{i\in\mathbb{Z}}, \boldsymbol{y}_o = (y_{o,i})_{i\in\mathbb{Z}} \in \ell^2$ and $\boldsymbol{g} = (g_i)_{i\in\mathbb{Z}}$, $\boldsymbol{h} = (h_i)_{i\in\mathbb{Z}} \in \ell^2$, let $\boldsymbol{x}^{(m)}(t,\omega)$, $\boldsymbol{y}^{(m)}(t,\omega)$ be two solutions of the equation (13.8) with initial data $\boldsymbol{x}_o$, $\boldsymbol{y}_o$ and $\boldsymbol{g}$ being replaced by $\boldsymbol{g}$, $\boldsymbol{h}$, respectively. Set $\boldsymbol{z}^{(m)}(t,\omega) = \boldsymbol{x}^{(m)}(t,\omega) - \boldsymbol{y}^{(m)}(t,\omega)$, then $\boldsymbol{z}^{(m)}(t,\omega) \in \ell^2 \subset \ell_\rho^2$ for $\omega \in \Omega_0$. By Assumptions 12.1–12.3 and 13.2 we have

$$\begin{aligned}\left|\langle \mathrm{A}_m(t,\omega)\boldsymbol{z}^{(m)}, \boldsymbol{z}^{(m)}\rangle_\rho\right| &\leq \tilde{p}\eta(\vartheta_t\omega)\left\|\boldsymbol{z}^{(m)}\right\|_\rho^2 \\ &\text{with} \quad \tilde{p} := p + 1 + \gamma + \gamma^2 + \cdots + \gamma^p,\end{aligned}$$

$$\begin{aligned}
&\langle F(e^{\mathcal{O}(\vartheta_t\omega)}\boldsymbol{x}^{(m)}) - F(e^{\mathcal{O}(\vartheta_t\omega)}\boldsymbol{y}^{(m)}), \boldsymbol{z}^{(m)}\rangle_\rho \\
&= \sum_{i\in\mathbb{Z}} \rho_i z_i^{(m)} \cdot \left(f_i(e^{\mathcal{O}(\vartheta_t\omega)}x_i^{(m)}) - f_i(e^{\mathcal{O}(\vartheta_t\omega)}y_i^{(m)})\right) \\
&\geq -\alpha e^{\mathcal{O}(\vartheta_t\omega)} \left\|\boldsymbol{z}^{(m)}\right\|_\rho^2,
\end{aligned}$$

$$\begin{aligned}
&\langle F(e^{\mathcal{O}(\vartheta_t\omega)}\boldsymbol{x}^{(m)}) - F(e^{\mathcal{O}(\vartheta_t\omega)}\boldsymbol{y}^{(m)}), \boldsymbol{z}^{(m)}\rangle_\rho \\
&\leq R\left(e^{\mathcal{O}(\vartheta_t\omega)}(\|\boldsymbol{x}^{(m)}\| + \|\boldsymbol{y}^{(m)}\|)\right) e^{\mathcal{O}(\vartheta_t\omega)} \|\boldsymbol{z}^{(m)}\|_\rho^2,
\end{aligned}$$

and

$$\langle \boldsymbol{g} - \boldsymbol{h}, \boldsymbol{z}^{(m)}\rangle_\rho \leq \|\boldsymbol{g} - \boldsymbol{h}\|_\rho^2 \cdot \|\boldsymbol{z}^{(m)}\|_\rho^2.$$

It then follows that

$$\begin{aligned}
\frac{\mathrm{d}\|\boldsymbol{z}^{(m)}\|_\rho^2}{\mathrm{d}t} &= 2\Big\{-\lambda\|\boldsymbol{z}^{(m)}\|_\rho^2 + \langle \mathrm{A}_m(t,\omega)\boldsymbol{z}^{(m)}, \boldsymbol{z}^{(m)}\rangle_\rho + \mathcal{O}(\vartheta_t\omega)\|\boldsymbol{z}^{(m)}\|_\rho^2 \\
&\quad - e^{-\mathcal{O}(\vartheta_t\omega)}\left(F(e^{\mathcal{O}(\vartheta_t\omega)}\boldsymbol{x}^{(m)}) - F(e^{\mathcal{O}(\vartheta_t\omega)}\boldsymbol{y}^{(m)}), \boldsymbol{z}^{(m)}\right)_\rho \\
&\quad + e^{-\mathcal{O}(\vartheta_t\omega)}\langle \boldsymbol{g} - \boldsymbol{h}, \boldsymbol{z}^{(m)}\rangle_\rho\Big\} \\
&\leq (-\lambda + 2\alpha + 2\mathcal{O}(\vartheta_t\omega) + 2\tilde{p}\eta(\vartheta_t\omega)) \cdot \|\boldsymbol{z}^{(m)}\|_\rho^2 \\
&\quad + \frac{1}{\lambda} e^{-2\mathcal{O}(\vartheta_t\omega)} \|\boldsymbol{h} - \boldsymbol{g}\|_\rho^2,
\end{aligned} \tag{13.16}$$

where $\tilde{p} = p + \sum_{k=0}^p \gamma^k$. Applying Gronwall's inequality to (13.16) on $[0, T)$, we obtain

$$\|\boldsymbol{x}^{(m)}(t) - \boldsymbol{y}^{(m)}(t)\|_\rho \leq C_T\left(\|\boldsymbol{x}_o - \boldsymbol{y}_o\|_\rho + \|\boldsymbol{g} - \boldsymbol{h}\|_\rho\right) \quad \forall t \in [0, T) \tag{13.17}$$

for some constant $= C_T$ depending on T. By similar arguments to the proof of Lemma 13.3, there exist $m_k \to \infty$ such that

$$\lim_{k\to\infty} \boldsymbol{x}^{(m_k)}(t) = \boldsymbol{v}(t;\omega,\boldsymbol{x}_o,\boldsymbol{g}), \quad \lim_{k\to\infty} \boldsymbol{y}^{(m_k)}(t) = \boldsymbol{v}(t;\omega,\boldsymbol{y}_o,\boldsymbol{h}) \quad \forall\, t \in [0, T),$$

which, together with (13.17) imply that

$$\|\boldsymbol{v}(t;\omega,\boldsymbol{x}_o,\boldsymbol{g}) - \boldsymbol{v}(t;\omega,\boldsymbol{y}_o,\boldsymbol{h})\|_\rho \leq C_T\left(\|\boldsymbol{x}_o - \boldsymbol{y}_o\|_\rho + \|\boldsymbol{g} - \boldsymbol{h}\|_\rho\right). \tag{13.18}$$

Similar to the arguments in the proof of Theorem 12.1, let $\tilde{\ell}^2$ be the space ℓ^2 equipped with the weighted norm $\|\cdot\|_\rho$. Then it follows by (13.18) that for given $T > 0$, there exists a mapping $\mathcal{S} \in \mathcal{C}(\tilde{\ell}^2 \times \tilde{\ell}^2, \mathcal{C}([0,T), \ell_\rho^2))$ such that $\mathcal{S}(\boldsymbol{v}_o, \boldsymbol{g}) = v(t;\omega,\boldsymbol{v}_o,\boldsymbol{g})$, where $\boldsymbol{v}(t;\omega,\boldsymbol{v}_o,\boldsymbol{g})$ is the solution of (13.5) on $[0,T)$ with $\boldsymbol{v}(0;\omega,\boldsymbol{v}_o,\boldsymbol{g}) = \boldsymbol{v}_o$. Since $\tilde{\ell}^2 \times \tilde{\ell}^2$ is dense in $\ell_\rho^2 \times \ell_\rho^2$, the mapping $\mathcal{S}$ can be extended uniquely to a continuous mapping $\widetilde{\mathcal{S}} : \ell_\rho^2 \times \ell_\rho^2 \to \mathcal{C}([0,T], \ell_\rho^2)$.

For given $\boldsymbol{v}_o \in \ell_\rho^2$ and $\boldsymbol{g} \in \ell_\rho^2$, we have $\boldsymbol{v}(\cdot;\omega,\boldsymbol{v}_o,\boldsymbol{g}) := \widetilde{\mathcal{S}}(\boldsymbol{v}_o,\boldsymbol{g}) \in \mathcal{C}([0,T),\ell_\rho^2)$ for $T > 0$. It remains to show that $\boldsymbol{v}(t;\omega,\boldsymbol{v}_o,\boldsymbol{g})$ satisfies the integral equation (13.14) for $t \in [0, T)$. In fact, since $\tilde{\ell}^2 \times \tilde{\ell}^2$ is dense in $\ell^2_\rho \times \ell^2_\rho$, there exist two sequences $\{\boldsymbol{v}_o^{(n)}\} \subset \tilde{\ell}^2$, $\{\boldsymbol{g}^{(n)}\} \subset \tilde{\ell}^2$ such that

$$\|\boldsymbol{v}_o^{(n)} - \boldsymbol{v}_o\|_\rho \to 0 \quad \text{and} \quad \|\boldsymbol{g}^{(n)} - \boldsymbol{g}\|_\rho \to 0 \quad \text{as} \quad n \to \infty.$$

Let $\boldsymbol{v}_n^{(m)}(t) = (v_{n,i}^{(m)}(t))_{i\in\mathbb{Z}} = \boldsymbol{v}^{(m)}(t;\omega, v_o^{(n)}, \boldsymbol{g}^{(n)}) \in \tilde{\ell}^2$ be the solution of the equation (13.8). Then it satisfies the integral equation

$$v_{n,i}^{(m)}(t) = v_{o,i}^{(m)} + \int_0^t \left(-\lambda v_{n,i}^{(m)}(s) + \mathrm{A}_m(s,\omega) v_{n,i}^{(m)}(s) + \mathcal{O}(\vartheta_s\omega) v_{n,i}^{(m)}(s)\right) \mathrm{d}s$$
$$+ \int_0^t \left(-e^{-\mathcal{O}(\vartheta_s\omega)} f_i(e^{\mathcal{O}(\vartheta_s\omega)} v_{n,i}^{(m)}(s)) + e^{-\mathcal{O}(\vartheta_s\omega)} g_i^{(n)}\right) \mathrm{d}s.$$

Moreover, $v_{n,i}^{(m)}(t)$ satisfies the differential equation (13.9) for each $i \in \mathbb{Z}$.

Multiplying (13.9) by $\rho_i v_{n,i}^{(m)}(t)$, and summing them over $i \in \mathbb{Z}$, we obtain that for $t \in [0, T)$,

$$\frac{\mathrm{d}\|\boldsymbol{v}_n^{(m)}(t)\|_\rho^2}{\mathrm{d}t} \leq (-\lambda + 2\alpha + 2\mathcal{O}(\vartheta_t\omega) + 2\tilde{p}\eta(\vartheta_t\omega)) \cdot \|\boldsymbol{v}_n^{(m)}(t)\|_\rho^2 + \frac{1}{\lambda} e^{-2\mathcal{O}(\vartheta_t\omega)} \|\boldsymbol{g}\|_\rho^2,$$

which can be integrated to give

$$\|\boldsymbol{v}_n^{(m)}(t)\|_\rho^2 \leq e^{2\alpha t + \int_0^t \left(|2\mathcal{O}(\vartheta_s\omega)| + 2\tilde{p}\eta(\vartheta_s\omega)\right)\mathrm{d}s} \|\boldsymbol{v}_o\|_\rho^2$$
$$+ \frac{\|\boldsymbol{g}\|^2}{\lambda} \left(e^{(2\alpha-\lambda)t + \int_0^t \left(2\mathcal{O}(\vartheta_s\omega) + 2\tilde{p}\eta(\vartheta_s\omega)\right)\mathrm{d}s}\right)$$
$$\cdot \left(\int_0^t e^{(\lambda-2\alpha)s - 2\mathcal{O}(\vartheta_s\omega) + \int_0^s \left(2\mathcal{O}(\vartheta_r\omega) + 2\tilde{p}\eta(\vartheta_r\omega)\right)\mathrm{d}r} \mathrm{d}s\right). \quad (13.19)$$

Noticing that the terms on the right hand of (13.19) are independent of m, n, by the arguments similar to those above, we have that $\boldsymbol{v}(t;\omega, \boldsymbol{v}_o, \boldsymbol{g}) \in \ell^2_\rho$ is the limit function of a subsequence of $\{\boldsymbol{v}^{(m)}(t;\omega, \boldsymbol{v}_o^{(n)}, \boldsymbol{g}^{(n)})\}$ in ℓ^2_ρ. Moreover, $\boldsymbol{v}(t;\omega, \boldsymbol{v}_o, \boldsymbol{g})$ satisfies the integral equation (13.14) for $t \in [0, T)$. The proof is complete. □

13.1.3 *Random dynamical systems generated by random LDS*

It will be shown here that the solution to (13.5) generates a random dynamical system with state space ℓ^2_ρ. Throughout the rest of this section, if no confusion occurs, we write $\boldsymbol{v}(t;\omega, \boldsymbol{v}_o, \boldsymbol{g})$ as $\boldsymbol{v}(t;\omega, \boldsymbol{v}_o)$, where $\boldsymbol{v}(t;\omega, \boldsymbol{v}_o, \boldsymbol{g})$ is shown to exist in Theorems 13.1.

Theorem 13.2. *Let Assumptions 12.1–12.3 and 13.1–13.2 hold. Then the solution of the equation* (13.5) *generates a continuous random dynamical system* $\{\varphi(t,\omega)\}_{t\geq 0, \omega\in\Omega_0}$ *over* $(\Omega_0, \mathcal{F}, \mathbb{P}, (\vartheta_t)_{t\in\mathbb{R}})$ *with state space* ℓ^2_ρ*:*

$$\varphi(t,\omega,\boldsymbol{v}_o) := \boldsymbol{v}(t;\omega,\boldsymbol{v}_o) \quad \text{for} \quad \boldsymbol{v}_o \in \ell^2_\rho,\ t \geq 0,\ \omega \in \Omega_0.$$

Proof. We first show that $\boldsymbol{v}(t;\omega,\boldsymbol{v}_o)=\boldsymbol{v}(t;\omega,\boldsymbol{v}_o,\boldsymbol{g})$ is measurable in $(t,\omega,\boldsymbol{v}_o)$. In fact, firstly, by the proof of Lemma 13.3, given any $\boldsymbol{v}_o\in\ell^2$ and $\boldsymbol{g}\in\ell^2$, the solution $\boldsymbol{v}(t;\omega,\boldsymbol{v}_o,\boldsymbol{g})\in\ell^2$ for $t\in[0,\infty)$. In this case, the function $\mathfrak{F}(\boldsymbol{v},t,\omega,\boldsymbol{g})=\mathfrak{F}(\boldsymbol{v},\vartheta_t\omega)$ is continuous in $\boldsymbol{v}$, $\boldsymbol{g}$ and measurable in t, ω, which implies that the mapping $\boldsymbol{v}:[0,\infty)\times\Omega_0\times\ell^2\times\ell^2\to\ell^2$, $(t;\omega,\boldsymbol{v}_o,\boldsymbol{g})\mapsto\boldsymbol{v}(t;\omega,\boldsymbol{v}_o,\boldsymbol{g})$ is $(\mathcal{B}([0,\infty))\times\mathcal{F}\times\mathcal{B}(\ell^2)\times\mathcal{B}(\ell^2),\mathcal{B}(\ell^2))$-measurable (see [Arnold (1998)]).

Secondly, by Theorem 13.1 for any $\boldsymbol{v}_o\in\ell^2_\rho$ and $\boldsymbol{g}\in\ell^2_\rho$, the solution $\boldsymbol{v}(t;\omega,\boldsymbol{v}_o,\boldsymbol{g})\in\ell^2_\rho$ for $t\in[0,\infty)$. Given any $n>0$, let the truncation operator $\Xi_n:\ell^2_\rho\to\ell^2$ be the same as defined in (12.26), i.e., Ξ_n is continuous operator satisfying

$$(\Xi_n\boldsymbol{v})_j=\begin{cases}v_j & \text{if } |j|\le n\\ 0 & \text{if } |j|>n\end{cases},$$

and for any $\boldsymbol{v}_o\in\ell^2_\rho$ and $\boldsymbol{g}\in\ell^2_\rho$

$$\boldsymbol{v}(t;\omega,\boldsymbol{v}_o,\boldsymbol{g})=\lim_{n\to\infty}\boldsymbol{v}_n\quad\text{with}\quad\boldsymbol{v}_n=\boldsymbol{v}(t;\omega,\Xi_n\boldsymbol{v}_o,\Xi_n\boldsymbol{g}).$$

We know that $\boldsymbol{v}:[0,\infty)\times\Omega_0\times\ell^2\times\ell^2\to\ell^2$ is $\mathcal{B}([0,\infty))\times\mathcal{F}\times\mathcal{B}(\ell^2)\times\mathcal{B}(\ell^2),\mathcal{B}(\ell^2))$-measurable and that $\mathrm{Id}:\ell^2\to\ell^2_\rho$ is continuous. Hence $\boldsymbol{v}:[0,\infty)\times\Omega_0\times\ell^2\times\ell^2\to\ell^2_\rho$ is $(\mathcal{B}([0,\infty))\times\mathcal{F}\times\mathcal{B}(\ell^2)\times\mathcal{B}(\ell^2),\mathcal{B}(\ell^2_\rho))$-measurable. Observe also that $(\mathrm{Id},\mathrm{Id},\Xi_n,\Xi_n)$: $[0,\infty)\times\Omega_0\times\ell^2_\rho\times\ell^2_\rho\to[0,\infty)\times\Omega_0\times\ell^2\times\ell^2$ is $(\mathcal{B}([0,\infty))\times\mathcal{F}\times\mathcal{B}(\ell^2_\rho)\times\mathcal{B}(\ell^2_\rho),\mathcal{B}([0,\infty))\times\mathcal{F}_0\times\mathcal{B}(\ell^2)\times\mathcal{B}(\ell^2))$-measurable. Hence

$$\boldsymbol{v}_n=\boldsymbol{v}\circ(\mathrm{Id},\mathrm{Id},\Xi_n,\Xi_n):[0,\infty)\times\Omega_0\times\ell^2_\rho\times\ell^2_\rho\to\ell^2_\rho$$

is $(\mathcal{B}([0,\infty))\times\mathcal{F}\times\mathcal{B}(\ell^2_\rho)\times\mathcal{B}(\ell^2_\rho),\mathcal{B}(\ell^2_\rho))$-measurable.

It then follows from (12.27) that $\boldsymbol{v}:[0,\infty)\times\Omega_0\times\ell^2_\rho\times\ell^2_\rho\to\ell^2_\rho$ is $(\mathcal{B}([0,\infty))\times\mathcal{F}\times\mathcal{B}(\ell^2_\rho)\times\mathcal{B}(\ell^2_\rho),\mathcal{B}(\ell^2_\rho))$-measurable. Therefore, fixing $\boldsymbol{g}\in\ell^2_\rho$, we have that $\boldsymbol{v}(t;\omega,\boldsymbol{v}_o)$ $=\boldsymbol{v}(t;\omega,\boldsymbol{v}_o,\boldsymbol{g})$ is measurable in $(t,\omega,\boldsymbol{v}_o)$. The other conditions of the RDS follow directly.

As a direct consequence of Theorem 13.2 and Lemma 11.1, the mapping

$$\tilde{\varphi}(t,\omega,\boldsymbol{u}_o):=\mathcal{T}(\vartheta_t\omega)\psi(t,\omega)\left(\mathcal{T}^{-1}(\omega)\boldsymbol{u}_o\right)\quad\text{for}\quad\boldsymbol{u}_o\in\ell^2_\rho,\,t\ge0,\,\omega\in\Omega_0,\quad(13.20)$$

where $\mathcal{T}$ is defined in Section 13.1.1, defines a continuous RDS $\{\tilde{\varphi}(t,\omega)\}_{t\ge0,\omega\in\Omega_0}$ over $(\Omega_0,\mathcal{F},\mathbb{P},(\vartheta_t)_{t\in\mathbb{R}})$ associated with (13.3). □

From now on, the RDS $\{\tilde{\varphi}(t,\omega)\}_{t\ge0,\omega\in\Omega_0}$ defined by (13.20) is referred to as the RDS associated with the problem (13.3).

13.2 Existence of global random attractors in weighted space

The existence of a global random attractor will be now established for the random dynamical system $\{\tilde{\varphi}(t,\omega)\}_{t\ge0,\omega\in\Omega_0}$ generated by the solution to (13.3) defined by (13.20) in Theorem 13.1 in weighted space ℓ^2_ρ. To that end, we first prove the existence of a tempered random bounded absorbing set and then verify the asymptotic nullness in Definition 11.6 for the RDS $\{\tilde{\varphi}(t,\omega)\}_{t\geq 0,\omega\in\Omega_0}$ in ℓ^2_ρ. The domain of attraction considered here is the set of all tempered random sets of ℓ^2_ρ, denoted by $\mathscr{D}(\ell^2_\rho)$.

13.2.1 *Existence of tempered random bounded absorbing sets*

In addition to the assumptions required for the existence of solutions, the following assumption on $\eta(\omega)$ defined in Assumption 13.2 is needed.

Assumption 13.3. $\lambda > \tilde{p}\,\mathbb{E}|\eta(\omega)|$, where

$$\tilde{p} = p + \sum_{k=0}^{p} \gamma^k, \quad \mathbb{E}|\eta(\omega)| = \lim_{t\to\pm\infty} \frac{1}{t}\int_0^t \eta(\vartheta_t\omega)\mathrm{d}s.$$

Lemma 13.4. *Assumptions 12.1–12.3 and 13.1–13.3 hold. Then the random dynamical system $\{\tilde{\varphi}(t,\omega)\}_{t\geq 0,\omega\in\Omega_0}$ possesses a tempered bounded closed random absorbing set $\mathcal{Q} = \{Q(\omega)\}_{\omega\in\Omega_0} \in \mathscr{D}(\ell^2_\rho)$ of such that for any $\mathcal{D} = \{D(\omega)\}_{\omega\in\Omega_0} \in \mathscr{D}(\ell^2_\rho)$ and each $D(\omega) \in \mathcal{D}$, there exists $T_{\mathcal{D}}(\omega) > 0$ such that $\tilde{\varphi}(t,\vartheta_{-t}\omega, D(\vartheta_{-t}\omega)) \subset Q(\omega)$ for all $t \geq T_{\mathcal{D}}(\omega)$. In particular, there exists $T_{\mathcal{Q}}(\omega) > 0$ such that $\tilde{\varphi}(t,\vartheta_{-t}\omega, Q(\vartheta_{-t}\omega)) \subset Q(\omega)$ for all $t \geq T_{\mathcal{Q}}(\omega)$.*

Proof. First we consider the initial condition $\boldsymbol{u}_o$ and $\boldsymbol{g}$ in ℓ^2. For $\omega \in \Omega_0$ and $\boldsymbol{g} \in \ell^2$, let $\boldsymbol{v}^{(m)}(t,\omega) = \boldsymbol{v}^{(m)}(t;\omega,\boldsymbol{v}_o(\omega),\boldsymbol{g})$ be a solution of equation (13.8) with $\boldsymbol{v}_o(\omega) = e^{-\mathcal{O}(\omega)}\boldsymbol{u}_o \in \ell^2$. Then $\boldsymbol{v}^{(m)}(t,\omega) \in \ell^2$ all $t \geq 0$. By Assumption 13.3 there exists $\epsilon > 0$ such that

$$2\lambda - \epsilon > 2\tilde{p}\mathbb{E}|\eta(\omega)|. \tag{13.21}$$

Using Assumptions 12.2 and 13.1, we have

$$\begin{aligned}
\frac{\mathrm{d}\|\boldsymbol{v}^{(m)}(t,\omega)\|_\rho^2}{\mathrm{d}t} &= 2\Big(- \lambda\|\boldsymbol{v}^{(m)}\|_\rho^2 + \langle \mathrm{A}_m(t,\omega)\boldsymbol{v}^{(m)}, \boldsymbol{v}^{(m)}\rangle_\rho + \mathcal{O}(\vartheta_t\omega)\|\boldsymbol{v}^{(m)}\|_\rho^2 \\
&\quad - e^{-2\mathcal{O}(\vartheta_t\omega)} \sum_{i\in\mathbb{Z}} \rho_i(e^{\mathcal{O}(\vartheta_t\omega)}v_i^{(m)}) \cdot f_i(e^{\mathcal{O}(\vartheta_t\omega)}v_i^{(m)}) \\
&\quad + \langle e^{-\mathcal{O}(\vartheta_t\omega)}\boldsymbol{g}, \boldsymbol{v}^{(m)}\rangle_\rho\Big) \\
&\leq \big(- (2\lambda - \epsilon) + 2\mathcal{O}(\vartheta_t\omega) + 2\tilde{p}\eta(\vartheta_t\omega)\big) \cdot \|\boldsymbol{v}^{(m)}\|_\rho^2 \\
&\quad + e^{-2\mathcal{O}(\vartheta_t\omega)}\big(2\|\boldsymbol{\beta}\|_\rho^2 + \frac{1}{\epsilon}\|\boldsymbol{g}\|_\rho^2\big).
\end{aligned} \tag{13.22}$$

Applying Gronwall's inequality to (13.22) gives that for $t > 0$,

$$\begin{aligned}
\|\boldsymbol{v}^{(m)}(t,\omega)\|_\rho^2 &\leq e^{-(2\lambda-\epsilon)t + \int_0^t \left(2\mathcal{O}(\vartheta_s\omega) + 2\tilde{p}\eta(\vartheta_s\omega)\right)\mathrm{d}s}\|\boldsymbol{v}_o\|_\rho^2 \\
&\quad + \big(2\|\boldsymbol{\beta}\|_\rho^2 + \frac{1}{\epsilon}\|\boldsymbol{g}\|_\rho^2\big)e^{-(2\lambda-\epsilon)t + \int_0^t \left(2\mathcal{O}(\vartheta_s\omega) + 2\tilde{p}\eta(\vartheta_s\omega)\right)\mathrm{d}s} \\
&\quad \cdot \int_0^t e^{(2\lambda-\epsilon)s - 2\mathcal{O}(\vartheta_s\omega) - \int_0^s \left(2\mathcal{O}(\vartheta_r\omega) + 2\tilde{p}\eta(\vartheta_r\omega)\right)\mathrm{d}r}\mathrm{d}s.
\end{aligned} \tag{13.23}$$

By the arguments in Theorem 13.1, there exists $m_k \to \infty$ such that $\boldsymbol{v}^{(m_k)}(t,\omega) \to \boldsymbol{v}(t,\omega) = \varphi(t,\omega,\boldsymbol{v}_o(\omega)) = \boldsymbol{v}(t;\omega,\boldsymbol{v}_o(\omega),\boldsymbol{g})$, where $\boldsymbol{v}(t;\omega,\boldsymbol{v}_o(\omega),\boldsymbol{g})$ is the solution of equation (13.5) with $\boldsymbol{v}_o(\omega) = e^{-\mathcal{O}(\omega)}\boldsymbol{u}_o \in \ell^2$ in the sense of Definition 13.1. Therefore, the equation (13.23) still holds with $\boldsymbol{v}^{(m)}(t,\omega)$ being replaced by $\boldsymbol{v}(t;\omega)$.

Now for any $\boldsymbol{v}_o \in \ell^2_\rho$ and $\boldsymbol{g} \in \ell^2_\rho$, let $\{\boldsymbol{v}_o^{(n)}\} \subset \ell^2$ and $\{\boldsymbol{g}^{(n)}\} \subset \ell^2$ be two sequences such that $\|\boldsymbol{v}_o^{(n)} - \boldsymbol{v}_o\|_\rho \to 0$ and $\|\boldsymbol{g}^{(n)} - \boldsymbol{g}\|_\rho \to 0$ as $n \to \infty$. Following the arguments in Theorem 13.1 again, $\boldsymbol{v}(t;\omega,\boldsymbol{v}_o^{(n)},\boldsymbol{g}^{(n)}) \to \boldsymbol{v}(t;\omega,\boldsymbol{v}_o,\boldsymbol{g})$ as $n \to \infty$ in ℓ^2_ρ. Hence (13.23) also holds in ℓ^2_ρ for $\boldsymbol{v}^{(m)}(t,\omega)$ being replaced by $\boldsymbol{v}(t;\omega,\boldsymbol{v}_o,\boldsymbol{g})$, and consequently

$$\begin{aligned}\|\boldsymbol{v}(t;\vartheta_{-t}\omega,\boldsymbol{v}_o(\vartheta_{-t}\omega),\boldsymbol{g})\|^2_\rho \le{}& e^{-(2\lambda-\epsilon)t+\int_0^t \left(2\mathcal{O}(\vartheta_{s-t}\omega)+2\tilde{p}\eta(\vartheta_{s-t}\omega)\right)\mathrm{d}s}\|\boldsymbol{v}_o(\vartheta_{-t}\omega)\|^2_\rho \\ &+\frac{1}{2}R^2(\omega)e^{-2\mathcal{O}(\omega)},\end{aligned}$$

where

$$\begin{aligned}R^2(\omega) ={}& 4e^{2\mathcal{O}(\omega)}(\|\boldsymbol{\beta}\|^2_\rho + \frac{1}{2\epsilon}\|\boldsymbol{g}\|^2_\rho)\int_{-\infty}^0 e^{(2\lambda-\epsilon)s-2\mathcal{O}(\vartheta_s\omega)+\int_s^0\left(2\mathcal{O}(\vartheta_r\omega)+2\tilde{p}\eta(\vartheta_r\omega)\right)\mathrm{d}r}\mathrm{d}s \\ &< \infty.\end{aligned}$$

Now consider the RDS $\tilde{\varphi}(t,\omega,\boldsymbol{u}_o) = \boldsymbol{u}(t,\omega)$ defined by the solution to the equation (13.3). Then for any $\mathcal{D} = \{D(\omega)\}_{\omega\in\Omega_0} \in \mathscr{D}(\ell^2_\rho)$ and any $\boldsymbol{u}_o \in D(\vartheta_{-t}\omega)$,

$$\begin{aligned}\|\tilde{\varphi}(t,\vartheta_{-t}\omega,\boldsymbol{u}_o)\|^2_\rho ={}& \left\|e^{\mathcal{O}(\omega)}\varphi(t,\vartheta_{-t}\omega,e^{-\mathcal{O}(\vartheta_{-t}\omega)}\boldsymbol{u}_o)\right\|^2_\rho \\ \le{}& e^{2\mathcal{O}(\omega)}e^{-(2\lambda-\epsilon)t-\mathcal{O}(\vartheta_{-t}\omega)+\int_{-t}^0\left(2\mathcal{O}(\vartheta_s\omega)+2\tilde{p}\eta(\vartheta_s\omega)\right)\mathrm{d}s}\sup_{\boldsymbol{u}\in D(\vartheta_{-t}\omega)}\|\boldsymbol{u}\|^2_\rho + \frac{1}{2}R^2(\omega).\end{aligned}$$

Let $\tilde{\lambda} = (2\lambda-\epsilon) - 2\tilde{p}\,\mathbb{E}|\eta(\omega)|$. Then $\tilde{\lambda} > 0$ due to (13.21), and thus by properties of $\eta(\vartheta_{\pm t}\omega)$ and that $\mathcal{D} \in \mathscr{D}(\ell^2_\rho)$, we have

$$\begin{aligned}&\lim_{t\to+\infty} e^{2\mathcal{O}(\omega)}e^{-(2\lambda-\epsilon)t-\mathcal{O}(\vartheta_{-t}\omega)+\int_{-t}^0\left(2\mathcal{O}(\vartheta_s\omega)+2\tilde{p}\eta(\vartheta_s\omega)\right)\mathrm{d}s}\sup_{\boldsymbol{u}\in D(\vartheta_{-t}\omega)}\|\boldsymbol{u}\|^2_\rho \\ &= \lim_{t\to+\infty} e^{2\mathcal{O}(\omega)}e^{-(\frac{\tilde{\lambda}}{2}+2\tilde{p}\mathbb{E}|\eta(\omega)|)t-\mathcal{O}(\vartheta_{-t}\omega)+\int_{-t}^0\left(2\mathcal{O}(\vartheta_s\omega)+2\tilde{p}\eta(\vartheta_s\omega)\right)\mathrm{d}s}e^{-\frac{\tilde{\lambda}}{2}t}\|D(\vartheta_{-t}\omega)\|^2_\rho \\ &= 0, \quad \text{with} \quad \|D(\vartheta_{-t}\omega)\|^2_\rho := \sup_{\boldsymbol{u}\in D(\vartheta_{-t}\omega)}\|\boldsymbol{u}\|^2_\rho.\end{aligned}$$

Denote by $\mathbb{B}_{\ell^2_\rho}(0,r)$ the ball in ℓ^2_ρ with center 0 and radius r. It then follows that

$$Q(\omega) = \overline{\{\boldsymbol{u} \in \ell^2_\rho : \|\boldsymbol{u}\|_\rho \le R(\omega)\}} = \overline{\mathbb{B}_{\ell^2_\rho}(0,R(\omega))} \subset \ell^2_\rho \tag{13.24}$$

forms a bounded closed random absorbing set for $\{\tilde{\varphi}(t,\omega)\}_{t\ge0,\omega\in\Omega_0}$.

It remains to show that $\mathcal{Q} = \{Q(\omega)\}_{\omega\in\Omega_0} \in \mathcal{D}(\ell^2_\rho)$. In fact, for every $c > 0$,

$$\left(e^{-ct}R(\vartheta_{-t}\omega)\right)^2 \left(\|\boldsymbol{\beta}\|^2_\rho + \frac{1}{2\epsilon}\|\boldsymbol{g}\|^2_\rho\right)^{-1}$$
$$= 4e^{-2ct+\mathcal{O}(\vartheta_{-t}\omega)}\int_{-\infty}^{-t} e^{(2\lambda-\epsilon)(s+t)-2\mathcal{O}(\vartheta_s\omega)+\int_s^{-t}\left(2\mathcal{O}(\vartheta_r\omega)+2\tilde{p}\eta(\vartheta_r\omega)\right)\mathrm{d}r}\mathrm{d}s$$
$$\longrightarrow 0 \quad \text{as} \quad t\to+\infty,$$

i.e., $\mathcal{Q} \in \mathcal{D}(\ell^2_\rho)$. The proof is complete. □

13.2.2 *Existence of global random attractors*

In this subsection, we show that the RDS $\{\tilde{\varphi}(t,\omega)\}_{t\geq 0,\omega\in\Omega_0}$ has a global random attractor in ℓ^2_ρ. Due to Theorem 11.4 it suffices to show that the RDS $\{\tilde{\varphi}(t,\omega)\}_{t\geq 0,\omega\in\Omega_0}$ is aymptotically null on $\mathcal{Q} = \{Q(\omega)\}_{\omega\in\Omega_0}$, where $Q(\omega)$ is defined as in (13.24).

Lemma 13.5. *Let Assumptions 12.1–12.3 and 13.1–13.3 hold. Then for any $\varepsilon > 0$, there exist $T(\varepsilon,\omega,Q)$ and $I(\varepsilon,\omega)\in\mathbb{N}$ such that the RDS $\{\tilde{\varphi}(t,\omega)\}_{t\in\mathbb{R},\omega\in\Omega_0}$ associated with the equation* (13.3) *satisfies*

$$\sum_{|i|\geq I(\varepsilon,\omega)} \rho_i|\tilde{\varphi}_i(t,\vartheta_{-t}\omega,\boldsymbol{u}_o)|^2 \leq \varepsilon, \quad \boldsymbol{u}_o\in Q(\vartheta_{-t}\omega), \quad t\geq T_Q(\omega),$$

where $\tilde{\varphi}_i(t,\vartheta_{-t}\omega,\boldsymbol{u}_o)$ is the ith component of $\tilde{\varphi}(t,\vartheta_{-t}\omega,\boldsymbol{u}_o)$.

Proof. Let $\xi \in C^1(\mathbb{R}^+,\mathbb{R})$ be the cut-off function defined as in (3.8), and recall that there exists $C_0 > 0$ such that $|\xi'(s)| \leq C_0$ for $s\in\mathbb{R}^+$. Let k be a suitable large integer (to be specified later) and define $\xi_k(s) := \xi(s/k)$ for $s\in\mathbb{R}$.

Let $\boldsymbol{u}(t;\omega,\boldsymbol{u}_o,\boldsymbol{g}) = (u_i(t;\omega,\boldsymbol{u}_o,\boldsymbol{g}))_{i\in\mathbb{Z}}$ be the solution of the equation (13.3) with initial data $\boldsymbol{u}_o$. Then

$$\boldsymbol{v}(t;\omega,\boldsymbol{v}_o(\omega),\boldsymbol{g}) = (v_i(t;\omega,\boldsymbol{v}_o(\omega),\boldsymbol{g}))_{i\in\mathbb{Z}} = e^{-\mathcal{O}(\vartheta_t\omega)}\boldsymbol{u}(t;\omega,\boldsymbol{u}_o,\boldsymbol{g})$$

is the solution of equation (13.5) with $\boldsymbol{v}_o(\omega) = e^{-\mathcal{O}(\omega)}\boldsymbol{u}_o \in \ell^2_\rho$. In addition, for $n\geq 1$ let $\boldsymbol{u}_o^{(n)} = \Xi_n\boldsymbol{u}_o$ and $\boldsymbol{g}^{(n)} = \Xi_n\boldsymbol{g}$, where Ξ_n is as defined in (12.26). Then $\boldsymbol{u}_o^{(n)}$, $\boldsymbol{g}^{(n)}$ $\in \ell^2$ and $\boldsymbol{u}(t;\omega,u_o^{(n)},\boldsymbol{g}^{(n)}) \to \boldsymbol{u}(t;\omega,\boldsymbol{u}_o,\boldsymbol{g})$ in ℓ^2_ρ.

Let $\boldsymbol{v}_n^{(m)}(t) = \boldsymbol{v}^{(m)}(t;\omega,\boldsymbol{u}_o^{(n)}(\omega),\boldsymbol{g}^{(n)})$ be the solution of (13.8) with $\boldsymbol{g} = \boldsymbol{g}^{(n)}$ satisfying $\boldsymbol{v}_n^{(m)}(0) = \boldsymbol{v}_o^{(n)}(\omega) = e^{-\mathcal{O}(\omega)}\boldsymbol{u}_o^{(n)}$. For each $n,m\in\mathbb{N}$, set $z_{n,i}^{(m)} = \xi_k(|i|)v_{n,i}^{(m)}$ for all $i\in\mathbb{Z}$, and write $\boldsymbol{z}_n^{(m)} = (z_{n,i}^{(m)})_{i\in\mathbb{Z}}$.

Noticing that $\boldsymbol{v}_n^{(m)}(\cdot) \in \mathcal{C}([0,\infty),\ell^2)\cap\mathcal{C}^1((0,\infty),\ell^2)$, we can then take the inner product $\langle\cdot,\cdot\rangle_\rho$ of (13.8) with $\boldsymbol{z}_n^{(m)}$ to obtain

$$\begin{aligned}\left\langle \frac{\mathrm{d}\boldsymbol{v}_n^{(m)}(t)}{\mathrm{d}t}, \boldsymbol{z}_n^{(m)}\right\rangle_\rho &= -\lambda\langle \boldsymbol{v}_n^{(m)}, \boldsymbol{z}_n^{(m)}\rangle_\rho + \langle \mathrm{A}_m(t,\omega)\boldsymbol{v}_n^{(m)}, \boldsymbol{z}_n^{(m)}\rangle_\rho \\ &\quad + \mathcal{O}(\vartheta_t\omega)\langle \boldsymbol{v}_n^{(m)}, \boldsymbol{z}_n^{(m)}\rangle_\rho - \left\langle e^{-\mathcal{O}(\vartheta_t\omega)}F(e^{\mathcal{O}(\vartheta_t\omega)}\boldsymbol{v}_n^{(m)}), \boldsymbol{z}_n^{(m)}\right\rangle_\rho \\ &\quad + \langle e^{-\mathcal{O}(\vartheta_t\omega)}\boldsymbol{g}^{(n)}, \boldsymbol{z}_n^{(m)}\rangle_\rho. \end{aligned} \tag{13.25}$$

The terms in (13.25) satisfy, respectively,

$$\begin{aligned}\left\langle \frac{\mathrm{d}\boldsymbol{v}_n^{(m)}(t)}{\mathrm{d}t}, \boldsymbol{z}_n^{(m)}\right\rangle_\rho &= \frac{1}{2}\frac{\mathrm{d}}{\mathrm{d}t}\sum_{i\in\mathbb{Z}} \xi_k(|i|)\rho_i (v_{n,i}^{(m)})^2, \\ \langle \boldsymbol{v}_n^{(m)}, \boldsymbol{z}_n^{(m)}\rangle_\rho &= \sum_{i\in\mathbb{Z}} \xi_k(|i|)\rho_i (v_{n,i}^{(m)})^2, \\ |\langle A_m(t,\omega)\boldsymbol{v}_n^{(m)}, \boldsymbol{z}_n^{(m)}\rangle_\rho| &\le (p + \sum_{k=0}^{p}\gamma^k)\eta(\vartheta_t\omega)\cdot \sum_{i\in\mathbb{Z}} \xi_k(|i|)\rho_i (v_{n,i}^{(m)})^2 \\ &\quad + \eta(\vartheta_t\omega)\cdot\frac{\gamma C_0}{k}\|\boldsymbol{v}_n^{(m)}\|_\rho^2, \\ \langle (e^{-\mathcal{O}(\vartheta_t\omega)}F(e^{\mathcal{O}(\vartheta_t\omega)}\boldsymbol{v}_n^{(m)}), \boldsymbol{z}_n^{(m)}\rangle_\rho &\ge e^{-2\mathcal{O}(\vartheta_t\omega)}\sum_{|i|\ge k}\rho_i\beta_i^2, \end{aligned}$$

and for some $\epsilon > 0$

$$\langle e^{-\mathcal{O}(\vartheta_t\omega)}\boldsymbol{g}^{(n)}, \boldsymbol{z}_n^{(m)}\rangle_\rho \le \frac{\epsilon}{2}\sum_{i\in\mathbb{Z}} \xi_k(|i|)\rho_i (v_{n,i}^{(m)})^2 + \frac{1}{2\epsilon}e^{-2\mathcal{O}(\vartheta_t\omega)}\sum_{|i|\ge k}\rho_i g_i^2.$$

Putting the above equations and inequalities into (13.25) gives

$$\begin{aligned}&\frac{\mathrm{d}}{\mathrm{d}t}\sum_{i\in\mathbb{Z}} \xi_k(|i|)\rho_i (v_{n,i}^{(m)})^2 + ((2\lambda-\epsilon) - 2\mathcal{O}(\vartheta_t\omega) - 2\tilde{p}|\eta(\vartheta_t\omega)|)\sum_{i\in\mathbb{Z}} \xi_k(|i|)\rho_i (v_{n,i}^{(m)})^2 \\ &\le \frac{1}{\epsilon}e^{-2\mathcal{O}(\vartheta_t\omega)}\sum_{|i|\ge k}\rho_i(g_i^2+\beta_i^2) + \eta(\vartheta_t\omega)\cdot\frac{\gamma C_0}{k}\|\boldsymbol{v}_n^{(m)}\|_\rho^2. \end{aligned} \tag{13.26}$$

Then applying Gronwall's inequality to (13.26) from $T_Q = T_Q(\omega)$ to t with ω being replaced by $\vartheta_{-t}\omega$ gives that for $\boldsymbol{u}_o \in Q(\vartheta_{-t}\omega)$

$$\begin{aligned}&\sum_{i\in\mathbb{Z}} \xi_k(|i|)\rho_i |v_{n,i}^{(m)}(t;\vartheta_{-t}\omega, e^{-\mathcal{O}(\vartheta_{-t}\omega)}\boldsymbol{u}_o^{(n)}, \boldsymbol{g}^{(n)})|^2 \\ &\le \mathcal{J}_1(t) + \frac{1}{\epsilon}\mathcal{J}_2(t) + \frac{\gamma C_0}{k}\int_{T_Q}^{t}\mathcal{J}_3(\tau)\mathrm{d}\tau, \end{aligned} \tag{13.27}$$

where

$$\mathcal{J}_1(t) = e^{-(2\lambda-\epsilon)(t-T_Q)+\int_{T_Q}^t \left(2\mathcal{O}(\vartheta_{s-t}\omega)+2\tilde{p}\eta(\vartheta_{s-t}\omega)\right)\mathrm{d}s} \cdot \left\| \boldsymbol{v}_n^{(m)}(T_Q;\vartheta_{-t}\omega, e^{-\mathcal{O}(\vartheta_{-t}\omega)}\boldsymbol{u}_o^{(n)}, \boldsymbol{g}^{(n)}) \right\|_\rho^2,$$

$$\mathcal{J}_2(t) = \sum_{|i|\geq k} \rho_i \left(g_i^2+\beta_i^2\right) \int_{T_Q}^t e^{-(2\lambda-\epsilon)(t-\tau)+\int_\tau^t \left(2\mathcal{O}(\vartheta_{s-t}\omega)+2\tilde{p}\eta(\vartheta_{s-t}\omega)\right)\mathrm{d}s - 2\mathcal{O}(\vartheta_{\tau-t}\omega)}\mathrm{d}\tau,$$

$$\mathcal{J}_3(\tau) = e^{-(2\lambda-\epsilon)(t-\tau)+\int_\tau^t \left(2\mathcal{O}(\vartheta_{s-t}\omega)+2\tilde{p}\eta(\vartheta_{s-t}\omega)\right)\mathrm{d}s - 2\mathcal{O}(\vartheta_{\tau-t}\omega)}\eta(\vartheta_{\tau-t}\omega) \cdot \| \boldsymbol{v}_n^{(m)}(\tau;\vartheta_{-t}\omega, e^{-\mathcal{O}(\vartheta_{-t}\omega)}\boldsymbol{u}_o^{(n)}, \boldsymbol{g}^{(n)}) \|_\rho^2.$$

We next estimate each term on the right-hand side of (13.27) for $\boldsymbol{u}_o \in Q(\vartheta_{-t}\omega) \subset \ell_\rho^2$. First, by (13.24)

$$\begin{aligned}\mathcal{J}_1(t) &\leq e^{-(2\lambda-\epsilon)t-2\mathcal{O}(\vartheta_{-t}\omega)+\int_0^t(2\mathcal{O}(\vartheta_{s-t}\omega)+2\tilde{p}\eta(\vartheta_{s-t}\omega))\mathrm{d}s}\|\boldsymbol{u}_o\|_\rho^2 \\ &\quad + \frac{1}{2}R^2(\omega)e^{-2\mathcal{O}(\omega)}\ e^{-(2\lambda-\epsilon)(t-T_Q)+\int_{T_Q}^t(2\mathcal{O}(\vartheta_{s-t}\omega)+2\tilde{p}\eta(\vartheta_{s-t}\omega))\mathrm{d}s} \\ &\to 0 \quad \text{as} \quad t \to +\infty,\end{aligned}$$

which implies that for all $\varepsilon > 0$ that there exists $T_1(\varepsilon, \omega, Q) \geq T_Q$ such that

$$\mathcal{J}_1(t) \leq \frac{\varepsilon}{3}e^{-2\mathcal{O}(\omega)}, \quad t \geq T_1(\varepsilon, \omega, Q). \tag{13.28}$$

Then by

$$\begin{aligned}&\int_{T_Q}^t e^{-(2\lambda-\epsilon)(t-\tau)+\int_\tau^t(2\mathcal{O}(\vartheta_{s-t}\omega)+2\tilde{p}\eta(\vartheta_{s-t}\omega))\mathrm{d}s-2\mathcal{O}(\vartheta_{\tau-t}\omega)}\mathrm{d}\tau \\ &\quad \leq \int_{-\infty}^0 e^{(2\lambda-\epsilon)\tau-\int_\tau^0(\mathcal{O}(\vartheta_s\omega)+2\tilde{p}\eta(\vartheta_s\omega))\mathrm{d}s-2\mathcal{O}(\vartheta_\tau\omega)}\mathrm{d}\tau < \infty,\end{aligned}$$

and the fact that $\boldsymbol{g}$, $\boldsymbol{\beta} \in \ell_\rho^2$, there exists $I_1(\varepsilon, \omega) \in \mathbb{N}$ such that

$$\frac{1}{\epsilon}\mathcal{J}_2(t) \leq \frac{\varepsilon}{3}e^{-2\mathcal{O}(\omega)}, \quad k > I_1(\varepsilon, \omega). \tag{13.29}$$

Recall that $\tilde{\lambda} = (2\lambda - \epsilon) - 2\tilde{p}\mathbb{E}\,|\eta(\omega)| > 0$. Then by Assumption 13.2, $\eta(\omega)$ is tempered, so thus there exists a $T_2 > 0$ such that

$$\eta(\vartheta_{\tau-t}\omega) \leq e^{\frac{\tilde{\lambda}}{3}(t-\tau)}, \quad \forall\, t-\tau \geq T_2. \tag{13.30}$$

By Assumption 13.2 and Lemma 12.2, there exists $0 < T_3 = T_3(\omega) < \infty$ such that

$$\frac{1}{\tau}\int_0^\tau (2\mathcal{O}(\vartheta_s\omega) + 2\tilde{p}\eta(\vartheta_s\omega))\,\mathrm{d}s < (2\lambda-\epsilon) - \frac{2\tilde{\lambda}}{3}, \quad \tau \geq T_3. \tag{13.31}$$

Let $T_4 = \max\{T_2, T_3\} < \infty$. Then it follows from (13.30) and (13.31) that for all $t > T_Q + T_4$ it holds

$$\begin{aligned}
&\int_{T_4}^{t-T_Q} \mathcal{J}_3(\tau)\mathrm{d}\tau \\
&\leq \|\boldsymbol{u}_o\|_\rho^2 e^{-((2\lambda-\epsilon)-\frac{\varepsilon_2}{3})t-4\mathcal{O}(\vartheta_{-t}\omega)+2\int_{-t}^0(\mathcal{O}(\vartheta_s\omega)+\tilde{p}\eta(\vartheta_s\omega))\mathrm{d}s} \int_{T_4}^{t-T_Q} e^{-\frac{\tilde{\lambda}}{3}\tau}\mathrm{d}\tau \\
&\quad + \frac{1}{2}R^2(\omega)e^{-2\mathcal{O}(\omega)} \int_{T_4}^{t-T_Q} e^{-((2\lambda-\epsilon)-\frac{\tilde{\lambda}}{3})\tau+\int_0^\tau(2\mathcal{O}(\vartheta_s\omega)+2\tilde{p}\eta(\vartheta_s\omega))\mathrm{d}s}\mathrm{d}\tau \\
&\leq \|\boldsymbol{u}_o\|_\rho^2 e^{-((2\lambda-\epsilon)-\frac{\tilde{\lambda}}{3})t-4\mathcal{O}(\vartheta_{-t}\omega)+\int_{-t}^0(2\mathcal{O}(\vartheta_s\omega)+2\tilde{p}\eta(\vartheta_s\omega))\mathrm{d}s} \frac{3}{\tilde{\lambda}} e^{-\frac{\tilde{\lambda}}{3}T_4} \\
&\quad + \frac{1}{2}R^2(\omega)e^{-2\mathcal{O}(\omega)}\frac{3}{\tilde{\lambda}}e^{-\frac{\tilde{\lambda}}{3}T_4},
\end{aligned}$$

where

$$\|\boldsymbol{u}_o\|_\rho^2 e^{-((2\lambda-\epsilon)-\frac{\tilde{\lambda}}{3})t-4\mathcal{O}(\vartheta_{-t}\omega)+\int_{-t}^0(2\mathcal{O}(\vartheta_s\omega)+2\tilde{p}\eta(\vartheta_s\omega))\mathrm{d}s} \frac{3}{\tilde{\lambda}} e^{-\frac{\tilde{\lambda}}{3}T_4} \to 0 \quad as \quad t \to \infty.$$

In addition,

$$\int_0^{T_4} \mathcal{J}_3(\tau)\mathrm{d}\tau < \infty.$$

Therefore there exist $T_5(\varepsilon, \omega, Q) \geq T_4$ and $I_2(\varepsilon, \omega) \in \mathbb{N}$ such that for $k > I_2(\varepsilon, \omega)$, $t \geq T_5(\varepsilon, \omega, Q)$,

$$\frac{\gamma C_0}{k}\int_{T_Q}^t \mathcal{J}_3(\tau)\mathrm{d}\tau \leq \frac{\varepsilon}{3}e^{-2\mathcal{O}(\omega)}. \tag{13.32}$$

In summary, let

$$T(\varepsilon,\omega,Q) = \max\{T_4(\varepsilon,\omega,Q), T_5(\varepsilon,\omega,Q)\}, \quad I(\varepsilon,\omega) = \max\{I_1(\varepsilon,\omega), I_2(\varepsilon,\omega)\}.$$

Then by inserting (13.28), (13.29) and (13.32) into (13.27) we obtain that for $t \geq T(\varepsilon, \omega, Q)$ and $k \geq I(\varepsilon, \omega)$,

$$\sum_{|i|\geq 2k} \rho_i |v_{n,i}^{(m)}(t;\vartheta_{-t}\omega, \boldsymbol{v}_o^{(n)}(\vartheta_{-t}\omega), \boldsymbol{g}^{(n)})|^2 \leq \varepsilon e^{-2\mathcal{O}(\omega)}. \tag{13.33}$$

Since $v_{n,i}^{(m_k)}(t;\vartheta_{-t}\omega, \boldsymbol{v}_o^{(n)}(\vartheta_{-t}\omega), \boldsymbol{g}^{(n)}) \to v_{n,i}(t;\vartheta_{-t}\omega, \boldsymbol{v}_o^{(n)}(\vartheta_{-t}\omega), \boldsymbol{g}^{(n)})$ as $m_k \to \infty$, by (13.33),

$$\begin{aligned}
&\sum_{|i|\geq 2k} \rho_i |u_{n,i}(t;\vartheta_{-t}\omega, \boldsymbol{u}_o^{(n)}(\vartheta_{-t}\omega), \boldsymbol{g}^{(n)})|^2 \\
&\quad = \sum_{|i|\geq 2k} \rho_i e^{2\mathcal{O}(\omega)} |v_{n,i}(t;\vartheta_{-t}\omega, \boldsymbol{v}_o^{(n)}(\vartheta_{-t}\omega), \boldsymbol{g}^{(n)})|^2 \leq \varepsilon
\end{aligned} \tag{13.34}$$

hold for every $n \geq 1$. Letting $n \to \infty$ in (13.34) then gives the desired assertion. □

The following theorem follows immediately from Lemma 13.4, Lemma 13.5, and Theorem 11.4.

Theorem 13.3. *Let Assumptions 12.1–12.3 and 13.1–13.3 hold. Then the random dynamical system* $\{\tilde{\varphi}(t,\omega)\}_{t\geq 0,\omega\in\Omega_0}$ *associated with the equation* (13.3) *possesses a unique global random* $\mathcal{D}$ *attractor* $\mathcal{A}=\{A(\omega)\}_{\omega\in\Omega_0}$ *with* $A(\omega)$ *given by*

$$A(\omega)=\bigcap_{\tau\geq T_Q(\omega)}\overline{\bigcup_{t\geq\tau}\tilde{\varphi}(t,\vartheta_{-t}\omega,Q(\vartheta_{-t}\omega))}\subset\ell^2_\rho.$$

13.3 End notes

This chapter is based on Han, Shen and Zhou [Han *et al.* (2011)]. Random attractors and pullback attractors for stochastic LDS with a multiplicative noise and non-Lipschitz nonlinearities were studied in [Caraballo *et al.* (2011)] and [Caraballo *et al.* (2012)], respectively.

Random attractors for second order stochastic LDS with multiplicative noise in weighted spaces were studied in [Han (2012)] and [Wang *et al.* (2010)] independently. Random attractors for stochastic sine-Gordon lattice system with multiplicative white noise were studied in [Han (2011a)]. Exponential stability of non-autonomous stochastic LDS with multiplicative noise and delays was investigated in [Wang *et al.* (2016)]. [Chen and Wang (2022)] investigated the asymptotic behavior of non-autonomous stochastic lattice systems with multiplicative noise and a discretised fractional Laplacian operator. Random attractors for stochastic non-autonomous p-Laplace lattice equation with delays and multiplicative noise were studied in [Wang and Li (2021)]. Random attractors of p-Laplacian lattice models with linear noise are investigated by Song, Li and Wang [Song *et al.* (2022)]. Invariant measures of stochastic sine-Gordon lattices with nonlinear noise are studied in [Yang and Li (2021)].

Random exponential attractors for non-autonomous first and second order stochastic LDS with multiplicative noise were studied in [Zhou (2017)] and [Su *et al.* (2019)], respectively. Random uniform exponential attractors for non-autonomous stochastic LDS, FitzHugh-Nagumo lattice systems and Schrödinger lattice systems with quasi-periodic forces and multiplicative noise were studied in [Han and Zhou (2020)] and [Zhang and Zhou (2022)].

All the above works considered one or a finite sum of linear multiplicative noise. Studies on stochastic LDS with infinite multiplicative noise require different techniques. Stochastic LDS with infinite linear multiplicative noise were studied in [Caraballo *et al.* (2016)]. See Chapter 20 for LDS with nonlinear noise coefficients.

13.4 Problems

Problem 13.1. Show that the random attractors of finite dimensional approximations of the lattice system (13.1) converge upper semi-continuously to the random attractor of (13.1).

Problem 13.2. Investigate the existence and uniqueness of solutions to the system (13.1) in ℓ^2_ρ when the Wiener process is state dependent, i.e., $W(t)$ is replaced by $W_i(t)$ where $\{W_i(t)\}_{i\in\mathbb{Z}}$ are mutually independent two-sided Wiener processes. What is the major difference in the analysis compare to that of the system (13.1) itself?

Problem 13.3. Investigate the existence of random attractors for the stochastic lattice system

$$\mathrm{d}u_i = \Big(-\lambda u_i - f_i(u_i) + g_i + \sum_{j=-p}^{p} \eta_{i,j}(\vartheta_t\omega)u_{i+j}\Big)\,\mathrm{d}t + \sum_{j=-n}^{n} u_{i+j}\circ \mathrm{d}W_{i+j}(t), i\in\mathbb{Z},$$

where $\{W_i(t)\}_{i\in\mathbb{Z}}$ are mutually independent two-sided Wiener processes. Where do the attractors approach when $n\to\infty$?

Problem 13.4. Consider the stochastic lattice system (13.1) with noise of affine structure, i.e.,

$$\mathrm{d}u_i = \Big(-\lambda u_i - f_i(u_i) + g_i + \sum_{j=-p}^{p} \eta_{i,j}(\vartheta_t\omega)u_{i+j}\Big)\,\mathrm{d}t + (\mathfrak{a}_i + \mathfrak{b}_i u_i)\circ \mathrm{d}W(t), i\in\mathbb{Z}.$$

Do the methods used in this and the previous chapter still apply?

Chapter 14

Stochastic lattice models with fractional Brownian motions

A fractional Brownian motion (fBm), usually denoted by $B^H(t)$, is a generalisation of Brownian motion. It is a continuous-time centered Gaussian process with a special covariance function that depends on the Hurst parameter $H \in (0, 1)$:

$$\mathrm{Cov}(s, t) = \mathbb{E}\left[B^H(t)B^H(s)\right] = \frac{1}{2}\left(|t|^{2H} + |s|^{2H} - |t - s|^{2H}\right). \tag{14.1}$$

For $H = 1/2$, $B^{1/2}(t)$ is the Brownian motion for which the generalised temporal derivative is the white noise. For $H \neq 1/2$, $B^H(t)$ is no longer a semi-martingale, and as a consequence, classical techniques of stochastic analysis are not applicable. In particular, the fBm with a Hurst parameter $H \in (1/2, 1)$ enjoys the property of a long range memory, which implies that the decay of stochastic dependence with respect to the past is only sub-exponentially slow. This long-range dependence property of the fBm makes it a realistic choice of noise for problems with long memory in the applied sciences.

In this chapter we consider the following stochastic LDS with the discrete Laplacian operator, a dissipative nonlinear reaction term, and an fBm at each node:

$$du_i(t) = \big(\nu(u_{i-1} - 2u_i + u_{i+1}) - \lambda u_i + f_i(u_i)\big)dt + \mathfrak{a}_i g_i(u_i) dB_i^H(t), \; i \in \mathbb{Z}, \tag{14.2}$$

with initial condition $u_i(0) = u_{o,i}$, where ν and λ are positive constants, u_i, $\mathfrak{a}_i \in \mathbb{R}$, each $B_i^H(t)$ is a one-dimensional two-sided fBm with Hurst parameter $H \in (1/2, 1)$, and f_i and g_i are smooth functions satisfying appropriate conditions to be specified later.

Different from previous chapters on LDS with Brownian motion, here we will study the existence of a unique mild solution for system (14.2) and analyse the exponential stability of the trivial solution when it exists. Due to the difference in properties of the noise process, techniques different from those used in the previous two chapters are needed. More precisely, the stochastic LDS (14.2) will not be transformed into a RODE, because the norm of any non-trivial solution depends on the magnitude of the norm of the noisy input. Instead, a cut-off argument will be used, in which the functions appeared in the stochastic LDS only need to be defined in a small time interval $[-\delta, \delta]$.

14.1 Preliminaries

Given any $T_1 < T_2$, denote by $\mathcal{C}^\beta([T_1,T_2];\ell^2)$ the Banach space of ℓ^2-valued Hölder continuous functions with exponent $0<\beta<1$ equipped with the norm

$$\|\boldsymbol{u}\|_{\beta,\gamma,T_1,T_2} = \|\boldsymbol{u}\|_{\infty,\gamma,T_1,T_2} + \left\vert\!\left\vert\!\left\vert \boldsymbol{u} \right\vert\!\right\vert\!\right\vert_{\beta,\gamma,T_1,T_2},$$

where $\gamma \geq 0$ and

$$\|\boldsymbol{u}\|_{\infty,\gamma,T_1,T_2} = \sup_{s\in[T_1,T_2]} e^{-\gamma(s-T_1)}\|\boldsymbol{u}(s)\|,$$

$$\left\vert\!\left\vert\!\left\vert \boldsymbol{u} \right\vert\!\right\vert\!\right\vert_{\beta,\gamma,T_1,T_2} = \sup_{T_1\leq s<t\leq T_2} e^{-\gamma(t-T_1)}\frac{\|\boldsymbol{u}(t)-\boldsymbol{u}(s)\|}{(t-s)^\beta}.$$

Notice that the corresponding norms are equivalent for $\gamma > 0$ and $\gamma = 0$. The index γ in the notation will be omitted if $\gamma = 0$, and the indices T_1, T_2 in the notations will be omitted for the special case $T_1 = 0$ and $T_2 = 1$.

To define integrals with Hölder-continuous integrators, we first recall the definition of Weyl fractional derivatives of functions on separable Hilbert spaces (see e.g., Samko, Kilbas & Marichev [Samko *et al.* (2014)]).

Definition 14.1. Let $\mathfrak{H}_1$ and $\mathfrak{H}_2$ be separable Hilbert spaces and let $0<\alpha<1$. The Weyl fractional derivatives of general measurable functions $Z:[s,t]\to\mathfrak{H}_1$ and $\omega:[s,t]\to\mathfrak{H}_2$, of order α and $1-\alpha$ respectively, are defined for $s<r<t$ by

$$\mathrm{D}^\alpha_{s+}Z[r] = \frac{1}{\Gamma(1-\alpha)}\left(\frac{Z(r)}{(r-s)^\alpha} + \alpha\int_s^r \frac{Z(r)-Z(\tau)}{(r-\tau)^{1+\alpha}}\mathrm{d}\tau\right)\in\mathfrak{H}_1,$$

$$\mathrm{D}^{1-\alpha}_{t-}\omega_{t-}[r] = \frac{(-1)^\alpha}{\Gamma(\alpha)}\left(\frac{\omega(r)-\omega(t-)}{(t-r)^{1-\alpha}} + (1-\alpha)\int_r^t \frac{\omega(r)-\omega(\tau)}{(\tau-r)^{2-\alpha}}\mathrm{d}\tau\right)\in\mathfrak{H}_2,$$

where Γ is the gamma function, and

$$\omega_{t-}(r) = \omega(r) - \omega(t-),$$

and $\omega(t-)$ is the left side limit of ω at t.

It is straightforward to show Weyl fractional derivatives are well-posed for Hölder-continuous functions with suitable Hölder exponents, presented in the following lemma. See [Bessaih *et al.* (2017)] for the proof.

Lemma 14.1. *Suppose that* $Z\in\mathcal{C}^\beta([T_1,T_2];\mathfrak{H}_1)$, $\omega\in\mathcal{C}^{\beta'}([T_1,T_2];\mathfrak{H}_2)$, *and* $1-\beta'<\alpha<\beta$. *Then* $\mathrm{D}^\alpha_{s+}Z$ *and* $\mathrm{D}^{1-\alpha}_{t-}\omega_{t-}$ *are well-defined for* $T_1\leq s<t\leq T_2$.

For the special case when $\mathfrak{H}_1=\mathfrak{H}_2=\mathbb{R}$, the fractional integral is defined by [Zähle (1998)]

$$\int_s^t Z(r)\mathrm{d}\omega(r) = (-1)^\alpha\int_s^t \mathrm{D}^\alpha_{s+}Z(r)\mathrm{D}^{1-\alpha}_{t-}\omega_{t-}(r)\mathrm{d}r.$$

The above definition of a fractional integral in $\mathbb{R}$ can be extended to a fractional integral in the separable Hilbert space ℓ^2, following the construction carried out in Chen, Gao, Atienza-Garrido & Schmalfuss [Chen *et al.* (2014)]. More precisely, consider the separable Hilbert space $\mathfrak{L}_2(\ell^2)$ of Hilbert–Schmidt operators from ℓ^2 into ℓ^2, with the usual norm $\|\cdot\|_{\mathfrak{L}_2(\ell^2)}$ defined by

$$\|F\|^2_{\mathfrak{L}_2(\ell^2)} = \sum_{i\in\mathbb{Z}} \|F\boldsymbol{e}^i\|^2, \quad F \in \mathfrak{L}_2(\ell^2),$$

where $\boldsymbol{e}^i$ denotes the element in ℓ^2 having 1 at position i and 0 elsewhere.

For any $\mathcal{Z} \in C^\beta([T_1,T_2];\mathfrak{L}_2(\ell^2))$, $\omega \in C^{\beta'}([T_1,T_2];\ell^2)$ with $\beta+\beta' > 1$, and $T_1 \le s < t \le T_2$, define the ℓ^2-valued integral from s to t as

$$\int_s^t \mathcal{Z}\mathrm{d}\omega := (-1)^\alpha \sum_{j\in\mathbb{Z}} \left(\sum_{i\in\mathbb{Z}} \int_s^t \mathrm{D}^\alpha_{s+}\langle \boldsymbol{e}^j, \mathcal{Z}(\cdot)\boldsymbol{e}^i\rangle[r] \mathrm{D}^{1-\alpha}_{t-}\langle \boldsymbol{e}^i, \omega(\cdot)\rangle_{t-}[r]\mathrm{d}r \right) \boldsymbol{e}^j, \tag{14.3}$$

for $1-\beta' < \alpha < \beta$, whose norm fulfills

$$\left\| \int_s^t \mathcal{Z}\mathrm{d}\omega \right\| \le \int_s^t \|\mathrm{D}^\alpha_{s+}\mathcal{Z}[r]\|_{\mathfrak{L}_2(\ell^2)} \|\mathrm{D}^{1-\alpha}_{t-}\omega_{t-}[r]\|\mathrm{d}r.$$

Notice that in (14.3) the integrals under the sums are one-dimensional fractional integrals. In particular, the following result from [Chen *et al.* (2014)] holds.

Theorem 14.1. *Suppose that* $\mathcal{Z} \in C^\beta([T_1,T_2];\mathfrak{L}_2(\ell^2))$ *and* $\omega \in C^{\beta'}([T_1,T_2];\ell^2)$ *where* $\beta+\beta' > 1$. *Then there exists* $\alpha \in (0,1)$ *such that* $1-\beta' < \alpha < \beta$ *and the integral* (14.3) *is well-defined. In addition,*

(i) there exists a positive constant $C_{\beta,\beta'}$ *such that for* $T_1 \le s < t \le T_2$

$$\left| \int_s^t \mathcal{Z}\mathrm{d}\omega \right| \le C_{\beta,\beta'}(1+(t-s)^\beta)(t-s)^{\beta'} \|\mathcal{Z}\|_{\beta,T_1,T_2} \|\omega\|_{\beta',T_1,T_2},$$

(ii) the following additive properties hold

$$\int_s^t (\mathcal{Z}_1+\mathcal{Z}_2)\mathrm{d}\omega = \int_s^t \mathcal{Z}_1\mathrm{d}\omega + \int_s^t \mathcal{Z}_2\, d\omega,$$
$$\int_s^t \mathcal{Z}\mathrm{d}(\omega_1+\omega_2) = \int_s^t \mathcal{Z}\mathrm{d}\omega_1 + \int_s^t \mathcal{Z}\mathrm{d}\omega_2,$$
$$\int_s^t \mathcal{Z}\mathrm{d}\omega = \int_s^\tau \mathcal{Z}\mathrm{d}\omega + \int_\tau^t \mathcal{Z}\mathrm{d}\omega, \quad \tau \in [s,t],$$

(iii) for any $\tau \in \mathbb{R}$

$$\int_s^t \mathcal{Z}(r)\mathrm{d}\omega(r) = \int_{s-\tau}^{t-\tau} \mathcal{Z}(r+\tau)\mathrm{d}\vartheta_\tau\omega(r), \tag{14.4}$$

where $\vartheta_\tau\omega(\cdot) = \omega(\cdot+\tau) - \omega(\tau)$,

(iv) for any sequence $\{\omega_n\}_{n\in\mathbb{N}}$ *converging to* ω *in* $C^{\beta'}([T_1,T_2];\mathbb{R})$

$$\lim_{n\to\infty} \left\| \int_{T_1}^{\cdot} \mathcal{Z}\mathrm{d}\omega_n - \int_{T_1}^{\cdot} \mathcal{Z}\mathrm{d}\omega \right\|_{\beta,T_1,T_2} = 0.$$

The following estimates of the integral with respect to the Hölder norms depending on γ is crucial for the analysis in the sequel.

Lemma 14.2. *Under the assumptions of Theorem 14.1, for $\beta' > \beta$ there exists a constant C depending on T_1, T_2, β, β' such that for $T_1 \le s < t \le T_2$*

$$e^{-\gamma t}\left\|\int_s^t \mathcal{Z}\mathrm{d}\omega\right\| \le C\kappa(\gamma)\|\mathcal{Z}\|_{\beta,\gamma,s,t}\,\|\omega\|_{\beta',s,t}(t-s)^{\beta},$$

where

$$\kappa(\gamma) = \sup_{0\le s<t\le T}\int_s^t e^{-\gamma(t-\tau)}(t-\tau)^{\alpha+\beta'-\beta-1}(\tau-s)^{-\alpha}\mathrm{d}\tau$$

is such that $\lim_{\gamma\to\infty}\kappa(\gamma) = 0$.

A detailed proof of Lemma 14.2 can be found in [Chen *et al.* (2014)].

Throughout this chapter $\kappa(\gamma)$ will be used to denote a generic function with the properties in Lemma 14.2, i.e., using $-b = \alpha < 1$ and $a = \alpha + \beta' - \beta - 1$,

$$\kappa(\gamma) = \sup_{0\le s<t\le T}\int_s^t e^{-\gamma(t-\tau)}(t-\tau)^{a}(\tau-s)^{b}\mathrm{d}\tau$$

for $T > 0$, $a, b > -1$ with $a+b+1 > 0$. Notice that the constraints in Lemma 14.2 imply that $\beta' > 1/2$.

To study the stochastic LDS system with fBm, here we will consider a particular case of Hölder-continuous integrator that is an fBm with values in ℓ^2 and Hurst-parameter $H > 1/2$. Let $(\Omega_0, \mathcal{F}, \mathbb{P})$ be a probability space and $(B_i^H(t))_{i\in\mathbb{Z}}$ be an i.i.d.-sequence of fBm with the same Hurst parameter $H > 1/2$ over this probability space, i.e., each $B_i^H(t)$ is a centered Gaussian process on $\mathbb{R}$ with the covariance defined by (14.1).

Similar to Lemma 12.1, Let $\mathfrak{Q}$ be a linear operator on ℓ^2 such that $\mathfrak{Q}e^i = \mathfrak{a}_i^2 e^i$. Then $\mathfrak{Q}$ is a non-negative and symmetric trace-class operator. A continuous ℓ^2-valued fractional Brownian motion $\boldsymbol{B}^H$ with covariance operator $\mathfrak{Q}$ and Hurst parameter H is defined by

$$\boldsymbol{B}^H(t) = \sum_{i\in\mathbb{Z}}(\mathfrak{a}_i B_i^H(t))e^i \tag{14.5}$$

with the covariance

$$\mathrm{Cov}_{\mathfrak{Q}}(s,t) = \frac{1}{2}\mathfrak{Q}(|s|^{2H} + |t|^{2H} - |t-s|^{2H}) \quad \text{for } s,\, t\in\mathbb{R}. \tag{14.6}$$

In fact, $\boldsymbol{B}^H(t)$ is a Gaussian process provided $\mathfrak{a} := (\mathfrak{a}_i)_{i\in\mathbb{Z}} \in \ell^2$ with

$$\begin{aligned}\mathbb{E}\left\|\boldsymbol{B}^H(t) - \boldsymbol{B}^H(s)\right\|^2 &= \sum_{i\in\mathbb{Z}}\mathfrak{a}_i^2\mathbb{E}\left[(B_i^H(t) - B_i^H(s))^2\right]\\ &= \sum_{i\in\mathbb{Z}}\mathfrak{a}_i^2|t-s|^{2H} = \|\mathfrak{a}\|^2|t-s|^{2H},\\ \mathbb{E}\left\|\boldsymbol{B}^H(t) - \boldsymbol{B}^H(s)\right\|^{2n} &\le C_n|t-s|^{2Hn}.\end{aligned}$$

Therefore, applying Theorem 1.4.1 in [Kunita (1990)], $\boldsymbol{B}^H(t)$ has a continuous version and also a Hölder-continuous version with exponent $\beta' < H$, see [Bauer (1996)] Chapter 39. Notice that $\boldsymbol{B}^H(0) = \boldsymbol{0}$, almost surely.

Let $\Omega := \{\omega \in \mathcal{C}(\mathbb{R}, \ell^2) : \omega(0) = \boldsymbol{0}\}$ be the space of continuous functions on $\mathbb{R}$ with values in ℓ^2 which are zero at zero, equipped with the compact open topology. Consider the canonical space for the fBm $(\Omega, \mathcal{B}(\Omega), \mathbb{P}_H)$, where $\mathbb{P}_H$ denotes the measure of the fBm with Hurst–parameter H. Then we introduce the Wiener shift ϑ on Ω given by the measurable flow

$$\vartheta : (\mathbb{R} \times \Omega, \mathcal{B}(\mathbb{R}) \otimes \mathcal{B}(\Omega)) \to (\Omega, \mathcal{B}(\Omega))$$

such that

$$\vartheta(t, \omega)(\cdot) = \vartheta_t\omega(\cdot) = \omega(\cdot + t) - \omega(t). \tag{14.7}$$

By [Mishura (2008)], Page 8, $\mathbb{P}_H$ is invariant under ϑ_t. In addition, $t \mapsto \vartheta_t\omega$ is continuous. Furthermore, thanks to [Bauer (1996)] Chapter 39, we can also conclude that the set $\Omega^{\beta'}$ defined by continuous functions which have a finite β'-Hölder-seminorm on any compact interval and which are zero at zero has $\mathbb{P}_H$-measure one for $\beta' < H$. This set is ϑ-invariant.

14.2 Existence of solutions

In this section, the existence of a unique mild solution will be shown by a fixed point argument, along with estimates satisfied by the stochastic integral with an fBm as integrator. To that end, we first rewrite the lattice system (14.2) as a stochastic evolution equation in ℓ^2.

Let Λ be the discrete Laplacian operator defined as in Section 3.1, and consider the linear bounded operator $\mathrm{A}_\lambda : \ell^2 \to \ell^2$ given by

$$\mathrm{A}_\lambda \boldsymbol{u} = \Lambda \boldsymbol{u} + \lambda \boldsymbol{u}, \quad \boldsymbol{u} \in \ell^2. \tag{14.8}$$

Then it is straightforward to check that

$$\langle \mathrm{A}_\lambda \boldsymbol{u}, \boldsymbol{u} \rangle \geq \lambda \|\boldsymbol{u}\|^2, \quad \forall \boldsymbol{u} \in \ell^2,$$

and hence $-\mathrm{A}_\lambda$ is a negative-defined bounded operator that generates a uniformly continuous semi-group $\mathcal{S}_\lambda := e^{-\mathrm{A}_\lambda t}$ on ℓ^2, for which the following estimates hold.

Lemma 14.3. *The uniformly continuous semigroup $\mathcal{S}_\lambda$ is exponentially stable, i.e.,*

$$\|\mathcal{S}_\lambda(t)\|_{\mathfrak{L}(\ell^2)} \leq e^{-\lambda t}, \quad \text{for} \quad t \geq 0.$$

In addition, for $0 \leq s \leq t$

$$\|\mathcal{S}_\lambda(t-s) - \mathrm{Id}\|_{\mathfrak{L}(\ell^2)} \leq \|\mathrm{A}_\lambda\|_{\mathfrak{L}(\ell^2)} \cdot (t-s),$$

$$\|\mathcal{S}_\lambda(t) - \mathcal{S}_\lambda(s)\|_{\mathfrak{L}(\ell^2)} \leq \|\mathrm{A}_\lambda\|_{\mathfrak{L}(\ell^2)} \cdot (t-s)e^{-\lambda s},$$

where $\mathfrak{L}(\ell^2)$ denotes the space of linear continuous operator from ℓ^2 into itself.

Proof. The proof of the first property is a direct consequence of the energy inequality, while the two last estimates follow easily by the mean value theorem. As a straightforward consequence, we also obtain that for $0 < s < t$,

$$\begin{aligned}\|\mathcal{S}_\lambda(t-\cdot)\|_{\beta,0,t} &= \sup_{0\le r_1<r_2\le t} \frac{\|\mathcal{S}_\lambda(t-r_2)-\mathcal{S}_\lambda(t-r_1)\|_{\mathfrak{L}(\ell^2)}}{(r_2-r_1)^\beta} \\ &\le \|\mathrm{A}_\lambda\|_{\mathfrak{L}(\ell^2)} t^{1-\beta},\end{aligned}$$

and

$$\begin{aligned}&\|\mathcal{S}_\lambda(t-\cdot)-\mathcal{S}_\lambda(s-\cdot)\|_{\beta,0,s} \\ &= \sup_{0\le r_1<r_2\le s} \frac{\|(\mathcal{S}_\lambda(t-s)-\mathrm{id})(\mathcal{S}_\lambda(s-r_2)-\mathcal{S}_\lambda(s-r_1))\|_{\mathfrak{L}(\ell^2)}}{(r_2-r_1)^\beta} \\ &\le \|\mathrm{A}_\lambda\|^2_{\mathfrak{L}(\ell^2)}(t-s)s^{1-\beta}.\end{aligned}$$

□

14.2.1 *Standing assumptions*

Recall that in Chapter 12, the process ω represents a (canonical) continuous Brownian motion with values in ℓ^2. Here in this chapter, the process ω is a (canonical) ℓ^2-valued continuous fBm with Hurst parameter H given by (14.5) and covariance given by (14.6). In addition, one or more the following assumptions will be needed.

Assumption 14.1. The parameters α, β and β' satisfy

$$\frac{1}{2} < \beta < \beta' < H \quad \text{and} \quad 1-\beta' < \alpha < \beta.$$

Assumption 14.2. For each $i \in \mathbb{Z}$, $f_i \in C^1(\mathbb{R},\mathbb{R})$, $\sum_{i\in\mathbb{Z}} f_i^2(0) < \infty$, and there exists a constant $L_f \ge 0$ such that

$$|f_i'(s)| \le L_f, \quad s\in\mathbb{R},\ i\in\mathbb{Z}.$$

Assumption 14.3. For each $i \in \mathbb{Z}$, $g_i \in C^2(\mathbb{R},\mathbb{R})$, $\sum_{i\in\mathbb{Z}} g_i^2(0) < \infty$, and there exist constants L_g, $D_g \ge 0$ such that

$$|g_i'(s)| \le L_g, \quad |g_i''(s)| \le D_g, \quad s\in\mathbb{R},\ i\in\mathbb{Z}.$$

14.2.2 *Properties of operators*

Under the Assumption 14.2 and Assumption 14.3, we can define the operators $F : \ell^2 \to \ell^2$, and $G(\boldsymbol{u}) \in \mathfrak{L}(\ell^2)$ by

$$F(\boldsymbol{u}) := (f_i(u_i))_{i\in\mathbb{Z}}, \quad G(\boldsymbol{u})\boldsymbol{v} = (g_i(u_i)v_i)_{i\in\mathbb{Z}} \in \ell^2 \tag{14.9}$$

for $\boldsymbol{u} = (u_i)_{i\in\mathbb{Z}}$, $\boldsymbol{v} = (v_i)_{i\in\mathbb{Z}} \in \ell^2$.

Lemma 14.4. *Let Assumptions 14.2–14.3 hold. Then*

(i) the operator $F : \ell^2 \to \ell^2$ defined in (14.9) *is well-defined and is Lipschitz continuous with Lipschitz constant L_f;*

(ii) the operator $\boldsymbol{u} \mapsto G(\boldsymbol{u}) \in \mathfrak{L}_2(\ell^2)$ defined by 14.9 is well-defined and continuously differentiable;

(iii) the operator $G(\boldsymbol{u})$ is Lipschitz continuous in $\boldsymbol{u}$ with Lipschitz constant L_g, and moreover for any $\boldsymbol{u}$, $\boldsymbol{v}$, $\boldsymbol{w}$, $\boldsymbol{z} \in \ell^2$,

$$\begin{aligned}&\|(G(\boldsymbol{u}) - G(\boldsymbol{v})) - (G(\boldsymbol{w}) - G(\boldsymbol{z}))\|_{\mathfrak{L}_2(\ell^2)}\\&\le \sqrt{2}L_g\|\boldsymbol{u} - \boldsymbol{v} - (\boldsymbol{w} - \boldsymbol{z})\| + 2D_g\|\boldsymbol{u} - \boldsymbol{w}\|(\|\boldsymbol{u} - \boldsymbol{v}\| + \|\boldsymbol{w} - \boldsymbol{z}\|).\end{aligned}$$

Proof. (i) By the definition of F, for $\boldsymbol{u} = (u_i)_{i\in\mathbb{Z}} \in \ell^2$ we have

$$\|F(\boldsymbol{u})\|^2 \le 2\sum_{i\in\mathbb{Z}} f_i^2(0) + 2L_f^2\|\boldsymbol{u}\|^2 < \infty,$$

hence it is well-posed. Moreover, for $\boldsymbol{v} = (v_i)_{i\in\mathbb{Z}} \in \ell^2$ there holds

$$\|F(\boldsymbol{u}) - F(\boldsymbol{v})\|^2 = \sum_{i\in\mathbb{Z}} |f_i(u_i) - f_i(v_i)|^2 \le L_f^2 \sum_{i\in\mathbb{Z}} |u_i - v_i|^2 = L_f^2\|\boldsymbol{u} - \boldsymbol{v}\|^2,$$

i.e., F is Lipschitz continuous with Lipschitz constant L_f.
(ii) for any $\boldsymbol{u} \in \ell^2$, the operator $G(\boldsymbol{u})$ is well-defined as a Hilbert-Schmidt operator, because

$$\begin{aligned}\|G(\boldsymbol{u})\|^2_{\mathfrak{L}_2(\ell^2)} &= \sum_{i\in\mathbb{Z}} \|G(\boldsymbol{u})\boldsymbol{e}^i\|^2 = \sum_{i,j\in\mathbb{Z}} |(G(\boldsymbol{u})\boldsymbol{e}^i)_j|^2 = \sum_{i\in\mathbb{Z}} |g_i(u_i)|^2\\&\le 2\sum_{i\in\mathbb{Z}} g_i^2(0) + 2L_g^2\|\boldsymbol{u}\|^2 < \infty.\end{aligned}$$

(iii) $G(\boldsymbol{u})$ is Lipschitz continuous in $\boldsymbol{u}$, since

$$\|G(\boldsymbol{u}) - G(\boldsymbol{v})\|^2_{\mathfrak{L}_2(\ell^2)} \le L_g^2\|\boldsymbol{u} - \boldsymbol{v}\|^2, \quad \text{for } \boldsymbol{u},\, \boldsymbol{v} \in \ell^2.$$

Next, notice that

$$\begin{aligned}&\|(G(\boldsymbol{u}) - G(\boldsymbol{v})) - (G(\boldsymbol{w}) - G(\boldsymbol{z}))\|^2_{\mathfrak{L}_2(\ell^2)} &&(14.10)\\&= \sum_{i\in\mathbb{Z}} |g_i(u_i) - g_i(v_i) - (g_i(w_i) - g_i(z_i))|^2\\&= \sum_{i\in\mathbb{Z}} |g_i'(\tilde{u}_i)(u_i - v_i) - g_i'(\tilde{w}_i)(w_i - z_i)|^2 &&(14.11)\end{aligned}$$

for some $\tilde{u}_i$ between u_i and v_i, and some $\tilde{w}_i$ between w_i and z_i. Further, for some ζ_i between $\tilde{u}_i$ and $\tilde{w}_i$ we have

$$\begin{aligned}&|g_i'(\tilde{u}_i)(u_i - v_i) - g_i'(\tilde{w}_i)(w_i - z_i)|^2\\&\le 2|g_i'(\tilde{u}_i)(u_i - v_i - (w_i - z_i))|^2 + 2|(g_i'(\tilde{u}_i) - g_i'(\tilde{w}_i))(w_i - z_i)|^2\\&\le 2L_g^2|u_i - v_i - (w_i - z_i)|^2 + 2|g_i''(\zeta_i)|^2|(\tilde{u}_i - \tilde{w}_i)(w_i - z_i)|^2,\end{aligned}$$

from which it follows that

$$
\begin{aligned}
&|g_i'(\tilde{u}_i)(u_i - v_i) - g_i'(\tilde{w}_i)(w_i - z_i)|^2 \\
&\le 2L_g^2|u_i - v_i - (w_i - z_i)|^2 + 4D_g^2|u_i - w_i|^2(|u_i - v_i|^2 + |w_i - z_i|^2). \quad (14.12)
\end{aligned}
$$

Inserting (14.12) into (14.11) immediately gives the desired assertion. □

14.2.3 *Existence of mild solutions*

Lemma 14.4 ensures that the lattice system (14.2) can be reformulated as the following stochastic evolution equation with values in ℓ^2,

$$
\mathrm{d}\boldsymbol{u}(t) = (-\mathrm{A}_\lambda \boldsymbol{u}(t) + F(\boldsymbol{u}(t)))\mathrm{d}t + G(\boldsymbol{u}(t))\mathrm{d}\omega(t), \quad (14.13)
$$

where A_λ, F, and $G(\boldsymbol{u})$ are defined by (14.8) and (14.9), respectively.

The stability result we are seeking is based on the exponential stability of the semigroup $\mathcal{S}_\lambda$ generated by the negative-definite bounded operator $-\mathrm{A}_\lambda$. Hence here we investigate the existence of a mild solution of the equation (14.13) defined as follows.

Definition 14.2. A mild solution $\boldsymbol{u}(t) = (u_i(t))_{i\in\mathbb{Z}} \in \ell^2$ to the equation (14.13) is a solution of the operator equation

$$
\boldsymbol{u}(t) = \mathcal{S}_\lambda(t)\boldsymbol{u}_o + \int_0^t \mathcal{S}_\lambda(t-s)F(\boldsymbol{u}(s))\mathrm{d}s + \int_0^t \mathcal{S}_\lambda(t-s)G(\boldsymbol{u}(s))\mathrm{d}\omega(s), \quad (14.14)
$$

where the initial condition $\boldsymbol{u}_o \in \ell^2$, and the last stochastic integral is defined in the sense of (14.3).

To obtain the existence and uniqueness of solutions to (14.14), estimates of the stochastic integral appearing on its right hand side are first presented in the following lemma.

Lemma 14.5. *Under Assumptions 14.1–14.3, the stochastic integral satisfies*

$$
\left\| \int_0^{\cdot} \mathcal{S}_\lambda(\cdot - s)G(\boldsymbol{u}(s))\mathrm{d}\omega(s) \right\|_{\beta,\gamma,0,T} \le C_1\kappa(\gamma)\,|\!|\!|\omega|\!|\!|_{\beta',0,T}\,\|G(\boldsymbol{u}(\cdot))\|_{\beta,\gamma,0,T},
$$

$$
\left\| \int_0^{\cdot} \mathcal{S}_\lambda(\cdot - s)G(\boldsymbol{u}(s)\mathrm{d}\omega(s) \right\|_{\infty,0,T} \le C_2(1 + \|\mathrm{A}_\lambda\|)\,|\!|\!|\omega|\!|\!|_{\beta',0,T}\,\|G(\boldsymbol{u}(\cdot))\|_{\beta,0,T},
$$

$$
\left|\!\left|\!\left| \int_0^{\cdot} \mathcal{S}_\lambda(\cdot - s)G(\boldsymbol{u}(s))\mathrm{d}\omega(s) \right|\!\right|\!\right|_{\beta,0,T} \le C_2(1 + \|\mathrm{A}_\lambda\|)^2\,|\!|\!|\omega|\!|\!|_{\beta',0,T}\,\|G(\boldsymbol{u}(\cdot))\|_{\beta,0,T},
$$

where C_1 may depend on β, β', T, $\|\mathrm{A}_\lambda\| = \|\mathrm{A}_\lambda\|_{\mathfrak{L}(\ell^2)}$, $\kappa(\gamma)$ is given in Lemma 14.2, and C_2 may depend on β, β' and T.

The proof of Lemma 14.5 can be found in Bessaih, Garrido-Atienza, Han & Schmalfuss [Bessaih *et al.* (2017)].

Theorem 14.2. *Let Assumptions 14.1–14.3 hold. Then for every* $T > 0$ *and* $\boldsymbol{u}_o \in \ell^2$ *the problem* (14.14) *has a unique solution* $\boldsymbol{u}(\cdot) = \boldsymbol{u}(\cdot;\omega,\boldsymbol{u}_o) \in \mathcal{C}^\beta([0,T];\ell^2)$ *in the sense of Definition 14.2.*

Proof. For $t \in [0,T]$, define the operator $\mathcal{T}$ by

$$\mathcal{T}_{\boldsymbol{u}_o,\omega}(\boldsymbol{u}(t)) = \mathcal{S}_\lambda(t)\boldsymbol{u}_o + \int_0^t \mathcal{S}_\lambda(t-s)F(\boldsymbol{u}(s))\mathrm{d}s + \int_0^t \mathcal{S}_\lambda(t-s)G(\boldsymbol{u}(s))\mathrm{d}\omega(s).$$

We next show that $\mathcal{T}$ has a unique fixed point in $\mathcal{C}^\beta([0,T];\ell^2)$ by applying the Banach fixed point theorem. To that end, we first show that there exists a closed centered ball with respect to the norm $\|\cdot\|_{\beta,\gamma,0,T}$ which is mapped by $\mathcal{T}_{\boldsymbol{u}_o,\omega}$ into itself. In fact, by Lemma 14.3

$$\|\mathcal{S}_\lambda(\cdot)\boldsymbol{u}_o\|_{\beta,\gamma,0,T} \le (1 + \|\mathrm{A}_\lambda\|T^{1-\beta})\|\boldsymbol{u}_o\|.$$

Then the Lebesgue integral within $\mathcal{T}_{\boldsymbol{u}_o,\omega}$ satisfies

$$\begin{aligned}
&\left\| \int_0^t \mathcal{S}_\lambda(t-s)F(\boldsymbol{u}(s))\mathrm{d}s \right\|_{\beta,\gamma,0,T} \\
&\le \sup_{t\in[0,T]} e^{-\gamma t}\left\| \int_0^t \mathcal{S}_\lambda(t-s)F(\boldsymbol{u}(s))\mathrm{d}s \right\| + \sup_{0\le s<t\le T} e^{-\gamma t}\frac{\left\| \int_s^t \mathcal{S}_\lambda(t-r)F(\boldsymbol{u}(r))\mathrm{d}r \right\|}{(t-s)^\beta} \\
&\quad + \sup_{0\le s<t\le T} e^{-\gamma t}\frac{\left\| \int_0^s (\mathcal{S}_\lambda(t-r)-\mathcal{S}_\lambda(s-r))F(\boldsymbol{u}(r))\mathrm{d}r \right\|}{(t-s)^\beta} \\
&\le \tilde{\kappa}(\gamma)\|F(\boldsymbol{u}(\cdot))\|_{\infty,\gamma,0,T}.
\end{aligned} \tag{14.15}$$

The next step is to show that $\lim_{\gamma\to\infty}\tilde{\kappa}(\gamma) = 0$ with

$$\tilde{\kappa}(\gamma) = \frac{1}{\gamma} + C_\beta\frac{1}{\gamma^{1-\beta}} + \frac{1}{\gamma}T^{1-\beta}\|\mathrm{A}_\lambda\|, \tag{14.16}$$

for some positive constant C_β depending on β. In fact, the first and second terms on the right hand side of (14.15) satisfy, respectively,

$$\sup_{t\in[0,T]}\int_0^t e^{-\gamma(t-s)}\mathrm{d}s\|F(\boldsymbol{u}(\cdot)\|_{\infty,\gamma,0,T} \le \frac{1}{\gamma}\|F(\boldsymbol{u}(\cdot))\|_{\infty,\gamma,0,T},$$

$$\frac{\int_s^t e^{-\gamma(t-r)}\mathrm{d}r}{(t-s)^\beta} \le \frac{1}{\gamma^{1-\beta}}\frac{1-e^{-\gamma(t-s)}}{\gamma^\beta(t-s)^\beta} \le \frac{1}{\gamma^{1-\beta}}\sup_{x>0}\frac{1-e^{-x}}{x^\beta} =: \frac{1}{\gamma^{1-\beta}}C_\beta.$$

The estimate of the last term on the right hand side of (14.15) follows by Lemma 14.3, because

$$\mathcal{S}_\lambda(t-r) - \mathcal{S}_\lambda(s-r) = (\mathcal{S}_\lambda(t-s) - \mathrm{Id})\mathcal{S}_\lambda(s-r).$$

On the other hand, since

$$\begin{aligned}\|F(\boldsymbol{u}(\cdot))\|_{\infty,\gamma,0,T} &\leq \sup_{0\leq t\leq T} e^{-\gamma t}\|F(\boldsymbol{u}_o)\| + \sup_{0\leq t\leq T} e^{-\gamma t}\|F(\boldsymbol{u}(t)) - F(\boldsymbol{u}_o)\| \\ &\leq \|F(\boldsymbol{u}_o)\| + L_f T^\beta \|\boldsymbol{u}\|_{\beta,\gamma,0,T},\end{aligned}$$

then

$$\left\| \int_0^{\cdot} \mathcal{S}_\lambda(\cdot - s)F(\boldsymbol{u}(s))\mathrm{d}s \right\|_{\beta,\gamma,0,T} \leq \hat{\kappa}(\gamma)(1 + \|\boldsymbol{u}\|_{\beta,\gamma,0,T}),$$

where $\hat{\kappa}(\gamma) = \max\{\|F(\boldsymbol{u}_o)\|, L_f T^\beta\}\tilde{\kappa}(\gamma)$, with $\tilde{\kappa}(\gamma)$ defined by (14.16).

In addition, noticing that

$$\begin{aligned}\|G(\boldsymbol{u}(\cdot))\|_{\beta,\gamma,0,t} &= \sup_{s\in[0,t]} e^{-\gamma s}\|G(\boldsymbol{u}(s))\|_{\mathfrak{L}_2(\ell^2)} \\ &\quad + \sup_{0\leq \tau < s\leq t} \frac{e^{-\gamma s}\|G(\boldsymbol{u}(s)) - G(\boldsymbol{u}(\tau))\|_{\mathfrak{L}_2(\ell^2)}}{(s-\tau)^\beta} \\ &\leq \|G(\boldsymbol{u}_o)\| + L_g(1 + T^\beta)\|\boldsymbol{u}\|_{\beta,\gamma,0,T},\end{aligned}$$

it follows from 14.15 that

$$\left\| \int_0^{\cdot} \mathcal{S}_\lambda(\cdot - s)G(\boldsymbol{u}(s))\mathrm{d}\omega(s) \right\|_{\beta,\gamma,0,T} \leq C_1\kappa(\gamma)\, \|\omega\|_{\beta',0,T}\, (1 + \|\boldsymbol{u}\|_{\beta,\gamma,0,T}),$$

where C_1 may depend on β, β', T, $\|\mathrm{A}_\lambda\|$, $\|G(\boldsymbol{u}_o)\|$ and L_g.

Summarising the above we have obtained

$$\|\mathcal{T}_{\boldsymbol{u}_o,\omega}(\boldsymbol{u})\|_{\beta,\gamma,0,T} \leq (1 + \|\mathrm{A}_\lambda\|T^{1-\beta})\|\boldsymbol{u}_o\| + K(\gamma)(1 + \|\omega\|_{\beta',0,T})(1 + \|\boldsymbol{u}\|_{\beta,\gamma,0,T})$$

where $\lim_{\gamma\to\infty} K(\gamma) = 0$. Notice that $K(\gamma)$ may also depend on the parameters related to F and H, the initial condition $\boldsymbol{u}_o$, $\|\mathrm{A}_\lambda\|$ and T.

Now choose γ to be sufficiently large, such that $K(\gamma)(1+\|\omega\|_{\beta',0,T}) \leq 1/2$. Then the ball

$$\mathbb{B} = \mathbb{B}(0, R(\boldsymbol{u}_o, \gamma)) = \{\boldsymbol{u} \in \mathcal{C}^\beta([0,T];\ell^2) \,:\, \|\boldsymbol{u}\|_{\beta,\gamma,0,T} \leq R\}$$

with

$$R = R(\boldsymbol{u}_o, \gamma) = 2(1 + \|\mathrm{A}_\lambda\|T^{1-\beta})\|\boldsymbol{u}_o\| + 1,$$

is mapped into itself because

$$\|\mathcal{T}_{\boldsymbol{u}_o,\omega}(\boldsymbol{u})\|_{\beta,\gamma,0,T} \leq (1 + \|\mathrm{A}_\lambda\|T^{1-\beta})\|\boldsymbol{u}_o\| + \frac{1}{2}(1 + R) = R.$$

It remains to show the contraction property of the operator $\mathcal{T}_{\boldsymbol{u}_o,\omega}$ with respect to the norm $\|\cdot\|_{\beta,\bar{\gamma},0,T}$ where the $\bar{\gamma}$ may be different from γ considered above. However, recalling that all these norms are equivalent for different $\gamma \geq 0$, the set $\mathbb{B}$ remains a complete space with respect to any $\|\cdot\|_{\beta,\bar{\gamma},0,T}$.

Consider any $\boldsymbol{u}$ and $\boldsymbol{v} \in \ell^2$. Similar to above, for the Lebesgue integral we obtain the estimate

$$\|F(\boldsymbol{u}(\cdot)) - F(\boldsymbol{v}(\cdot))\|_{\beta,\bar{\gamma},0,T} \leq \tilde{k}(\bar{\gamma})L_f\|\boldsymbol{u} - \boldsymbol{v}\|_{\beta,\bar{\gamma},0,T},$$

where $\tilde{k}(\gamma)$ is defined by (14.16) replacing γ by $\bar{\gamma}$.

Also, from part (iv) of Lemma 14.4 it follows immediately that

$$\begin{aligned}\|G(\boldsymbol{u}(\cdot)) - G(\boldsymbol{v}(\cdot))\|_{\beta,\bar{\gamma},0,T} &\leq L_g(\|\boldsymbol{u} - \boldsymbol{v}\|_{\infty,\bar{\gamma},0,T} + \sqrt{2}\,|\!|\!|\boldsymbol{u} - \boldsymbol{v}|\!|\!|_{\beta,\bar{\gamma},0,T}) \\ &\quad + 2D_g(\|\boldsymbol{u}\|_{\infty,0,T} + \|\boldsymbol{v}\|_{\infty,0,T})\|\boldsymbol{u} - \boldsymbol{v}\|_{\infty,\bar{\gamma},0,T}.\end{aligned}$$

Since $\|\boldsymbol{u}\|_{\infty,0,T}, \|\boldsymbol{v}\|_{\infty,0,T} \leq e^{\gamma T}R(\boldsymbol{u}_o, \gamma)$, then

$$\begin{aligned}&\|\mathcal{T}_{\boldsymbol{u}_o,\omega}(\boldsymbol{u}) - \mathcal{T}_{\boldsymbol{u}_o,\omega}(\boldsymbol{v})\|_{\beta,\bar{\gamma},0,T} \\ &\quad \leq K(\bar{\gamma})(1 + |\!|\!|\omega|\!|\!|_{\beta',0,T})(1 + \|\boldsymbol{u}\|_{\infty,0,T} + \|\boldsymbol{v}\|_{\infty,0,T}) \cdot \|\boldsymbol{u} - \boldsymbol{v}\|_{\infty,\bar{\gamma},0,T},\end{aligned}$$

where again $\lim_{\bar{\gamma}\to 0} K(\bar{\gamma}) = 0$. Now choosing $\bar{\gamma}$ sufficiently large gives

$$\|\mathcal{T}_{\boldsymbol{u}_o,\omega}(\boldsymbol{u}) - \mathcal{T}_{\boldsymbol{u}_o,\omega}(\boldsymbol{v})\|_{\beta,\bar{\gamma},0,T} \leq \frac{1}{2}\|\boldsymbol{u} - \boldsymbol{v}\|_{\beta,\bar{\gamma},0,T},$$

which implies the contraction property of the map $T_{\boldsymbol{u}_o,\omega}$. Therefore the equation (14.14) has a unique solution $\boldsymbol{u} \in \mathcal{C}^\beta([0,T];\ell^2)$. The proof is complete. □

14.3 Generation of an RDS

Similar to Chapters 12 and 13, where the metric dynamical system is the Brownian motion, here the metric dynamical system is the fBm. More precisely, consider the quadruple $(\Omega, \mathcal{F}, \mathbb{P}, \vartheta) = (\Omega, \mathcal{B}(\Omega), \mathbb{P}_H, \vartheta)$, where $\mathbb{P}_H$ denotes the measure of the fBm with Hurst parameter H, and ϑ is given by the Wiener flow introduced in (14.7).

Theorem 14.3. *The solution to the equation* (14.14) *generates a random dynamical system* $\varphi : \mathbb{R}^+ \times \Omega \times \ell^2 \to \ell^2$ *given by*

$$\varphi(t, \omega, \boldsymbol{u}_o) = \boldsymbol{u}(t; \omega, \boldsymbol{u}_o),$$

where $\boldsymbol{u}(t; \omega, \boldsymbol{u}_o)$ *is the unique solution to* (14.14) *corresponding to* $\omega \in \Omega$ *and initial condition* $\boldsymbol{u}_o \in \ell^2$.

Proof. The cocycle property is a direct consequence of the additivity of the stochastic integral as well as the property of the stochastic integral when performing a change of variable given by (14.4). More specifically,

$$\begin{aligned}
\varphi(t+\tau,\omega,\boldsymbol{u}_o) &= \mathcal{S}_\lambda(t+\tau)\boldsymbol{u}_o + \int_0^{t+\tau} \mathcal{S}_\lambda(t+\tau-s)F(\boldsymbol{u}(s))\mathrm{d}s \\
&\quad + \int_0^{t+\tau} \mathcal{S}_\lambda(t+\tau-s)G(\boldsymbol{u}(s))\mathrm{d}\omega(s) \\
&= \mathcal{S}_\lambda(t)\left(\mathcal{S}_\lambda(\tau)\boldsymbol{u}_o + \int_0^{\tau} \mathcal{S}_\lambda(\tau-s)F(\boldsymbol{u}(s))\mathrm{d}s \right.\\
&\quad \left. + \int_0^{\tau} \mathcal{S}_\lambda(\tau-s)G(\boldsymbol{u}(s))\mathrm{d}\omega(s)\right) \\
&\quad + \int_\tau^{t+\tau} \mathcal{S}_\lambda(t+\tau-s)F(\boldsymbol{u}(s))\mathrm{d}s + \int_\tau^{t+\tau} \mathcal{S}_\lambda(t+\tau-s)G(\boldsymbol{u}(s))\mathrm{d}\omega(s) \\
&= \mathcal{S}(t)\boldsymbol{u}(\tau) + \int_0^t \mathcal{S}_\lambda(t-s)F(\boldsymbol{u}(s+\tau))\mathrm{d}s \\
&\quad + \int_0^t \mathcal{S}_\lambda(t-s)G(\boldsymbol{u}(s+\tau))\mathrm{d}\vartheta_\tau\omega(s).
\end{aligned}$$

Set $\tilde{\boldsymbol{u}}(\cdot) = \boldsymbol{u}(\cdot+\tau)$. Then the inequality above becomes

$$\begin{aligned}
\varphi(t+\tau,\omega,\boldsymbol{u}_o) &= \mathcal{S}_\lambda(t)\tilde{\boldsymbol{u}}(0) + \int_0^t \mathcal{S}_\lambda(t-s)F(\tilde{\boldsymbol{u}}(s))\mathrm{d}s \\
&\quad + \int_0^t \mathcal{S}_\lambda(t-s)G(\tilde{\boldsymbol{u}}(s))\mathrm{d}\vartheta_\tau\omega(s) \\
&= \varphi(t,\vartheta_\tau\omega,\varphi(\tau,\omega,\boldsymbol{u}_o)).
\end{aligned}$$

The measurability of the mapping φ follows from its continuity with respect to ω (that implies measurability with respect to ω), along with its continuity with respect to (t,x) and the separability of ℓ^2, see Lemma III.14 in [Castaing and Valadier (1977)]. The proof is complete. □

14.4 Exponential stability of the trivial solution

In this section we show that the trivial solution of (14.2), if exists, is exponential stable. The key idea is to consider the composition of the functions defined locally with a cut-off function depending on a random variable $\mathcal{R}$. With these compositions, we construct a sequence $\{\boldsymbol{u}^{(n)}\}_{n\in\mathbb{N}}$ such that each element $\boldsymbol{u}^{(n)}$ is a solution of a modified stochastic LDS on the time interval $[0,1]$ driven by a path of the fBm depending also on n. It is easily conceived that $\boldsymbol{u}^{(n)}(0) = \boldsymbol{u}^{(n-1)}(1)$ will be required.

The norm of each $\boldsymbol{u}^{(n)}$ depends on the magnitude of the corresponding driving noise and a new random variable $\hat{\mathcal{R}}$ related to the aforementioned $\mathcal{R}$. By a suitable choice of these random variables, we can apply a discrete Gronwall–like lemma to obtain a subexponential estimate of every element of the sequence. Finally, temperedness comes into play in order to ensure that $\{\boldsymbol{u}^{(n)}\}_{n\in\mathbb{N}}$ describes the solution of our SLDS on the positive real line, and such a solution converges to the equilibrium given by the trivial solution exponentially fast.

First of all, the following definition of exponential stability is adapted to the stochastic lattice systems using the concept of random dynamical systems.

Definition 14.3. The trivial solution of the stochastic lattice system (14.2) is said to be exponential stable with a rate $\mu > 0$, if for almost every $\omega \in \Omega$ there exists a random variable $\mathfrak{r}(\omega) > 0$ and a random neighborhood $\mathcal{N}(\omega)$ of zero such that for all $\omega \in \Omega$ and $t \in \mathbb{R}^+$

$$\sup_{\boldsymbol{u}_o \in \mathcal{N}(\omega)} \|\varphi(t, \omega, \boldsymbol{u}_o)\| \leq \mathfrak{r}(\omega) e^{-\mu t},$$

where $\varphi : \mathbb{R}^+ \times \Omega \times \ell^2 \to \ell^2$ is the cocycle mapping given in Theorem 14.3.

The aim is to prove that the trivial solution of the stochastic lattice system (14.2) is exponentially stable with a rate μ less than λ. First, an alternative definition of temperedness to Def.11.3 in Chapter 11 is recalled below.

Definition 14.4. A random variable $\mathcal{R} \in (0, \infty)$ is said to be tempered from above with respect to the metric dynamical system $(\Omega, \mathcal{F}, \mathbb{P}, \vartheta)$ if

$$\limsup_{t \to \pm\infty} \frac{\ln^+ \mathcal{R}(\vartheta_t \omega)}{t} = 0 \quad \text{with probability 1},$$

where $\ln^+ x = \max\{0, \ln x\}$ for $x > 0$.

In other words, temperedness from above describes the sub-exponential growth of a stochastic stationary process $(t, \omega) \mapsto \mathcal{R}(\vartheta_t \omega)$. A random variable $\mathcal{R}$ is said to be tempered from below if $\mathcal{R}^{-1}$ is tempered from above. In particular, if a random variable $\mathcal{R}$ is tempered from below and $t \mapsto \mathcal{R}(\vartheta_t \omega)$ is continuous, then for any $\epsilon > 0$ there exists a random variable $\mathcal{R}_\epsilon(\omega) > 0$ such that

$$\mathcal{R}(\vartheta_t \omega) \geq \mathcal{R}_\epsilon(\omega) e^{-\epsilon |t|} \quad \text{with probability 1}.$$

A sufficient condition for temperedness with respect to an ergodic metric dynamical system is that (see [Arnold (1998)], Page 165)

$$\mathbb{E}\left[\sup_{t \in [0,1]} \ln^+ \mathcal{R}(\vartheta_t \omega)\right] < \infty.$$

Hence, by Theorem 1.4.1 in [Kunita (1990)] we obtain that $\mathcal{R}(\omega) = \|\omega\|_{\beta',0,1}$ is tempered from above because $\ln^+ r \leq r$ for $r > 0$, and trivially $\sup_{t \in [0,1]} \|\vartheta_t \omega\|_{\beta,0,1} \leq \|\omega\|_{\beta,0,2}$. Furthermore, the set of all ω satisfying Definition 14.4 is invariant with respect to the flow ϑ.

14.4.1 *Existence of trivial solutions*

The following assumptions imply that the system (14.2) has a unique trivial solution.

Assumption 14.4. Each f_i is defined on $[-\delta, \delta]$ for some $\delta > 0$. In addition, $f_i(0) = f_i'(0) = 0$, $f_i \in \mathcal{C}^2([-\delta, \delta]), \mathbb{R})$ and there exists a positive constant D_f such that

$$|f_i''(s)| \leq D_f, \quad s \in [-\delta, \delta],\ i \in \mathbb{Z}.$$

Assumption 14.5. For each $i \in \mathbb{Z}$, g_i is defined on $[-\delta, \delta]$ with $g_i(0) = g_i'(0) = 0$.

The operators F, G are then defined on $\bar{\mathbb{B}}_{\ell^2}(0, \delta)$. In particular, from Assumption 14.4 we can show that F is Fréchet differentiable and its derivative $F' : \ell^2 \mapsto \mathfrak{L}(\ell^2)$ is continuous. Indeed, for $\boldsymbol{u}, \boldsymbol{v} \in \ell^2$ we have

$$\|F(\boldsymbol{u}+\boldsymbol{v}) - F(\boldsymbol{u}) - F'(\boldsymbol{u})\boldsymbol{v}\|^2 \le \frac{1}{4} D_f^2 \|\boldsymbol{v}\|^4,$$

and

$$\|F'(\boldsymbol{u}) - F'(\boldsymbol{v})\|_{\mathfrak{L}(\ell^2)}^2 = \sup_{\|\boldsymbol{z}\|=1} \|F'(\boldsymbol{u})\boldsymbol{z} - F'(\boldsymbol{v})\boldsymbol{z}\|^2 \le D_f^2 \|\boldsymbol{u} - \boldsymbol{v}\|^2.$$

Thus Assumptions 14.4–14.5 ensure that (14.2) has a unique trivial solution.

14.4.2 *The cut–off strategy*

Let ζ be a twice continuously differentiable cut–off function defined by

$$\zeta : \ell^2 \to \bar{\mathbb{B}}_{\ell^2}(0,1), \qquad \zeta(\boldsymbol{u}) = \begin{cases} \boldsymbol{u} : \|\boldsymbol{u}\| \le \frac{1}{2} \\ 0 \; : \|\boldsymbol{u}\| \ge 1 \end{cases}$$

such that the norm of $\zeta(\boldsymbol{u})$ is bounded by 1. In addition, ζ is chosen such that the first and second derivatives ζ' and ζ'' of ζ are bounded by L_ζ, D_ζ, respectively. Next for some random variable $\mathcal{R} \in (0, \delta]$ define

$$\zeta_{\mathcal{R}}(\boldsymbol{u}) = \mathcal{R} \cdot \zeta\Big(\frac{\boldsymbol{u}}{\mathcal{R}}\Big) \in \bar{\mathbb{B}}_{\ell^2}(0, \mathcal{R}), \quad \boldsymbol{u} \in \ell^2. \tag{14.17}$$

Then it is straightforward to check that the first derivative $\zeta_{\mathcal{R}}'$ of $\zeta_{\mathcal{R}}$ is bounded by L_ζ, while the second derivative $\zeta_{\mathcal{R}}''$ is bounded by $D_\zeta/\mathcal{R}$.

We now modify the operators F and G by considering their compositions with the cut–off function defined in (14.17). More precisely, set

$$F_{\mathcal{R}} := F \circ \zeta_{\mathcal{R}} : \ell^2 \to \ell^2, \qquad G_{\mathcal{R}} := G \circ \zeta_{\mathcal{R}} : \ell^2 \to \mathfrak{L}_2(\ell^2).$$

Consider again the equation (14.14) with F replaced by $F_{\mathcal{R}}$ and G replaced by $G_{\mathcal{R}}$, and the sequence $\{\boldsymbol{u}^{(n)}\}_{n \in \mathbb{N}}$ defined by

$$\begin{aligned} \boldsymbol{u}^{(n)}(t) = \mathcal{S}_\lambda(t)\boldsymbol{u}^{(n)}(0) &+ \int_0^t \mathcal{S}_\lambda(t-s) F_{\mathcal{R}(\vartheta_n \omega)}(\boldsymbol{u}^{(n)}(s)) \mathrm{d}s \\ &+ \int_0^t \mathcal{S}_\lambda(t-s) G_{\mathcal{R}(\vartheta_n \omega)}(\boldsymbol{u}^{(n)}(s)) \mathrm{d}\vartheta_n \omega, \qquad t \in [0,1], \end{aligned} \tag{14.18}$$

where $\boldsymbol{u}^{(0)}(0) = \boldsymbol{u}_o$ and $\boldsymbol{u}^{(n)}(0) = \boldsymbol{u}^{(n-1)}(1)$. Notice that the modified coefficients satisfy assumptions in Theorem 14.2 for any $n \in \mathbb{N}$, then the equation (14.18) has a unique solution $\boldsymbol{u}^{(n)}$ on $[0,1]$.

14.4.3 *Preliminary estimates*

The next result is crucial for the exponential stability of the trivial solution.

Lemma 14.6. *For every $\hat{\mathcal{R}} > 0$ there exists a positive $\mathcal{R} \leq \delta$ such that for all $\boldsymbol{u}$, $\boldsymbol{v} \in \ell^2$*

$$\|F_{\mathcal{R}}(\boldsymbol{u})\| \leq \hat{\mathcal{R}} L_\zeta \|\boldsymbol{u}\|, \tag{14.19}$$

$$\|G_{\mathcal{R}}(\boldsymbol{u})\| \leq \hat{\mathcal{R}} L_\zeta \|\boldsymbol{u}\|, \tag{14.20}$$

$$\|G_{\mathcal{R}}(\boldsymbol{u}) - G_{\mathcal{R}}(\boldsymbol{v})\| \leq \hat{\mathcal{R}} L_\zeta \|\boldsymbol{u} - \boldsymbol{v}\|. \tag{14.21}$$

Proof. By $F'(0) = 0$ and the continuity of F', for any $\hat{\mathcal{R}} > 0$ we can choose an $\mathcal{R} \leq \delta$ such that

$$\sup_{\|\boldsymbol{z}\| \leq \mathcal{R}} \|F'(\boldsymbol{z})\|_{\mathfrak{L}(\ell^2)} \leq \hat{\mathcal{R}}.$$

Then for $\boldsymbol{u} \in \ell^2$, since $F(0) = 0$ from the mean value theorem we have

$$\begin{aligned}
\|F_{\mathcal{R}}(\boldsymbol{u})\| &\leq \sup_{\boldsymbol{z} \in \ell^2} \|F'(\zeta_{\mathcal{R}}(\boldsymbol{z}))\| \|\boldsymbol{u}\| \\
&\leq \sup_{\|\boldsymbol{z}\| \leq \mathcal{R}} \|F'(\boldsymbol{z})\|_{\mathfrak{L}(\ell^2)} \sup_{\boldsymbol{z} \in \ell^2} \|\zeta'_{\mathcal{R}}(\boldsymbol{z})\| \|\boldsymbol{u}\| \leq \hat{\mathcal{R}} L_\zeta \|\boldsymbol{u}\|,
\end{aligned}$$

which implies (14.19). The assertion (14.20) can be shown following similar arguments.

Finally, by the regularity of G',

$$\begin{aligned}
\|G_{\mathcal{R}}(\boldsymbol{u}) - G_{\mathcal{R}}(\boldsymbol{v})\| &\leq \sup_{\|\boldsymbol{z}\| \leq \mathcal{R}} \|G'(\boldsymbol{z})\|_{\mathfrak{L}(\ell^2, \mathfrak{L}_2(\ell^2))} \|\zeta_{\mathcal{R}}(\boldsymbol{u}) - \zeta_{\mathcal{R}}(\boldsymbol{v})\| \\
&\leq L_\zeta \sup_{\|\boldsymbol{z}\| \leq \mathcal{R}} \|G'(\boldsymbol{z})\|_{\mathfrak{L}(\ell^2, \mathfrak{L}_2(\ell^2))} \|\boldsymbol{u} - \boldsymbol{v}\| \\
&\leq \hat{\mathcal{R}} L_\zeta \|\boldsymbol{u} - \boldsymbol{v}\|.
\end{aligned}$$

The proof is complete. □

Three technical results needed in the sequel are summarised in the following lemma without proofs.

Lemma 14.7.

(i) Let $(\mathfrak{V}, \|\cdot\|_{\mathfrak{V}})$ and $(\mathfrak{E}, \|\cdot\|_{\mathfrak{E}})$ be Banach spaces. For $\delta > 0$, let $U : \bar{\mathbb{B}}_{\mathfrak{V}}(0, \delta) \to \mathfrak{E}$ be a continuously differentiable function with $U(0) = 0$ and $U \not\equiv 0$ and

$$\sup_{\boldsymbol{z} \in \bar{\mathbb{B}}_{\mathfrak{V}}(0,\delta)} \|U'(\boldsymbol{z})\|_{\mathfrak{L}(\mathfrak{V}, \mathfrak{E})} := L_U < \infty.$$

For $R > 0$, let $\hat{R} := \sup_{r>0}\{r : \mathbb{B}_{\mathfrak{V}}(0, r) \subset F^{-1}(\mathbb{B}_{\mathfrak{E}}(0, R))\}$. Then

$$\sup_{\boldsymbol{z} \in \bar{\mathbb{B}}_{\mathfrak{V}}(0,\hat{R})} \|U(\boldsymbol{z})\|_{\mathfrak{E}} \leq R, \quad \frac{\hat{R}}{R} \geq \frac{1}{L_U} \quad \text{for} \quad 0 < R < \sup_{\boldsymbol{z} \in \bar{\mathbb{B}}_{\mathfrak{V}}(0,\delta)} \|U(\boldsymbol{z})\|_{\mathfrak{E}}.$$

(ii) Let $\{x_n\}$ and $\{y_n\}$ be non-negative sequences. Then for $c > 0$

$$x_n \leq c + \sum_{i=0}^{n-1} x_i y_i \quad \Longrightarrow \quad x_n \leq c \prod_{i=1}^{n-1}(1 + y_i).$$

(iii) Suppose that $R_i \geq C_\epsilon e^{-\epsilon i}$ for every $0 < \epsilon < M$ and $i \in \mathbb{N}$ and some $C_\epsilon > 0$. Let $\{x_i\}_{i \in \mathbb{N}}$ be a sequence with $v_i \leq v_0 e^{-Mi}$. Then $v_i \leq R_i$ for all $i \in \mathbb{N}$ for v_0 sufficiently small.

14.4.4 *Exponential stability*

For $n \in \mathbb{Z}^+$ set

$$\boldsymbol{u}(t) = \boldsymbol{u}^{(n)}(t-n) \quad \text{for } t \in [n, n+1]. \tag{14.22}$$

Notice that the function $\boldsymbol{u}$ in (14.22) is defined on the whole positive real line, and moreover, is Hölder continuous on any interval $[n, n+1]$. However, we cannot claim yet that $\boldsymbol{u}$ defined by (14.22) is exactly the mild solution obtained in Theorem 14.2. The reason is that every $\boldsymbol{u}^{(n)}$ is a solution of a modified lattice problem depending of the cut–off function $\zeta_{\mathcal{R}}$ and driven by a path $\vartheta_n\omega$. But as we will show below, using the additivity of the integrals, the estimates of the functions $F_{\mathcal{R}}$ and $G_{\mathcal{R}}$ given in Lemma 14.6, and a suitable choice of the random variables $\hat{\mathcal{R}}$ and $\mathcal{R}$, we will end up proving that not only $\boldsymbol{u}$ given by (14.22) is the solution of our original stochastic lattice system (14.14), but also that it is locally exponential stable with a certain decay rate μ.

In order to prove the previous assertions, we first express $\boldsymbol{u}$ given by (14.22) for $t \in [n, n+1]$ as follows

$$\begin{aligned}
\boldsymbol{u}(t) &= \mathcal{S}_\lambda(t-n)\boldsymbol{u}(n) + \int_n^t \mathcal{S}_\lambda(t-r)F_{\mathcal{R}(\vartheta_n\omega)}(\boldsymbol{u}(r))\mathrm{d}r \\
&\quad + \int_n^t \mathcal{S}_\lambda(t-r)G_{\mathcal{R}(\vartheta_n\omega)}(\boldsymbol{u}(r))\mathrm{d}\omega(r) \\
&= \mathcal{S}_\lambda(t)\boldsymbol{u}_o + \sum_{j=0}^{n-1} \mathcal{S}_\lambda(t-j-1)\Big(\int_0^1 \mathcal{S}_\lambda(1-r)F_{\mathcal{R}(\vartheta_j\omega)}(\boldsymbol{u}^{(j)}(r))\mathrm{d}r \\
&\quad + \int_0^1 \mathcal{S}_\lambda(1-r)G_{\mathcal{R}(\vartheta_j\omega)}(\boldsymbol{u}^{(j)}(r))\mathrm{d}\vartheta_j\omega(r)\Big) \\
&\quad + \int_0^{t-n} \mathcal{S}_\lambda(t-n-r)F_{\mathcal{R}(\vartheta_n\omega)}(\boldsymbol{u}^{(n)}(r))\mathrm{d}r \\
&\quad + \int_0^{t-n} \mathcal{S}_\lambda(t-n-r)G_{\mathcal{R}(\vartheta_n\omega)}(\boldsymbol{u}^{(n)}(r))\mathrm{d}\vartheta_n\omega(r), \quad t \in [n, n+1],
\end{aligned} \tag{14.23}$$

where this splitting is a consequence of the additivity of the integrals, Theorem 14.1 and (14.4).

Notice that, in all the integrals on the right hand side of the expression (14.23), the time varies in the interval $[0,1]$ (in the last two integrals, $[0,t-n]$ is contained in $[0,1]$). Hence, we can focus on estimating the Hölder-norm of all the terms in (14.23) with $T_1 = 0$ and $T_2 = 1$. Also, due to the presence of the semigroup $\mathcal{S}_\lambda$ as a factor in all terms under the sum, it suffices to consider the β-norm instead of the β,γ-norm, i.e., $\gamma = 0$ in all estimates in the sequel.

Lemma 14.8. *Let Assumptions 14.1–14.5 hold. Then there exists a zero neighborhood $\mathcal{N}(\omega)$ depending on ω and a tempered random variable $\mathcal{R}(\omega)$ such that for $\boldsymbol{u}_o \in \mathcal{N}(\omega)$, the unique solution $\boldsymbol{u}^{(n)}$ to the equation (14.18) satisfies*

$$\|\boldsymbol{u}^{(n)}\|_\beta \leq \frac{\mathcal{R}(\vartheta_n\omega)}{2} \quad \text{for all } n \in \mathbb{Z}^+,$$

Proof. First, notice that by (14.19) we have

$$\left\| \int_0^\cdot \mathcal{S}_\lambda(\cdot - r) F_{\mathcal{R}(\vartheta_n\omega)}(\boldsymbol{u}^{(n)}(r))\mathrm{d}r \right\|_\infty \leq \hat{\mathcal{R}}(\vartheta_n\omega) L_\varsigma \|\boldsymbol{u}^{(n)}\|_\infty.$$

Then for the Hölder-seminorm, thanks to Lemma 14.3,

$$\begin{aligned}
&\left|\left|\left| \int_0^\cdot \mathcal{S}_\lambda(\cdot - r) F_{\mathcal{R}(\vartheta_n\omega)}(\boldsymbol{u}^{(n)}(r))\mathrm{d}r \right|\right|\right|_\beta \\
&= \sup_{0\leq s<t\leq 1} \frac{\left\| \int_s^t \mathcal{S}_\lambda(t-r) F_{\mathcal{R}(\vartheta_n\omega)}(\boldsymbol{u}^{(n)}(r))\mathrm{d}r + \int_0^s (\mathcal{S}_\lambda(t-r) - \mathcal{S}_\lambda(s-r)) F_{\mathcal{R}(\vartheta_n\omega)}(\boldsymbol{u}^{(n)}(r))\mathrm{d}r \right\|}{(t-s)^\beta} \\
&\leq \sup_{0\leq s<t\leq 1} \left((t-s)^{1-\beta} \sup_{r\in[s,t]} (\|\mathcal{S}_\lambda(t-r)\|_{\mathfrak{L}(\ell^2)} \|F_{\mathcal{R}(\vartheta_n\omega)}(\boldsymbol{u}^{(n)}(r))\|) \right) \\
&\quad + \sup_{0\leq s<t\leq 1} \left(\frac{s}{(t-s)^\beta} \sup_{r\in[0,s]} (\|\mathcal{S}_\lambda(t-r) - \mathcal{S}_\lambda(s-r)\|_{\mathfrak{L}(\ell^2)} \|F_{\mathcal{R}(\vartheta_n\omega)}(\boldsymbol{u}^{(n)}(r))\|) \right) \\
&\leq \hat{\mathcal{R}}(\vartheta_n\omega) L_\varsigma \|\boldsymbol{u}^{(n)}\|_\infty + \|\mathrm{A}_\lambda\| \hat{\mathcal{R}}(\vartheta_n\omega) L_\varsigma \|\boldsymbol{u}^{(n)}\|_\infty.
\end{aligned}$$

Thus

$$\left\| \int_0^\cdot \mathcal{S}_\lambda(\cdot - r) F_{\mathcal{R}(\vartheta_n\omega)}(\boldsymbol{u}^{(n)}(r))\mathrm{d}r \right\|_\beta \leq (2 + \|\mathrm{A}_\lambda\|) \hat{\mathcal{R}}(\vartheta_n\omega) L_\varsigma \|\boldsymbol{u}^{(n)}\|_\beta.$$

On the other hand, since $G(0) = 0$, by (14.20) and (14.21) we get

$$\begin{aligned}
\|G_{\mathcal{R}(\vartheta_n\omega)}(\boldsymbol{u}(\cdot))\|_\beta &= \sup_{t\in[0,1]} \|G_{\mathcal{R}(\vartheta_n\omega)}(\boldsymbol{u}(t))\| \\
&\quad + \sup_{0\leq r<\tau\leq 1} \frac{\|G_{\mathcal{R}(\vartheta_n\omega)}(\boldsymbol{u}(r)) - G_{\mathcal{R}(\vartheta_n\omega)}(\boldsymbol{u}(\tau))\|}{(r-\tau)^\beta} \\
&\leq L_\varsigma \hat{\mathcal{R}}(\vartheta_n\omega) \|\boldsymbol{u}\|_\beta.
\end{aligned}$$

For the stochastic integral, thanks to Lemma 14.5 we obtain

$$\begin{aligned}&\left\| \int_0^{\cdot} \mathcal{S}_\lambda(\cdot - r) G_{\mathcal{R}(\vartheta_n\omega)}(\boldsymbol{u}^{(n)}(r)) \mathrm{d}\vartheta_n\omega(r) \right\|_\beta \\ &\quad \leq C_{\beta,\beta'} \left\|\vartheta_n\omega\right\|_{\beta'} (1 + \|\mathrm{A}_\lambda\|)(2 + \|\mathrm{A}_\lambda\|) \|G_{\mathcal{R}(\vartheta_n\omega)}(\boldsymbol{u}^{(n)}(\cdot))\|_\beta \\ &\quad \leq L_\varsigma C_{\beta,\beta'} \left\|\vartheta_n\omega\right\|_{\beta'} (1 + \|\mathrm{A}_\lambda\|)(2 + \|\mathrm{A}_\lambda\|) \hat{\mathcal{R}}(\vartheta_n\omega) \|\boldsymbol{u}^{(n)}\|_\beta.\end{aligned}$$

For the terms under the sum we have

$$\begin{aligned}&\left\| \mathcal{S}_\lambda(\cdot - j - 1) \int_0^1 \mathcal{S}_\lambda(1 - r) F_{\mathcal{R}(\vartheta_j\omega)}(\boldsymbol{u}^{(j)}(r)) \mathrm{d}r \right\|_{\beta,n,n+1} \\ &\quad = \|\mathcal{S}_\lambda(\cdot - j - 1)\|_{\beta,n,n+1} \left\| \int_0^1 \mathcal{S}_\lambda(1 - r) F_{\mathcal{R}(\vartheta_j\omega)}(\boldsymbol{u}^{(j)}(r)) \mathrm{d}r \right\| \\ &\quad \leq \|\mathcal{S}_\lambda(\cdot - j - 1)\|_{\beta,n,n+1} \left\| \int_0^{\cdot} \mathcal{S}_\lambda(\cdot - r) F_{\mathcal{R}(\vartheta_j\omega)}(\boldsymbol{u}^{(j)}(r)) \mathrm{d}r \right\|_\infty,\end{aligned}$$

and from Lemma 14.3,

$$\|\mathcal{S}_\lambda(\cdot - j - 1)\|_{\beta,n,n+1} \leq (1 + \|\mathrm{A}_\lambda\|) e^{-\lambda(n-j-1)},$$

such that

$$\begin{aligned}&\left\| \mathcal{S}_\lambda(\cdot - j - 1) \int_0^1 \mathcal{S}_\lambda(1 - r) F_{\mathcal{R}(\vartheta_j\omega)}(\boldsymbol{u}^{(j)}(r)) \mathrm{d}r \right\|_{\beta,n,n+1} \\ &\quad \leq (1 + \|\mathrm{A}_\lambda\|) e^{-\lambda(n-j-1)} \hat{\mathcal{R}}(\vartheta_j\omega) L_\varsigma \|\boldsymbol{u}^{(j)}\|_\beta.\end{aligned}$$

Following similar steps, and thanks to (14.15) we have

$$\begin{aligned}&\left\| \mathcal{S}_\lambda(\cdot - j - 1) \int_0^1 \mathcal{S}_\lambda(1 - r) H_{\mathcal{R}(\vartheta_j\omega)}(\boldsymbol{u}^{(j)}(r)) \mathrm{d}\vartheta_j\omega(r) \right\|_{\beta,n,n+1} \\ &\quad \leq L_\varsigma c_{\beta,\beta'} \left\|\vartheta_j\omega\right\|_{\beta'} (1 + \|\mathrm{A}_\lambda\|)^2 e^{-\lambda(n-j-1)} \hat{\mathcal{R}}(\vartheta_j\omega) \|\boldsymbol{u}^{(j)}\|_\beta.\end{aligned}$$

Therefore, taking the $\|\cdot\|_{\beta,n,n+1}$ norm of the different terms in (14.23), applying the triangle inequality and in view of the above estimates, we obtain

$$\begin{aligned}\|\boldsymbol{u}^{(n)}\|_\beta &\leq \|\mathcal{S}_\lambda\|_{\beta,n,n+1} \|\boldsymbol{u}_o\| + C \sum_{j=0}^{n-1} e^{-\lambda(n-j-1)} \hat{\mathcal{R}}(\vartheta_j\omega)(1 + \left\|\vartheta_j\omega\right\|_{\beta'}) \|\boldsymbol{u}^{(j)}\|_\beta \\ &\quad + C\hat{\mathcal{R}}(\vartheta_n\omega)(1 + \left\|\vartheta_n\omega\right\|_{\beta'}) \|\boldsymbol{u}^{(n)}\|_\beta, \qquad (14.24)\end{aligned}$$

where $C = \max\{1, c_{\beta,\beta'}\} L_\varsigma (1 + \|\mathrm{A}_\lambda\|)(2 + \|\mathrm{A}_\lambda\|)$.

Now set $\epsilon \in (0, 1)$, that will be determined later more precisely. Define the variables $\hat{\mathcal{R}}$ and $\mathcal{R}$ as follows:

$$\hat{\mathcal{R}}(\omega) = \frac{\epsilon}{2C(1 + \left\|\omega\right\|_{\beta'})}$$

and

$$\mathcal{R}(\omega) = \sup\left\{r \in [0,\delta] : \|F'(\boldsymbol{z})_{\mathfrak{L}(\ell^2)} + \|G'(\boldsymbol{z})_{\mathfrak{L}(\ell^2,\mathfrak{L}_2(\ell^2))} \leq \hat{\mathcal{R}}(\omega)\ \forall\ \boldsymbol{v} \in \bar{\mathbb{B}}(0,r)\right\}.$$

Then $\mathcal{R}(\omega)$ is a random variable [Garrido-Atienza (2018)]. In addition, since $\|\omega\|_{\beta'}$ is tempered from above then $\hat{\mathcal{R}}$ is tempered from below. According to Lemma 14.7-(i), the random variable $\mathcal{R}$ is tempered from below. In fact, suppose (for contradiction) that $\mathcal{R}$ is not tempered from below. Then there exist $\omega \in \Omega$, $\mu \in \mathbb{R}^+ \setminus \{0\} \cup \{+\infty\}$ and a sequence $(t_i)_{i\in\mathbb{N}}$ approaching $+\infty$ or $-\infty$ such that

$$\mathcal{R}(\vartheta_{t_i}\omega) \leq e^{-\mu|t_i|}.$$

Applying Lemma 14.7-(i) again with $\mathfrak{V} = \ell^2$ and $\mathfrak{E} = \mathfrak{L}(\ell^2) \times \mathfrak{L}(\ell^2, \mathfrak{L}_2(\ell^2))$ we have

$$\hat{\mathcal{R}}(\vartheta_{t_i}\omega) \leq ce^{-\mu|t_i|} \quad \text{for } i \text{ large enough},$$

which contradicts the temperedness of $\hat{\mathcal{R}}$.

With the above choice of $\hat{\mathcal{R}}$, coming back to (14.24), since $\epsilon < 1$ we obtain

$$\frac{1}{2}\|\boldsymbol{u}^{(n)}\|_\beta \leq \|\mathcal{S}_\lambda\|_{\beta,n,n+1}\|\boldsymbol{u}_o\| + \frac{\epsilon}{2}\sum_{j=0}^{n-1} e^{-\lambda(n-j-1)}\|\boldsymbol{u}^{(j)}\|_\beta,$$

and thus

$$\|\boldsymbol{u}^{(n)}\|_\beta \leq 2(1+\|\mathrm{A}_\lambda\|)\|\boldsymbol{u}_o\|e^{-\lambda n} + \epsilon\sum_{j=0}^{n-1} e^{-\lambda(n-j-1)}\|\boldsymbol{u}^{(j)}\|_\beta.$$

Letting $x_n = e^{\lambda n}\|\boldsymbol{u}^{(n)}\|_\beta$, and $y_n = \epsilon e^\lambda$, Lemma 14.7-(ii) then implies that

$$x_n \leq 2(1+\|\mathrm{A}_\lambda\|)\|\boldsymbol{u}_o\|\prod_{i=0}^{n-1}(1+\epsilon e^\lambda) = 2(1+\|\mathrm{A}_\lambda\|)\|\boldsymbol{u}_o\|(1+\epsilon e^\lambda)^n.$$

It follows that

$$\|\boldsymbol{u}^{(n)}\|_\beta \leq 2(1+\|\mathrm{A}_\lambda\|)\|\boldsymbol{u}_o\|e^{-n(\lambda-\log(1+\epsilon e^\lambda))}. \tag{14.25}$$

On the other hand, due to Lemma 14.7-(iii), since $\mathcal{R}(\omega)/2$ is tempered from below, we can find a zero neighborhood $\mathcal{N}$ of $\boldsymbol{u}_o$ depending on ω, such that the assertion of this lemma holds. □

Lemma 14.8 above is crucial in proving that the sequence of truncated solutions $\{\boldsymbol{u}^{(n)}\}_{n\in\mathbb{N}}$ defines a solution of (14.14) on $\mathbb{R}^+$. We are now ready to prove the main result on exponential stability of the trivial solution.

Theorem 14.4. *Let Assumptions 14.1–14.5 hold. Consider* $\epsilon(\lambda) = \epsilon \in (0, 1-e^{-\lambda})$. *Then the trivial solution to* (14.14) *is exponentially stable with an exponential rate less than or equal to* $\mu < \lambda - \log(1+\epsilon e^\lambda)$.

Proof. First, we show that $\boldsymbol{u}$ given by (14.22) is a solution of the original stochastic lattice system (14.14). Indeed, it follows by Lemma 14.8 that

$$\|\boldsymbol{u}^{(n)}(t)\| \leq \frac{\mathcal{R}(\vartheta_n\omega)}{2}, \quad n \in \mathbb{Z}^+, \quad t \in [0,1].$$

Consequently $\zeta_{\mathcal{R}(\vartheta_n\omega)}(\boldsymbol{u}^{(n)}(t)) = \boldsymbol{u}^{(n)}(t)$, and hence

$$F_{\mathcal{R}(\vartheta_n\omega)}(\boldsymbol{u}^{(n)}(t)) = F(\boldsymbol{u}^{(n)}(t)), \quad G_{\mathcal{R}(\vartheta_n\omega)}(\boldsymbol{u}^{(n)}(t)) = G(\boldsymbol{u}^{(n)}(t))$$

for any $t \in [0,1]$ and $n \in \mathbb{Z}^+$. Thus $\boldsymbol{u}$ given by (14.22), where $\boldsymbol{u}^{(n)}$ solves (14.18), is a solution of the equation (14.14) on $\mathbb{R}^+$.

Now we show the exponential stability of the trivial solution, according to the Definition 14.3. Take $\epsilon \in (0, 1-e^{-\lambda})$, $\hat{\epsilon} > 0$ small enough and $\mu < \lambda - \log(1+\epsilon e^{\lambda}) - \hat{\epsilon}$. From (14.25) we derive that there exist $T_0(\omega,\epsilon) \in \mathbb{N}$ and a neighborhood of zero $\mathcal{N}(\omega)$ such that if $\boldsymbol{u}_o \in \mathcal{N}(\omega)$, for $t \geq T_0(\omega,\epsilon)$

$$\|\varphi(t,\omega,\boldsymbol{u}_o)\| \leq \alpha_1(\omega) e^{-(\lambda - \log(1+\epsilon e^{\lambda}) - \hat{\epsilon})t} \leq \mathfrak{r}_1(\omega) e^{-\mu t},$$

where the positive random variable $\mathfrak{r}_1(\omega)$ depends on the coefficients κ and $\|\mathrm{A}_\lambda\|$.

On the other hand, from Lemma 14.8 we derive

$$\sup_{\boldsymbol{u}_o \in \mathcal{N}(\omega), t \in [0, T_0(\omega,\epsilon)]} \|\varphi(t,\omega,\boldsymbol{u}_o)\| \leq \sup_{\boldsymbol{u}_o \in \mathcal{N}(\omega), 0 \leq n \leq T_0(\omega,\epsilon)} \|\boldsymbol{u}^{(n)}\|_\beta \leq \mathfrak{r}_2(\omega) e^{-\mu t},$$

where

$$\mathfrak{r}_2(\omega) = \sup_{0 \leq n \leq T_0(\omega,\epsilon)} \frac{\mathcal{R}(\vartheta_n\omega)}{2} e^{\mu T_0}.$$

Therefore,

$$\begin{aligned}\sup_{\boldsymbol{u}_o \in \mathcal{N}(\omega)} \|\varphi(t,\omega,\boldsymbol{u}_o)\| &\leq \sup_{\boldsymbol{u}_o \in \mathcal{N}(\omega), t \in [0, T_0(\omega,\epsilon)]} \|\varphi(t,\omega,\boldsymbol{u}_o)\| \\ &\quad + \sup_{\boldsymbol{u}_o \in \mathcal{N}(\omega), t \geq T_0(\omega,\epsilon)} \|\varphi(t,\omega,\boldsymbol{u}_o)\| \\ &\leq (\mathfrak{r}_1(\omega) + \mathfrak{r}_2(\omega)) e^{-\mu t}.\end{aligned}$$

The proof is complete. □

14.5 End notes

The chapter is based on Bessaih, Garrido-Atienza, Han & Schmalfuss [Bessaih *et al.* (2017)]. See also [Garrido-Atienza (2018); Nualart and Rascanu (2022)].

Information about fractional Brownian motion and related fractional calculus can be found in [Mishura (2008); Samko *et al.* (2014); Zähle (1998)].

The exponential stability of systems driven by continuous semi-martingales is investigated in the monograph [Mao (1994)].

14.6 Problems

Problem 14.1. Investigate the existence of a random attractor in a dissipative RODE driven by fractional Brownian motion. How do these attractors depend on the Hurst parameter H?

Problem 14.2. What happens to the solutions of the stochastic lattice system (14.2) as $H \to \frac{1}{2}^+$? Do they approach the solution to an otherwise identical stochastic lattice system with a Brownian motion?

Problem 14.3. For the special case $g_i \equiv 1$, show that the random dynamical system generated by the stochastic lattice system (14.2) has a global random attractor under appropriate conditions on f_i.

PART 5

Hopfield Lattice Models

Chapter 15

Hopfield neural network lattice model

15.1 Introduction

One of the most popular mathematical models for artificial neural networks was proposed by John J. Hopfield in 1984 and is now known as the Hopfield neural network model [Hopfield (1984)]. It is described by the following system of ordinary differential equations

$$\mu_i \frac{\mathrm{d}u_i(t)}{\mathrm{d}t} = -\frac{u_i(t)}{\varkappa_i} + \sum_{j=1}^{n} \lambda_{i,j} f_j(u_j(t)) + g_i, \qquad i = 1, ..., n, \tag{15.1}$$

where u_i represents the voltage on the input of the ith neuron at time t, $\mu_i > 0$ and $\varkappa_i > 0$ represent the neuron amplifier input capacitance and resistance of the ith neuron, respectively, and g_i is the external forcing on the ith neuron.

The system (15.1) describes the dynamics of a total number of n neurons coupled by the $n \times n$ matrix $(\lambda_{i,j})_{1 \leq i,j \leq n}$, where the parameter $\lambda_{i,j}$ represents the connection strength between the ith and the jth neuron. More precisely, for each pair of $i, j \in \{1, \dots, n\}$, the parameter $\lambda_{i,j}$ represents the synapse efficacy between neurons i and j, and thus $\lambda_{i,j} > 0$ ($\lambda_{i,j} < 0$, resp.) means the output of neuron j excites (inhibits, resp.) neuron i. The term $\lambda_{i,j} f_j(u_j(t))$ represents the electric current input to neuron i due to the present potential of neuron j, in which the function f_j is the neuron activation function.

When the size of an underlying neural network becomes increasingly large, and approaching infinity, the finite dimensional system (15.1) can be extended to the following lattice dynamical system, which will be referred to as the Hopfield lattice model:

$$\mu_i \frac{\mathrm{d}u_i(t)}{\mathrm{d}t} = -\frac{u_i(t)}{\varkappa_i} + \sum_{j=i-\mathfrak{n}}^{i+\mathfrak{n}} \lambda_{i,j} f_j(u_j(t)) + g_i, \qquad i \in \mathbb{Z}. \tag{15.2}$$

In this lattice model, the total number of neurons are assumed to be infinitely large, but each neuron is interacting only with the neurons within its $\mathfrak{n}$-neighbourhood. More precisely, the ith neuron interacts with the $(i - \mathfrak{n})$th through to the $(i + \mathfrak{n})$th neuron.

Notice that the operator $\sum_{j=i-\mathfrak{n}}^{i+\mathfrak{n}} \lambda_{i,j} f_j(u_j(t))$ describing interactions among neurons is nonlinear when f_j is nonlinear. This makes the lattice system (15.2) different from most of the lattice systems with linear operators, such as the discrete Laplacian operator, considered in the earlier chapters of this book.

15.2 Formulation as an ODE

The lattice system (15.2) can be written as an ODE on the sequence space ℓ^2 under the following assumptions.

Assumption 15.1. The efficacy between each pair of neurons is finite, i.e., there exists $M_\lambda > 0$ such that

$$\sup_{i,j\in\mathbb{Z}} |\lambda_{i,j}| \leq M_\lambda.$$

Assumption 15.2. The neuron amplifier input capacitance and resistance of neurons are uniformly bounded from above and below, i.e., there exist positive constants m_μ, M_μ, $m_\varkappa$, and $M_\varkappa$, such that

$$m_\mu \leq \mu_i \leq M_\mu, \quad m_\varkappa \leq \varkappa_i \leq M_\varkappa \quad \text{for all } i \in \mathbb{Z}.$$

Assumption 15.3. The forcing $\boldsymbol{g} := (g_i)_{i\in\mathbb{Z}}$ on the whole network is square summable, i.e., $\boldsymbol{g} \in \ell^2$ with

$$\|\boldsymbol{g}\| = \left(\sum_{i\in\mathbb{Z}} g_i^2\right)^{1/2} < \infty.$$

Assumption 15.4. For all $i \in \mathbb{Z}$, the neuron activation function f_i is globally Lipschitz with Lipschitz constant L and satisfies $f_i(0) = 0$.

Given any $\boldsymbol{u} = (u_i)_{i\in\mathbb{Z}}$, define the operators $\mathrm{A}\boldsymbol{u} = ((\mathrm{A}\boldsymbol{u})_i)_{i\in\mathbb{Z}}$, $F(\boldsymbol{u}) = ((F(\boldsymbol{u}))_i)_{i\in\mathbb{Z}}$ component wise by

$$(\mathrm{A}\boldsymbol{u})_i = \frac{u_i}{\varkappa_i \mu_i}, \qquad (F(\boldsymbol{u}))_i = \sum_{j=i+\mathfrak{n}}^{i+\mathfrak{n}} \frac{\lambda_{i,j}}{\mu_i} f_j(u_j),$$

and write $\tilde{\boldsymbol{g}} = \left(\frac{g_i}{\mu_i}\right)_{i\in\mathbb{Z}}$. Then define the operator $\mathfrak{F}$ to be

$$\mathfrak{F}(\boldsymbol{u}) = -\mathrm{A}\boldsymbol{u} + F(\boldsymbol{u}) + \tilde{\boldsymbol{g}}.$$

Lemma 15.1. *Let the Assumptions 15.1–15.4 hold. Then the operator $\mathfrak{F} : \ell^2 \to \ell^2$ is globally Lipschitz with*

$$\|\mathfrak{F}(\boldsymbol{u}) - \mathfrak{F}(\boldsymbol{v})\| \leq \frac{\sqrt{2}}{m_\mu}\left(\frac{1}{m_\varkappa^2} + M_\lambda^2(2\mathfrak{n}+1)^2 L^2\right)^{1/2} \|\boldsymbol{u} - \boldsymbol{v}\| \quad \textit{for all } \boldsymbol{u}, \boldsymbol{v} \in \ell^2.$$

Proof. First given any $\boldsymbol{u}, \boldsymbol{v} \in \ell^2$,

$$\|\mathrm{A}\boldsymbol{u} - \mathrm{A}\boldsymbol{v}\|^2 = \sum_{i\in\mathbb{Z}} \frac{1}{\mu_i^2 \varkappa_i^2}(u_i - v_i)^2 \le \frac{1}{m_\mu^2 m_\varkappa^2}\|\boldsymbol{u} - \boldsymbol{v}\|^2. \tag{15.3}$$

Second, by the inequality $\left(\sum_{j=i-\mathfrak{n}}^{i+\mathfrak{n}} a_i\right)^2 \le 2^{2\mathfrak{n}} \sum_{j=i-\mathfrak{n}}^{i+\mathfrak{n}} a_i^2$ and the Assumptions 15.1 and 15.4,

$$\begin{aligned}
\|F(\boldsymbol{u}) - F(\boldsymbol{v})\|^2 &= \sum_{i\in\mathbb{Z}} \frac{1}{\mu_i^2}\left(\sum_{j=i-\mathfrak{n}}^{i+\mathfrak{n}} \lambda_{i,j}\left[f_j(u_j) - f_j(v_j)\right]\right)^2 \\
&\le \frac{M_\lambda^2}{m_\mu^2}\sum_{i\in\mathbb{Z}} 2^{2\mathfrak{n}} \sum_{j=i-\mathfrak{n}}^{i+\mathfrak{n}} \left[f_j(u_j) - f_j(v_j)\right]^2 \\
&\le \frac{M_\lambda^2}{m_\mu^2} 2^{2\mathfrak{n}} \sum_{i\in\mathbb{Z}} \sum_{j=i-\mathfrak{n}}^{i+\mathfrak{n}} L^2 |u_j - v_j|^2 \\
&= \frac{M_\lambda^2}{m_\mu^2}(2\mathfrak{n}+1)^2 2^{2\mathfrak{n}} L^2 \|\boldsymbol{u} - \boldsymbol{v}\|^2.
\end{aligned} \tag{15.4}$$

Hence summing (15.3) and (15.4) yields

$$\begin{aligned}
\|\mathfrak{F}(\boldsymbol{u}) - \mathfrak{F}(\boldsymbol{v})\|^2 &\le 2\|A\boldsymbol{u} - A\boldsymbol{v}\|^2 + 2\|F(\boldsymbol{u}) - F(\boldsymbol{v})\|^2 \\
&\le \frac{2}{m_\mu^2}\left(\frac{1}{m_\varkappa^2} + M_\lambda^2 (2\mathfrak{n}+1)^2 2^{2\mathfrak{n}} L^2\right)\|\boldsymbol{u} - \boldsymbol{v}\|^2,
\end{aligned}$$

which implies that $\mathfrak{F}(\boldsymbol{u})$ is globally Lipschitz in $\boldsymbol{u}$ for every fixed $\mathfrak{n} \in \mathbb{N}$.

Setting $\boldsymbol{v} = \mathbf{0}$ in (15.3) and (15.4) and using $f_i(0) = 0$ gives immediately

$$\begin{aligned}
\|\mathfrak{F}(\boldsymbol{u}))\| &\le \|\mathrm{A}\boldsymbol{u}\| + \|F(\boldsymbol{u})\| + \frac{1}{m_\mu}\left(\sum_{i\in\mathbb{Z}} g_i^2\right)^{1/2} \\
&\le \frac{1}{m_\mu}\left(\frac{1}{m_\varkappa} + (2\mathfrak{n}+1)2^{\mathfrak{n}} M_\lambda L\right)\|\boldsymbol{u}\| + \frac{1}{m_\mu}\|\boldsymbol{g}\|,
\end{aligned}$$

which implies that F maps ℓ^2 to ℓ^2 for every fixed $\mathfrak{n} \in \mathbb{N}$. □

Lemma 15.1 ensures that the Hopfield lattice system (15.2) can be rewritten as the ODE

$$\frac{\mathrm{d}\boldsymbol{u}(t)}{\mathrm{d}t} = \mathfrak{F}(\boldsymbol{u}) = -\mathrm{A}\boldsymbol{u} + F(\boldsymbol{u}) + \tilde{\boldsymbol{g}}, \quad \boldsymbol{u} \in \ell^2, \tag{15.5}$$

on ℓ^2 with a globally Lipschitz vector field. Hence the standard theory of ODEs on Banach spaces (see, e.g., [Deimling (1977)]) gives the existence and uniqueness of solutions.

Theorem 15.1. *Let Assumptions 15.1–15.4 hold. Then for any initial data* $\boldsymbol{u}_o = (u_{o,i})_{i\in\mathbb{Z}} \in \ell^2$, *the LDS* (15.5) *has a unique solution* $\boldsymbol{u}(\cdot;\boldsymbol{u}_o) \in \mathcal{C}([0,\infty),\ell^2)$ *with* $\boldsymbol{u}(0;\boldsymbol{u}_o) = \boldsymbol{u}_o$. *In addition, the solution* $\boldsymbol{u}(\cdot;\boldsymbol{u}_o)$ *is continuous in* $\boldsymbol{u}_o \in \ell^2$.

15.3 Existence of attractors

It follows by Theorem 15.1 that the solutions of the ODE (15.5) defines an autonomous semi-dynamical system $\{\varphi(t)\}_{t\geq 0}$ on ℓ^2 to ℓ^2 by

$$\varphi(t, \boldsymbol{u}_o) = \boldsymbol{u}(t; \boldsymbol{u}_o) \quad \text{for all } t \geq 0, \ \boldsymbol{u}_o \in \ell^2.$$

The semi-dynamical system $\{\varphi(t)\}_{t\geq 0}$ has an absorbing set if the parameters satisfy the following assumption

Assumption 15.5. $\delta := \frac{1}{M_\mu M_\varkappa} - 2(2\mathfrak{n}+1)\frac{LM_\lambda}{m_\mu} > 0.$

Lemma 15.2. *Let Assumptions 15.1–15.5 hold. Then the continuous semi-dynamical system $\{\varphi(t)\}_{t\geq 0}$ generated by solutions to the system (15.5) has a positively invariant closed and bounded absorbing set.*

Proof. First multiplying both sides of the equation (15.2) by $u_i(t)$ and summing over all $i \in \mathbb{Z}$ to obtain

$$\frac{1}{2}\frac{\mathrm{d}\,\|\boldsymbol{u}(t)\|^2}{\mathrm{d}t} = -\sum_{i\in\mathbb{Z}} \frac{u_i^2}{\mu_i \varkappa_i} + \sum_{i\in\mathbb{Z}} \frac{u_i}{\mu_i} \sum_{j=i-\mathfrak{n}}^{i+\mathfrak{n}} \lambda_{i,j} f_j(u_j) + \sum_{i\in\mathbb{Z}} \frac{u_i}{\mu_i} g_i. \tag{15.6}$$

By Assumption 15.2,

$$-\sum_{i\in\mathbb{Z}} \frac{u_i^2}{\mu_i \varkappa_i} \leq -\frac{\|\boldsymbol{u}\|^2}{M_\mu M_\varkappa}. \tag{15.7}$$

Using the fact that $xy \leq \frac{1}{2}(ax^2 + \frac{1}{a}y^2)$, for some $a > 0$ (to be determined later),

$$\begin{aligned}\sum_{i\in\mathbb{Z}} \frac{u_i}{\mu_i} g_i &\leq \sum_{i\in\mathbb{Z}} \frac{1}{2}\left(a u_i^2 + \frac{g_i^2}{a\mu_i^2}\right) \leq \frac{a}{2}\|\boldsymbol{u}\|^2 + \frac{1}{2a}\sum_{i\in\mathbb{Z}} \frac{g_i^2}{\mu_i^2} \\ &\leq \frac{a}{2}\|\boldsymbol{u}\|^2 + \frac{1}{2am_\mu^2}\|\boldsymbol{g}\|^2. \end{aligned} \tag{15.8}$$

Then by Assumption 15.4

$$\lambda_{i,j} u_i f_j(u_j) \leq M_\lambda L |u_i||u_j| \leq \frac{LM_\lambda}{2}(u_i^2 + u_j^2), \tag{15.9}$$

which after a shift of index implies that

$$\sum_{i\in\mathbb{Z}} \frac{u_i}{\mu_i} \sum_{j=i-\mathfrak{n}}^{i+\mathfrak{n}} \lambda_{i,j} f_j(u_j) \leq \frac{LM_\lambda}{m_\mu}(2\mathfrak{n}+1)\|\boldsymbol{u}\|^2. \tag{15.10}$$

Now inserting the inequality (15.7), the inequality (15.8) with $a = \frac{1}{M_\mu M_\varkappa}$, and the inequality (15.10) into (15.6) results in

$$\frac{\mathrm{d}\|\boldsymbol{u}(t)\|^2}{\mathrm{d}t} \leq -\delta\|\boldsymbol{u}(t)\|^2 + \frac{M_\mu M_\varkappa}{m_\mu^2}\|\boldsymbol{g}\|^2, \tag{15.11}$$

where δ is defined as in Assumption 15.4. Integrating the differential inequality (15.11) from 0 to t gives

$$\begin{aligned}\|\varphi(t,\boldsymbol{u}_o)\|^2 &\le e^{-\delta t}\|\boldsymbol{u}_o\|^2 + \frac{M_\mu M_\varkappa}{m_\mu^2}\|\boldsymbol{g}\|^2 \int_0^t e^{-\delta(t-s)}\mathrm{d}s \\ &\le e^{-\delta t}\|\boldsymbol{u}_o\|^2 + \frac{M_\mu M_\varkappa}{\delta m_\mu^2}\|\boldsymbol{g}\|^2(1-e^{-\delta t}).\end{aligned}$$

Define Q to be the closed and bounded subset

$$Q := \left\{\boldsymbol{u} \in \ell^2 : \|\boldsymbol{u}\| \le \frac{\sqrt{M_\mu M_\varkappa}}{\sqrt{\delta}}\frac{\|\boldsymbol{g}\|^2}{m_\mu} + 1\right\}. \tag{15.12}$$

Then given any bounded set $B \in \ell^2$, there exists T_B such that

$$\varphi(t,B) \subset Q \quad \text{for all } t \ge T_B,$$

i.e., Q is a bounded absorbing set for the semi-dynamical system φ.

As a particular case, there exists T_Q such that

$$\varphi(t,Q) \subset Q \quad \text{for all } t > T_Q,$$

i.e., Q absorbs itself under $\{\varphi(t)\}_{t\ge 0}$. Then the set $\tilde{Q} := \bigcup_{0\le t\le T_Q}\varphi(t,Q)$ is closed, bounded, absorbing and positive invariant. □

The asymptotic compactness of $\{\varphi(t)\}_{t\ge0}$ will also be proved by using a tail estimate of solutions, but a cut-off function different from (3.8) that was used in previous chapters is adopted. Given a fixed positive integer k let $\eta_k : \mathbb{Z}_+ \to [0,1]$ be a continuous, increasing and sub-additive function such that

$$\eta_k(|i|) \in \begin{cases} \{0\}, & |i| < k, \\ [0,1], & |i| \in [k,2k] \\ \{1\}, & |i| \ge 2k, \end{cases} \qquad i \in \mathbb{Z}. \tag{15.13}$$

Lemma 15.3. *Let* η_k *be a function defined as in* (15.13). *Then, for any* $k \ge \mathfrak{n}$,

$$\eta_k(|j|) = \eta_k(|i|) \quad \textit{for all } j = i-\mathfrak{n},\ldots,i,\ldots,\mathfrak{n}.$$

Proof. Note that $|i-j| \le \mathfrak{n}$ for any $j = i-\mathfrak{n},\ldots,i+\mathfrak{n}$, so $\eta_k(|i-j|) = 0$. Then, by the increasingness and sub-additivity of η_k,

$$\eta_k(|i|) \le \eta_k(|i-j| + |j|) \le \eta_k(|i-j|) + \eta_k(|j|) = \eta_k(|j|),$$

$$\eta_k(|j|) \le \eta_k(|i| + |j-i|) \le \eta_k(|i|) + \eta_k(|j-i|) = \eta_k(|i|),$$

i.e., $\eta_k(|j|) = \eta_k(|i|)$. □

Remark 15.1. The cut-off function η_k is not assumed to be smooth as in earlier chapters, but instead is assumed to satisfy a sub-additivity property to handle the nonlinear interaction term. A possible example is with $\eta_k(s) = \frac{1}{k}s - 1$ for $s \in [k,2k]$.

The following asymptotic tails property holds.

Lemma 15.4. *Let the Assumptions 15.1–15.5 hold. Then for any $\varepsilon > 0$ there exist $T_\varepsilon = T(\varepsilon, Q) \geq T_Q$ and $I_\varepsilon \in \mathbb{N}$ such that the solution $\varphi(t, \boldsymbol{u}_o) = \boldsymbol{u}(t; \boldsymbol{u}_o) = (u_i(t; \boldsymbol{u}_o))_{i\in\mathbb{Z}}$ of the equation (15.2) with $\boldsymbol{u}_o = (u_{o,i})_{i\in\mathbb{Z}} \in Q$ satisfies*

$$\sum_{|i|\geq I_\varepsilon} |(\varphi(t, \boldsymbol{u}_o))_i|^2 = \sum_{|i|\geq I_\varepsilon} |u_i(t; \boldsymbol{u}_o)|^2 \leq \varepsilon \quad \textit{for all } t \geq T_\varepsilon.$$

Proof. The proof uses the same idea as in Chapter 3, but with more technical analysis due to the nonlinear structure of the vector field operator $\mathfrak{F}$.

For any solution $\boldsymbol{u}(t; \boldsymbol{u}_o) = (u_i(t; \boldsymbol{u}_o))_{i\in\mathbb{Z}}$ of (15.2), let $v_i := \eta_k(|i|)u_i$ where η_k is the continuous, increasing and sub-additive function as defined in (15.13) with $k \geq \mathfrak{n}$. Then $\boldsymbol{v} := (v_i)_{i\in\mathbb{Z}} \in \ell^2$. Taking the inner product of $\boldsymbol{v}$ with (15.2) gives

$$\frac{1}{2}\frac{\mathrm{d}}{\mathrm{d}t}\sum_{i\in\mathbb{Z}} \eta_k(|i|)u_i^2(t; \boldsymbol{u}_o) = -\sum_{i\in\mathbb{Z}} \eta_k(|i|)\frac{u_i^2}{\mu_i \varkappa_i} + \sum_{i\in\mathbb{Z}} \eta_k(|i|)u_i \frac{1}{\mu_i} g_i$$

$$+ \sum_{i\in\mathbb{Z}} \frac{1}{\mu_i} \sum_{j=i-\mathfrak{n}}^{i+\mathfrak{n}} \lambda_{i,j}\eta_k(|i|)u_i f_j(u_j). \tag{15.14}$$

First by Assumption 15.2,

$$-\sum_{i\in\mathbb{Z}} \eta_k(|i|)\frac{u_i^2}{\mu_i \varkappa_i} \leq -\frac{1}{M_\mu M_\varkappa}\sum_{i\in\mathbb{Z}} \eta_k(|i|)u_i^2. \tag{15.15}$$

Second, similar to (15.8) there exists some $a > 0$ such that

$$\sum_{i\in\mathbb{Z}} \eta_k(|i|)\frac{u_i}{\mu_i} g_i \leq \frac{a}{2}\sum_{i\in\mathbb{Z}} \eta_k(|i|)u_i^2 + \frac{1}{2am_\mu^2}\sum_{i\in\mathbb{Z}} \eta_k(|i|)g_i^2. \tag{15.16}$$

It remains to estimate the last term of (15.14). To this end, multiplying (15.9) by $\frac{1}{\mu_i}\eta_k(|i|)$ and summing over $i \in \mathbb{Z}$ to get

$$\sum_{i\in\mathbb{Z}} \eta_k(|i|)\frac{1}{\mu_i} \sum_{j=i-\mathfrak{n}}^{i+\mathfrak{n}} \lambda_{i,j}u_i f_j(u_j) \leq \frac{LM_\lambda}{2m_\mu}\sum_{i\in\mathbb{Z}} \eta_k(|i|) \sum_{j=i-\mathfrak{n}}^{i+\mathfrak{n}} (u_i^2 + u_j^2).$$

Notice that by a shift of index and Lemma 15.3,

$$\sum_{i\in\mathbb{Z}} \eta_k(|i|) \sum_{j=i-\mathfrak{n}}^{i+\mathfrak{n}} u_j^2 = \sum_{i\in\mathbb{Z}} \sum_{j=i-\mathfrak{n}}^{i+\mathfrak{n}} \eta_k(|j|)u_i^2 = (2\mathfrak{n}+1)\sum_{i\in\mathbb{Z}} \eta_k(|i|)u_i^2. \tag{15.17}$$

Inserting estimates (15.15), (15.16) with $a = \frac{1}{M_\mu M_\varkappa}$, and (15.17) in (15.14) results in

$$\frac{\mathrm{d}}{\mathrm{d}t}\sum_{i\in\mathbb{Z}} \eta_k(|i|)u_i^2(t) \leq -\delta\sum_{i\in\mathbb{Z}} \eta_k(|i|)u_i^2 + \frac{M_\mu M_\varkappa}{m_\mu^2}\sum_{i\in\mathbb{Z}} \eta_k(|i|)g_i^2,$$

where δ is as defined as in Assumption 15.5.

Integrating the above inequality from 0 to t and using the fact that $\int_0^t e^{-\delta(t-s)}\mathrm{d}s \leq \frac{1}{\delta}$ gives

$$\sum_{i\in\mathbb{Z}} \eta_k(|i|)u_i^2(t) \leq e^{-\delta t}\sum_{i\in\mathbb{Z}} \eta_k(|i|)u_{o,i}^2 + \frac{M_\mu M_\varkappa}{\delta m_\mu^2}\sum_{i\in\mathbb{Z}} \eta_k(|i|)g_i^2. \tag{15.18}$$

First notice that $\sum_{i\in\mathbb{Z}} \eta_k(|i|)u_{o,i}^2 \leq \|\boldsymbol{u}_o\|^2$. Then for any $\varepsilon > 0$ and $\boldsymbol{u}_o \in Q$, there exists $T_\varepsilon > 0$ such that

$$e^{-\delta t}\sum_{i\in\mathbb{Z}} \eta_k(|i|)u_{o,i}^2 < \frac{\varepsilon}{2} \quad \text{for all } t \geq T_\varepsilon. \tag{15.19}$$

Second, since $\boldsymbol{g} = (g_i)_{i\in\mathbb{Z}} \in \ell^2$, for any $\varepsilon > 0$ there exists $I(\varepsilon) > 0$ such that

$$\sum_{|i|\geq I(\varepsilon)} g_i^2 < \frac{\delta m_\mu^2}{2M_\mu M_\varkappa}\varepsilon. \tag{15.20}$$

Therefore letting $k = k(\varepsilon) = \max\{I(\varepsilon), \mathfrak{n}\}$ and applying (15.19) and (15.20) to (15.18) we obtain

$$\sum_{i\in\mathbb{Z}} \eta_k(|i|)u_i^2(t; \boldsymbol{u}_o) < \epsilon, \quad \text{for alll } t \geq T_\varepsilon.$$

This implies that with $I_\varepsilon = 2k$,

$$\sum_{|i|\geq I_\varepsilon} u_i^2(t, \boldsymbol{u}_o) \leq \sum_{i\in\mathbb{Z}} \eta_k(|i|)u_i^2(t, \boldsymbol{u}_o) < \varepsilon, \quad \text{for all } t \geq T_\varepsilon.$$

The proof is complete. □

Theorem 15.2. *Let the Assumptions 15.1–15.5 hold. Then, the semi-dynamical system $\{\varphi(t)\}_{t\geq 0}$ generated by solutions to the system (15.5) possesses a global attractor $\mathcal{A}$ which is minimal and connected.*

Proof. The existence of a bounded absorbing set for $\{\varphi(t)\}_{t\geq 0}$ has been established in Lemma 15.2, so it remains to prove that the absorbing set Q defined in (15.12) is asymptotically compact under φ. To this end consider a sequence of time $\{t_m\}$ with $\lim_{m\to\infty} t_m = \infty$ and let $\{\boldsymbol{u}_o^{(m)}\}$ be a sequence in Q. For $m = 1, 2, \ldots$, let

$$\boldsymbol{u}^{(m)} = (u_i^{(m)})_{i\in\mathbb{Z}} \quad \text{where} \quad u_i^{(m)} = (\varphi(t_m, \boldsymbol{u}^{(m)}))_i \quad \text{for } i \in \mathbb{Z}.$$

First since $\lim_{m\to\infty} t_m = \infty$ there exists $M_1(Q) \in \mathbb{N}$ such that $t_m \geq T_Q$ when $m \geq M_1(Q)$ and hence

$$\boldsymbol{u}^{(m)} = \varphi(t_m, \boldsymbol{u}_o^{(m)}) \in Q \quad \text{for alll } m \geq M_1(Q).$$

Then there is a subsequence of $\{\boldsymbol{u}^{(m)}\}$ (still denoted by $\{\boldsymbol{u}^{(m)}\}$), and $\boldsymbol{u}^* \in \ell^2$ such that

$$\boldsymbol{u}^{(m)} = \varphi(t_m, \boldsymbol{u}_o^{(m)}) \rightharpoonup \boldsymbol{u}^* \text{ weakly in } \ell^2.$$

It will now be shown that this convergence is actually strong. Given any $\varepsilon > 0$, by Lemma 15.4 there exists $I_1(\varepsilon) > 0$ and $M_2(\varepsilon) > 0$ such that

$$\sum_{|i|\geq I_1(\varepsilon)} \left|\left(\varphi(t_m, \boldsymbol{u}^{(m)})\right)_i\right|^2 \leq \frac{\varepsilon^2}{8} \quad \text{for all } m \geq M_2(\varepsilon). \tag{15.21}$$

Also, since $\boldsymbol{u}^* = (u_i^*)_{i\in\mathbb{Z}} \in \ell^2$, there exists $I_2(\varepsilon) > 0$ such that

$$\sum_{|i|\geq I_2(\varepsilon)} |u_i^*|^2 \leq \frac{\varepsilon^2}{8}. \tag{15.22}$$

Set $I(\varepsilon) := \max\{I_1(\varepsilon), I_2(\varepsilon)\}$. Since $\varphi(t_m, \boldsymbol{u}^{(m)}) \rightharpoonup \boldsymbol{u}^*$ in ℓ^2 then the components

$$\varphi_i\left(t_m, \boldsymbol{u}^{(m)}\right) \longrightarrow u_i^* \quad \text{for} \quad |i| \leq I(\varepsilon), \quad \text{as } m \to \infty.$$

Therefore there exists $M_3(\varepsilon) > 0$ such that

$$\sum_{|i|\leq I(\varepsilon)} \left|\varphi_i\left(t_m, \boldsymbol{u}^{(m)}\right) - u_i^*\right|^2 \leq \frac{\varepsilon^2}{2} \quad \text{for all } m \geq M_3(\varepsilon). \tag{15.23}$$

Set $M(\varepsilon) := \max\{M_1(\varepsilon), M_2(\varepsilon), M_3(\varepsilon)\}$, then by using (15.21)–(15.23) we have

$$\begin{aligned}\left\|\varphi(t_m, \boldsymbol{u}^{(m)}) - \boldsymbol{u}^*\right\|^2 &= \sum_{|i|\leq I(\varepsilon)} \left|\varphi_i(t_m, \boldsymbol{u}^{(m)}) - u_i^*\right|^2 + \sum_{|i|>I(\varepsilon)} \left|\varphi_i(t_m, \boldsymbol{u}^{(m)}) - u_i^*\right|^2 \\ &\leq \frac{\varepsilon^2}{2} + 2\sum_{|i|>I(\varepsilon)} \left|\varphi_i(t_m, \boldsymbol{u}^{(m)})\right|^2 + |u_i^*|^2 \leq \varepsilon^2 \text{ for all } m \geq M(\epsilon).\end{aligned}$$

Hence $\boldsymbol{u}^{(m)}$ (the subsequence) converges to $\boldsymbol{u}^*$ strongly in ℓ^2, and therefore Q is asymptotically compact. The desired assertion follows immediately from Theorem 2.3. □

The invariance $\varphi(t, \mathcal{A}) = \mathcal{A}$ of the global attractor $\mathcal{A}$ holds for all $t \geq 0$. In fact $\mathcal{A}$ consists of all bounded solutions defined on the whole $\mathbb{R}$, i.e., the bounded entire solutions. The next theorem shows the backward uniqueness of solutions to (15.5) on $\mathcal{A}$, and hence φ is indeed a group rather than just a semi-group on $\mathcal{A}$.

Theorem 15.3. *Let the Assumptions 15.1–15.5 hold. For any two solutions* $\boldsymbol{u}(t) = \varphi(t, \boldsymbol{u}_o)$ *and* $\boldsymbol{v}(t) = \varphi(t, \boldsymbol{v}_o)$ *of* (15.5), *if* $\boldsymbol{u}(T) = \boldsymbol{v}(T)$ *for some* $T > 0$, *then* $\boldsymbol{u}(t) = \boldsymbol{v}(t)$ *for all* $t \in [0, T]$.

Proof. Let $\boldsymbol{w}(t) = \boldsymbol{u}(t) - \boldsymbol{v}(t)$ and (for the sake of contradiction) suppose there exists $t_0 \in [0, T)$ such that $\boldsymbol{w}(t_0) \neq 0$. Then by the continuity of $\boldsymbol{w}$, there exists $t_1 \in [t_0, T)$ such that $\|\boldsymbol{w}(t)\| \neq 0$ for all $t \in [t_0, t_1)$ and

$$0 < \|\boldsymbol{w}(t_1)\| < \epsilon \quad \text{for any } \epsilon > 0. \tag{15.24}$$

Write $\boldsymbol{w} = (w_i)_{i\in\mathbb{Z}}$ then $\boldsymbol{w}(t)$ satisfies

$$\frac{\mathrm{d}}{\mathrm{d}t} w_i(t) = -\frac{w_i}{\varkappa_i \mu_i} + \sum_{j=i-\mathfrak{n}}^{i+\mathfrak{n}} \frac{\lambda_{i,j}}{\mu_i} \left(f_j(u_j) - f_j(v_j)\right) \quad i \in \mathbb{Z}.$$

Multiplying the above equation by $-w_i$ and summing over $i \in \mathbb{Z}$ we get

$$-\frac{1}{2}\frac{\mathrm{d}}{\mathrm{d}t}\|\boldsymbol{w}(t)\|^2 = \sum_{i\in\mathbb{Z}} \frac{w_i^2}{\varkappa_i \mu_i} - \sum_{i\in\mathbb{Z}} \frac{w_i}{\mu_i} \sum_{j=i-\mathfrak{n}}^{i+\mathfrak{n}} \lambda_{i,j}\left(f_j(v_j) - f_j(u_j)\right)$$
$$\leq \left(\frac{1}{M_\varkappa M_\mu} + \frac{M_\lambda L}{m_\mu}(2\mathfrak{n}+1)\right)\|\boldsymbol{w}\|^2 := \hat{\delta}\|\boldsymbol{w}\|^2 \quad \text{for all } t \in [0,T)$$

which implies that

$$\frac{\mathrm{d}}{\mathrm{d}t}\ln\left(\frac{1}{\|\boldsymbol{w}(t)\|}\right) \leq \hat{\delta} \quad \text{for all } t \in [0,T) \text{ and } \boldsymbol{w}(t) \neq 0.$$

Integrating the above differential inequality from t_0 to t_1 results in

$$\ln\left(\frac{1}{\|\boldsymbol{w}(t_1)\|}\right) \leq \ln\left(\frac{1}{\|\boldsymbol{w}(t_0)\|}\right) + \hat{\delta}(t_1 - t_0) \leq \ln\left(\frac{1}{\|w(t_0)\|}\right) + \hat{\delta}T. \tag{15.25}$$

On the other side, due to (15.24), $\ln\left(\frac{1}{\|\boldsymbol{w}(t_1)\|}\right) > \ln\frac{1}{\epsilon}$. Now picking $\epsilon = \|\boldsymbol{w}(t_0)\|/e^{2\hat{\delta}T}$, then

$$\ln\left(\frac{1}{\|\boldsymbol{w}(t_1)\|}\right) > \ln\frac{1}{\epsilon} = \ln\left(\frac{1}{\|\boldsymbol{w}(t_0)\|}\right) + 2\hat{\delta}T,$$

which contradicts with (15.25). Thus $\boldsymbol{w}(t) = 0$ for all $t \in [t_0, T)$. □

15.4 Finite dimensional approximations

Finite dimensional approximations of the lattice reaction diffusion equation were considered in Chapter 4. Here they will be investigated for Hopfield lattice system (15.2) under various boundary conditions.

To this end, first consider the following $(2N+1)$-dimensional system of ODEs obtained from directly truncating the lattice system (15.2):

$$\mu_i \frac{\mathrm{d}}{\mathrm{d}t}u_i(t) = -\frac{u_i(t)}{\varkappa_i} + \sum_{j=i-\mathfrak{n}}^{i+\mathfrak{n}} \lambda_{i,j} f_j(u_j(t)) + g_i, \qquad i = -N, \ldots, 0, \ldots, N. \tag{15.26}$$

Since every neuron is interacting with $2\mathfrak{n}$ neurons in the neighborhood ordered by their indices, throughout this section it is assumed that $N \geq \mathfrak{n}$ to capture enough dynamics of the network. Note that the sum term in the above system, involves $u_{-N-\mathfrak{n}}, \ldots, u_{-N-1}$ and $u_{N+1}, \ldots, u_{N+\mathfrak{n}}$, which are not governed by any equation within the system. Thus in order for the system (15.26) to be well-defined, some boundary conditions need to be imposed.

There are multiple ways to construct such boundary conditions. Here three most intuitive constructions will be considered in terms of additional assumptions.

Assumption 15.6. (A Dirichlet type boundary condition with vanishing tails) The neuron activation functions satisfy $f_i(u_i) = 0$ for $i = -\mathfrak{n} - N, \ldots, -N-1$ and $i = N+1, \ldots, N+\mathfrak{n}$.

Assumption 15.7. (A periodic type boundary condition with period $2\mathfrak{n}+1$) The neuron activation functions and connection strength among neurons satisfy

$$f_i(u_i) = f_{i+2\mathfrak{n}+1}(u_{i+2\mathfrak{n}+1}), \quad \lambda_{i,j} = \lambda_{i,j+2\mathfrak{n}+1}$$

for $i = -\mathfrak{n} - N, \ldots, -N-1$ and $i = N - 2\mathfrak{n}, \ldots, N - \mathfrak{n} - 1$.

Assumption 15.8. (A periodic type boundary condition with period $2N+1$) The neuron activation functions and connection strength among neurons satisfy

$$f_i(u_i) = f_{i+2N+1}(u_{i+2N+1}), \quad \lambda_{i,j} = \lambda_{i,j+2N+1}$$

for $i = -\mathfrak{n} - N, \ldots, -N-1$ and $i = -N, \ldots, \mathfrak{n} - N - 1$.

Remark 15.2. Assumption 15.6 assumes the neurons out of the index domain $i = -N, \ldots, N$ do not excite or inhibit those within the index domain. Assumptions 15.7 and 15.8 ensure that the activation functions for those neurons out of the index domain $i = -N, \ldots, N$ can be replicated by neurons inside the index domain.

In particular, in Assumption 15.7 each neuron outside the index domain is substituted by the neuron $2\mathfrak{n}+1$ index away within the domain (see Figures 15.1 and 15.2 below); and in Assumption 15.8 each neuron outside the index domain is substituted by the neuron $2N+1$ indices away within the domain (see Figures 15.3 and 15.4 below).

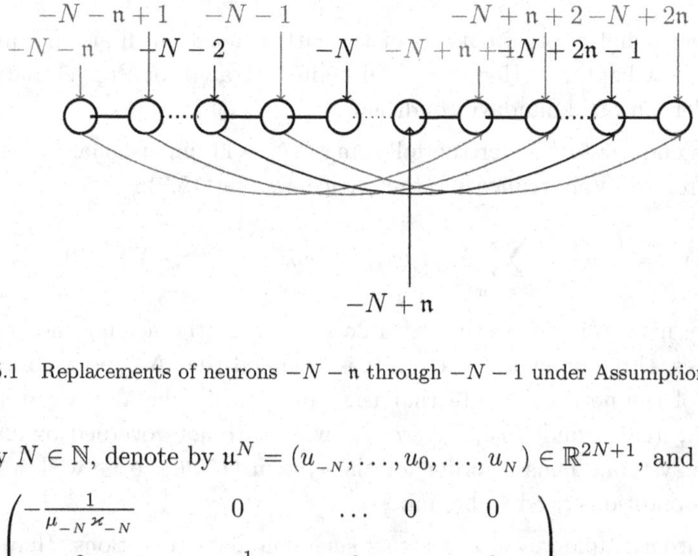

Fig. 15.1 Replacements of neurons $-N-\mathfrak{n}$ through $-N-1$ under Assumption 15.7.

For any $N \in \mathbb{N}$, denote by $\mathfrak{u}^N = (u_{-N}, \ldots, u_0, \ldots, u_N) \in \mathbb{R}^{2N+1}$, and

$$\mathrm{A}^N := \begin{pmatrix} -\frac{1}{\mu_{-N}\varkappa_{-N}} & 0 & \cdots & 0 & 0 \\ 0 & -\frac{1}{\mu_{-N+1}\varkappa_{-N+1}} & 0 & \cdots & 0 \\ \vdots & \vdots & & \ddots & \vdots \\ 0 & 0 & \cdots & 0 & -\frac{1}{\mu_N\varkappa_N} \end{pmatrix} \in \mathbb{R}^{(2N+1)\times(2N+1)}.$$

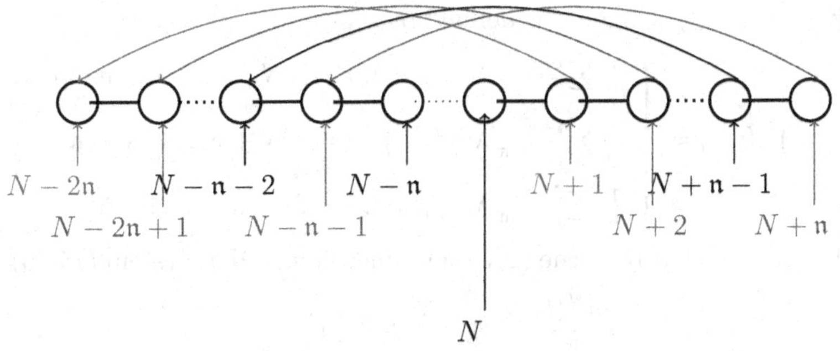

Fig. 15.2 Replacements of neurons $N+1$ through $N+\mathfrak{n}$ under Assumption 15.7.

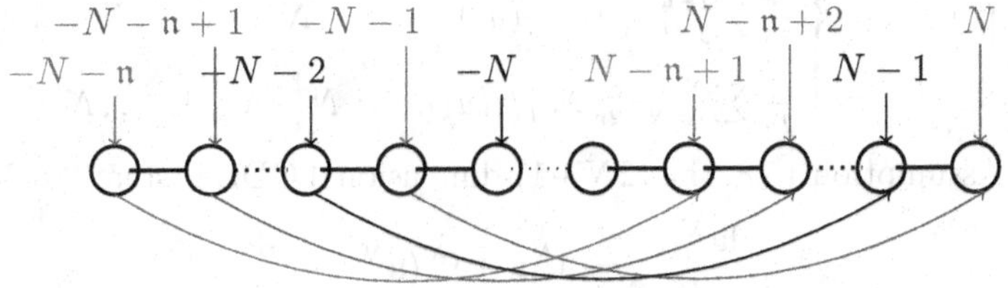

Fig. 15.3 Replacements of neurons $-N-\mathfrak{n}$ through $-N-1$ under Assumption 15.8.

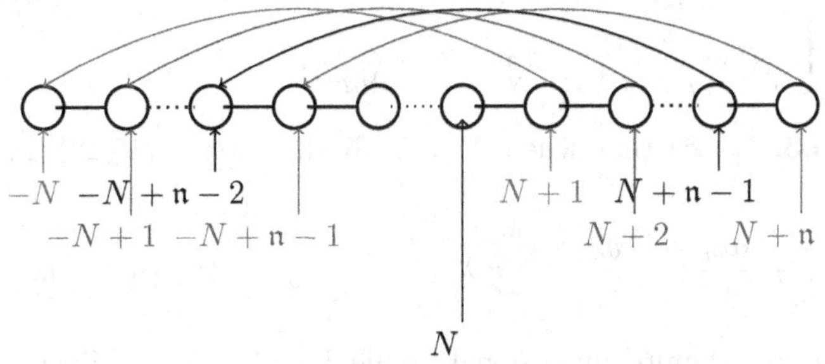

Fig. 15.4 Replacements of neurons $N+1$ through $N+\mathfrak{n}$ under Assumption 15.8.

In addition, let

$$\tilde{\mathfrak{g}}^N := \left(\frac{g_{-N}}{\mu_{-N}}, \dots, \frac{g_N}{\mu_N}\right) \in \mathbb{R}^{2N+1}.$$

Under Assumption 15.6, the $(2N+1)$-dimensional ODE system (15.26) becomes

$$\frac{\mathrm{d}\mathfrak{u}^N(t)}{\mathrm{d}t} = A^N \mathfrak{u}^N + \mathfrak{f}^N(\mathfrak{u}^N) + \tilde{\mathfrak{g}}^N, \tag{15.27}$$

where $\mathfrak{f}^N(\cdot) = (\mathfrak{f}_i^N(\cdot))_{i=-N,\dots N}$ is defined by

$$\mathfrak{f}_i^N(\mathfrak{u}^N) = \begin{cases} \frac{1}{\mu_i}\sum_{j=-N}^{i+\mathfrak{n}} \lambda_{i,j} f_j(u_j),\ i = -N, \dots, -N+\mathfrak{n}-1 \\ \frac{1}{\mu_i}\sum_{j=i-\mathfrak{n}}^{i+\mathfrak{n}} \lambda_{i,j} f_j(u_j),\ i = -N+\mathfrak{n}, \dots, N-\mathfrak{n} \\ \frac{1}{\mu_i}\sum_{j=i-\mathfrak{n}}^{N} \lambda_{i,j} f_j(u_j),\ i = N-\mathfrak{n}+1, \dots, N \end{cases}$$

Under Assumption 15.7, the $(2N+1)$-dimensional ODE system (15.26) becomes

$$\frac{\mathrm{d}\mathfrak{u}^N(t)}{\mathrm{d}t} = A^N\mathfrak{u}^N + \mathfrak{f}^N(\mathfrak{u}^N) + \tilde{\mathfrak{g}}^N, \tag{15.28}$$

where $\mathfrak{f}^N(\cdot) = (\mathfrak{f}_i^N(\cdot))_{i=-N,\dots N}$ is defined by

$$\mathfrak{f}_i^N(\mathfrak{u}^N) = \begin{cases} \frac{1}{\mu_i}\sum_{j=-N}^{-N+2\mathfrak{n}} \lambda_{i,j} f_j(u_j),\quad i = -N, \dots, -N+\mathfrak{n}-1 \\ \frac{1}{\mu_i}\sum_{j=i-\mathfrak{n}}^{i+\mathfrak{n}} \lambda_{i,j} f_j(u_j),\quad i = -N+\mathfrak{n}, \dots, N-\mathfrak{n} \\ \frac{1}{\mu_i}\sum_{j=N-2\mathfrak{n}}^{N} \lambda_{i,j} f_j(u_j),\ i = N-\mathfrak{n}+1, \dots, N \end{cases}$$

Under Assumption 15.8, the $(2N+1)$-dimensional ODE system (15.26) becomes

$$\frac{\mathrm{d}\mathfrak{u}^N(t)}{\mathrm{d}t} = A^N\mathfrak{u} + \mathfrak{f}^N(\mathfrak{u}^N) + \tilde{\mathfrak{g}}^N, \tag{15.29}$$

where $\mathfrak{f}^N(\cdot) = (\mathfrak{f}_i^N(\cdot))_{i=-N,\dots N}$ is defined by

$$\mathfrak{f}_i^N(\mathfrak{u}^N) = \begin{cases} \frac{1}{\mu_i}\left(\sum_{j=-N}^{i+\mathfrak{n}} + \sum_{j=i-\mathfrak{n}+2N+1}^{N}\right) \lambda_{i,j} f_j(u_j),\ i = -N, \dots, -N+\mathfrak{n}-1 \\ \frac{1}{\mu_i}\sum_{j=i-\mathfrak{n}}^{i+\mathfrak{n}} \lambda_{i,j} f_j(u_j), \qquad i = -N+\mathfrak{n}, \dots, N-\mathfrak{n} \\ \frac{1}{\mu_i}\left(\sum_{j=i-\mathfrak{n}}^{N} + \sum_{-N}^{i+\mathfrak{n}-2N-1}\right) \lambda_{i,j} f_j(u_j) \qquad i = N-\mathfrak{n}+1, \dots, N \end{cases}$$

Remark 15.3. Notice that when $N = \mathfrak{n}$, both systems (15.28) and (15.29) are reduced to

$$\mu_i \frac{\mathrm{d}u_i}{\mathrm{d}t} = -\frac{u_i}{\varkappa_i} + \sum_{i=-\mathfrak{n}}^{\mathfrak{n}} \lambda_{i,j} f_j(u_j) + g_i, \quad i = -\mathfrak{n}, \dots, \mathfrak{n}, \tag{15.30}$$

which is the classical finite dimensional Hopfield model (15.1). When $N = \mathfrak{n}$, system (15.27) also resembles (15.30) with $\lambda_{i,j} = 0$ for some i, j.

The existence of global attractors for each of the systems (15.27), (15.28) and (15.29) will now be investigated and will then be shown that they converge to the global attractor for the lattice system (15.2) when $N \to \infty$.

The following Lemma provides critical estimations to be used later.

Lemma 15.5. *Given any* $\mathfrak{u}^N = (u_{-N}, \dots, u_0, \dots, u_N) \in \mathbb{R}^{2N+1}$,

$$\sum_{i=-N}^{N} u_i \mathfrak{f}_i^N(\mathfrak{u}^N) \le \frac{LM_\lambda}{m_\mu}(7\mathfrak{n}+2)\sum_{i=-N}^{N} u_i^2.$$

Proof. By Assumptions 15.1, 15.2 and 15.4, and the definitions of $\mathfrak{f}_i^N$, $\mathfrak{f}_i^N$ and $\mathfrak{f}_i^N$

$$\sum_{i=-N}^{N} u_i\mathfrak{f}_i^N \le \frac{LM_\lambda}{m_\mu}\Big(\underbrace{\sum_{i=-N}^{-N+\mathfrak{n}-1}\Big(\sum_{j=-N}^{i+\mathfrak{n}}+\sum_{j=i-\mathfrak{n}+2N+1}^{N}\Big)|u_iu_j|}_{\textbf{(ii)}}+\underbrace{\sum_{-N+\mathfrak{n}}^{N-\mathfrak{n}}\sum_{j=i-\mathfrak{n}}^{i+\mathfrak{n}}|u_iu_j|}_{\textbf{(i)}}\Big)$$

$$+\frac{LM_\lambda}{m_\mu}\underbrace{\sum_{i=N-\mathfrak{n}+1}^{N}\Big(\sum_{j=i-\mathfrak{n}}^{N}+\sum_{-N}^{i+\mathfrak{n}-2N-1}\Big)|u_iu_j|}_{\textbf{(iii)}}. \qquad (15.31)$$

Then using $|u_iu_j| \le \frac{1}{2}(u_i^2+u_j^2)$ and shifting of indices repeatedly gives

$$\textbf{(i)} \le \frac{1}{2}\sum_{i=-N+\mathfrak{n}}^{N-\mathfrak{n}} u_i^2 \sum_{j=i-\mathfrak{n}}^{i+\mathfrak{n}} 1 + \frac{1}{2}\sum_{k=0}^{\mathfrak{n}}\sum_{i=-N+k}^{N-2\mathfrak{n}+k} u_i^2$$

$$\le \frac{2\mathfrak{n}+1}{2}\sum_{i=-N+\mathfrak{n}}^{N-\mathfrak{n}} u_i^2 + \frac{\mathfrak{n}+1}{2}\sum_{i=-N}^{N} u_i^2 \le \frac{3\mathfrak{n}+2}{2}\sum_{i=-N}^{N} u_i^2;$$

$$\textbf{(ii)} \le \frac{1}{2}\sum_{i=-N}^{-N+2\mathfrak{n}-1} u_i^2\Big(\sum_{j=-N}^{i+\mathfrak{n}} 1+\sum_{j=i-\mathfrak{n}+2N+1}^{N} 1\Big)+\frac{1}{2}\sum_{i=-N}^{-N+2\mathfrak{n}-1}\Big(\sum_{j=-N}^{i+\mathfrak{n}} u_j^2+\sum_{j=i-\mathfrak{n}+2N+1}^{N} u_j^2\Big)$$

$$\le \frac{2\mathfrak{n}+1}{2}\sum_{i=-N}^{-N+2\mathfrak{n}-1} u_i^2 + \mathfrak{n}\sum_{i=-N}^{N} u_i^2 \le \frac{5\mathfrak{n}+1}{2}\sum_{i=-N}^{N} u_i^2;$$

$$\textbf{(iii)} \le \frac{1}{2}\sum_{i=N-\mathfrak{n}+1}^{N} u_i^2\Big(\sum_{j=i-\mathfrak{n}}^{N} 1+\sum_{-N}^{i+\mathfrak{n}-2N-1} 1\Big)+\frac{1}{2}\sum_{i=N-\mathfrak{n}+1}^{N}\Big(\sum_{j=i-\mathfrak{n}}^{N} u_j^2+\sum_{-N}^{i+\mathfrak{n}-2N-1} u_j^2\Big)$$

$$\le \frac{2\mathfrak{n}+1}{2}\sum_{i=N-\mathfrak{n}+1}^{N} u_i^2 + \mathfrak{n}\sum_{i=-N}^{N} u_i^2 \le \frac{5\mathfrak{n}+1}{2}\sum_{i=-N}^{N} u_i^2.$$

Inserting estimations **(i)**–**(iii)** in (15.31) implies the assertion of the Lemma. □

Let the initial condition for systems (15.27), (15.28) and (15.29) be $\mathfrak{u}^N(0) = \mathfrak{u}_o^N$, where $\mathfrak{u}_o^N$ is the $(2N+1)$-dimensional truncation of the initial value $\boldsymbol{u}_o$ of the infinite dimensional system (15.5), i.e., $\mathfrak{u}_o^N = (u_{o,i})_{i=-N,\dots,N}$. Then for every $\mathfrak{u}_o^N \in \mathbb{R}^{2N+1}$, each of the systems (15.27), (15.28) and (15.29) has a unique solution that generates a dynamical system from $\mathbb{R}^{2N+1}$ to $\mathbb{R}^{2N+1}$.

More precisely, denote by π^N, $\mathcal{S}^N$ and φ^N the semi-dynamical systems generated by solutions of the system (15.27), (15.28) and (15.29), respectively, i.e.,

$$\pi^N(t,\mathfrak{u}_o^N) = \mathfrak{u}^N(t;\mathfrak{u}_o^N) \text{ is the solution to (15.27) with } \mathfrak{u}^N(0) = \mathfrak{u}_o^N,$$

$$\mathcal{S}^N(t,\mathfrak{u}_o^N) = \mathfrak{u}^N(t;\mathfrak{u}_o^N) \text{ is the solution to (15.28) with } \mathfrak{u}^N(0) = \mathfrak{u}_o^N,$$

$$\varphi^N(t,\mathfrak{u}_o^N) = \mathfrak{u}^N(t;\mathfrak{u}_o^N) \text{ is the solution to (15.29) with } \mathfrak{u}^N(0) = \mathfrak{u}_o^N.$$

Remark 15.4. The proofs are given for the case of a periodic boundary condition with period $2N+1$ under Assumption 15.8. The proofs for the other two cases are similar and can be found in Han, Usman & Kloeden [Han *et al.* (2020b)].

In the calculations below $\|\cdot\|$ denotes the norm of ℓ^2 and $|\cdot|_N$ denotes the Euclidean norm of $\mathbb{R}^{2N+1}$. Moreover, for $i=-N,\dots,N$ denote by $u_i^{N,\pi}(t)$, $u_i^{N,\mathcal{S}}(t)$, and $u_i^{N,\varphi}(t)$ the ith component of $\pi^N(t,\mathrm{u}_o^N)$, $\mathcal{S}^N(t,\mathrm{u}_o^N)$ and $\varphi^N(t,\mathrm{u}_o^N)$, respectively.

Lemma 15.6. *Let Assumptions 15.1–15.5 hold. In addition, assume that*

$$\tilde{\delta} := \frac{1}{M_\mu M_\varkappa} - 2\frac{LM_\lambda}{m_\mu}(7\mathfrak{n}+2) > 0. \tag{15.32}$$

Then each of the semi-dynamical systems π^N under 15.6, $\mathcal{S}^N$ under 15.7 and φ^N under 15.8 has an absorbing set in $\mathbb{R}^{2N+1}$.

Proof. First taking the inner product on $\mathbb{R}^{2N+1}$ of equation (15.29) with $\varphi^N(t,\mathrm{u}_o^N)$ gives

$$\begin{aligned}\frac{1}{2}\frac{\mathrm{d}}{\mathrm{d}t}|\varphi^N(t,\mathrm{u}_o^N)|_N^2 &= -\sum_{i=-N}^{N}\frac{(u_i^{N,\varphi})^2}{\mu_i\varkappa_i}\\ &\quad+\sum_{i=-N}^{N}u_i^{N,\varphi}\mathfrak{f}_i^N(\varphi^N(t,\mathrm{u}_o^N))+\sum_{i=-N}^{N}\frac{g_iu_i^{N,\varphi}}{\mu_i}.\end{aligned} \tag{15.33}$$

Then by Assumption 15.2,

$$-\sum_{i=-N}^{N}\frac{(u_i^{N,\varphi})^2}{\mu_i\varkappa_i} \le -\frac{1}{M_\mu M_\varkappa}|\varphi^N(t)\mathrm{u}_o^N|^2, \tag{15.34}$$

$$\sum_{i=-N}^{N}\frac{g_iu_i^{N,\varphi}}{\mu_i} \le \frac{1}{2M_\mu M_\varkappa}|\varphi^N(t)\mathrm{u}_o^N|^2 + \frac{M_\mu M_\varkappa}{2m_\mu^2}\|g\|^2. \tag{15.35}$$

In addition, it follows immediately from Lemma 15.5 that

$$\sum_{i=-N}^{N}\varphi_i^N\mathfrak{f}_i^N(\varphi^N(t,\mathrm{u}_o^N)) \le \frac{LM_\lambda}{m_\mu}(7\mathfrak{n}+2)|\varphi^N(t,\mathrm{u}_o^N)|_N^2. \tag{15.36}$$

Collecting (15.33)–(15.36) results in

$$\frac{\mathrm{d}}{\mathrm{d}t}|\varphi^N(t,\mathrm{u}_o^N)|_N^2 \le -\frac{1}{M_\mu M_\varkappa} + \frac{LM_\lambda}{m_\mu}(7\mathfrak{n}+2)|\varphi^N(t,\mathrm{u}_o^N)|_N^2 + \frac{M_\mu M_\varkappa}{m_\mu^2}\|g\|^2.$$

Integrating the above inequality from 0 to t gives immediately

$$|\varphi^N(t,\mathrm{u}_o^N)|_N^2 \le e^{-\tilde{\delta}t}|\mathrm{u}_o^N|_N^2 + \frac{M_\mu M_\varkappa}{\tilde{\delta}m_\mu^2}\|g\|^2(1-e^{-\tilde{\delta}t})$$

where $\tilde{\delta}$ is defined as in (15.32). Therefore,

$$Q^N = \left\{\mathrm{u}^N \in \mathbb{R}^{2N+1} : |\mathrm{u}^N|_N \le \frac{M_\mu M_\varkappa}{\sqrt{\tilde{\delta}}}\frac{\|g\|^2}{m_\mu} + 1\right\} \tag{15.37}$$

is an absorbing set for the semi-dynamical systems φ^N (as well as π^N and $\mathcal{S}^N$). This completes the proof. □

It follows immediately from Lemma 15.6 that each of the the semi-dynamical systems π^N, $\mathcal{S}^N$ and φ^N possesses a global attractor, denoted by $\mathcal{A}_\pi^N$, $\mathcal{A}_\mathcal{S}^N$ and $\mathcal{A}_\varphi^N$, respectively.

Recall that every $\mathfrak{a} = (a_i)_{i=-N,\dots,N} \in \mathbb{R}^{2N+1}$, can be extended naturally to the element $\boldsymbol{a} = (\tilde{a}_i)_{i\in\mathbb{Z}} \in \ell^2$ such that

$$\tilde{a}_i = \begin{cases} a_i, & i = -N, \dots, N, \\ 0, & \text{otherwise} \end{cases}.$$

It will now be shown that natural embeddings of the global attractors $\mathcal{A}_\phi^N$, $\mathcal{A}_\mathcal{S}^N$ and $\mathcal{A}_\varphi^N$ into ℓ^2 all approach the global attractor $\mathcal{A}$ for the lattice system (15.2) upper semi-continuously as $N \to \infty$.

Lemma 15.7. *Let Assumptions 15.1–15.5 hold. Then given any $\varepsilon > 0$ there exists $I(\varepsilon) > 0$ depending on $\boldsymbol{g}$, m_μ, M_λ, M_μ, $M_\varkappa$, but independent of N, such that every $\mathfrak{a} = (a_i)_{i=-N,\dots,N} \in \mathcal{A}_\pi^N, \mathcal{A}_\mathcal{S}^N, \mathcal{A}_\varphi^N$ under Assumptions 15.6, 15.7, 15.8 (resp.) satisfies*

$$\sum_{I(\varepsilon)\le|i|\le N} |a_i|^2 < \varepsilon.$$

Proof. The proof is similar to that of Lemma 15.4 with the details given here for the system φ. Just as φ is a group on its attractor $\mathcal{A}$, so are the other semi-dynamical systems on their attractors. Hence, in particular,

$$\varphi^N(t, \mathcal{A}_\varphi^N) = \mathcal{A}_\varphi^N \quad \text{for all } t \in \mathbb{R}.$$

Thus given any $\mathfrak{u}_o^{N,\varphi} \in \mathcal{A}_\varphi^N$, the solution $\varphi^N(t, \mathfrak{u}_o^{N,\varphi})$ of (15.29) are defined for all $t \in \mathbb{R}$ and

$$\varphi^N(t, \mathfrak{u}_o^{N,\varphi}) \subset Q^N \quad \text{for all } t \in \mathbb{R},$$

where Q^N is defined as in (15.37).

For simplicity, denote by $\left(u_i^{N,\varphi}(t)\right)_{i=-N,\dots,N}$ the component wise solutions of (15.29). Given a fixed $k(\le N)$ which will be determine later, define

$$\boldsymbol{v}^\varphi = (v_i^\varphi)_{|i|\le N}, \quad \text{with} \quad v_i^\varphi = \eta_k(|i|)u_i^{N,\varphi}, \quad i = -N, \dots, N,$$

and η_k is as defined as in (15.13).

Taking the inner product of (15.29) with $\boldsymbol{v}^\varphi$ in $\mathbb{R}^{2N+1}$, results in

$$\begin{aligned}\frac{1}{2}\frac{\mathrm{d}}{\mathrm{d}t}\sum_{|i|\le N}\eta_k(|i|)(u_i^{N,\varphi})^2 \le &-\sum_{|i|\le N}\eta_k(|i|)\frac{(u_i^{N,\varphi})^2}{\varkappa_i\mu_i} + \sum_{|i|\le N}\eta_k(|i|)\frac{\mathfrak{f}_i^N(\mathfrak{u}^{N,\varphi})}{\mu_i} \\ &+ \sum_{|i|\le N}\eta_k(|i|)\frac{u_i^{N,\varphi}g_i}{\mu_i}.\end{aligned}$$

Then by using Lemma 15.5 and proceeding as in the proof of Lemma 15.4 yields

$$\sum_{|i|\le N} \eta_k(|i|)(u_i^{N,\varphi})^2 \le e^{-\tilde{\delta}t} \sum_{|i|\le N} \eta_k(|i|)(u_{o,i}^{N,\varphi})^2 + \frac{M_\mu m_\varkappa}{\delta m_\mu^2} \sum_{|i|\le N} \eta_k(|i|) g_i.$$

Therefore as in the proof of Lemma 15.4, given any $\varepsilon > 0$ there exists $T_\varepsilon > 0$ and $I(\varepsilon) > 0$ such that with $\frac{k}{2} \ge I(\varepsilon)$,

$$\sum_{I(\varepsilon)\le|i|\le N} (u_i^\varphi)^2(t) \le \sum_{|i|\le N} \eta_k(|i|)(u_i^\varphi)^2(t) < \varepsilon \qquad \text{for all } t \ge T_\varepsilon. \tag{15.38}$$

Consider $\mathfrak{a} = (a_i)_{i=-N,\dots,N} \in \mathcal{A}_\varphi^N$. Then by the fact that $\varphi^N(T_\varepsilon, \mathcal{A}_\varphi^N) = \mathcal{A}_\varphi^N$ there exists $\mathfrak{u}_o^\varphi \in \mathcal{A}_\varphi^N$, such that $\mathfrak{a} = \varphi^N(T_\varepsilon, \mathfrak{u}_o^\varphi) = (u_i^\varphi(T_\varepsilon))_{|i|\le N}$. It then follows from (15.38) that

$$\sum_{I(\varepsilon)\le|i|\le N} |a_i|^2 = \sum_{I(\varepsilon)\le|i|\le N} (u_i^\varphi)^2(t) \le \varepsilon \qquad \text{for all } \mathfrak{a} \in \mathcal{A}_\varphi^N. \qquad \square$$

Lemma 15.8. *Let Assumptions 15.1–15.5, and 15.8 hold. Then there exists a subsequence* $\{\mathfrak{a}^{N_k,\varphi}\}_{k\in\mathbb{N}}$ *of* $\{\mathfrak{a}^{N,\varphi}\}_{N\in\mathbb{N}}$ *and* $\boldsymbol{a}^{*,\varphi} \in \mathcal{A}$ *such that* $\{\mathfrak{a}^{N_k,\varphi}\}$ *converges to* $\boldsymbol{a}^{*,\varphi}$ *in* ℓ^2 *as* $k \to \infty$.

Similar results holds for $\mathfrak{a}^{N,\pi} \in \mathcal{A}_\pi^N$ *under Assumption 15.6, and for* $\mathfrak{a}^{N,S} \in \mathcal{A}_S^N$ *under Assumption 15.7.*

Proof. Let $\mathfrak{a}^{N,\varphi}(t) = \varphi^N(t, \mathfrak{a}_o^{N,\varphi})$ be the solution of (15.29). Since $\mathfrak{a}_o^{N,\varphi} \in \mathcal{A}_\varphi^N$, then $\mathfrak{a}^{N,\varphi}(t) \in \mathcal{A}_\varphi^N$ for all $t \in \mathbb{R}$. For $N \in \mathbb{N}$, let $\boldsymbol{a}^{N,\varphi}(t)$ be the natural extension of $\mathfrak{a}^{N,\varphi}(t)$ in ℓ^2. Then due to Lemma 15.6

$$\left\|\boldsymbol{a}^{N,\varphi}(t)\right\| \le \frac{M_\mu M_\varkappa}{\sqrt{\delta}} \frac{\|\boldsymbol{g}\|^2}{m_\mu} + \varepsilon := R_\varepsilon \quad \text{for} \quad t \in \mathbb{R}, \quad N \in \mathbb{N}. \tag{15.39}$$

In addition, by (15.29)

$$\left\|\frac{\mathrm{d}\boldsymbol{a}^{N,\varphi}(t)}{\mathrm{d}t}\right\| \le \left|\mathrm{A}^N \mathfrak{a}^{N,\varphi}(t)\right|_N + \left|\mathfrak{f}^N(\mathfrak{a}^{N,\varphi}(t))\right|_N + \left|\mathfrak{g}^N\right|_N.$$

First notice that

$$\left|\mathfrak{g}^N\right|_N \le \|\boldsymbol{g}\|, \qquad \left|\mathrm{A}^N \mathfrak{a}^{N,\varphi}(t)\right|_N \le \frac{1}{m_\mu m_\varkappa} \left\|\boldsymbol{a}^{N,\varphi}(t)\right\|.$$

Then by definitions of $\mathfrak{f}^N$, $\mathfrak{f}^N$ and $\mathfrak{f}^N$, and following computations similar to those in the proof of Lemma 15.5 to obtain

$$\begin{aligned}
\left|\mathfrak{f}^N(\mathfrak{a}^{N,\varphi}(t))\right|_N^2 \le \frac{M_\lambda^2 L^2}{m_\mu^2}\Bigg(&\sum_{i=-N}^{-N+\mathfrak{n}-1} \Big(\sum_{j=-N}^{i+\mathfrak{n}} + \sum_{j=i-\mathfrak{n}+2N+1}^{N} \Big) \mathfrak{a}_j^{N,\varphi}(t)^2 \\
&+ \sum_{i=-N+\mathfrak{n}}^{N-\mathfrak{n}} \sum_{j=i-\mathfrak{n}}^{i+\mathfrak{n}} \mathfrak{a}_j^{N,\varphi}(t)^2 \Bigg) \\
&+ \frac{M_\lambda^2 L^2}{m_\mu^2} \sum_{i=N-\mathfrak{n}+1}^{N} \Big(\sum_{j=i-\mathfrak{n}}^{N} + \sum_{-N}^{i+\mathfrak{n}-2N-1} \Big) \mathfrak{a}_j^{N,\varphi}(t)^2 \\
\le (5\mathfrak{n}+1) &\left\|\boldsymbol{a}^{N,\varphi}(t)\right\|^2.
\end{aligned}$$

In summary, there exists $\tilde{R}_\varepsilon$ depending on L, M_λ, m_μ, $m_\varkappa$, $\mathfrak{n}$, $\|\boldsymbol{g}\|$ but independent of N such that

$$\left\| \frac{\mathrm{d}\boldsymbol{a}^{N,\varphi}(t)}{\mathrm{d}t} \right\| \leq \tilde{R}_\varepsilon, \quad t \in \mathbb{R}, \quad N \in \mathbb{N}. \tag{15.40}$$

Let J_k be a sequence of compact intervals in $\mathbb{R}$ such that $J_k \subset J_{k+1}$ and $\bigcup_k J_k = \mathbb{R}$. Then it follows from (15.40) that $\{\boldsymbol{a}^{N,\varphi}\}_{N\in\mathbb{N}}$ is equicontinuous in $\mathcal{C}(J_k, \ell^2)$, that is, for each $t \in J_k$ and $\varepsilon > 0$, there is a neighborhood $\mathcal{N}$ of $\mathcal{A}$ such that

$$\|\boldsymbol{a}^{N,\varphi}(t) - \boldsymbol{a}^{N,\varphi}(s)\| \leq \varepsilon, \qquad \text{for all } s \in \mathcal{N},\ N \in \mathbb{N}.$$

Next it will be shown that for each fixed $t \in J_k$, the set $\{\boldsymbol{a}^{N,\varphi}\}_{N\in\mathbb{N}}$ is precompact in ℓ^2. First of all by (15.39) the sequence $\{\boldsymbol{a}^{N,\varphi}\}_{N\in\mathbb{N}}$ is bounded in ℓ^2, and hence there is a subsequence of $\{\boldsymbol{a}^{N,\varphi}\}_{N\in\mathbb{N}}$, still denoted by $\{\boldsymbol{a}^{N,\varphi}\}_{N\in\mathbb{N}}$, and $\alpha_t^\varphi \in \ell^2$ such that

$$\boldsymbol{a}^{N,\varphi}(t) \rightharpoonup \alpha_t^\varphi \qquad \text{weakly in } \ell^2.$$

Then, by a similar argument as in Theorem 15.2, it can be shown that the above convergence is in fact strong and hence by the Arzelà-Ascoli theorem $\{\boldsymbol{a}^{N,\varphi}(t)\}_{N\in\mathbb{N}}$ is precompact in $\mathcal{C}(J_k, \ell^2)$ for each $k = 1, 2, \ldots$. Therefore there exist a subsequence $\{\boldsymbol{a}^{N_1,\varphi}\}$ of $\{\boldsymbol{a}^{N,\varphi}\}$ and $\boldsymbol{a}^{\varphi,*} \in \mathcal{C}(J_1, \ell^2)$ such that $\{\boldsymbol{a}^{N_1,\varphi}\}$ converges to $\boldsymbol{a}^{\varphi,*}$ in $\mathcal{C}(J_1, \ell^2)$.

Using the Arzelà-Ascoli theorem it can be shown inductively that there is a subsequence $\{\boldsymbol{a}^{N_{k+1},\varphi}\}$ of $\{\boldsymbol{a}^{N_k,\varphi}\}$ such that $\{\boldsymbol{a}^{N_{k+1},\varphi}\}$ converges to $\boldsymbol{a}^{*,\varphi}$ in $\mathcal{C}(J_{k+1}, \ell^2)$. Taking a diagonal sequence in usual way gives a subsequence $\{\boldsymbol{a}^{N_N,\varphi}\}$ of $\{\boldsymbol{a}^{N,\varphi}\}$ and $\boldsymbol{a}^{*,\varphi} \in \mathcal{C}(\mathbb{R}, \ell^2)$ such that

$$\boldsymbol{a}^{N_N,\varphi} \overset{N\to\infty}{\longrightarrow} \boldsymbol{a}^{*,\varphi} \quad \text{in} \ \ \mathcal{C}(J, \ell^2) \ \ \text{for any compact interval } J \subset \mathbb{R}. \tag{15.41}$$

It follows immediately from (15.39) that

$$\|\boldsymbol{a}^{*,\varphi}(t)\| \leq R_\varepsilon \quad \text{for all } t \in \mathbb{R}.$$

Now it will be shown that that $\boldsymbol{a}^{\varphi,*}(t)$ is a solution of the original lattice system (15.2) and that $\boldsymbol{a}^{*,\varphi}(0) \in \mathcal{A}$, the global attractor for (15.2). To simplify notation, denote the diagonal subsequence in (15.41) by $\{\boldsymbol{a}^{N,\varphi}\}$. Then, by (15.40) and the Banach-Alaoglu theorem, Theorem 2.10,

$$\dot{\boldsymbol{a}}^{N,\varphi} \overset{N\to\infty}{\longrightarrow} \dot{\boldsymbol{a}}^{*,\varphi} \qquad \text{weak star in } \mathcal{L}^\infty(\mathbb{R}, \ell^2).$$

For $i \in \mathbb{Z}$ fixed, let $N > |i| + \mathfrak{n}$. Since $\mathfrak{a}^{N,\varphi}(t)$ is a solution of (15.29), then

$$\mu_i \frac{\mathrm{d}}{\mathrm{d}t} a_i^{N,\varphi}(t) = -\frac{1}{\varkappa_i} a_i^{N,\varphi}(t) + \sum_{j=i-\mathfrak{n}}^{i+\mathfrak{n}} \lambda_{i,j} f_j(a_j^{N,\varphi}) + g_i, \quad i = -N+\mathfrak{n}, \ldots, N-\mathfrak{n}.$$

Consider $\phi \in C_0^\infty(J)$, then for each $i = -N + \mathfrak{n}, \ldots, N - \mathfrak{n}$

$$\int_J \frac{\mathrm{d}}{\mathrm{d}t} a_i^{N,\varphi}(t)\phi(t)\mathrm{d}t = -\frac{1}{\mu_i \varkappa_i} \int_J a_i^{N,\varphi}(t)\phi(t)\mathrm{d}t + \frac{1}{\mu_i} \int_J \sum_{i-\mathfrak{n}}^{i+\mathfrak{n}} \lambda_{i,j} f_j(a_j^{N,\varphi})\phi(t)\mathrm{d}t + \frac{1}{\mu_i} \int_J g_i \phi(t)\mathrm{d}t,$$

and using Assumption 15.4 and (15.41) to obtain

$$\begin{aligned} &\left| \int_J \sum_{j=i-\mathfrak{n}}^{i+\mathfrak{n}} \lambda_{i,j} f_j(a_j^{N,\varphi})\phi(t)\mathrm{d}t - \int_J \sum_{j=i-\mathfrak{n}}^{i+\mathfrak{n}} \lambda_{i,j} f_j(a_j^{*,\varphi})\phi(t)\mathrm{d}t \right| \\ &\le M_\lambda \int_J \sum_{j=i-\mathfrak{n}}^{i+\mathfrak{n}} \left| f_j(a_j^{N,\varphi}) - f_j(a_j^{*,\varphi}) \right| |\phi(t)|\mathrm{d}t \\ &\le M_\lambda L \sup_{t\in J} \max_{j=i-\mathfrak{n},\ldots,i+\mathfrak{n}} |\boldsymbol{a}_j^{N,\varphi} - a_j^{*,\varphi}| \int_J |\phi(t)|\mathrm{d}t \stackrel{N\to\infty}{\longrightarrow} 0. \end{aligned}$$

As a result

$$\frac{\mathrm{d}}{\mathrm{d}t} a_i^{*,\varphi}(t) = -\frac{a_i^{*,\varphi}(t)}{\varkappa_i \mu_i} + \sum_{j=i-\mathfrak{n}}^{i+\mathfrak{n}} \frac{\lambda_{i,j}}{\mu_i} f_j(a_j^{*,\varphi}(t)) + \frac{g_i}{\mu_i} \quad \text{for all } i \in \mathbb{Z}.$$

Since J was arbitrarily chosen the above equation holds for all $t \in \mathbb{R}$, and hence $\boldsymbol{a}^{*,\varphi}$ is a bounded solution of (15.2) for all $t \in \mathbb{R}$. Therefore $\boldsymbol{a}^{*,\varphi}(t) \in \mathcal{A}$ for all $t \in \mathbb{R}$. Finally, by (15.41),

$$\boldsymbol{a}^{N_N,\varphi}(0) \stackrel{N\to\infty}{\longrightarrow} \boldsymbol{a}^{*,\varphi}(0),$$

which completes the proof. □

Theorem 15.4. *Let Assumptions 15.1–15.5 hold. Then*

$$\lim_{N\to\infty} dist_{\ell^2}(\mathcal{A}_\varphi^N, \mathcal{A}) = 0 \qquad \textit{(under 15.8)}$$

and similarly

$$\lim_{N\to\infty} dist_{\ell^2}(\mathcal{A}_\phi^N, \mathcal{A}) = 0 \qquad \textit{(under 15.6)},$$

$$\lim_{N\to\infty} dist_{\ell^2}(\mathcal{A}_\S^N, \mathcal{A}) = 0 \qquad \textit{(under 15.7)}.$$

Proof. Suppose (for contradiction) that the first assertion is not true. Then there exist a subsequence $\boldsymbol{a}^{N_k,\varphi} \in \mathcal{A}_\varphi^{N_k}$ and $\varepsilon_0 > 0$ such that

$$\mathrm{dis}_{\ell^2}(\boldsymbol{a}^{N_k,\varphi}, \mathcal{A}) \ge \varepsilon_0 > 0. \tag{15.42}$$

On the other hand due to Lemma 15.8 there exists a subsequence $\{\mathfrak{a}^{N_{k_l},\varphi}\}$ of $\{\boldsymbol{a}^{N_k,\varphi}\}$ such that

$$\mathrm{dist}_{\ell^2}(\mathfrak{a}^{N_{k_l},\varphi}, \mathcal{A}) \longrightarrow 0,$$

which contradicts (15.42) and completes the proof. □

15.5 End notes

There is an extensive literature on Hopfield neural network models, see for example [Van den Driessche and Zou (1998)]. This chapter is based mainly on Han, Kloeden & Usman [Han *et al.* (2020b)]. See also Han, Usman & Kloeden [Han *et al.* (2019)] and Wang, Kloeden & Han [Wang *et al.* (2020)].

15.6 Problems

Problem 15.1. Does the Hopfield lattice model have steady state solutions in the space ℓ^2 for appropriate coefficient functions? When are there at most countably many steady states?

Problem 15.2. Can the Hopfield lattice model considered here have periodic solutions?

Problem 15.3. The original Hopfield model used two-state threshold neurons following a stochastic algorithm. Each neuron i has two states, characterised by the output V_i of the neuron having the values V_i^0 or V_i^1. Each neuron changes the value of its output or leaves it fixed according to a threshold rule with thresholds U_i:

$$V_i \longrightarrow \begin{cases} V_i^0, & \text{if } \sum_{j\neq i} \lambda_{ij} V_j + g_i < U_i \\ V_i^1, & \text{if } \sum_{j\neq i} \lambda_{ij} V_j + g_i < U_i \end{cases}$$

where λ_{ij} and g_i have the same interpretation as in (15.1). Investigate the relation between the above model and the ODE model (15.1).

Chapter 16

The Hopfield lattice model in weighted spaces

The Hopfield lattice neural model

$$\mu_i \frac{\mathrm{d}u_i(t)}{\mathrm{d}t} = -\frac{u_i(t)}{\varkappa_i} + \sum_{j=i-\mathfrak{n}}^{i+\mathfrak{n}} \lambda_{i,j}^{(\mathfrak{n})} f_j(u_j(t)) + g_i, \qquad i \in \mathbb{Z}, \tag{16.1}$$

was investigated in Chapter 15 on the sequence space ℓ^2 for $\lambda_{i,j}^{(\mathfrak{n})}$ independent of $\mathfrak{n}$. Here we are interested in what happens as the number $\mathfrak{n} \in \mathbb{N}$ in (16.1) increases without bound, leading to the lattice system

$$\mu_i \frac{\mathrm{d}u_i(t)}{\mathrm{d}t} = -\frac{u_i(t)}{\varkappa_i} + \sum_{j\in\mathbb{Z}} \lambda_{i,j} f_j(u_j(t)) + g_i, \qquad i \in \mathbb{Z}. \tag{16.2}$$

This will be formulated in this chapter as an LDS on the weighted space ℓ^2_ρ. The aim is to show that the system (16.2) has an attractor $\mathcal{A}$ and that the attractors $\mathcal{A}_\mathfrak{n}$ of (16.1) converge upper semi-continuously to $\mathcal{A}$ as $\mathfrak{n} \to \infty$ when $\lambda_{i,j}^{(\mathfrak{n})} \to \lambda_{i,j}$ as $n \to \infty$.

The weights for the space ℓ^2_ρ are assumed to satisfy

Assumption 16.1. $\rho_i > 0$ for $i \in \mathbb{Z}$ and $\rho_\Sigma := \sum_{i\in\mathbb{Z}} \rho_i < \infty$.

Recall that ℓ^2_ρ is a separable Hilbert space and $\ell^2_\rho \supset \ell^2$ under Assumption 16.1.

The Assumption 15.2 on model parameters μ_i's and $\varkappa_i$'s is still assumed to hold here. It is repeated in Assumption 16.2 below for the reader's convenience.

Assumption 16.2. The neuron amplifier input capacitance and resistance of neurons are uniformly bounded, i.e., there exist positive constants m_μ, M_μ, $m_\varkappa$, and $M_\varkappa$, such that

$$m_\mu \le \mu_i \le M_\mu, \quad m_\varkappa \le \varkappa_i \le M_\varkappa \quad \text{for all} \quad i \in \mathbb{Z}.$$

Assumption 16.3. The reciprocal-weighted aggregate efficacy on each neuron is finite, in the sense that there exists $M_\lambda > 0$ such that

$$\sup_{i\in\mathbb{Z}} \sum_{j\in\mathbb{Z}} \frac{\lambda_{i,j}^2}{\rho_j^2} \le M_\lambda, \quad \text{and} \quad \sup_{i\in\mathbb{Z}} \sum_{j\in\mathbb{Z}} \frac{(\lambda_{i,j}^{(\mathfrak{n})})^2}{\rho_j^2} \le M_\lambda \text{ for all } \mathfrak{n} \in \mathbb{N}.$$

The forcing terms are assumed to satisfy

Assumption 16.4. $f_i(0) = 0$ for all $i \in \mathbb{Z}$.

Assumption 16.5. $f_i : \mathbb{R} \to \mathbb{R}$ is continuously differentiable with equi-bounded derivatives, i.e., there exist positive constants L_i with $\boldsymbol{L} := (L_i)_{i\in\mathbb{Z}} \in \ell^2_\rho$ and such that

$$|f_i'(s)| \le L_i \quad \text{for all } s \in \mathbb{R},\ i \in \mathbb{Z}.$$

Assumption 16.6. $\boldsymbol{g} := (g_i)_{i\in\mathbb{Z}} \in \ell^2_\rho$.

Note that Assumptions 16.4 and 16.5 together imply that f_i is locally Lipschitz:

$$|f_i(s) - f_i(r)| \le L_i|s - r| \quad \text{for all } s, r \in \mathbb{R},\ i \in \mathbb{Z}, \tag{16.3}$$

and satisfies the growth bound

$$|f_i(s)| \le L_i|s| \quad \text{for all } s \in \mathbb{R},\ i \in \mathbb{Z}. \tag{16.4}$$

16.1 Reformulation as an ODE on ℓ^2_ρ

Given any $\boldsymbol{u} = (u_i)_{i\in\mathbb{Z}} \in \ell^2_\rho$, define the operators $\mathrm{A}\boldsymbol{u} = ((A\boldsymbol{u})_i)_{i\in\mathbb{Z}}$, $F(\boldsymbol{u}) = (F_i(\boldsymbol{u}))_{i\in\mathbb{Z}}$ and $F^{\mathrm{n}}(\boldsymbol{u}) = (F_i^{\mathrm{n}}(\boldsymbol{u}))_{i\in\mathbb{Z}}$ component wise by

$$(\mathrm{A}\boldsymbol{u})_i = \frac{u_i}{\mu_i\varkappa_i}, \quad F_i(\boldsymbol{u}) = \sum_{j\in\mathbb{Z}} \frac{\lambda_{i,j}}{\mu_i} f_j(u_j), \quad (F_i^{\mathrm{n}}(\boldsymbol{u}))_{i\in\mathbb{Z}} = \sum_{j=i-\mathrm{n}}^{i+\mathrm{n}} \frac{\lambda_{i,j}^{(\mathrm{n})}}{\mu_i} f_j(u_j), \quad i \in \mathbb{Z}.$$

Lemma 16.1. *Let Assumptions 16.1–16.5 hold. Then the operators* A, F, F^{n} *map* ℓ^2_ρ *to* ℓ^2_ρ.

Proof. First by Assumption 16.2,

$$\sum_{i\in\mathbb{Z}} \rho_i(\mathrm{A}\boldsymbol{u})_i^2 = \sum_{i\in\mathbb{Z}} \frac{1}{\varkappa_i^2\mu_i^2}\rho_i u_i^2 \le \frac{1}{m_\varkappa^2 \cdot m_\mu^2} \sum_{i\in Z} \rho_i u_i^2 = \frac{1}{m_\varkappa m_\mu}\|\boldsymbol{u}\|_\rho^2,$$

which implies that $\mathrm{A}\boldsymbol{u} \in \ell^2_\rho$ for every $\boldsymbol{u} \in \ell^2_\rho$.

Then by Assumption 16.2, the growth bound (16.4), the Cauchy inequality $(\sum_i a_ib_i)^2 \le \sum_i a_i^2 \sum_i b_i^2$, and the inequality $\sum_i |a_ib_i| \le \sum_i |a_i| \sum_i |b_i|$ we have

$$\begin{aligned}
\Big(\sum_{j\in\mathbb{Z}} \frac{\lambda_{i,j}}{\mu_i} f_j(u_j)\Big)^2 &\le \frac{1}{m_\mu^2}\Big(\sum_{j\in\mathbb{Z}} |\lambda_{i,j}L_ju_j|\Big)^2 \\
&\le \frac{1}{m_\mu^2}\Big(\sum_{j\in\mathbb{Z}} \frac{\lambda_{i,j}^2}{\rho_j^2}\Big) \sum_{j\in\mathbb{Z}} \Big(\rho_jL_j^2\Big)\Big(\rho_ju_j^2\Big). \\
&\le \frac{1}{m_\mu^2}\Big(\sum_{j\in\mathbb{Z}} \frac{\lambda_{i,j}^2}{\rho_j^2}\Big)\Big(\sum_{j\in\mathbb{Z}} \rho_jL_j^2\Big)\Big(\sum_{j\in\mathbb{Z}} \rho_ju_j^2\Big), \quad i \in \mathbb{Z}.
\end{aligned}$$

Next, by Assumptions 16.1, 16.3 and 16.5

$$\sum_{i\in\mathbb{Z}} \rho_i F_i^2(\boldsymbol{u}) = \sum_{i\in\mathbb{Z}} \rho_i \Big(\sum_{j\in\mathbb{Z}} \frac{\lambda_{i,j}}{\mu_i} f_j(u_j)\Big)^2 \leq \frac{\rho_\Sigma}{m_\mu^2} M_\lambda \|\boldsymbol{L}\|_\rho^2 \|\boldsymbol{u}\|_\rho^2,$$

i.e., $F(\boldsymbol{u}) \in \ell_\rho^2$ for every $\boldsymbol{u} \in \ell_\rho^2$. Similarly, $F^{\mathrm{n}}(\boldsymbol{u}) \in \ell_\rho^2$ for every $\boldsymbol{u} \in \ell_\rho^2$. □

Lemma 16.2. *Let Assumptions 16.1–16.5 hold. Then F and F^{n} are globally Lipschitz with*

$$\|F(\boldsymbol{u}) - F(\boldsymbol{v})\|_\rho \leq \frac{\sqrt{\rho_\Sigma M_\lambda}\|\boldsymbol{L}\|_\rho}{m_\mu} \|\boldsymbol{u} - \boldsymbol{v}\|_\rho,$$

$$\|F^{\mathrm{n}}(\boldsymbol{u}) - F^{\mathrm{n}}(\boldsymbol{v})\|_\rho \leq \frac{\sqrt{\rho_\Sigma M_\lambda}\|\boldsymbol{L}\|_\rho}{m_\mu} \|\boldsymbol{u} - \boldsymbol{v}\|_\rho.$$

Proof. For each $i \in \mathbb{Z}$ and any $\boldsymbol{u} = (u_i)_{i\in\mathbb{Z}} \in \ell_\rho^2$, $\boldsymbol{v} = (v_i)_{i\in\mathbb{Z}} \in \ell_\rho^2$, following calculations similar to Lemma 16.1 gives

$$\begin{aligned}
\Big|F_i(\boldsymbol{u}) - F_i(\boldsymbol{v})\Big|^2 &\leq \frac{1}{m_\mu^2}\Big(\sum_{j\in\mathbb{Z}} |\lambda_{i,j}|\big|f_j(u_j) - f_j(v_j)\big|\Big)^2 \\
&\leq \frac{1}{m_\mu^2}\Big(\sum_{j\in\mathbb{Z}} \frac{\lambda_{i,j}^2}{\rho_j^2}\Big)\Big(\sum_{j\in\mathbb{Z}} \rho_j L_j^2\Big)\Big(\sum_{j\in\mathbb{Z}} \rho_j |u_j - v_j|^2\Big) \\
&\leq \frac{1}{m_\mu^2} M_\lambda \|\boldsymbol{L}\|_\rho^2 \|\boldsymbol{u} - \boldsymbol{v}\|_\rho^2,
\end{aligned}$$

and hence

$$\|F(\boldsymbol{u}) - F(\boldsymbol{v})\|_\rho \leq \frac{\sqrt{\rho_\Sigma M_\lambda}\|\boldsymbol{L}\|_\rho}{m_\mu} \|\boldsymbol{u} - \boldsymbol{v}\|_\rho.$$

The above estimates also hold uniformly in n for the finite sum operator F^{n} . □

Notice that due to Assumptions 16.2 and 16.6, $(g_i/\mu_i)_{i\in\mathbb{Z}} \in \ell_\rho^2$, and hence the lattice systems (16.1) and (16.2) can be rewritten as the following ODEs on ℓ_ρ^2, respectively:

$$\frac{\mathrm{d}\boldsymbol{u}(t)}{\mathrm{d}t} = -\mathrm{A}\boldsymbol{u} + F^{\mathrm{n}}(\boldsymbol{u}) + \Big(\frac{g_i}{\mu_i}\Big)_{i\in\mathbb{Z}} := \mathfrak{F}^{\mathrm{n}}(\boldsymbol{u}), \quad \boldsymbol{u} \in \ell_\rho^2, \tag{16.5}$$

$$\frac{\mathrm{d}\boldsymbol{u}(t)}{\mathrm{d}t} = -\mathrm{A}\boldsymbol{u} + F(\boldsymbol{u}) + \Big(\frac{g_i}{\mu_i}\Big)_{i\in\mathbb{Z}} := \mathfrak{F}(\boldsymbol{u}), \quad \boldsymbol{u} \in \ell_\rho^2. \tag{16.6}$$

16.2 Existence and uniqueness of solutions

It follows directly from Lemma 16.1 and Lemma 16.2 that $\mathfrak{F}^{\mathrm{n}}$ in (16.5) and $\mathfrak{F}$ in (16.6) both map ℓ_ρ^2 into ℓ_ρ^2. Also notice that due to the linearity of the operator A,

$$\|\mathrm{A}\boldsymbol{u} - \mathrm{A}\boldsymbol{v}\|_\rho \leq \frac{1}{m_\varkappa m_\mu}\|\boldsymbol{u} - \boldsymbol{v}\|_\rho \quad \text{for all } \boldsymbol{u}, \boldsymbol{v} \in \ell_\rho^2.$$

Thus

$$\|\mathfrak{F}^n(\boldsymbol{u})-\mathfrak{F}^n(\boldsymbol{v})\|_\rho \le \Big(\frac{1}{m_\varkappa m_\mu}+\frac{\sqrt{\rho_\Sigma M_\lambda}}{m_\mu}\|\boldsymbol{L}\|_\rho\Big)\|\boldsymbol{u}-\boldsymbol{v}\|_\rho,$$

$$\|\mathfrak{F}(\boldsymbol{u})-\mathfrak{F}(\boldsymbol{v})\|_\rho \le \Big(\frac{1}{m_\varkappa m_\mu}+\frac{\sqrt{\rho_\Sigma M_\lambda}}{m_\mu}\|\boldsymbol{L}\|_\rho\Big)\|\boldsymbol{u}-\boldsymbol{v}\|_\rho,$$

i.e., $\mathfrak{F}^n(\boldsymbol{u})$ and $\mathfrak{F}(\boldsymbol{u})$ are globally equi-Lipschitz on ℓ^2_ρ with the Lipschitz constant

$$M_L := \frac{1}{m_\varkappa m_\mu}+\frac{\sqrt{\rho_\Sigma M_\lambda}}{m_\mu}\|\boldsymbol{L}\|_\rho.$$

Standard existence and uniqueness theorems for ODEs in Banach spaces (see, e.g., [Deimling (1977)]) then give the global existence and uniqueness of solutions of equations (16.5) and (16.6) given an initial condition $\boldsymbol{u}(0)=\boldsymbol{u}_o\in\ell^2_\rho$.

Moreover, the solutions of each system depends continuously on initial data. In fact, given any $\boldsymbol{u}_o,\boldsymbol{v}_o\in\ell^2_\rho$, suppose that $\boldsymbol{u}$ and $\boldsymbol{v}$ are the solutions of the equation (16.6) with the initial conditions $\boldsymbol{u}_o$ and $\boldsymbol{v}_o$. Set $\boldsymbol{w}=\boldsymbol{u}-\boldsymbol{v}$, then $\boldsymbol{w}$ satisfies the following equation:

$$\frac{\mathrm{d}\boldsymbol{w}(t)}{\mathrm{d}t} = -\mathrm{A}\boldsymbol{w}+F(\boldsymbol{u})-F(\boldsymbol{v}), \quad \boldsymbol{w}\in\ell^2_\rho. \tag{16.7}$$

Multiply both sides of the ith of the equation (16.7) by $\rho_i w_i(t)$ and summarise over $i\in\mathbb{Z}$ to obtain

$$\begin{aligned}
\frac{1}{2}\frac{\mathrm{d}}{\mathrm{d}t}\|\boldsymbol{w}\|^2_\rho &= -\sum_{i\in\mathbb{Z}}\frac{\rho_i w_i^2}{\mu_i\varkappa_i}+\sum_{i\in\mathbb{Z}}\rho_i w_i\Big(\sum_{j\in\mathbb{Z}}\frac{\lambda_{i,j}}{\mu_i}(f_j(u_j)-f_j(v_j))\Big)\\
&\le -\sum_{i\in\mathbb{Z}}\frac{\rho_i w_i^2}{\mu_i\varkappa_i}+\sum_{i\in\mathbb{Z}}\rho_i w_i\Big(\sum_{j\in\mathbb{Z}}\frac{\lambda_{i,j}}{\mu_i}L_j|w_j|\Big)\\
&\le -\frac{1}{M_\mu M_\varkappa}\|\boldsymbol{w}\|^2_\rho+\Big(\sum_{i\in\mathbb{Z}}\rho_i w_i^2\Big)^{1/2}\Big(\sum_{i\in\mathbb{Z}}\rho_i\big(\sum_{j\in\mathbb{Z}}\frac{\lambda_{i,j}}{\mu_i}L_j|w_j|\big)^2\Big)^{1/2}\\
&\le -\frac{1}{M_\mu M_\varkappa}\|\boldsymbol{w}\|^2_\rho+\|\boldsymbol{w}\|_\rho\Big(\sum_{i\in\mathbb{Z}}\rho_i\big(\sum_{j\in\mathbb{Z}}\frac{\lambda^2_{i,j}}{\rho_j^2 m_\mu^2}\big)\big(\sum_{j\in\mathbb{Z}}\rho_j L_j^2\big)\big(\sum_{j\in\mathbb{Z}}\rho_j w_j^2\big)\Big)^{1/2}\\
&\le -\frac{1}{M_\mu M_\varkappa}\|\boldsymbol{w}\|^2_\rho+\|\boldsymbol{w}\|_\rho\Big(\frac{\rho_\Sigma}{m_\mu^2}M_\lambda\|\boldsymbol{w}\|^2_\rho\|\boldsymbol{L}\|^2_\rho\Big)^{1/2}\\
&\le \Big(-\frac{1}{M_\mu M_\varkappa}+\frac{\sqrt{\rho_\Sigma M_\lambda}}{m_\mu}\|\boldsymbol{L}\|_\rho\Big)\|\boldsymbol{w}\|^2_\rho.
\end{aligned}$$

It immediately follows

$$\|\boldsymbol{w}\|^2_\rho \le e^{2t\big(-\frac{1}{M_\mu M_\varkappa}+\frac{\sqrt{\rho_\Sigma M_\lambda}}{m_\mu}\|\boldsymbol{L}\|_\rho\big)}\|\boldsymbol{w}_o\|^2_\rho,$$

which implies that the solutions of (16.6) depend continuously on initial data. Similar arguments give the same conclusion for (16.5).

Theorem 16.1. *Let Assumptions 16.1–16.6 hold. Then the solution of system* (16.5) *generates a semi-dynamical system* $\{\varphi^{\mathrm{n}}(t)\}_{t\geq 0}$ *on* ℓ^2_ρ *defined by* $\varphi^{\mathrm{n}}(t,\boldsymbol{u}_o) = \boldsymbol{u}(t;\boldsymbol{u}_o)$*, where* $\boldsymbol{u}(t;\boldsymbol{u}_o)$ *is the solution of* (16.5) *satisfying* $\boldsymbol{u}(0) = \boldsymbol{u}_o \in \ell^2_\rho$*. Similarly the solution of system* (16.6) *generates a semi-dynamical system* $\{\varphi(t)\}_{t\geq 0}$ *on* ℓ^2_ρ *defined by* $\varphi(t,\boldsymbol{u}_o) = \boldsymbol{u}(t;\boldsymbol{u}_o)$ *where* $\boldsymbol{u}(t;\boldsymbol{u}_o)$ *is the solution of* (16.6) *satisfying* $\boldsymbol{u}(0) = \boldsymbol{u}_o \in \ell^2_\rho$.

16.3 Existence of attractors

The existence of attractors for the semi-dynamical systems $\{\varphi^{\mathrm{n}}(t)\}_{t\geq 0}$ and $\{\varphi(t)\}_{t\geq 0}$ will now be established under additional conditions to ensure the dissipativity of systems (16.1) and (16.2).

There are multiple ways to achieve this. In particular, an upper bound for $\|\boldsymbol{L}\|_\rho$ can be imposed to make sure that the growth of $F(\boldsymbol{u})$ would not overcome the dissipativity coming from the linear term $-\frac{u_i}{\mu_i \varkappa_i}$. This will essentially put a restriction on the growth rates of f_i's depending on the magnitudes of μ_i's and $\varkappa_i$'s.

Alternatively, an upper bound could be imposed for the magnitude of f_i's themselves, then the maximum growth rates of f_i's do not need to depend on the magnitudes of μ_i's and $\varkappa_i$'s. Another possibility is to impose dissipativity conditions on the functions f_i's, instead of assuming bounds on f_i's or f_i''s.

Since, in the context of Hopfield models, the f_i's are usually sigmoidal functions, it will be assumed here that the f_i's are bounded in the sense that

Assumption 16.7. there exists $\boldsymbol{K} := (K_i)_{i\in\mathbb{Z}} \in \ell^2_\rho$ such that
$$|f_i(s)| \leq |K_i| \quad \text{for all} \quad i \in \mathbb{Z},\ s \in \mathbb{R}.$$

16.3.1 *Existence of absorbing sets*

Absorbing sets for the semi-dynamical systems $\{\varphi^{\mathrm{n}}(t)\}_{t\geq 0}$ and $\{\varphi(t)\}_{t\geq 0}$ will now be established.

Lemma 16.3. *Let Assumptions 16.1–16.7 hold. Then the semi-dynamical system* $\{\varphi(t)\}_{t\geq 0}$ *generated by the equation* (16.2) *has a bounded absorbing set in* ℓ^2_ρ
$$Q := \left\{\boldsymbol{u} \in \ell^2_\rho : \|\boldsymbol{u}\|_\rho \leq R_Q := \frac{M_\mu M_\varkappa}{m_\mu}\sqrt{\rho_\Sigma^2 M_\lambda \|\boldsymbol{K}\|_\rho^2 + \|\boldsymbol{g}\|^2_\rho + 1}\right\}. \tag{16.8}$$

Proof. Multiply both sides of the equation (16.2) by $\rho_i u_i(t)$ and sum over $i \in \mathbb{Z}$ to obtain
$$\frac{1}{2}\frac{\mathrm{d}\,\|\boldsymbol{u}(t)\|_\rho^2}{\mathrm{d}t} = -\sum_{i\in\mathbb{Z}}\frac{\rho_i u_i^2}{\mu_i\varkappa_i} + \sum_{i\in\mathbb{Z}}\frac{\rho_i u_i}{\mu_i}\sum_{j\in\mathbb{Z}}\lambda_{i,j}f_j(u_j) + \sum_{i\in\mathbb{Z}}\rho_i\frac{u_i}{\mu_i}g_i. \tag{16.9}$$
First by Assumption 16.2, we have
$$-\sum_{i\in\mathbb{Z}}\frac{\rho_i u_i^2}{\mu_i\varkappa_i} \leq -\frac{\|\boldsymbol{u}\|_\rho^2}{M_\mu M_\varkappa}. \tag{16.10}$$

Then by Assumptions 16.1, 16.3 and 16.7

$$\sum_{i\in\mathbb{Z}}\rho_i\Big(\sum_{j\in\mathbb{Z}}\lambda_{i,j}f_j(u_j)\Big)^2\le\sum_{i\in\mathbb{Z}}\rho_i\Big(\sum_{j\in\mathbb{Z}}\frac{\lambda_{i,j}^2}{\rho_j^2}\Big)\Big(\sum_{j\in\mathbb{Z}}\rho_j\Big)\Big(\sum_{j\in\mathbb{Z}}\rho_jK_j^2\Big)\le\rho_\Sigma^2M_\lambda\|\boldsymbol{K}\|_\rho^2.$$

Thus, by Cauchy's inequality and Assumption 16.7,

$$\begin{aligned}\sum_{i\in\mathbb{Z}}\frac{\rho_iu_i}{\mu_i}\sum_{j\in\mathbb{Z}}\lambda_{i,j}f_j(u_j)&\le\frac{1}{m_\mu}\Big(\sum_{i\in\mathbb{Z}}\rho_iu_i^2\Big)^{1/2}\Big(\sum_{i\in\mathbb{Z}}\rho_i\Big(\sum_{j\in\mathbb{Z}}\lambda_{i,j}f_j(u_j)\Big)^2\Big)^{1/2}\\&\le\frac{\rho_\Sigma\sqrt{M_\lambda}\|\boldsymbol{K}\|_\rho}{m_\mu}\|\boldsymbol{u}\|_\rho\le\frac{a}{2}\|\boldsymbol{u}\|_\rho^2+\frac{1}{2a}\frac{\rho_\Sigma^2M_\lambda\|\boldsymbol{K}\|_\rho^2}{m_\mu^2}\end{aligned}$$

for some $a>0$. Also, for some $b>0$,

$$\sum_{i\in\mathbb{Z}}\frac{\rho_iu_i}{\mu_i}g_i\le\frac{b}{2}\|\boldsymbol{u}\|_\rho^2+\frac{1}{2b}\sum_{i\in\mathbb{Z}}\frac{\rho_ig_i^2}{\mu_i^2}\le\frac{b}{2}\|\boldsymbol{u}\|_\rho^2+\frac{1}{2bm_\mu^2}\|\boldsymbol{g}\|_\rho^2. \tag{16.11}$$

Insert estimates (16.10)–(16.11) into (16.9) to obtain

$$\frac{\mathrm{d}\,\|\boldsymbol{u}(t)\|_\rho^2}{\mathrm{d}t}\le2\left(-\frac{1}{M_\mu M_\varkappa}+\frac{a}{2}+\frac{b}{2}\right)\cdot\|\boldsymbol{u}\|_\rho^2+\frac{\rho_\Sigma^2M_\lambda\|\boldsymbol{K}\|_\rho^2}{am_\mu^2}+\frac{1}{bm_\mu^2}\|\boldsymbol{g}\|_\rho^2.$$

Letting $a=b=\frac{1}{2M_\mu M_\varkappa}$ in the above inequality leads to

$$\frac{\mathrm{d}\,\|\boldsymbol{u}(t)\|_\rho^2}{\mathrm{d}t}\le-\frac{1}{M_\mu M_\varkappa}\|\boldsymbol{u}\|_\rho^2+\frac{M_\mu M_\varkappa}{m_\mu^2}\left(\rho_\Sigma^2M_\lambda\|\boldsymbol{K}\|_\rho^2+\|\boldsymbol{g}\|_\rho^2\right),$$

which can be integrated from 0 to t to obtain

$$\|\boldsymbol{u}(t)\|_\rho^2\le e^{-\frac{t}{M_\mu M_\varkappa}}\|\boldsymbol{u}(0)\|_\rho^2+\frac{M_\mu^2M_\varkappa^2}{m_\mu^2}\left(\rho_\Sigma^2M_\lambda\|\boldsymbol{K}\|_\rho^2+\|\boldsymbol{g}\|_\rho^2\right)\left(1-e^{-\frac{t}{M_\mu M_\varkappa}}\right).$$

It follows immediately that the closed and bounded set Q defined in (16.8) is an absorbing set for the semi-dynamical system φ on ℓ_ρ^2. □

The following Lemma follows from similar analysis.

Lemma 16.4. *Let Assumptions 16.1–16.7 hold. Then the set Q defined in* (16.8) *is an absorbing set for the semi-dynamical system $\{\varphi^{\mathfrak{n}}(t)\}_{t\ge0}$ generated by equation* (16.1) *uniformly in $\mathfrak{n}$.*

16.3.2 *Asymptotic compactness*

To prove the asymptotic compactness of the semi-dynamical systems $\{\varphi(t)\}_{t\ge0}$ and $\{\varphi^{\mathfrak{n}}(t)\}_{t\ge0}$, it first needs to be shown that the solutions of systems (16.1) and (16.2) have asymptotic null tails at $|i|$ large by using a standard cut-off argument.

Lemma 16.5. *Let Assumptions 16.1–16.7 hold and let $\boldsymbol{u}_o\in Q$, where Q is the absorbing set defined in* (16.8). *Then for every $\varepsilon>0$ there exist $T(\varepsilon)$ and $I(\varepsilon)$ such that the solution of equation* (16.6) *with $\boldsymbol{u}(0)=\boldsymbol{u}_o$ satisfies*

$$\sum_{|i|\ge2I(\varepsilon)}\rho_iu_i(t;\boldsymbol{u}_o)^2\le\varepsilon\quad\textit{for all } t\ge T(\varepsilon).$$

Proof. Let $\xi : \mathbb{R}^+ \to [0,1]$ be the smooth cut-function defined in (3.8). Recall that there exists $C_0 > 0$ such that $|\xi'(s)| \le C_0$ for all $s \in \mathbb{R}^+$. Let k be a fixed (and large) natural number to be specified later, and set

$$v_i(t) = \xi_k(|i|)u_i(t) \quad \text{with} \quad \xi_k(|i|) = \xi\Big(\frac{|i|}{k}\Big), \quad i \in \mathbb{Z}.$$

Multiplying both sides of system (16.2) by $\rho_i v_i(t)$ and summing over $i \in \mathbb{Z}$ yields

$$\frac{1}{2}\frac{\mathrm{d}}{\mathrm{d}t}\sum_{i\in\mathbb{Z}} \rho_i \xi_k(|i|)u_i^2(t) = C_1 + C_2 + C_3$$

where

$$C_1 = -\sum_{i\in\mathbb{Z}} \rho_i\xi_k(|i|)\frac{u_i^2}{\mu_i \varkappa_i} \le -\frac{1}{M_\mu M_\varkappa}\sum_{i\in\mathbb{Z}}\rho_i\xi_k(|i|)u_i^2(t),$$

$$C_2 = \sum_{i\in\mathbb{Z}}\rho_i\xi_k(|i|)u_i\sum_{j\in\mathbb{Z}}\frac{\lambda_{i,j}}{\mu_i}f_j(u_j(t)),$$

$$C_3 = \sum_{i\in\mathbb{Z}}\rho_i\xi_k(|i|)\frac{u_i}{\mu_i}g_i \le \frac{b}{2}\sum_{i\in\mathbb{Z}}\rho_i\xi_k(|i|)u_i^2(t) + \frac{1}{2bm_\mu^2}\sum_{i\in\mathbb{Z}}\rho_i\xi_k(|i|)g_i^2.$$

Using the Cauchy inequality $(\sum_i a_i b_i)^2 \le \sum_i a_i^2 \sum_i b_i^2$ and the inequality $xy \le \frac{1}{2}(ax^2 + \frac{1}{a}y^2)$ (where a is positive and to be determined later), by Assumption 16.1, 16.2, 16.3, 16.7,

$$\begin{aligned}
C_2 &\le \frac{1}{m_\mu}\Big(\sum_{i\in\mathbb{Z}}\rho_i\xi_k(|i|)u_i^2(t)\Big)^{\frac{1}{2}}\Big(\sum_{i\in\mathbb{Z}}\rho_i\xi_k(|i|)\big(\sum_{j\in\mathbb{Z}}\lambda_{i,j}f_j(u_j(t))\big)^2\Big)^{\frac{1}{2}}\\
&\le \frac{1}{m_\mu}\Big(\sum_{i\in\mathbb{Z}}\rho_i\xi_k(|i|)u_i^2(t)\Big)^{\frac{1}{2}}\Big(\sum_{i\in\mathbb{Z}}\rho_i\xi_k(|i|)\big(\sum_{j\in\mathbb{Z}}\frac{\lambda_{i,j}^2}{\rho_j^2}\big)\big(\sum_{j\in\mathbb{Z}}\rho_j\big)\big(\sum_{j\in\mathbb{Z}}\rho_jK_j^2\big)\Big)^{\frac{1}{2}}\\
&\le \frac{1}{m_\mu}\Big(\sum_{i\in\mathbb{Z}}\rho_i\xi_k(|i|)u_i^2(t)\Big)^{\frac{1}{2}}\Big(\sum_{i\in\mathbb{Z}}\rho_i\xi_k(|i|)M_\lambda\rho_\Sigma\|\boldsymbol{K}\|_\rho^2\Big)^{\frac{1}{2}}\\
&\le \frac{1}{2cm_\mu}\sum_{i\in\mathbb{Z}}\rho_i\xi_k(|i|)u_i^2(t) + \frac{c}{2m_\mu}\sum_{i\in\mathbb{Z}}\rho_i\xi_k(|i|)M_\lambda\rho_\Sigma\|\boldsymbol{K}\|_\rho^2,
\end{aligned}$$

and as a result

$$\begin{aligned}
\frac{1}{2}\frac{\mathrm{d}}{\mathrm{d}t}\sum_{i\in\mathbb{Z}}\rho_i\xi_k(|i|)u_i^2(t) &\le \Big(-\frac{1}{M_\varkappa M_\mu} + \frac{b}{2} + \frac{1}{2cm_\mu}\Big)\sum_{i\in\mathbb{Z}}\rho_i\xi_k(|i|)u_i^2(t)\\
&\quad + \frac{c}{2m_\mu}\sum_{i\in\mathbb{Z}}\rho_i\xi_k(|i|)M_\lambda\rho_\Sigma\|\boldsymbol{K}\|_\rho^2 + \frac{1}{2bm_\mu^2}\sum_{i\in\mathbb{Z}}\rho_i\xi_k(|i|)g_i^2.
\end{aligned}$$

Choose

$$b = \frac{1}{2M_\varkappa M_\mu}, \qquad c = \frac{2M_\varkappa M_\mu}{m_\mu}.$$

Then

$$\frac{\mathrm{d}}{\mathrm{d}t}\sum_{i\in\mathbb{Z}}\rho_i\xi_k(|i|)u_i^2(t) \le -\frac{1}{M_\varkappa M_\mu}\sum_{i\in\mathbb{Z}}\rho_i\xi_k(|i|)u_i^2(t) + \frac{2M_\varkappa M_\mu}{m_\mu^2}\sum_{i\in\mathbb{Z}}\rho_i\xi_k(|i|)g_i^2$$
$$+\frac{2M_\mu M_\varkappa}{m_\mu^2}\sum_{i\in\mathbb{Z}}\rho_i\xi_k(|i|)M_\lambda\rho_\Sigma\|\boldsymbol{K}\|_\rho^2.$$

Since $\rho_\Sigma := \sum_{i\in\mathbb{Z}}\rho_i < \infty$, for every $\varepsilon > 0$, there exists $I_1(\varepsilon) > 0$ such that

$$\sum_{|i|\ge k}\rho_i \le \varepsilon \qquad \text{when} \qquad k \ge I_1(\varepsilon).$$

In addition, since $\boldsymbol{g} := (g_i)_{i\in\mathbb{Z}} \in \ell_\rho^2$, for every $\varepsilon > 0$, there exists $I_2(\varepsilon) > 0$ such that

$$\sum_{|i|\ge k}\rho_i g_i^2 \le \varepsilon \qquad \text{when} \qquad k \ge I_2(\varepsilon).$$

Consequently, when $k \ge I(\varepsilon) := \max\{I_1(\varepsilon), I_2(\varepsilon)\}$, we have

$$\frac{\mathrm{d}}{\mathrm{d}t}\sum_{i\in\mathbb{Z}}\rho_i\xi_k(|i|)u_i^2(t) \le -\frac{1}{M_\mu M_\varkappa}\sum_{i\in\mathbb{Z}}\rho_i\xi_k(|i|)u_i^2(t) + \frac{2M_\varkappa M_\mu}{m_\mu^2}(M_\lambda\rho_\Sigma\|\boldsymbol{K}\|_\rho^2+1)\varepsilon.$$

It follows immediately from Gronwall's lemma that

$$\sum_{i\in\mathbb{Z}}\rho_i\xi_k(|i|)u_i^2(t) \le e^{-\frac{t}{M_\mu M_\varkappa}}\sum_{i\in\mathbb{Z}}\rho_i\xi_k(|i|)u_i^2(0) + \frac{2M_\varkappa^2 M_\mu^2}{m_\mu^2}(M_\lambda\rho_\Sigma\|\boldsymbol{K}\|_\rho^2+1)\varepsilon$$
$$\le e^{-\frac{t}{M_\mu M_\varkappa}}\|\boldsymbol{u}_o\|_\rho^2 + \frac{2M_\varkappa^2 M_\mu^2}{m_\mu^2}(M_\lambda\rho_\Sigma\|\boldsymbol{K}\|_\rho^2+1)\varepsilon$$

which implies that there exists $T = T(\varepsilon)$ such that

$$\sum_{|i|\ge 2k}\rho_i u_i^2(t) \le \sum_{i\in\mathbb{Z}}\rho_i\xi_k(|i|)u_i^2(t) \le \frac{4M_\varkappa^2 M_\mu^2}{m_\mu^2}(M_\lambda\rho_\Sigma\|\boldsymbol{K}\|_\rho^2+1)\varepsilon,$$

for all $k \ge I(\varepsilon), t \ge T(\varepsilon)$. □

Lemma 16.6. *Suppose that Assumptions 16.1–16.7 hold. Then the semi-dynamical system $\{\varphi(t)\}_{t\ge 0}$ is asymptotically compact, i.e., if $\boldsymbol{u}_o^{(m)}$ is bounded in ℓ_ρ^2 and $t_m \to \infty$, then every sequence $\boldsymbol{u}^{(m)}(t_m) := \varphi(t_m, \boldsymbol{u}_o^{(m)})$, $m \in \mathbb{N}$, is precompact in ℓ_ρ^2.*

Proof. For every $\boldsymbol{u}_o^{(m)} \in \ell_\rho^2$, by the existence of the absorbing set, there exists $N_1 > 0$ such that

$$\|\boldsymbol{u}^{(m)}(t_m)\|_\rho^2 \le R_Q^2, \qquad m > N_1, \tag{16.12}$$

where R_Q is defined by (16.8). By (16.12) there exist $\boldsymbol{u}^* \in \ell_\rho^2$ and a subsequence of $\{\boldsymbol{u}^{(m)}(t_m)\}$ (still denoted by $\{\boldsymbol{u}^{(m)}(t_m)\}$) such that

$$\boldsymbol{u}^{(m)}(t_m) \rightharpoonup \boldsymbol{u}^* \quad \text{(weak convergence) in} \quad \ell_\rho^2. \tag{16.13}$$

It will be shown next that this weak convergence is actually strong convergence, in particular that for every $\varepsilon > 0$, there exists $N_0(\varepsilon) > 0$ such that when $m \geq N_0(\varepsilon)$,

$$\|\boldsymbol{u}^{(m)}(t_m) - \boldsymbol{u}^*\|_\rho^2 \leq \varepsilon.$$

Indeed, there exists $N_2(\varepsilon) > 0$ such that $t_m \geq T(\varepsilon)$ when $m \geq N_2(\varepsilon)$ since $t_m \to \infty$ (where $T(\varepsilon)$ is the constant in Lemma 16.5) and by the tail estimate there exists $M_1(\varepsilon) > 0$ such that

$$\sum_{|i| \geq 2M_1(\varepsilon)} \rho_i |u_i^{(m)}(t_m)|^2 \leq \frac{\varepsilon}{8} \quad \text{for all} \quad m \geq \max\{N_2(\varepsilon), N_1(\varepsilon)\}. \tag{16.14}$$

On the other hand, since $\boldsymbol{u}^* \in \ell_\rho^2$, there exists $M_2(\varepsilon)$ such that

$$\sum_{|i| \geq 2M_2(\varepsilon)} \rho_i (u_i^*)^2 \leq \frac{\varepsilon}{8}. \tag{16.15}$$

Let $M_0(\varepsilon) := \max\{M_1(\varepsilon), M_2(\varepsilon)\}$, by the weak convergence (16.13) there exists $N_3(\varepsilon) > N_1(\varepsilon)$ such that when $m \geq N_3(\varepsilon)$

$$\sum_{|i| \leq 2M_0(\varepsilon)} \rho_i |u_i^{(m)}(t_m) - u_i^*|^2 \leq \frac{\varepsilon}{2}. \tag{16.16}$$

Setting $N_0(\varepsilon) := \max\{N_2(\varepsilon), N_3(\varepsilon)\}$, from (16.14), (16.15) and (16.16), it follows for $m \geq N_0(\varepsilon)$ that

$$\begin{aligned} \|\boldsymbol{u}^{(m)}(t_m) - \boldsymbol{u}^*\|_\rho^2 &= \sum_{|i| \leq 2M_0(\varepsilon)} \rho_i |u_i^{(m)}(t_m) - u_i^*|^2 + \sum_{|i| > 2M_0(\varepsilon)} \rho_i |u_i^{(m)}(t_m) - u_i^*|^2 \\ &\leq \frac{\varepsilon}{2} + 2 \sum_{|i| > 2M_0(\varepsilon)} \rho_i (|u_i^{(m)}(t_m)|^2 + |u_i^*|^2) \\ &\leq \varepsilon \end{aligned}$$

Hence, $\{\boldsymbol{u}^{(m)}(t_m)\}_{m \in \mathbb{N}}$ is precompact in ℓ_ρ^2. □

The above results give the following Theorem.

Theorem 16.2. *Let Assumptions 16.1–16.7 hold. Then system* (16.2) *has a global attractor $\mathcal{A}$ and the system* (16.1) *has a global attractor $\mathcal{A}_\mathfrak{n}$ for every $\mathfrak{n} \geq 1$. Moreover the attractors $\mathcal{A}$ and $\mathcal{A}_\mathfrak{n}$ are minimal and connected.*

16.4 Upper semi-continuity of attractors in $\lambda_{i,j}$

In this section it will be shown that the attractors $\mathcal{A}_\mathfrak{n}$ for the semi-dynamical systems $\{\varphi^\mathfrak{n}(t)\}_{t \geq 0}$ generated by system (16.1) converge upper semi-continuously to the attractor for the semi-dynamical system $\{\varphi(t)\}_{t \geq 0}$ generated by system (16.2) as $\mathfrak{n} \to \infty$. To this end a series of results that are crucial for the proof of the upper semi-continuity are first presented.

In particular, Lemma 16.7 shows that the global attractors $\mathcal{A}_\mathfrak{n}$ and $\mathcal{A}$ consist of all bounded entire solutions of systems (16.1) and (16.2), respectively, defined on the whole $\mathbb{R}$, while Lemma 16.8 compares the semi-dynamical systems $\{\varphi^{\mathfrak{n}}(t)\}_{t\geq 0}$ and $\{\varphi(t)\}_{t\geq 0}$ and Lemma 16.9 shows that any sequence of elements picked from $\mathcal{A}_{\mathfrak{n}}$ converges to an element in $\mathcal{A}$ as $\mathfrak{n} \to \infty$.

Lemma 16.7. *The global attractors $\mathcal{A}_{\mathfrak{n}}$ and $\mathcal{A}$ are strictly invariant for all $t \in \mathbb{R}$, i.e.,*

$$\varphi(t, \mathcal{A}) = \mathcal{A}, \quad \varphi^{\mathfrak{n}}(t, \mathcal{A}_{\mathfrak{n}}) = \mathcal{A}_{\mathfrak{n}} \textit{ for every } \mathfrak{n} \geq 1 \quad \textit{for all } t \in \mathbb{R}.$$

Proof. Note that the global attractors $\mathcal{A}$ and $\mathcal{A}_{\mathfrak{n}}$ obtained in Theorem 16.2 are strictly invariant for $t \geq 0$, i.e.,

$$\varphi(t, \mathcal{A}) = \mathcal{A}, \quad \varphi^{\mathfrak{n}}(t, \mathcal{A}_{\mathfrak{n}}) = \mathcal{A}_{\mathfrak{n}} \text{ for every } \mathfrak{n} \geq 1 \quad \text{for all } t \geq 0.$$

Thus only the backward uniqueness of solutions to each of the systems (16.1) and (16.2) needs to be shown. To this end for any $\boldsymbol{u}_o \in \ell^2_\rho$ let $\boldsymbol{u}(t) = \varphi(t, \boldsymbol{u}_o)$ and $\boldsymbol{v}(t) = \varphi(t, \boldsymbol{v}_o)$ be two solutions of the system (16.2) such that $\boldsymbol{u}(T) = \boldsymbol{v}(T)$ for some $T > 0$. It will be shown that $\boldsymbol{u}(t) = \boldsymbol{v}(t)$ for all $0 \leq t \leq T$.

Define $\boldsymbol{w}(t) := \boldsymbol{u}(t) - \boldsymbol{v}(t)$ then by equation (16.2)

$$\mu_i \frac{\mathrm{d}w_i(t)}{\mathrm{d}t} = -\frac{1}{\varkappa_i} w_i(t) + \sum_{j\in\mathbb{Z}} \lambda_{i,j} \left(f_j(u_j) - f_j(v_j)\right).$$

Multiplying the above equation by $-\frac{\rho_i}{\mu_i} w_i$ and summing over $i \in \mathbb{Z}$ results in

$$-\sum_{i\in\mathbb{Z}} \rho_i w_i \frac{\mathrm{d}w_i(t)}{\mathrm{d}t} = \sum_{i\in\mathbb{Z}} \frac{\rho_i}{\mu_i \varkappa_i} w_i^2(t) - \sum_{i\in\mathbb{Z}} \Big(\frac{\rho_i}{\mu_i} w_i \sum_{j\in\mathbb{Z}} \lambda_{i,j} \left(f_j(u_j) - f_j(v_j)\right) \Big),$$

in which

$$\sum_{i\in\mathbb{Z}} \frac{\rho_i}{\mu_i \varkappa_i} w_i^2(t) \leq \frac{1}{m_\mu m_\varkappa} \|\boldsymbol{w}(t)\|_\rho^2,$$

and

$$\begin{aligned}
&\sum_{i\in\mathbb{Z}} \Big(\frac{\rho_i}{\mu_i} w_i \sum_{j\in\mathbb{Z}} \lambda_{i,j} \left(f_j(u_j) - f_j(v_j)\right) \Big) \\
&\leq \frac{1}{m_\mu} \Big(\sum_{i\in\mathbb{Z}} \rho_i w_i^2 \Big)^{1/2} \Big(\sum_{i\in\mathbb{Z}} \rho_i \sum_{j\in\mathbb{Z}} \frac{\lambda_{i,j}^2}{\rho_j^2} \sum_{j\in\mathbb{Z}} \rho_j L_j^2 \sum_{j\in\mathbb{Z}} \rho_j w_j^2 \Big)^{1/2} \\
&\leq \frac{1}{m_\mu} \sqrt{\rho_\Sigma} \sqrt{M_\lambda} \|\boldsymbol{L}\|_\rho \|\boldsymbol{w}(t)\|_\rho^2.
\end{aligned}$$

Therefore

$$-\sum_{i\in\mathbb{Z}} \rho_i w_i \frac{\mathrm{d}w_i(t)}{\mathrm{d}t} \leq \frac{1}{m_\mu m_\varkappa} \|\boldsymbol{w}(t)\|_\rho^2 + \frac{\sqrt{\rho_\Sigma}}{m_\mu} \sqrt{M_\lambda} \|\boldsymbol{L}\|_\rho \|\boldsymbol{w}(t)\|_\rho^2 \leq c \|\boldsymbol{w}(t)\|_\rho^2, \tag{16.17}$$

where

$$c = \frac{1}{m_\mu} \left(\frac{1}{m_\varkappa} + \sqrt{\rho_\Sigma M_\lambda \|\boldsymbol{L}\|_\rho^2} \right).$$

Now suppose (for contradiction) that there exists $t_0 \in [0, T]$ such that $\boldsymbol{w}(t_0) \neq 0$. Then by continuity of solutions there exists $\tau \in [t_0, T]$ such that

$$\boldsymbol{w}(\tau) = 0 \quad \text{but } \boldsymbol{w}(t) \neq 0 \quad \text{for all } t \in [t_0, \tau).$$

Thus

$$\lim_{t \to \tau^-} \ln \frac{1}{\|\boldsymbol{w}(t)\|_\rho} = \infty. \tag{16.18}$$

On the other hand it follows from (16.17) that

$$\frac{\mathrm{d}}{\mathrm{d}t} \ln \frac{1}{\|\boldsymbol{w}(t)\|_\rho} = -\frac{1}{2}\frac{\mathrm{d}}{\mathrm{d}t} \ln \|\boldsymbol{w}(t)\|_\rho^2 = -\frac{\sum_{i\in\mathbb{Z}} \rho_i w_i \frac{\mathrm{d}w_i(t)}{\mathrm{d}t}}{\|\boldsymbol{w}(t)\|_\rho^2} \leq c.$$

Integrating the above inequality from t_0 and $t \in [t_0, \tau)$ gives

$$\ln \frac{1}{\|\boldsymbol{w}(t)\|_\rho} \leq \ln \frac{1}{\|\boldsymbol{w}(t_0)\|_\rho} + c(t - t_0) \leq \ln \frac{1}{\|\boldsymbol{w}(t_0)\|_\rho} + cT,$$

which contradicts with (16.18). Therefore the solutions to system (16.2) is backward unique, consequently $\varphi(t, \mathcal{A}) = \mathcal{A}$ for all $t \in \mathbb{R}$. By a similar procedure it follows that solutions of system (16.1) are backward unique, and that $\varphi^{\mathfrak{n}}(t, \mathcal{A}_{\mathfrak{n}}) = \mathcal{A}_{\mathfrak{n}}$ for all $t \in \mathbb{R}$. □

The rest of this section requires the convergence rate from $\lambda_{i,j}^{(\mathfrak{n})} \to \lambda_{i,j}$ to be specified. In particular it is assumed that

Assumption 16.8. $\lambda_{i,j}^{(\mathfrak{n})} \to \lambda_{i,j}$ as $\mathfrak{n} \to \infty$ in the sense that for every $\varepsilon > 0$ there exists $N(\varepsilon) \in \mathbb{N}$ such that

$$\sum_{j\in\mathbb{Z}} \frac{(\lambda_{i,j}^{(\mathfrak{n})} - \lambda_{i,j})^2}{\rho_j} \leq \varepsilon^2 \quad \text{for all } \mathfrak{n} \geq N(\epsilon), \qquad i \in \mathbb{Z}.$$

Lemma 16.8. *Let Assumptions 16.1–16.8 hold. Then for every $\boldsymbol{u}_o \in Q$ and $\varepsilon > 0$, there exists $N(\varepsilon) > 0$ such that*

$$\|\varphi^{\mathfrak{n}}(t, \boldsymbol{u}_o) - \varphi(t, \boldsymbol{u}_o)\|_\rho < \varepsilon \quad \textit{for all } \mathfrak{n} \geq N(\varepsilon) \quad \textit{for each fixed } t.$$

Proof. Write $\boldsymbol{u}^{\mathfrak{n}}(t) = (u_i^{\mathfrak{n}}(t))_{i\in\mathbb{Z}} = \varphi^{\mathfrak{n}}(t, \boldsymbol{u}_o)$ and $\boldsymbol{v}^{\mathfrak{n}}(t) = (v_i^{\mathfrak{n}}(t))_{i\in\mathbb{Z}} = \varphi(t, \boldsymbol{u}_o)$. Then $\boldsymbol{u}^{\mathfrak{n}}(t)$ and $\boldsymbol{v}^{\mathfrak{n}}(t)$ satisfy systems (16.1) and (16.2) respectively. Define

$$\boldsymbol{w}^{\mathfrak{n}}(t) = \boldsymbol{u}^{\mathfrak{n}}(t) - \boldsymbol{v}^{\mathfrak{n}}(t) = (u_i^{\mathfrak{n}}(t) - v_i^{\mathfrak{n}}(t))_{i\in\mathbb{Z}} = (w_i^{\mathfrak{n}}(t))_{i\in\mathbb{Z}}.$$

Then $\boldsymbol{w}^{\mathfrak{n}}(t)$ satisfies the equations

$$\mu_i \frac{\mathrm{d}w_i^{\mathfrak{n}}(t)}{\mathrm{d}t} = -\frac{1}{\varkappa_i} w_i^{\mathfrak{n}}(t) + \sum_{j=i-\mathfrak{n}}^{i+\mathfrak{n}} \lambda_{i,j}^{(\mathfrak{n})} f_j(u_j^{\mathfrak{n}}(t)) - \sum_{j\in\mathbb{Z}} \lambda_{i,j} f_j(v_j^{\mathfrak{n}}(t)).$$

Multiplying both sides of the above equation by $2\rho_i w_i^{\mathfrak{n}}(t)/\mu_i$ and summing over $i \in \mathbb{Z}$ gives

$$\begin{aligned}\frac{\mathrm{d}}{\mathrm{d}t}\|\boldsymbol{w}^{\mathfrak{n}}(t)\|_\rho^2 = &-\sum_{i\in\mathbb{Z}} \frac{2}{\varkappa_i\mu_i}\rho_i (w_i^{\mathfrak{n}})^2 - \sum_{i\in\mathbb{Z}}\Big(\frac{2}{\mu_i}\rho_i w_i^{\mathfrak{n}} \sum_{|j-i|>n} \lambda_{i,j} f_j(v_j^{\mathfrak{n}})\Big)\\ &+\sum_{i\in\mathbb{Z}}\Big(\frac{2}{\mu_i}\rho_i w_i^{\mathfrak{n}} \sum_{j=i-\mathfrak{n}}^{i+\mathfrak{n}} (\lambda_{i,j}^{(\mathfrak{n})} f_j(u_j^{\mathfrak{n}}) - \lambda_{i,j} f_j(v_j^{\mathfrak{n}}))\Big). \end{aligned} \tag{16.19}$$

To simplify notations, for each $i \in \mathbb{Z}$, set

$$X_i := \sum_{|j-i|>n} \big|\lambda_{i,j} f_j(v_j^{\mathfrak{n}})\big|$$

$$Y_i := \sum_{j=i-n}^{i+n} \Big|\lambda_{i,j}^{(\mathfrak{n})} (f_j(u_j^{\mathfrak{n}}(t)) - f_j(v_j^{\mathfrak{n}}(t)))\Big|$$

$$Z_i := \sum_{j=i-n}^{i+n} \Big|(\lambda_{i,j}^{(\mathfrak{n})} - \lambda_{i,j}) f_j(v_j^{\mathfrak{n}}(t))\Big|.$$

Then by Assumption 16.1, Cauchy's inequality and the triangle inequality it follows from equation (16.19) that

$$\begin{aligned}\frac{\mathrm{d}}{\mathrm{d}t}\|\boldsymbol{w}^{\mathfrak{n}}(t)\|_\rho^2 \le &-\frac{2}{M_\mu M_\varkappa}\|\boldsymbol{w}^{\mathfrak{n}}(t)\|_\rho^2\\ &+\frac{2}{m_\mu}\|\boldsymbol{w}^{\mathfrak{n}}(t)\|_\rho\Big(\big(\sum_{i\in\mathbb{Z}}\rho_i X_i^2\big)^{1/2} + \big(\sum_{i\in\mathbb{Z}}\rho_i Y_i^2\big)^{1/2} + \big(\sum_{i\in\mathbb{Z}}\rho_i Z_i^2\big)^{1/2}\Big). \end{aligned} \tag{16.20}$$

Each term on the right hand side of (16.20) will now be estimated.

First, since $\sum_{i\in\mathbb{Z}}\rho_i = \rho_\Sigma < \infty$, for every $\varepsilon > 0$ there exists $I(\varepsilon) > 0$ such that $\sum_{|i|>I(\varepsilon)}\rho_i < \varepsilon^2$. Pick $N_1(\varepsilon) = 2I(\varepsilon)$, then $|j| > I(\varepsilon)$ if $|j-i| > N_1(\varepsilon)$ and $|i| \le I(\varepsilon)$, and thus

$$\sum_{|j-i|>\mathfrak{n}} \rho_j < \varepsilon^2 \quad \text{for all } \mathfrak{n} \ge N_1(\varepsilon).$$

Consequently,

$$\begin{aligned}\sum_{i\in\mathbb{Z}}\rho_i X_i^2 &\le \sum_{i\in\mathbb{Z}}\rho_i\Big(\sum_{|j-i|>\mathfrak{n}}\frac{\lambda_{i,j}^2}{\rho_j}\sum_{|j-i|>\mathfrak{n}}\rho_j^2\sum_{|j-i|>\mathfrak{n}}\rho_j K_j^2\Big)\\ &\le \sum_{|i|\le I(\varepsilon)}\rho_i M_\lambda \varepsilon^2\|\boldsymbol{K}\|_\rho^2 + \sum_{|i|>I(\varepsilon)}\rho_i M_\lambda\rho_\Sigma\|\boldsymbol{K}\|_\rho^2\\ &\le 2M_\lambda\rho_\Sigma\|\boldsymbol{K}\|_\rho^2\varepsilon^2 \quad \text{for all } \mathfrak{n} \ge N_1(\varepsilon). \end{aligned} \tag{16.21}$$

Then, by the local Lipschitz property of f_i's in (16.3)

$$\begin{aligned}\sum_{i\in\mathbb{Z}}\rho_i Y_i^2 &\le \sum_{i\in\mathbb{Z}}\rho_i\Big(\sum_{j=i-n}^{i+n}\frac{(\lambda_{i,j}^{(\mathfrak{n})})^2}{\rho_j^2}\sum_{j=i-\mathfrak{n}}^{i+\mathfrak{n}}\rho_j L_j^2\sum_{j=i-n}^{i+n}\rho_j(w_j^{\mathfrak{n}})^2\Big)\\ &\le \rho_\Sigma M_\lambda\|\boldsymbol{L}\|_\rho^2\cdot\|\boldsymbol{w}^{\mathfrak{n}}(t)\|_\rho^2. \end{aligned} \tag{16.22}$$

Now by Assumption 16.8, for every $\varepsilon > 0$ there exists $N_2(\varepsilon) > 0$ such that $\sum_{j=i-\mathfrak{n}}^{i+\mathfrak{n}} (\lambda_{i,j}^{(\mathfrak{n})} - \lambda_{i,j})^2/\rho_j < \varepsilon^2$ for all $\mathfrak{n} \geq N_2(\varepsilon)$ and $i \in \mathbb{Z}$. Therefore

$$\begin{aligned}\sum_{i\in\mathbb{Z}} \rho_i Z_i^2 &\leq \sum_{i\in\mathbb{Z}} \rho_i \Big(\sum_{j=i-\mathfrak{n}}^{i+\mathfrak{n}} \frac{(\lambda_{i,j}^{(\mathfrak{n})} - \lambda_{i,j})^2}{\rho_j} \sum_{j=i-\mathfrak{n}}^{i+\mathfrak{n}} \rho_j K_j^2 \Big) \\ &\leq \varepsilon^2 \rho_\Sigma \|\boldsymbol{K}\|_\rho^2 \quad \text{for all } \mathfrak{n} \geq N_2(\varepsilon). \end{aligned} \tag{16.23}$$

Inserting estimations (16.21)–(16.23) in (16.20) gives

$$\begin{aligned}\frac{\mathrm{d}}{\mathrm{d}t}\|\boldsymbol{w}^{\mathfrak{n}}(t)\|_\rho^2 &\leq \Big(-\frac{2}{M_\mu M_\varkappa} + \frac{2}{m_\mu}\sqrt{\rho_\Sigma M_\lambda}\|\boldsymbol{L}\|_\rho \Big) \|\boldsymbol{w}^{\mathfrak{n}}(t)\|_\rho^2 \\ &\quad + \frac{2\|\boldsymbol{K}\|_\rho \sqrt{\rho_\Sigma}}{m_\mu} \varepsilon \left(\sqrt{2M_\lambda} + 1 \right) \|\boldsymbol{w}^{\mathfrak{n}}(t)\|_\rho \\ &:= a\|\boldsymbol{w}^{\mathfrak{n}}(t)\|_\rho^2 + b\|\boldsymbol{w}^{\mathfrak{n}}(t)\|_\rho. \end{aligned}$$

Divide the above inequality by $2\|\boldsymbol{w}^{\mathfrak{n}}(t)\|_\rho$ and integrate the resulting inequality from 0 to t with $\boldsymbol{w}^{\mathfrak{n}}(0) = 0$ to obtain

$$\|\boldsymbol{w}^{\mathfrak{n}}(t)\|_\rho \leq \frac{b}{|a|}\varepsilon \cdot \max\{1, e^{|a|t}\} \quad \text{for all } t \geq 0 \text{ fixed } \mathfrak{n} \geq \max\{N_1(\varepsilon), N_2(\varepsilon)\}.$$

The desired assertion then follows immediately after minor adjustment in ε. □

The following important property can be proved with the above lemma.

Lemma 16.9. *Let Assumptions 16.1–16.8 hold. Then for every sequence $\{\boldsymbol{a}_{\mathfrak{n}}\}_{\mathfrak{n}\geq 1}$ with $\boldsymbol{a}_{\mathfrak{n}} \in \mathcal{A}_{\mathfrak{n}}$ there exist a subsequence $\{\boldsymbol{a}_{\mathfrak{n}_k}\}_{k\geq 1}$ of $\{\boldsymbol{a}_{\mathfrak{n}}\}_{\mathfrak{n}\geq 1}$ and $\boldsymbol{a} \in \mathcal{A}$ such that $\boldsymbol{a}_{\mathfrak{n}_k} \to \boldsymbol{a}$ in ℓ_ρ^2.*

Proof. Let $\boldsymbol{u}^{\mathfrak{n}}(t; \boldsymbol{a}_{\mathfrak{n}}) = \varphi^{\mathfrak{n}}(t, \boldsymbol{a}_{\mathfrak{n}})$ be the solution of system (16.1) with $\boldsymbol{u}^{\mathfrak{n}}(0) = \boldsymbol{a}_{\mathfrak{n}}$. Let $\boldsymbol{v}^{\mathfrak{n}}(t) = \varphi(t, \boldsymbol{a}_{\mathfrak{n}})$. Then it follows from Lemma 16.8 above that $\boldsymbol{u}^{\mathfrak{n}}(t) \to \boldsymbol{v}^{\mathfrak{n}}(t)$ in ℓ_ρ^2 for every fixed $t \in \mathbb{R}$.

Since $\boldsymbol{a}_{\mathfrak{n}} \in \mathcal{A}_{\mathfrak{n}} \subset Q$, the absorbing set defined in (16.8), then $\boldsymbol{v}^{\mathfrak{n}}(t) \in Q$ for all $t \in \mathbb{R}$ with

$$\|\boldsymbol{v}^{\mathfrak{n}}(t)\|_\rho \leq R_Q \quad \text{for all } t \in \mathbb{R} \text{ and } \mathfrak{n} = 1, 2, \ldots,$$

where R_Q is the radius of the absorbing set defined in (16.8).

One the one hand, by equation (16.6),

$$\left\|\frac{\mathrm{d}\boldsymbol{v}^{\mathfrak{n}}(t)}{\mathrm{d}t}\right\|_\rho \leq \left\|A\boldsymbol{v}^{\mathfrak{n}}(t)\right\|_\rho + \left\|A\boldsymbol{v}^{\mathfrak{n}}(t)\right\|_\rho + \frac{1}{m_\mu}\|\boldsymbol{g}\|_\rho,$$

and then by Lemmas 16.1 and 16.3

$$\begin{aligned}\left\|\frac{\mathrm{d}\boldsymbol{v}^{\mathfrak{n}}(t)}{\mathrm{d}t}\right\|_\rho &\leq \frac{1}{m_\mu m_\varkappa}\left\|\boldsymbol{v}^{\mathfrak{n}}(t)\right\|_\rho + \frac{\sqrt{\rho_\Sigma M_\lambda}}{m_\mu}\|\boldsymbol{L}\|_\rho \left\|\boldsymbol{v}^{\mathfrak{n}}(t)\right\|_\rho + \frac{1}{m_\mu}\|\boldsymbol{g}\|_\rho \\ &\leq \frac{1}{m_\mu}\Big(\frac{1}{m_\varkappa} + \sqrt{\rho_\Sigma M_\lambda}\|\boldsymbol{L}\|_\rho\Big) R_Q + \frac{1}{m_\mu}\|\boldsymbol{g}\|_\rho := M_{\dot{v}}. \end{aligned}$$

It follows directly that

$$\left\|v^{\mathfrak{n}}(t)-v^{\mathfrak{n}}(s)\right\|_{\rho} \leq \left\|\frac{dv^{\mathfrak{n}}(t)}{dt}\right\|_{\rho}|s-t| \leq M_{\dot{v}}|s-t| \quad \text{for all } s,t \in \mathbb{R},$$

i.e., $\{v^{\mathfrak{n}}(t)\}_{\mathfrak{n}\geq 1}$ is equicontinuous on $\mathcal{C}(\mathbb{R},\ell^2_\rho)$.

On the other hand following analysis similar to those in the proof of Lemma 16.6 we are able to show that $\{v^{\mathfrak{n}}(t)\}_{\mathfrak{n}\geq 1}$ is precompact in ℓ^2_ρ. Then due to the Ascoli-Arzelà Theorem, $\{v^{\mathfrak{n}}(t)\}_{\mathfrak{n}\geq 1}$ is precompact on $\mathcal{C}(J,\ell^2_\rho)$ for any compact interval $J\subset\mathbb{R}$.

Notice that $v^{\mathfrak{n}}(t)$ are solutions of (16.2) which is bounded and defined on the whole $\mathbb{R}$. Thus $v^{\mathfrak{n}}(t)\in\mathcal{A}$ for all $t\in\mathbb{R}$. Then by the compactness of $\mathcal{A}$ and the Ascoli-Arzelà Theorem again, there is a subsequence $\{v^{\mathfrak{n}_k}(t)\}_{k\geq 1}$ of $\{v^{\mathfrak{n}}(t)\}_{\mathfrak{n}\geq 1}$ and $v^*(t)\in\mathcal{A}$ such that $v^{n_k}(t)\to v^*(t)$ in $\mathcal{C}(J,\ell^2_\rho)$ for any compact interval $J\subset\mathbb{R}$.

Summarizing the above, $u^{\mathfrak{n}_k}(t)\to v^*(t)$ in $\mathcal{C}(J,\ell^2_\rho)$ for any compact interval $J\subset\mathbb{R}$. And it follows directly by Lemma 16.7 that

$$a_{\mathfrak{n}_k} = u^{\mathfrak{n}_k}(0) \to v^*(0) := a \in \mathcal{A} \quad \text{as } k\to\infty.$$

□

Recall that $\text{dist}_{\ell^2_\rho}$ denotes the Hausdorff semi distance between any two subsets of ℓ^2_ρ.

Theorem 16.3. *Let Assumptions 16.1–16.8 hold. Then the global attractors for system* (16.1) *converge to the global attractor for system* (16.2) *upper semi-continuously, i.e.,*

$$\lim_{\mathfrak{n}\to\infty} \text{dist}_{\ell^2_\rho}(\mathcal{A}_{\mathfrak{n}},\mathcal{A}) = 0.$$

Proof. Suppose (for contradiction) that $\lim_{n\to\infty}\text{dist}_{\ell^2_\rho}(\mathcal{A}_{\mathfrak{n}},\mathcal{A})\neq 0$, then by the definition of $\text{dist}_{\ell^2_\rho}$ there exist sequences $\mathfrak{n}_k\to\infty$, $a_{\mathfrak{n}_k}\in\mathcal{A}_{\mathfrak{n}_k}$ and $\epsilon_0>0$ such that

$$\text{dist}_{\ell^2_\rho}(a_{\mathfrak{n}_k},\mathcal{A}) \geq \epsilon_0 > 0.$$

On the other hand, by Lemma 16.9 there exists a subsequence $\{a_{\mathfrak{n}_{k_m}}\}$ of $\{a_{\mathfrak{n}_k}\}$ such that

$$\text{dist}_{\ell^2_\rho}(a_{\mathfrak{n}_{k_m}},\mathcal{A}) \to 0 \quad \text{as } m\to\infty.$$

This contradiction completes the proof. □

16.5 End notes

This chapter is based on the paper by Wang, Han & Kloeden [Wang *et al.* (2020)].

16.6 Problems

Problem 16.1. Can the Hopfield lattice model have steady state solutions in the space ℓ^2_ρ that are not in the space ℓ^2? Give an example.

Problem 16.2. Construct an absorbing set $\tilde{Q}$ in ℓ^2_ρ for the semi-dynamical system $\{\varphi(t)\}_{t\geq 0}$ generated by the equation (16.2) under appropriate dissipative conditions on the function f_i's instead of Assumption 16.7 and show that $\{\varphi(t)\}_{t\geq 0}$ is asymptotically compact on $\tilde{Q}$.

Chapter 17

A random Hopfield lattice model

The goal of this chapter is to investigate the long term dynamics, in particular, the existence of a random attractor for the random Hopfield neural lattice model

$$\mu_i \frac{du_i(t)}{dt} = -\frac{u_i(t)}{\varkappa_i} + \sum_{j=i-n}^{i+n} \lambda_{i,j} f_j(u_j(t)) + g_i(\vartheta_t \omega), \quad i \in \mathbb{Z} \tag{17.1}$$

in the sequence space ℓ^2.

Here the noise term is represented in canonical form by a measure preserving dynamical system $\vartheta = \{\vartheta_t\}_{t\in\mathbb{R}}$ acting on a probability space $(\Omega, \mathcal{F}, \mathbb{P})$ rather than as a specific noise process, see Definition 11.2 in Chapter 11.

17.1 Basic properties of solutions

For $\boldsymbol{u} = (u_i)_{i\in\mathbb{Z}} \in \ell^2$, define the operators $\mathrm{A}\boldsymbol{u} = ((A\boldsymbol{u})_i)_{i\in\mathbb{Z}}$ and $F(\boldsymbol{u}) = (F_i(\boldsymbol{u}))_{i\in\mathbb{Z}}$ by

$$(\mathrm{A}\boldsymbol{u})_i = \frac{1}{\mu_i \varkappa_i} u_i, \qquad F_i(\boldsymbol{u}) = \frac{1}{\mu_i} \sum_{j=i-n}^{i+n} \lambda_{i,j} f_j(u_j).$$

Assumptions on model parameters and activation functions similar to Chapter 16 are assumed here. For the reader's convenience, they are stated below.

Assumption 17.1. The efficacy among neurons is finite, i.e., there exists $M_\lambda > 0$ such that

$$\sup_{i,j\in\mathbb{Z}} |\lambda_{i,j}| \le M_\lambda.$$

Assumption 17.2. There exist $0 < m_\mu \le M_\mu$ and $0 < m_\varkappa \le M_\varkappa$ such that $\mu_i \in [m_\mu, M_\mu]$, $\varkappa_i \in [m_\varkappa, M_\varkappa]$ for each $i \in \mathbb{Z}$.

Assumption 17.3. For each $i \in \mathbb{Z}$, $f_i \in \mathcal{C}^1(\mathbb{R}, \mathbb{R})$, $f_i(0) = 0$, and there exists a continuous function $L(r) \in \mathcal{C}(\mathbb{R}_+, \mathbb{R}_+)$ such that

$$\sup_{i\in\mathbb{Z}} \max_{s\in[-r,r]} |f_i'(s)| \le L(r) \quad \text{for all } r \in \mathbb{R}_+.$$

It follows immediately from the Assumption 17.2 that $A\boldsymbol{u} \in \ell^2$ for every $\boldsymbol{u} \in \ell^2$. In addition, notice that the Assumption 17.3 implies that given any $\boldsymbol{u} = (u_i)_{i\in\mathbb{Z}} \in \ell^2$, for each $i \in \mathbb{Z}$, by the Mean Value Theorem, there exists $s_i \in \mathbb{R}$ with $|s_i| \le |u_i|$ such that

$$|f_i(u_i)| = |f_j'(s_i)u_i| \le L(|u_i|)|u_i| \le L(\|\boldsymbol{u}\|)|u_i|.$$

Then by the Assumptions 17.1 and 17.3

$$\begin{aligned}\sum_{i\in\mathbb{Z}} F_i(\boldsymbol{u})^2 &\le \frac{(2n+1)}{m_\mu^2}\sum_{i\in\mathbb{Z}}\sum_{j=i-n}^{i+n} \lambda_{i,j}^2 f_j^2(u_j)\\ &\le \frac{(2n+1)}{m_\mu^2} M_\lambda^2 L^2(\|\boldsymbol{u}\|)\sum_{i\in\mathbb{Z}}\sum_{j=i-n}^{i+n} u_j^2\\ &\le \frac{(2n+1)^2}{m_\mu^2} M_\lambda^2 \cdot L^2(\|\boldsymbol{u}\|)\cdot\|\boldsymbol{u}\|^2,\end{aligned}$$

and thus $F(\boldsymbol{u}) \in \ell^2$ for every $\boldsymbol{u} \in \ell^2$.

The forcing terms $g_i(\vartheta_t\omega)$ are stochastic processes. In particular, it is assumed that

Assumption 17.4. $t \mapsto g_i(\vartheta_t\omega)$ is continuous for each $i \in \mathbb{Z}$ and satisfies

$$\sum_{i\in\mathbb{Z}} g_i^2(\vartheta_t\omega) < \infty \quad \text{for all } t \in \mathbb{R}.$$

Define $\tilde{\boldsymbol{g}}(\vartheta_t\omega) = (g_i(\vartheta_t\omega)/\mu_i)_{i\in\mathbb{Z}}$. Then by Assumption 17.4, $\tilde{\boldsymbol{g}}(\vartheta_t\omega) \in \mathcal{C}(\mathbb{R}, \ell^2)$, and the lattice system (17.1) can be rewritten as the random ordinary differential equation (RODE):

$$\frac{d\boldsymbol{u}(t,\omega)}{dt} = -A\boldsymbol{u} + F(\boldsymbol{u}) + \tilde{\boldsymbol{g}}(\vartheta_t\omega). \tag{17.2}$$

Theorem 17.1 below states the existence and uniqueness of a global solution to the RODE (17.2). The proof needs the following general Gronwall-like Lemma.

Lemma 17.1. *Let $a(t)$ be non-negative and non-decreasing, $b(t)$ be non-negative and continuous, and $c(t)$ be non-negative and integrable on $[t_0, T]$, and assume that the function $w : [t_0, T] \to [0, \infty)$ satisfies*

$$w(t) \le a(t) + \int_{t_0}^t b(s)w(s)ds + \int_{t_0}^t c(s)\ln\big(w(s)+1\big)w(s)ds, \quad \textit{for all } t \in [t_0, T]$$

Then

$$w(t) \le \max\left\{2a(t)e^{2\int_{t_0}^t b(s)ds}, 2^{e^{2\int_{t_0}^t c(s)ds}} - 1\right\} \quad \textit{for all } t \in [t_0, T].$$

Proof. Since all functions and integrals are non-negative,

$$w(t) \leq 2\max\left\{a(t) + \int_{t_0}^{t} b(s)w(s)\mathrm{d}s, \int_{t_0}^{t} c(s)\ln(w(s)+1)w(s)\mathrm{d}s\right\} \tag{17.3}$$

for all $t \in [t_0, T]$.

With $w(t) \leq 2a(t) + 2\int_{t_0}^{t} b(s)w(s)\mathrm{d}s$ and the assumption that $a(t)$ is non-decreasing it follows directly from Gronwall's inequality that

$$w(t) \leq 2a(t)e^{2\int_{t_0}^{t} b(s)\mathrm{d}s}. \tag{17.4}$$

With $w(t) \leq 2\int_{t_0}^{t} c(s)\ln(w(s)+1)w(s)\mathrm{d}s$ and the assumption that $c(t)$ is non-negative

$$\begin{aligned} w(t) + 1 &\leq 2\int_{t_0}^{t} c(s)\ln\big(w(s)+1\big)(w(s)+1)\mathrm{d}s + 1 \\ &\leq 2\left(\int_{t_0}^{t} c(s)\ln[w(s)+1](w(s)+1)\mathrm{d}s + 1\right). \end{aligned}$$

Then, using a generalised Gronwall-like integral inequality [Engler (1989)], gives $w(t) + 1 \leq 2^{e^{2\int_{t_0}^{t} c(s)\mathrm{d}s}}$, i.e.,

$$w(t) \leq 2^{e^{2\int_{t_0}^{t} c(s)\mathrm{d}s}} - 1. \tag{17.5}$$

Inserting the inequalities (17.4) and (17.5) into (17.3) gives immediately the desired assertion. □

Theorem 17.1. *Let Assumptions 17.1–17.4 hold. In addition, assume that there exist* $\kappa_1, \kappa_2 > 0$ *such that*

$$L(r) \leq \kappa_1 \ln(r^2+1) + \kappa_2 \qquad \text{for all } r \geq 0. \tag{17.6}$$

Then for any $\omega \in \Omega$, $t_0 \in \mathbb{R}$ *and any initial data* $\boldsymbol{u}_o = (u_{o,i})_{i\in\mathbb{Z}} \in \ell^2$, *the RODE (17.2) has a unique solution* $\boldsymbol{u}(\cdot; t_0, \omega, \boldsymbol{u}_o) \in C([t_0,\infty), \ell^2)$ *with* $\boldsymbol{u}(t_0; t_0, \omega, \boldsymbol{u}_o) = \boldsymbol{u}_o$. *In addition, the solution* $\boldsymbol{u}(\cdot; t_0, \omega, \boldsymbol{u}_o)$ *is continuous in* $\boldsymbol{u}_o \in \ell^2$.

Proof. Given $\omega \in \Omega$ and $\boldsymbol{u} \in \ell^2$, let

$$\mathfrak{F}(\boldsymbol{u}, \omega) = -\mathrm{A}\boldsymbol{u} + F(\boldsymbol{u}) + \tilde{\boldsymbol{g}}(\omega).$$

Then $\mathfrak{F}$ is continuous in $\boldsymbol{u}$ and measurable in ω from $\ell^2 \times \Omega$ into ℓ^2. First for any $\boldsymbol{u}, \boldsymbol{v} \in \ell^2$,

$$\|\mathrm{A}\boldsymbol{u} - \mathrm{A}\boldsymbol{v}\|^2 = \sum_{i\in\mathbb{Z}} \frac{1}{\mu_i^2 \varkappa_i^2}(u_i - v_i)^2 \leq \frac{1}{m_\mu^2 m_\varkappa^2}\|\boldsymbol{u} - \boldsymbol{v}\|^2. \tag{17.7}$$

Second, by using Assumptions 17.2 and 17.3

$$\begin{aligned}\|F(\boldsymbol{u})-F(\boldsymbol{v})\|^2 &= \sum_{i\in\mathbb{Z}}\frac{1}{\mu_i^2}\left(\sum_{j=i-\mathfrak{n}}^{i+\mathfrak{n}}\lambda_{i,j}\left(f_j(u_j)-f_j(v_j)\right)\right)^2\\ &\le \frac{1}{b_\mu^2}(2\mathfrak{n}+1)\sum_{i\in\mathbb{Z}}\sum_{j=i-\mathfrak{n}}^{i+\mathfrak{n}}\lambda_{i,j}^2\left(f_j(u_j)-f_j(v_j)\right)^2\\ &\le \frac{M_\lambda^2}{b_\mu^2}(2\mathfrak{n}+1)\sum_{i\in\mathbb{Z}}\sum_{j=i-\mathfrak{n}}^{i+\mathfrak{n}}L^2(2\max\{|u_j|,|v_j|\})|u_j-v_j|^2\\ &\le \frac{M_\lambda^2}{b_\mu^2}(2\mathfrak{n}+1)^2L^2(2\max\{\|\boldsymbol{u}\|,\|\boldsymbol{v}\|\})\|\boldsymbol{u}-\boldsymbol{v}\|^2. \end{aligned}\tag{17.8}$$

Hence by summing (17.7) and (17.8)

$$\begin{aligned}\|\mathfrak{F}(\boldsymbol{u},\omega)-\mathfrak{F}(\boldsymbol{v},\omega)\|^2 &\le 2\|\mathrm{A}\boldsymbol{u}-\mathrm{A}\boldsymbol{v}\|^2+2\|F(\boldsymbol{u})-F(\boldsymbol{v})\|^2\\ &\le \frac{2}{m_\mu^2}\left(\frac{1}{m_\varkappa^2}+M_\lambda^2(2\mathfrak{n}+1)^2L^2(2\max\{\|\boldsymbol{u}\|,\|\boldsymbol{v}\|\})\right)\|\boldsymbol{u}-\boldsymbol{v}\|^2,\end{aligned}$$

which implies that $\mathfrak{F}(\boldsymbol{u},\omega)$ is locally Lipschitz in $\boldsymbol{u}$ uniformly in ω. Therefore the RODE (17.2) has a unique local solution.

It will now be shown that for any given $T>t_0$ the solution $\boldsymbol{u}(t)$ is bounded for all $t\in[t_0,T]$. To this end, multiply both sides of the equation (17.1) by $u_i(t)$ and sum over all $i\in\mathbb{Z}$ to give

$$\frac{1}{2}\frac{d\,\|\boldsymbol{u}(t,\omega)\|^2}{dt}=-\sum_i\frac{u_i^2}{\mu_i\varkappa_i}+\sum_i\frac{1}{\mu_i}\sum_{j=i-\mathfrak{n}}^{i+\mathfrak{n}}\lambda_{i,j}u_if_j(u_j)+\sum_{i\in\mathbb{Z}}\frac{u_i}{\mu_i}g_i(\vartheta_t\omega).\tag{17.9}$$

First by Assumption 17.2,

$$-\sum_{i\in\mathbb{Z}}\frac{u_i^2}{\mu_i\varkappa_i}\le-\frac{\|\boldsymbol{u}\|^2}{M_\mu M_\varkappa}.\tag{17.10}$$

Then by using the fact that $xy\le\frac{1}{2}(ax^2+\frac{1}{a}y^2)$, for some $a>0$ (to be determined later),

$$\begin{aligned}\sum_{i\in\mathbb{Z}}\frac{u_i}{\mu_i}g_i(\vartheta_t\omega) &\le \sum_{i\in\mathbb{Z}}\frac{1}{2}\left(au_i^2+\frac{g_i^2(\vartheta_t\omega)}{a\mu_i^2}\right)\le\frac{a}{2}\|\boldsymbol{u}\|^2+\frac{1}{2a}\sum_{i\in\mathbb{Z}}\frac{g_i^2(\vartheta_t\omega)}{\mu_i^2}\\ &\le \frac{a}{2}\|\boldsymbol{u}\|^2+\frac{1}{2am_\mu^2}\sum_{i\in\mathbb{Z}}g_i^2(\vartheta_t\omega).\end{aligned}\tag{17.11}$$

Then by Assumptions 17.1 and 17.2,

$$\begin{aligned}\sum_i\frac{1}{\mu_i}\sum_{j=i-\mathfrak{n}}^{i+\mathfrak{n}}\lambda_{i,j}u_if_j(u_j) &\le \frac{M_\lambda}{m_\mu}\sum_i\sum_{j=i-\mathfrak{n}}^{i+\mathfrak{n}}|u_if_j(u_j)|\\ &= \frac{M_\lambda}{m_\mu}\sum_i\sum_{j=i-\mathfrak{n}}^{i+\mathfrak{n}}|u_iu_j|\cdot|f'(s_j)|\end{aligned}$$

for some s_j with $|s_j| \leq |u_j|$. Thus by Assumption 17.2, Assumption 17.3 and the fact that $xy \leq \frac{1}{2}(x^2+y^2)$ we obtain

$$\sum_{i\in\mathbb{Z}} \frac{1}{\mu_i} \sum_{j=i-\mathfrak{n}}^{i+\mathfrak{n}} \lambda_{i,j} u_i f_j(u_j) \leq \frac{M_\lambda}{2m_\mu} L(\|\boldsymbol{u}\|) \sum_{i\in\mathbb{Z}} \sum_{j=i-\mathfrak{n}}^{i+\mathfrak{n}} (u_j^2+u_i^2)$$
$$\leq (2\mathfrak{n}+1)\frac{M_\lambda}{m_\mu} L(\|\boldsymbol{u}\|)\|\boldsymbol{u}\|^2. \tag{17.12}$$

Inserting estimations (17.10)–(17.12) into (17.9) gives

$$\frac{\mathrm{d}\,\|\boldsymbol{u}(t)\|^2}{\mathrm{d}t} \leq 2\left(-\frac{1}{M_\mu M_\varkappa} + \frac{(2\mathfrak{n}+1)M_\lambda L(\|\boldsymbol{u}\|)}{m_\mu} + \frac{a}{2}\right)\|\boldsymbol{u}\|^2 + \frac{1}{ab_\mu^2}\sum_{i\in\mathbb{Z}} g_i^2(\vartheta_t\omega).$$

Now pick $a > 0$ such that $-\frac{1}{M_\mu M_\varkappa} + \frac{a}{2} < 0$. In particular, let $a = \frac{1}{M_\mu M_\varkappa}$, and for simplicity denote

$$K_1 := \frac{1}{M_\mu M_\varkappa}, \quad K_2 := \frac{2(2\mathfrak{n}+1)M_\lambda}{m_\mu}, \quad M(\vartheta_t\omega) := \frac{M_\mu M_\varkappa}{m_\mu^2}\sum_{i\in\mathbb{Z}} g_i^2(\vartheta_t\omega).$$

Then the above inequality becomes

$$\frac{\mathrm{d}\,\|\boldsymbol{u}(t)\|^2}{\mathrm{d}t} \leq -K_1\|\boldsymbol{u}\|^2 + K_2 L(\|\boldsymbol{u}\|)\|\boldsymbol{u}\|^2 + M(\vartheta_t\omega). \tag{17.13}$$

Multiplying both sides of (17.13) by $e^{K_1 t}$ gives

$$\frac{\mathrm{d}\,\left\|e^{K_1 t}\boldsymbol{u}(t)\right\|^2}{\mathrm{d}t} \leq K_2 e^{K_1 t} L(\|\boldsymbol{u}(t)\|)\|\boldsymbol{u}(t)\|^2 + M(\vartheta_t\omega)e^{K_1 t}.$$

Integrating the above inequality from t_0 to t results in

$$e^{K_1 t}\|\boldsymbol{u}(t)\|^2 \leq \|\boldsymbol{u}_o\|^2 e^{K_1 t_0} + K_2\int_{t_0}^t e^{K_1 s} L(\|\boldsymbol{u}(s)\|)\|\boldsymbol{u}(s)\|^2 \mathrm{d}s + \int_{t_0}^t M(\vartheta_s\omega)e^{K_1 s}\mathrm{d}s,$$

and hence

$$\|\boldsymbol{u}(t)\|^2 \leq \|\boldsymbol{u}_o\|^2 e^{-K_1(t-t_0)} + \int_{t_0}^t M(\vartheta_s\omega)e^{-K_1(t-s)}\mathrm{d}s$$
$$+K_2\int_{t_0}^t e^{-K_1(t-s)} L(\|\boldsymbol{u}(s)\|)\|\boldsymbol{u}(s)\|^2\mathrm{d}s. \tag{17.14}$$

First, notice that Assumption 17.4 implies that $\int_{t_0}^t M(\vartheta_s\omega)e^{-K_1(t-s)}\mathrm{d}s < \infty$ for all $t \geq t_0$. Then using the condition (17.6) in (17.14) gives

$$\|\boldsymbol{u}(t)\|^2 \leq \|\boldsymbol{u}_o\|^2 + \int_{t_0}^t M(\vartheta_s\omega)e^{-K_1(t-s)}\mathrm{d}s + K_2\kappa_2\int_{t_0}^t \|\boldsymbol{u}(s)\|^2\mathrm{d}s$$
$$+K_2\kappa_1\int_{t_0}^t \ln(\|\boldsymbol{u}(s)\|^2+1)\|\boldsymbol{u}(s)\|^2\mathrm{d}s.$$

Since $M(\vartheta_s\omega) \geq 0$ for all $s \in \mathbb{R}$, $\int_{t_0}^t M(\vartheta_s\omega)e^{-K_1(t-s)}\mathrm{d}s$ is non-negative and non-decreasing. It then follows directly from Lemma 17.1 that

$$\|\boldsymbol{u}(t)\|^2 \leq \max\left\{2(\|\boldsymbol{u}_o\|^2 + \hat{M}(\vartheta_t\omega))e^{2\kappa_2 K_2(t-t_0)}, 2^{e^{2\kappa_1 K_2(t-t_0)}} - 1\right\} \tag{17.15}$$

for all $t \in [t_0, T]$, where

$$\hat{M}(t,\omega) := \int_{t_0}^{t} M(\vartheta_s\omega)e^{-K_1(t-s)}\mathrm{d}s < \infty \qquad \text{for all } \ t \geq t_0.$$

This shows that the solution exists for all $t \geq t_0$.

It remains to show the continuous dependence of solutions on initial data. To this end, let $\boldsymbol{u}_o, \boldsymbol{v}_o \in \ell^2$ and consider two solutions of system (17.2) with initial value $\boldsymbol{u}(t_0) = \boldsymbol{u}_o$ and $\boldsymbol{u}(t_0) = \boldsymbol{v}_o$, respectively. Write

$$X(t) = (X_i(t))_{i\in\mathbb{Z}} := \boldsymbol{u}(t;t_0,\omega,\boldsymbol{u}_o), \qquad Y(t) = (Y_i(t))_{i\in\mathbb{Z}} = \boldsymbol{u}(t;t_0,\omega,\boldsymbol{v}_o),$$

and define $h(t) = (h_i(t))_{i\in\mathbb{Z}} = X(t) - Y(t)$. Then $h(t)$ satisfies the random lattice system

$$\frac{\mathrm{d}h_i(t)}{\mathrm{d}t} = -\frac{1}{\mu_i \varkappa_i} h_i + \frac{1}{\mu_i}\sum_{j=i-\mathfrak{n}}^{i+\mathfrak{n}} \lambda_{i,j}\left(f_j(X_j) - f_j(Y_j)\right), \quad i \in \mathbb{Z}. \tag{17.16}$$

Multiplying the equation (17.16) by $h_i(t)$ and summing over $i \in \mathbb{Z}$ gives

$$\begin{aligned}
\frac{1}{2}\frac{\mathrm{d}}{\mathrm{d}t}\|h(t)\|^2 &\leq -\sum_{i\in\mathbb{Z}} \frac{1}{\mu_i\varkappa_i}h_i^2 + \sum_{i\in\mathbb{Z}}\frac{1}{\mu_i}\sum_{j=i-\mathfrak{n}}^{i+\mathfrak{n}} \lambda_{i,j}\left(f_j(X_j) - f_j(Y_j)\right)h_i \\
&\leq -\frac{1}{M_\mu M_\varkappa}\|h(t)\|^2 + \frac{M_\lambda}{m_\mu}\sum_{i\in\mathbb{Z}}\sum_{j=i-\mathfrak{n}}^{i+\mathfrak{n}} |f_j(X_j) - f_j(Y_j)| \cdot |h_i|.
\end{aligned}$$

By Assumption 17.3 again,

$$|f_j(X_j) - f_j(Y_j)| \leq L(\max\{|X_j|,|Y_j|\})|h_j| \leq L_T|h_j|,$$

where L_T is a constant which depends on T as in (17.15). Therefore,

$$\begin{aligned}
\frac{\mathrm{d}}{\mathrm{d}t}\|h(t)\|^2 &\leq -\frac{2}{M_\mu M_\varkappa}\|h(t)\|^2 + \frac{4M_\lambda L_T}{m_\mu}\sum_{i\in\mathbb{Z}}\sum_{j=i-\mathfrak{n}}^{i+\mathfrak{n}} h_i^2 \\
&\leq \left(-\frac{2}{M_\mu M_\varkappa} + \frac{4(2\mathfrak{n}+1)M_\lambda L_T}{m_\mu}\right)\|h(t)\|^2.
\end{aligned} \tag{17.17}$$

Integrating the above inequality gives

$$\|h(t)\|^2 \leq e^{\left(-\frac{2}{M_\mu M_\varkappa} + \frac{4(2\mathfrak{n}+1)M_\lambda L_T}{m_\mu}\right)(t-t_0)} \cdot \|h(0)\|^2,$$

which implies that

$$\sup_{t\in[t_0,T]} \|X(t) - Y(t)\|^2 \leq \max\left\{ e^{2\left(-\frac{2}{M_\mu M_\varkappa} + \frac{4(2\mathfrak{n}+1)M_\lambda L_T}{m_\mu}\right)(T-t_0)}, 1\right\} \cdot \|\boldsymbol{u}_o - \boldsymbol{v}_o\|^2.$$

The solution depends continuously on the initial data, and this completes the proof of Theorem 17.1. □

It is straightforward to check that

$$u(t+t_0;t_0,\omega,u_o)=u(t;0,\vartheta_{t_0}\omega,u_o),\qquad \text{for all } t\geq 0,\ u_0\in\ell^2,\ \omega\in\Omega.$$

Thus a continuous random dynamical system $\{\varphi(t,\omega)\}_{t\geq 0,\,\omega\in\Omega}$ can be defined by

$$\varphi(t,\omega,u_o)=u(t;0,\omega,u_o),\qquad \text{for all } t\geq 0,\ u_o\in\ell^2,\ \omega\in\Omega.$$

Henceforth, write $u(t;\omega,u_o)$ instead of $u(t;0,\omega,u_o)$.

17.2 Existence of random attractors

The existence of attractors for the random dynamical system $\{\varphi(t,\omega)\}_{t\geq 0,\omega\in\Omega}$ defined by the solutions to the RODE (17.2) is established in this section. First, a closed and bounded absorbing set for $\varphi(t,\omega)$ will be constructed. Then the asymptotic compactness of the absorbing set will be proved.

For simplification of exposition, throughout this section it will be assumed that $L(r)\equiv L>0$ in Assumption 17.3. In addition, it will be assumed that the functions f_i also satisfy the following dissipative condition.

Assumption 17.5. There exist $\alpha\geq 0$ and $\beta=(\beta_i)_{i\in\mathbb{Z}}\in\ell^2$ such that

$$sf_i(s)\leq -\alpha s^2+\beta_i^2\quad \text{for all } i\in\mathbb{Z},\quad s\in\mathbb{R}.$$

Lemma 17.2. *Suppose that Assumptions 17.1–17.5 hold. Then the continuous random dynamical system $\{\varphi(t,\omega)\}_{t\geq 0,\omega\in\Omega}$ generated by the RODE (17.2) has a random absorbing set $\mathcal{Q}=\{Q(\omega)\}_{\omega\in\Omega}$ in ℓ^2 provided*

$$m_\lambda:=\inf_{i\in\mathbb{Z}}\min_{\substack{j=i-\mathfrak{n},\dots,i+\mathfrak{n}\\ \lambda_{i,j}\neq 0}}|\lambda_{i,j}|>0; \tag{17.18}$$

$$\delta:=\frac{1}{M_\mu M_\varkappa}+\frac{\alpha(2\mathfrak{n}+1)m_\lambda}{M_\mu}-(2\mathfrak{n}+1)\frac{LM_\lambda}{m_\mu}\left(\frac{9}{2}+\frac{L}{\alpha}\right)>0. \tag{17.19}$$

Proof. Start from the equation (17.9) and use the inequalities (17.10) and (17.11). It then remains to estimate $\sum_{i\in\mathbb{Z}}\frac{1}{\mu_i}\sum_{j=i-\mathfrak{n}}^{i+N}\lambda_{i,j}u_if_j(u_j)$. To this end first note that

$$\sum_{j=i-\mathfrak{n}}^{i+\mathfrak{n}}\lambda_{i,j}u_if_j(u_j)=\sum_{\substack{j=i-\mathfrak{n},\dots,i+\mathfrak{n}\\ \lambda_{i,j}>0}}\lambda_{i,j}u_if_j(u_j)+\sum_{\substack{j=i-\mathfrak{n},\dots,i+\mathfrak{n}\\ \lambda_{i,j}<0}}\lambda_{i,j}u_if_j(u_j). \tag{17.20}$$

Also, by Assumption 17.5,

$$u_jf_j(u_j)\leq -\alpha u_j^2+\beta_j^2,\qquad \text{for all } j\in\mathbb{Z}.$$

Then by Assumption 17.3 with $L(r)\equiv L$ and the fact that $xy\leq\frac{1}{2}\left(\frac{x^2}{a}+ay^2\right)$,

$$\begin{aligned}f_j(u_j)\cdot(u_i-u_j)&=f_j'(\zeta_j)u_j(u_i-u_j)\quad \text{for some } \zeta_j \text{ with } |\zeta_j|\leq|u_j|\\ &\leq L\cdot\left(\frac{u_i^2}{2a}+\frac{au_j^2}{2}+u_j^2\right)\end{aligned}$$

for some $a > 0$ (to be determined later). Therefore

$$\begin{aligned} u_i f_j(u_j) &= (u_i - u_j) f_j(u_j) + u_j f_j(u_j) \\ &\leq \left(\frac{a}{2}L - \alpha\right) u_j^2 + L u_j^2 + \frac{L}{2a} u_i^2 + \beta_j^2, \qquad \text{for all } i, j \in \mathbb{Z}. \quad (17.21) \end{aligned}$$

Pick a such that $\frac{a}{2}L - \alpha < 0$ in (17.21). In particular, let $a = \frac{\alpha}{L}$. Then multiplying (17.21) by $\lambda_{i,j} > 0$ and summing over all j with $\lambda_{i,j} > 0$ gives

$$\begin{aligned} &\sum_{\substack{j=i-\mathfrak{n},\ldots,i+\mathfrak{n} \\ \lambda_{i,j}>0}} \lambda_{i,j} u_i f_j(u_j) \\ &\leq -\frac{\alpha}{2} \sum_{\substack{j=i-\mathfrak{n},\ldots,i+\mathfrak{n} \\ \lambda_{i,j}>0}} \lambda_{i,j} u_j^2 + M_\lambda \left(L \sum_{j=i-\mathfrak{n}}^{i+\mathfrak{n}} u_j^2 + \frac{(2\mathfrak{n}+1)L^2}{2\alpha} u_i^2 + \sum_{j=i-\mathfrak{n}}^{i+\mathfrak{n}} \beta_j^2 \right) \\ &\leq -\frac{\alpha m_\lambda}{2} \sum_{j=i-\mathfrak{n}}^{i+\mathfrak{n}} u_j^2 + M_\lambda \left(L \sum_{j=i-\mathfrak{n}}^{i+\mathfrak{n}} u_j^2 + \frac{(2\mathfrak{n}+1)L^2}{2\alpha} u_i^2 + \sum_{j=i-\mathfrak{n}}^{i+\mathfrak{n}} \beta_j^2 \right). \quad (17.22) \end{aligned}$$

Next with $\lambda_{i,j} < 0$

$$\begin{aligned} \sum_{\substack{j=i-\mathfrak{n},\ldots,i+\mathfrak{n} \\ \lambda_{i,j}<0}} \lambda_{i,j} u_i f_j(u_j) &\leq M_\lambda \sum_{\substack{j=i-\mathfrak{n},\ldots,i+\mathfrak{n} \\ \lambda_{i,j}<0}} |u_i||f_j(u_j)| \\ &\leq M_\lambda \sum_{j=i-\mathfrak{n}}^{i+\mathfrak{n}} L |u_i||u_j| \\ &\leq M_\lambda L \sum_{j=i-\mathfrak{n}}^{i+\mathfrak{n}} u_j^2 + \frac{M_\lambda L}{4} (2\mathfrak{n}+1) u_i^2. \quad (17.23) \end{aligned}$$

Using (17.20), (17.22) and (17.23) then gives

$$\begin{aligned} \sum_{j=i-\mathfrak{n}}^{i+\mathfrak{n}} \lambda_{i,j} u_i f_j(u_j) \leq -\frac{\alpha m_\lambda}{2} \sum_{j=i-\mathfrak{n}}^{i+\mathfrak{n}} u_j^2 + 2M_\lambda L \sum_{j=i-\mathfrak{n}}^{i+\mathfrak{n}} u_j^2 \\ + M_\lambda \frac{(2\mathfrak{n}+1)L}{4\alpha} (2L+\alpha) u_i^2 + M_\lambda \sum_{j=i-\mathfrak{n}}^{i+\mathfrak{n}} \beta_j^2. \quad (17.24) \end{aligned}$$

Multiplying both sides of (17.24) by $\frac{1}{\mu_i}$ and summing over $i \in \mathbb{Z}$,

$$\begin{aligned} \sum_{i\in\mathbb{Z}} \frac{1}{\mu_i} \sum_{j=i-\mathfrak{n}}^{i+\mathfrak{n}} \lambda_{i,j} u_i f_j(u_j) \leq -\frac{\alpha m_\lambda}{2M_\mu} \sum_{i\in\mathbb{Z}} \sum_{j=i-\mathfrak{n}}^{i+\mathfrak{n}} u_j^2 + \frac{2M_\lambda}{m_\mu} L \sum_{i\in\mathbb{Z}} \sum_{j=i-\mathfrak{n}}^{i+\mathfrak{n}} u_j^2 \\ + \frac{M_\lambda}{m_\mu} \frac{(2\mathfrak{n}+1)L}{4\alpha} (2L+\alpha) \sum_{i\in\mathbb{Z}} u_i^2 + \frac{M_\lambda}{m_\mu} \sum_{i\in\mathbb{Z}} \sum_{j=i-\mathfrak{n}}^{i+\mathfrak{n}} \beta_j^2. \end{aligned}$$

It then follows from the Assumption (17.18) and a shift of index that

$$\sum_{i\in\mathbb{Z}}\frac{1}{\mu_i}\sum_{j=i-\mathfrak{n}}^{i+\mathfrak{n}}\lambda_{i,j}u_if_j(u_j)\le(2\mathfrak{n}+1)\cdot\left(-\frac{\alpha m_\lambda}{2M_\mu}+\frac{M_\lambda L}{4\alpha m_\mu}(2L+\alpha)\right.$$
$$\left.+\frac{2M_\lambda L}{m_\mu}\right)\|\boldsymbol{u}(t)\|^2+(2\mathfrak{n}+1)\frac{M_\lambda}{m_\mu}\|\boldsymbol{\beta}\|^2. \tag{17.25}$$

Now inserting the inequality (17.10), the inequality (17.11) with $a=\frac{1}{M_\mu M_\varkappa}$, and the inequality (17.25) into (17.9) results in

$$\frac{\mathrm{d}\|\boldsymbol{u}(t,\omega)\|^2}{\mathrm{d}t}\le-\delta\|\boldsymbol{u}(t,\omega)\|^2+2(2\mathfrak{n}+1)\frac{M_\lambda}{m_\mu}\|\boldsymbol{\beta}\|^2+M(\vartheta_t\omega), \tag{17.26}$$

where δ is defined as in (17.19) and $M(\vartheta_t\omega)=\frac{M_\mu M_\varkappa}{m_\mu^2}\sum_{i\in\mathbb{Z}}g_i^2(\vartheta_t\omega)$.

Integrating (17.26) from 0 to t gives

$$\begin{aligned}\|\varphi(t,\omega,\boldsymbol{u}_o)\|^2&\le e^{-\delta t}\|\boldsymbol{u}_o\|^2+e^{-\delta t}2(2\mathfrak{n}+1)\frac{M_\lambda}{m_\mu}\|\boldsymbol{\beta}\|^2\int_0^t e^{\delta s}\mathrm{d}s\\&\quad+\int_0^t M(\vartheta_s\omega)e^{-\delta(t-s)}\mathrm{d}s\\&\le e^{-\delta t}\|\boldsymbol{u}_o\|^2+2(2\mathfrak{n}+1)\frac{M_\lambda}{m_\mu\delta}\|\boldsymbol{\beta}\|^2+\int_0^t M(\vartheta_s\omega)e^{-\delta(t-s)}\mathrm{d}s.\end{aligned}$$

Then replacing ω by $\vartheta_{-t}\omega$ in the above inequality we finally obtain

$$\|\varphi(t,\vartheta_{-t}\omega,\boldsymbol{u}_o)\|^2\le e^{-\delta t}\|\boldsymbol{u}_o\|^2+2(2\mathfrak{n}+1)\frac{M_\lambda}{m_\mu\delta}\|\boldsymbol{\beta}\|^2+\int_{-t}^0 M(\vartheta_\tau\omega)e^{\delta\tau}\mathrm{d}\tau.$$

Note that $\int_{-t}^0 M(\vartheta_\tau\,\omega)e^{\delta\tau}\mathrm{d}\tau$ is a tempered random variable due to Assumption 17.4 and the integrability of $e^{\delta\tau}$ for $\tau<0$. For $\omega\in\Omega$, define

$$Q(\omega):=\left\{\boldsymbol{u}\in\ell^2:\|\boldsymbol{u}\|\le\left(2(2\mathfrak{n}+1)\frac{M_\lambda}{m_\mu\delta}\|\boldsymbol{\beta}\|^2+\int_{-\infty}^0 M(\vartheta_\tau\omega)e^{\delta\tau}\mathrm{d}\tau+1\right)^{1/2}\right\}. \tag{17.27}$$

Denote by $\mathfrak{D}(\ell^2)$ the collection of all tempered sets of ℓ^2. Then for any $\mathcal{D}=\{D(\omega):\omega\in\Omega\}\in\mathfrak{D}(\ell^2)$ and given $\boldsymbol{u}_o(\omega)\in D(\omega)$, there exists $T_D(\omega)>0$ such that

$$\|\varphi(t,\vartheta_{-t}\omega,\boldsymbol{u}_o(\vartheta_{-t}\omega))\|^2\le2(2\mathfrak{n}+1)\frac{M_\lambda}{m_\mu\delta}\|\boldsymbol{\beta}\|^2+\int_{-\infty}^0 M(\vartheta_\tau\omega)e^{\delta\tau}\mathrm{d}\tau+1$$

for all $t\ge T_D(\omega)$, i.e., $\varphi(t,\vartheta_{-t}\omega,D(\vartheta_{-t}\omega))\in Q(\omega)$. The proof is complete. □

Next it will be shown that the absorbing set defined in (17.27) is asymptotically compact under the RDS $\{\varphi(t,\omega)\}_{t\ge0,\omega\in\Omega}$ by following the techniques introduced in Chapter 3. This will be done by a tail estimate of solutions, presented in the lemma below. Note that as a particular case of Lemma 17.2, there exists $T_Q(\omega)$ such that

$$\varphi(t,\vartheta_{-t}\omega,Q(\vartheta_{-t}\omega))\in Q(\omega),\qquad\text{for all }\ t\ge T_Q(\omega).$$

Lemma 17.3. *Suppose that Assumptions 17.1–17.5 and 17.18–17.19 hold. Then for any* $\varepsilon > 0$ *there exist* $\hat{T}(\varepsilon, \omega, \mathcal{Q}) \geq T_{\mathcal{Q}}(\omega)$ *and* $I(\varepsilon, \omega) \in \mathbb{N}$ *such that the solution* $\varphi(t, \omega, \boldsymbol{u}_o) = \boldsymbol{u}(t; \omega, \boldsymbol{u}_o) = (u_i(t; \omega, \boldsymbol{u}_o))_{i\in\mathbb{Z}}$ *of the RODE* (17.2) *with* $\boldsymbol{u}_o = (u_{o,i})_{i\in\mathbb{Z}} \in Q(\vartheta_{-t}\omega)$ *satisfies*

$$\sum_{|i|\geq I(\varepsilon,\omega)} |(\varphi(t, \vartheta_{-t}\omega, \boldsymbol{u}_o))_i|^2 = \sum_{|i|\geq I(\varepsilon,\omega)} |u_i(t; \vartheta_{-t}\omega, \boldsymbol{u}_o)|^2 \leq \varepsilon, \ \forall t \geq \hat{T}(\varepsilon, \omega, \mathcal{Q}).$$

Proof. Given $k \in \mathbb{N}$ fixed (to be determine later), let $\eta_k : \mathbb{R}_+ \to [0, 1]$ be the continuous, increasing and sub-additive function defined in (15.13). For any solution $\boldsymbol{u}(t; \omega, \boldsymbol{u}_o) = (u_i(t; \omega, \boldsymbol{u}_o))_{i\in\mathbb{Z}}$ of (17.2), set $\boldsymbol{v} = (v_i)_{i\in\mathbb{Z}}$ where $v_i = \eta_k(|i|)u_i$, then $\boldsymbol{v} \in \ell^2$. Taking the inner product of $\boldsymbol{v}$ with (17.2) gives

$$\begin{aligned}\frac{1}{2}\frac{\mathrm{d}}{\mathrm{d}t}\sum_{i\in\mathbb{Z}} \eta_k(|i|)u_i^2(t; \omega, \boldsymbol{u}_o) &= -\sum_{i\in\mathbb{Z}} \eta_k(|i|)\frac{u_i^2}{\mu_i\varkappa_i} + \sum_{i\in\mathbb{Z}} \eta_k(|i|)u_i\frac{1}{\mu_i}g_i(\vartheta_t\omega) \\ &\quad + \sum_{i\in\mathbb{Z}} \frac{1}{\mu_i} \sum_{j=i-\mathfrak{n}}^{i+\mathfrak{n}} \lambda_{i,j}\eta_k(|i|)u_i f_j(u_j). \qquad (17.28)\end{aligned}$$

First by Assumption 17.2,

$$-\sum_{i\in\mathbb{Z}} \eta_k(|i|)\frac{u_i^2}{\mu_i\varkappa_i} \leq -\frac{1}{M_\mu M_\varkappa}\sum_{i\in\mathbb{Z}} \eta_k(|i|)u_i^2. \qquad (17.29)$$

Second, similar to (17.11) there exists some $a > 0$ such that

$$\sum_{i\in\mathbb{Z}} \eta_k(|i|)\frac{u_i}{\mu_i}g_i(\vartheta_t\omega) \leq \frac{a}{2}\sum_{i\in\mathbb{Z}} \eta_k(|i|)u_i^2 + \frac{1}{2am_\mu^2}\sum_{i\in\mathbb{Z}} \eta_k(|i|)g_i^2(\vartheta_t\omega). \qquad (17.30)$$

It remains to estimate the last term of (17.28). To this end, multiply (17.24) by $\frac{1}{\mu_i}\eta_k(|i|)$ and sum over $i \in \mathbb{Z}$ to obtain

$$\begin{aligned}&\sum_{i\in\mathbb{Z}} \eta_k(|i|)\frac{1}{\mu_i}\sum_{j=i-\mathfrak{n}}^{i+\mathfrak{n}} \lambda_{i,j}u_i f_j(u_j) \\ &\leq -\frac{\alpha m_\lambda}{2M_\mu}\sum_{i\in\mathbb{Z}} \eta_k(|i|)\sum_{j=i-\mathfrak{n}}^{i+\mathfrak{n}} u_j^2 + \frac{2M_\lambda}{m_\mu}L\sum_{i\in\mathbb{Z}} \eta_k(|i|)\sum_{j=i-\mathfrak{n}}^{i+\mathfrak{n}} u_j^2 \\ &\quad + \frac{M_\lambda L(2\mathfrak{n}+1)(2L+\alpha)}{4m_\mu\alpha}\sum_{i\in\mathbb{Z}} \eta_k(|i|)u_i^2 + \frac{M_\lambda}{m_\mu}\sum_{i\in\mathbb{Z}} \eta_k(|i|)\sum_{j=i-\mathfrak{n}}^{i+\mathfrak{n}} \beta_j^2. \qquad (17.31)\end{aligned}$$

Choose $k > \mathfrak{n}$. Recall that by Lemma 15.3

$$\eta_k(i) = \eta_k(j), \qquad \text{for all } |i-j| \leq \mathfrak{n}. \qquad (17.32)$$

Then using (17.32) in the inequality (17.31) results in

$$\begin{aligned}
&\sum_{i\in\mathbb{Z}}\eta_k(|i|)\frac{1}{\mu_i}\sum_{j=i-\mathfrak{n}}^{i+\mathfrak{n}}\lambda_{i,j}u_i f_j(u_j)\\
&\le -\frac{\alpha m_\lambda}{2M_\mu}\sum_{i\in\mathbb{Z}}\sum_{j=i-\mathfrak{n}}^{i+\mathfrak{n}}\eta_k(|j|)u_j^2+\frac{2M_\lambda}{m_\mu}L\sum_{i\in\mathbb{Z}}\sum_{j=i-\mathfrak{n}}^{i+\mathfrak{n}}\eta_k(|j|)u_j^2\\
&\quad+\frac{M_\lambda L(2\mathfrak{n}+1)(2L+\alpha)}{4m_\mu\alpha}\sum_{i\in\mathbb{Z}}\eta_k(|i|)u_i^2+\frac{M_\lambda}{m_\mu}\sum_{i\in\mathbb{Z}}\eta_k(|i|)\sum_{j=i-\mathfrak{n}}^{i+\mathfrak{n}}\beta_j^2.
\end{aligned}$$

Then it follows from a shift of index that

$$\begin{aligned}
\sum_{i\in\mathbb{Z}}\eta_k(|i|)\frac{1}{\mu_i}\sum_{j=i-\mathfrak{n}}^{i+\mathfrak{n}}\lambda_{i,j}u_i f_j(u_j) &\le \frac{M_\lambda}{m_\mu}\sum_{i\in\mathbb{Z}}\left(\sum_{j=i-\mathfrak{n}}^{i+\mathfrak{n}}\eta_k(|j|)\right)\beta_i^2\\
&+(2\mathfrak{n}+1)\left(-\frac{\alpha m_\lambda}{2M_\mu}+\frac{2M_\lambda L}{m_\mu}+\frac{M_\lambda(2L^2+\alpha L)}{4m_\mu\alpha}\right)\sum_{i\in\mathbb{Z}}\eta_k(|i|)u_i^2. \quad (17.33)
\end{aligned}$$

Inserting estimations (17.29), (17.30) with $a=\frac{1}{M_\mu M_\varkappa}$ and (17.33) in (17.28),

$$\begin{aligned}
\frac{\mathrm{d}}{\mathrm{d}t}\sum_{i\in\mathbb{Z}}\eta_k(|i|)u_i^2(t) &\le -\delta\sum_{i\in\mathbb{Z}}\eta_k(|i|)u_i^2+\frac{2M_\lambda}{m_\mu}\sum_{i\in\mathbb{Z}}\left(\sum_{j=i-\mathfrak{n}}^{i+\mathfrak{n}}\eta_k(|j|)\right)\beta_i^2\\
&\quad+\frac{M_\mu M_\varkappa}{m_\mu^2}\sum_{i\in\mathbb{Z}}\eta_k(|i|)g_i^2(\vartheta_t\omega),
\end{aligned}$$

where δ is as defined in (17.19). Integrating the above inequality from 0 to t, then replacing ω by $\vartheta_{-t}\omega$ gives

$$\begin{aligned}
&\sum_{i\in\mathbb{Z}}\eta_k(|i|)u_i^2(t;\vartheta_{-t}\omega,\boldsymbol{u}_o)\le e^{-\delta t}\sum_{i\in\mathbb{Z}}\eta_k(|i|)u_{o,i}^2(\vartheta_{-t}\omega)\\
&\quad+\frac{2M_\lambda}{\delta m_\mu}\sum_{i\in\mathbb{Z}}\left(\sum_{j=i-\mathfrak{n}}^{i+\mathfrak{n}}\eta_k(|j|)\right)\beta_i^2+\frac{M_\mu M_\varkappa}{m_\mu^2}\int_{-t}^0\sum_{i\in\mathbb{Z}}\eta_k(|i|)g_i^2(\vartheta_\tau\omega)e^{\delta\tau}d\tau. \quad (17.34)
\end{aligned}$$

First notice that $\sum_{i\in\mathbb{Z}}\eta_k(|i|)u_{o,i}^2(\vartheta_{-t}\omega)\le\|\boldsymbol{u}_o(\vartheta_{-t}\omega)\|^2$. Then for any $\varepsilon>0$, there exists $\hat{T}(\varepsilon,\omega)>0$ such that

$$e^{-\delta t}\sum_{i\in\mathbb{Z}}\eta_k(|i|)u_{o,i}^2(\vartheta_{-t}\omega)<\frac{\varepsilon}{3}\quad\text{for all } t\ge\hat{T}(\varepsilon,\omega). \quad (17.35)$$

Second, since $\boldsymbol{\beta}\in\ell^2$, given any $\varepsilon>0$ there exists $N_1(\varepsilon)>0$ such that

$$\sum_{|i|\ge N_1(\varepsilon)}\beta_i^2<\frac{\delta m_\mu}{6M_\lambda(2\mathfrak{n}+1)}\varepsilon.$$

Pick k such that $k>N_1(\varepsilon)+\mathfrak{n}$. Then

$$\sum_{|j|\ge k}\beta_i^2<\frac{\delta m_\mu}{6M_\lambda(2\mathfrak{n}+1)}\varepsilon,\quad\text{for each } j=i-\mathfrak{n},\dots,i,\dots,i+\mathfrak{n},$$

and consequently

$$\frac{2M_\lambda}{\delta m_\mu}\sum_{i\in\mathbb{Z}}\left(\sum_{j=i-\mathfrak{n}}^{i+\mathfrak{n}}\eta_k(|j|)\right)\beta_i^2 \le \frac{2M_\lambda}{\delta m_\mu}\left(\sum_{|i-\mathfrak{n}|\ge k}\beta_i^2+\cdots+\sum_{|i+\mathfrak{n}|\ge k}\beta_i^2\right)$$
$$\le \frac{2M_\lambda}{\delta m_\mu}(2\mathfrak{n}+1)\cdot\frac{\delta m_\mu}{6M_\lambda(2\mathfrak{n}+1)}\varepsilon = \frac{\varepsilon}{3}. \quad (17.36)$$

Next by using Assumption 17.4, for any $\varepsilon > 0$ there exists $N_2(\varepsilon,\omega) > 0$ such that

$$\sum_{i\in\mathbb{Z}}\eta_k(|i|)g_i^2(\vartheta_t\omega) \le \sum_{|i|\ge N_2} g_i^2(\vartheta_t\omega) < \frac{\delta m_\mu^2}{3M_\mu M_\varkappa}\varepsilon \quad \text{for all } \ t\in\mathbb{R}.$$

Therefore

$$\frac{M_\mu M_\varkappa}{m_\mu^2}\int_{-t}^{0}\sum_{i\in\mathbb{Z}}\eta_k(|i|)g_i^2(\vartheta_\tau\omega)e^{\delta\tau} \le \frac{M_\mu M_\varkappa}{m_\mu^2}\cdot\frac{\delta m_\mu^2}{3M_\mu M_\varkappa}\varepsilon\cdot\int_{-t}^{0}e^{\delta\tau}\mathrm{d}\tau < \frac{\varepsilon}{3}. \quad (17.37)$$

In summary, letting $k := \max\{N_1(\varepsilon)+\mathfrak{n}, N_2(\varepsilon,\omega)\}$, and applying (17.35)–(17.37) to (17.34),

$$\sum_{i\in\mathbb{Z}}\eta_k(|i|)u_i^2(t;\vartheta_{-t}\omega,\boldsymbol{u}_o) < \varepsilon \quad \text{for all } \ t\ge\hat{T}(\varepsilon,\omega),$$

which implies that

$$\sum_{|i|\ge 2k} u_i^2(t,\vartheta_{-t}\omega,\boldsymbol{u}_o) \le \sum_{i\in\mathbb{Z}}\eta_k(|i|)u_i^2(t,\vartheta_{-t}\omega,\boldsymbol{u}_o) < \varepsilon \quad \text{for all } t\ge\hat{T}(\varepsilon,\omega).$$

The proof is complete. □

Lemma 17.4. *Suppose that Assumptions 17.1–17.5 and 17.18–17.19 hold. Then the absorbing set $\mathcal{Q} = \{Q(\omega)\}_{\omega\in\Omega}$ defined in (17.27) is asymptotically compact under the RDS $\{\varphi(t,\omega)\}_{t\ge 0,\omega\in\Omega}$ defined by solutions of the RODE (17.2).*

Proof. For a sequence $\{t_m\}$ with $\lim_{m\to\infty} t_m = \infty$, let $\boldsymbol{u}_o^{(m)}(\omega) \in Q(\vartheta_{-t_m}\omega) \in \mathfrak{D}(\ell^2)$ and

$$\boldsymbol{u}^{(m)}(\omega) = \varphi(t_m, \vartheta_{-t_m}\,\omega, \boldsymbol{u}_o^{(m)}), \quad m = 1,2,\ldots,$$

where $u_i^{(m)} = \varphi_i(t_m,\vartheta_{-t_m}\,\omega,\boldsymbol{u}_o^{(m)})$ for $i\in\mathbb{Z}$.

First since $\lim_{m\to\infty} t_m = \infty$ there exists $N_1(\omega, Q)\in\mathbb{N}$ such that $t_m \ge T_{\mathcal{Q}}(\omega)$ if $m \ge N_1(\omega,\mathcal{Q})$ and hence

$$\boldsymbol{u}^{(m)}(\omega) = \varphi(t_m,\vartheta_{-t_m}\omega,\boldsymbol{u}_o^{(m)}) \in Q(\omega), \qquad \text{for all } \ m\ge N_1(\omega,\mathcal{Q}).$$

Then there is a subsequence of $\{\boldsymbol{u}^{(m)}\}$ (still denoted by $\{\boldsymbol{u}^{(m)}\}$), and $\boldsymbol{u}^*\in\ell^2$ such that

$$\boldsymbol{u}^{(m)} = \varphi(t_m,\vartheta_{-t_m}\omega,\boldsymbol{u}_o^{(m)}) \rightharpoonup \boldsymbol{u}^* \text{ weakly in } \ \ell^2.$$

It will now be shown that this convergence is actually strong. Given any $\varepsilon > 0$, by Lemma 17.3 there exist $I_1(\varepsilon,\omega) > 0$ and $N_2(\varepsilon,\omega) > 0$ such that

$$\sum_{|i|\geq I(\varepsilon,\omega)} \left|\varphi_i(t_m, \vartheta_{-t_m}\omega, \boldsymbol{u}_o^{(m)})\right|^2 \leq \frac{\varepsilon^2}{8}, \qquad \text{for all } n \geq N_2(\varepsilon,\omega). \tag{17.38}$$

Also, since $\boldsymbol{u}^* = (u_i^*)_{i\in\mathbb{Z}} \in \ell^2$, there exists $I_2(\varepsilon) > 0$ such that

$$\sum_{|i|\geq I_2(\varepsilon)} |u_i^*|^2 \leq \frac{\varepsilon^2}{8}. \tag{17.39}$$

Set $I(\varepsilon,\omega) := \max\{I_1(\varepsilon,\omega), I_2(\varepsilon,\omega)\}$ and since $\varphi(t_m, \vartheta_{-t_m}\omega, \boldsymbol{u}_o^{(m)}) \rightharpoonup \boldsymbol{u}^*$ in ℓ^2 then, componenet wise,

$$\varphi_i(t_m, \vartheta_{-t_m}\omega, \boldsymbol{u}_o^{(m)}) \longrightarrow u_i^* \quad \text{for} \quad |i| \leq I(\varepsilon,\omega), \quad \text{as } m \to \infty.$$

Therefore there exists $N_3(\varepsilon,\omega) > 0$ such that

$$\sum_{|i|\leq I(\varepsilon,\omega)} |\varphi_i(t_m, \vartheta_{-t_m}\omega, \boldsymbol{u}_o^{(m)}) - u_i^*|^2 \leq \frac{\varepsilon^2}{2}, \qquad \text{for all } m \geq N_3(\epsilon,\omega). \tag{17.40}$$

Set $\hat{N}(\varepsilon,\omega) := \max\{N_1(\varepsilon,\omega), N_2(\varepsilon,\omega), N_3(\epsilon,\omega)\}$, then using (17.38)–(17.40) we have

$$\begin{aligned}\left\|\varphi(t_m, \vartheta_{-t_m}\omega, \boldsymbol{u}_o^{(m)}) - \boldsymbol{u}^*\right\|^2 &= \sum_{|i|\leq I(\varepsilon,\omega)} |\varphi_i(t_m, \vartheta_{-t_m}\omega, \boldsymbol{u}_o^{(m)}) u_i^*|^2 \\ &\quad + \sum_{|i|> I(\varepsilon,\omega)} |\varphi_i(t_m, \vartheta_{-t_m}\omega, \boldsymbol{u}_o^{(m)}) - u_i^*|^2 \\ &\leq \frac{\varepsilon^2}{2} + 2\sum_{|i|> I(\varepsilon,\omega)} |\varphi_i(t_m, \vartheta_{-t_m}\omega, \boldsymbol{u}_o^{(m)})|^2 + |u_i^*|^2 \leq \varepsilon^2.\end{aligned}$$

Hence $\boldsymbol{u}^{(m)}$ (the subsequence) is strongly convergent in ℓ^2, and therefore $\mathcal{Q}$ is asymptotically compact. □

The following theorem follows directly from Lemma 17.2, Lemma 17.4 and Theorem 11.3.

Theorem 17.2. *Suppose that Assumptions 17.1–17.5 and 17.18–17.19 hold. The random dynamical system* $\{\varphi(t,\omega)\}_{t\geq 0,\omega\in\Omega}$ *generated by the RODE* (17.2) *possesses a unique global random attractor with component subsets*

$$\mathcal{A}(\omega) = \bigcap_{\tau \geq \hat{T}_{\mathcal{Q}}(\omega)} \overline{\bigcup_{t\geq\tau} \varphi(t, \vartheta_{-t}\omega, Q(\vartheta_{-t}\omega))},$$

where $\mathcal{Q} = \{Q(\omega)\}_{\omega\in\Omega}$ *is defined as in* (17.27).

Corollary 17.1. *Suppose that Assumptions 17.1–17.4 hold. In addition, assume that all the f_j's are uniformly globally Lipschitz with Lipschitz constant L_f. Then the random attractor for the RDS $\{\varphi(t,\omega)\}_{t\geq 0,\omega\in\Omega}$ consists of singleton component sets $\mathcal{A}(\omega)$ provided*

$$(2\mathfrak{n}+1)M_\lambda L_f < \frac{m_\mu}{2M_\mu M_\varkappa}.$$

Proof. Let $\boldsymbol{u}(t;\omega,\boldsymbol{u}_o) = (u_i(t;\omega,\boldsymbol{u}_o))_{i\in\mathbb{Z}}$ and $\boldsymbol{v}(t;\omega,\boldsymbol{v}_o) = (v_i(t;\omega,\boldsymbol{u}_o))_{i\in\mathbb{Z}}$ be two solutions of the RODE (17.2). Since the f_j's are uniformly globally Lipschitz,

$$|f_j(u_j) - f_j(v_j)| \leq L_f|u_j - v_j|.$$

Thanks to (17.17)

$$\frac{\mathrm{d}}{\mathrm{d}t}\|\boldsymbol{u}(t)-\boldsymbol{v}(t)\|^2 \leq \left(-\frac{2}{M_\mu M_\varkappa} + \frac{4(2\mathfrak{n}+1)M_\lambda L_f}{m_\mu}\right)\cdot\|\boldsymbol{u}(t)-\boldsymbol{v}(t)\|^2, \qquad \text{for all } t\geq 0,$$

which implies that

$$\begin{aligned}\|\boldsymbol{u}(t) - \boldsymbol{v}(t)\|^2 &\leq e^{-\left(\frac{2}{M_\mu M_\varkappa} - \frac{4(2\mathfrak{n}+1)M_\lambda L_f}{m_\mu}\right)t}\|\boldsymbol{u}_o - \boldsymbol{v}_o\|^2 \\ &\longrightarrow 0 \qquad \text{as } t\to\infty.\end{aligned}$$

The proof is complete. □

17.3 End notes

This chapter is based on Han, Usman & Kloeden [Han *et al.* (2019)]. Random ordinary differential equations (RODEs) are discussed in [Han and Kloeden (2017a)].

17.4 Problems

Problem 17.1. Do similar results hold in the weighted norm sequence space ℓ^2_ρ?

Problem 17.2. Give an example of functions f_i that satisfy the Assumption 17.3, with $L(r)$ satisfying (17.6).

Problem 17.3. How much must the proofs be changed when

(1) the randomness is in the weights $\lambda_{i,j}$ instead of in the external term g?
(2) the noise term $g_i(\vartheta_t\omega)$ in (17.1) is replaced by an additive or multiplicative white noise?

PART 6
LDS in Biology

Chapter 18

FitzHugh-Nagumo lattice model

The FitzHugh-Nagumo system is a system of two partial differential equations which arises as a model in neurobiology describing the signal transmission across axons. In this chapter a lattice version of the FitzHugh-Nagumo system is considered in a suitable weighted space of infinite sequences ℓ_ρ^2, since it includes traveling wave solutions.

The lattice FitzHugh-Nagumo system is defined for each $i \in \mathbb{Z}$ by

$$\frac{\mathrm{d}u_i}{\mathrm{d}t} = \nu(u_{i-1} - 2u_i + u_{i+1}) + f(u_i) - v_i, \tag{18.1}$$

$$\frac{\mathrm{d}v_i}{\mathrm{d}t} = \delta(u_i - \mu v_i), \tag{18.2}$$

where $\boldsymbol{u} = (u_i)_{i\in\mathbb{Z}}$, $\boldsymbol{v} = (v_i)_{i\in\mathbb{Z}}$ are real valued bi-infinite sequences of real numbers and ν, δ, μ are positive constants. The nonlinear term f is assumed to satisfy

Assumption 18.1. $f : \mathbb{R} \to \mathbb{R}$ is a continuously differentiable function that satisfies

$$f(0) = 0, \quad f(s)s \le -\alpha s^2 + \beta, \quad f'(s) \le D_f, \qquad \forall\, s \in \mathbb{R},$$

for some positive constants α, β, D_f.

The system (18.1)–(18.2) will be studied in the weighted space of infinite sequences ℓ_ρ^2 defined in Chapter 1. The sequence of positive weights $(\rho_i)_{i\in\mathbb{Z}}$ satisfy the Assumptions 1.1 and 1.2. For the reader's convenience they are restated here.

Assumption 18.2. $\rho_i > 0$ for all $i \in \mathbb{Z}$ and $\rho_\Sigma := \sum_{i\in\mathbb{Z}} \rho_i < \infty$.

Assumption 18.3. There exist positive constants γ_0 and γ_1 such that

$$\rho_{i\pm1} \le \gamma_0 \rho_i, \qquad |\rho_i - \rho_i| \le \gamma_1 \rho_i \quad \text{for all } i \in \mathbb{Z}.$$

Recall that ℓ_ρ^2 is a Hilbert space with the inner product on ℓ_ρ^2 defined by

$$\langle \boldsymbol{u}, \boldsymbol{v} \rangle_\rho := \sum_{i\in\mathbb{Z}} \rho_i u_i v_i \quad \text{for} \quad \boldsymbol{u} = (u_i)_{i\in\mathbb{Z}}, \ \boldsymbol{v} = (v_i)_{i\in\mathbb{Z}} \in \ell_\rho^2,$$

and the norm by

$$\|\boldsymbol{u}\|_\rho := \sqrt{\sum_{i\in\mathbb{Z}} \rho_i u_i^2}.$$

Also, under Assumptions 18.2 and 18.3, $\ell^2 \subset \ell^2_\rho$ and ℓ^2 is dense in ℓ^2_ρ with $\|\boldsymbol{u}\|_\rho \le \sqrt{\rho_\Sigma}\|\boldsymbol{u}\|$ for $\boldsymbol{u} \in \ell^2$.

18.1 Generation of a semi-dynamical system on $\ell^2_\rho \times \ell^2_\rho$

Under Assumption 18.1 the LDS (18.1)–(18.2) can be written as the infinite dimensional ordinary differential equation (ODE) in the sequence space $\ell^2 \times \ell^2$

$$\frac{d\boldsymbol{u}}{dt} = \nu\Lambda\boldsymbol{u} + F(\boldsymbol{u}) - \boldsymbol{v} =: \mathfrak{F}_1(\boldsymbol{u},\boldsymbol{v}), \tag{18.3}$$

$$\frac{d\boldsymbol{v}}{dt} = \delta(\boldsymbol{u} - \mu\boldsymbol{v}) =: \mathfrak{F}_2(\boldsymbol{u},\boldsymbol{v}), \tag{18.4}$$

where Λ is the discrete Laplacian operator, and the Nemytskii operator $F : \ell^2 \to \ell^2$ is defined componentwise by

$$F(\boldsymbol{u}) = (F_i(\boldsymbol{u}))_{i\in\mathbb{Z}} := (f(u_i))_{i\in\mathbb{Z}}, \qquad F_i(\boldsymbol{u}) = f(u_i), \quad i \in \mathbb{Z}.$$

Theorem 18.1. *Under Assumptions 18.1–18.3 one can associate the initial value problem for the lattice system* (18.3)–(18.4) *with a semi-group* $\{\varphi^\delta(t)\}_{t\ge0}$ *in* $\ell^2_\rho \times \ell^2_\rho$ *such that* $\varphi^\delta(t,\boldsymbol{u}_o,\boldsymbol{v}_o)$ *is the unique solution of* (18.3)–(18.4) *with the initial value* $(\boldsymbol{u}(0),\boldsymbol{v}(0)) = (\boldsymbol{u}_o,\boldsymbol{v}_o) \in \ell^2 \times \ell^2$.

Similarly to Chapter 5 the proof of Theorem 18.1 consists of several parts. First, the existence and uniqueness of solutions of (18.3)–(18.4) in $\ell^2 \times \ell^2$ will be established. Such solutions can be considered as functions in $\ell^2_\rho \times \ell^2_\rho$ since $\ell^2 \times \ell^2$ is contained in $\ell^2_\rho \times \ell^2_\rho$, which will be shown to be globally Lipschitz in the norm of $\ell^2_\rho \times \ell^2_\rho$. This allows the solution mapping to be extended from $\mathcal{C}([0,T],\ell^2\times\ell^2)$ to $\mathcal{C}([0,T],\ell^2_\rho\times\ell^2_\rho)$. Since the solutions are also continuous in the initial times and satisfy the semi-group property, this extended mapping in $\ell^2_\rho\times\ell^2_\rho$ forms a semi-group on the space $\ell^2_\rho \times \ell^2_\rho$.

18.1.1 *Existence and uniqueness of solutions in $\ell^2 \times \ell^2$*

As shown in Chapter 5, $F : \ell^2 \to \ell^2$ is Lipschitz on any bounded set $B \subset \ell^2$ with

$$\|F(\boldsymbol{u}) - F(\boldsymbol{v})\| \le L_B\|\boldsymbol{u} - \boldsymbol{v}\|$$

for some local Lipschitz constant $L_B > 0$.

Since the other terms in (18.3)–(18.4) are linear, the mapping $\mathfrak{F} := (\mathfrak{F}_1,\mathfrak{F}_2) : \ell^2 \times \ell^2 \to \ell^2 \times \ell^2$ is locally Lipschitz continuous in both variables. Therefore the system (18.3)–(18.4) has a local solution. Moreover, the solutions are bounded on bounded time intervals as shown in the following lemma.

Lemma 18.1. *Given any initial data* $(\boldsymbol{u}(0), \boldsymbol{v}(0)) = (\boldsymbol{u}_o, \boldsymbol{v}_o) \in \ell^2 \times \ell^2$, *the solution* $(\boldsymbol{u}(t; \boldsymbol{u}_o, \boldsymbol{v}_o), \boldsymbol{v}(t; \boldsymbol{u}_o, \boldsymbol{v}_o))$ *of the lattice system* (18.3)–(18.4) *satisfies*

$$\|\boldsymbol{u}(t)\|^2 + \frac{1}{\delta}\|\boldsymbol{v}(t)\|^2 \leq \left(\|\boldsymbol{u}_o\|^2 + \frac{1}{\delta}|\boldsymbol{v}_o|^2\right) e^{-\lambda_1 t} + \frac{2\beta}{\lambda_1}\left(1 - e^{-\lambda_1 t}\right), \qquad t \geq 0.$$

where $\lambda_1 = 2\min\{\alpha, \mu\} > 0$.

Proof. Taking the inner product in ℓ^2 of (18.3) with $2\boldsymbol{u}$, and (18.4) with $2\boldsymbol{v}$, respectively, gives

$$\begin{aligned}
\frac{\mathrm{d}}{\mathrm{d}t}\|\boldsymbol{u}\|^2 &= 2\nu\langle\Lambda\boldsymbol{u}, \boldsymbol{u}\rangle + 2\langle F(\boldsymbol{u}), \boldsymbol{u}\rangle - 2\langle\boldsymbol{v}, \boldsymbol{u}\rangle \\
&\leq -2\nu\|D^+\boldsymbol{u}\|^2 - 2\alpha\|\boldsymbol{u}\|^2 + 2\beta - 2\langle\boldsymbol{v}, \boldsymbol{u}\rangle, \\
\frac{1}{\delta}\frac{\mathrm{d}}{\mathrm{d}t}\|\boldsymbol{v}\|^2 &= 2\langle\boldsymbol{v}, \boldsymbol{u}\rangle - 2\mu\langle\boldsymbol{v}, \boldsymbol{v}\rangle = -2\mu\|\boldsymbol{v}\|^2 + 2\langle\boldsymbol{v}, \boldsymbol{u}\rangle.
\end{aligned}$$

Hence by properties of Λ and Assumption 18.1,

$$\begin{aligned}
\frac{\mathrm{d}}{\mathrm{d}t}\left(\|\boldsymbol{u}\|^2 + \frac{1}{\delta}\|\boldsymbol{v}\|^2\right) &\leq -2\alpha\|\boldsymbol{u}\|^2 - 2\mu\|\boldsymbol{v}\|^2 + 2\beta \\
&\leq -\lambda_1\left(\|\boldsymbol{u}\|^2 + \frac{1}{\delta}\|\boldsymbol{v}\|^2\right) + 2\beta,
\end{aligned}$$

where $\lambda_1 = 2\min\{\alpha, \mu\} > 0$. The result then follows by the Gronwall inequality. □

Given any initial data $(\boldsymbol{u}_o, \boldsymbol{v}_o) \in \ell^2 \times \ell^2$, the existence and uniqueness of a global solution $(\boldsymbol{u}(\cdot; \boldsymbol{u}_o \boldsymbol{v}_o), \boldsymbol{v}(\cdot; \boldsymbol{u}_o \boldsymbol{v}_o)) \in \mathcal{C}([t_0; \infty), \ell^2 \times \ell^2)$ of the system (18.3)–(18.4) in the Hilbert space $\ell^2 \times \ell^2$ follows by Lemma 18.1 and standard arguments. Moreover, these solutions are continuous in their initial data and satisfy the semigroup property due to the uniqueness of solutions.

18.1.2 *Lipschitz ℓ^2_ρ-continuity of solutions in initial data*

The next lemma establishes the Lipschitz continuity of these solutions in ℓ^2_ρ which is required for the extension of the solution operator from $\ell^2 \times \ell^2$ to $\ell^2_\rho \times \ell^2_\rho$. The proof uses the fact that the mapping F satisfies a one-sided Lipschitz condition in ℓ^2_ρ, see Lemma 5.6 in Chapter 5, i.e.,

$$\langle F(\boldsymbol{u}) - F(\boldsymbol{v}), \boldsymbol{u} - \boldsymbol{v}\rangle_\rho \leq D_f\|\boldsymbol{u} - \boldsymbol{v}\|^2_\rho, \qquad \boldsymbol{u}, \boldsymbol{v} \in \ell^2_\rho. \tag{18.5}$$

Lemma 18.2. *Let* $(\boldsymbol{u}(t; \boldsymbol{u}_o, \boldsymbol{v}_o), \boldsymbol{v}(t; \boldsymbol{u}_o, \boldsymbol{v}_o))$ *and* $(\tilde{\boldsymbol{u}}(t; \tilde{\boldsymbol{u}}_o, \tilde{\boldsymbol{v}}_o), \tilde{\boldsymbol{v}}(t; \tilde{\boldsymbol{u}}_o, \tilde{\boldsymbol{v}}_o))$ *be two solutions of* (18.3)–(18.4) *with the initial data* $(\boldsymbol{u}_o, \boldsymbol{v}_o)$ *and* $(\tilde{\boldsymbol{u}}_o, \tilde{\boldsymbol{v}}_o)$ *in* $\ell^2 \times \ell^2$, *respectively. Then for each $T > 0$ there exists a constant K_T depending on T such that*

$$\|\boldsymbol{u}(t;\boldsymbol{u}_o,\boldsymbol{v}_o)-\tilde{\boldsymbol{u}}(t;\tilde{\boldsymbol{u}}_o,\tilde{\boldsymbol{v}}_o)\|_\rho+\|\boldsymbol{v}(t;\boldsymbol{u}_o,\boldsymbol{v}_o)-\tilde{\boldsymbol{v}}(t;\tilde{\boldsymbol{u}}_o,\tilde{\boldsymbol{v}}_o)\|_\rho$$
$$\leq \frac{1}{\delta}K_T\left(\|\boldsymbol{u}_o-\tilde{\boldsymbol{u}}_o\|_\rho+\|\boldsymbol{v}_o-\tilde{\boldsymbol{v}}_o\|_\rho\right) \quad \text{for} \quad 0\leq t\leq T.$$

Proof. Define $\boldsymbol{h}_u(t) = \boldsymbol{u}(t;\boldsymbol{u}_o,\boldsymbol{v}_o) - \tilde{\boldsymbol{u}}(t;\tilde{\boldsymbol{u}}_o,\tilde{\boldsymbol{v}}_o)$ and $\boldsymbol{h}_v(t) = \boldsymbol{v}(t;\boldsymbol{u}_o,\boldsymbol{v}_o) - \tilde{\boldsymbol{v}}(t;\tilde{\boldsymbol{u}}_o,\tilde{\boldsymbol{v}}_o)$. Then, from equations(18.3)–(18.4)

$$\frac{\mathrm{d}\boldsymbol{h}_u}{\mathrm{d}t} = \nu\Lambda\boldsymbol{h}_u + F(\boldsymbol{u}) - F(\boldsymbol{v}) - \boldsymbol{h}_v.$$

Taking the inner product of (18.3) with $\boldsymbol{h}_u$ in ℓ^2_ρ gives

$$\frac{1}{2}\frac{\mathrm{d}}{\mathrm{d}t}\|\boldsymbol{h}_u\|^2_\rho = \nu\langle\Lambda\boldsymbol{h}_u,\boldsymbol{h}_u\rangle_\rho + \langle F(\boldsymbol{u})-F(\boldsymbol{v}),\boldsymbol{h}_u\rangle_\rho - \langle\boldsymbol{h}_v,\boldsymbol{h}_u\rangle_\rho$$
$$\leq \nu\langle\Lambda\boldsymbol{h}_u,\boldsymbol{h}_u\rangle_\rho + D_f\|\boldsymbol{h}_u\|^2_\rho - \langle\boldsymbol{h}_v,\boldsymbol{h}_u\rangle_\rho,$$

due to the local one-sided Lipschitz property (18.5) of the mapping F in ℓ^2_ρ.

Then using Lemma 5.5 in Chapter 5,

$$\frac{1}{2}\frac{\mathrm{d}}{\mathrm{d}t}\|\boldsymbol{h}_u\|^2_\rho \leq \nu\langle\Lambda\boldsymbol{h}_u,\boldsymbol{h}_u\rangle_\rho + D_f\|\boldsymbol{h}_u\|^2_\rho - \langle\boldsymbol{h}_v,\boldsymbol{h}_u\rangle_\rho$$
$$\leq -\nu\frac{1}{2}\left\|D^+\boldsymbol{h}_u\right\|^2_\rho + \frac{1}{2}\nu\gamma_0\gamma_1^2\|\boldsymbol{h}_u\|^2_\rho + D_f\|\boldsymbol{h}_u\|^2_\rho - \langle\boldsymbol{h}_v,\boldsymbol{h}_u\rangle_\rho,$$

from which it follows

$$\frac{\mathrm{d}}{\mathrm{d}t}\|\boldsymbol{h}_u\|^2_\rho \leq \left(\nu\gamma_0\gamma_1^2+2D_f\right)\|\boldsymbol{h}_u\|^2_\rho - 2\langle\boldsymbol{h}_v,\boldsymbol{h}_u\rangle_\rho. \tag{18.6}$$

Similarly, taking the inner product of (18.4) with $\boldsymbol{h}_v$ in ℓ^2_ρ gives

$$\frac{1}{\delta}\frac{\mathrm{d}}{\mathrm{d}t}\|\boldsymbol{h}_v\|^2_\rho = 2\langle\boldsymbol{h}_v,\boldsymbol{h}_u\rangle_\rho - 2\mu\langle\boldsymbol{h}_v,\boldsymbol{h}_v\rangle_\rho = -2\mu\|\boldsymbol{h}_v\|^2_\rho + 2\langle\boldsymbol{h}_v,\boldsymbol{h}_u\rangle_\rho,$$

which, along with (18.6) give

$$\frac{\mathrm{d}}{\mathrm{d}t}\left(\|\boldsymbol{h}_u\|^2_\rho + \frac{1}{\delta}\|\boldsymbol{h}_v\|^2_\rho\right) \leq \left(\nu\gamma_0\gamma_1^2+2D_f\right)\|\boldsymbol{h}_u\|^2_\rho - 2\mu\|\boldsymbol{h}_v\|^2_\rho$$
$$\leq \lambda_2\left(\|\boldsymbol{h}_u\|^2_\rho + \frac{1}{\delta}\|\boldsymbol{h}_v\|^2_\rho\right), \tag{18.7}$$

where $\lambda_2 = \nu\gamma_0\gamma_1^2 + 2D_f$.

Integrating the differential inequality (18.7) from 0 to t results in

$$\|\boldsymbol{h}_u(t)\|^2_\rho + \frac{1}{\delta}\|\boldsymbol{h}_v(t)\|^2_\rho \leq e^{\lambda_2 t}\left(\|\boldsymbol{h}_u(0)\|^2_\rho + \frac{1}{\delta}\|\boldsymbol{h}_v(0)\|^2_\rho\right), \quad t\geq 0,$$

which implies that

$$\|\boldsymbol{h}_u(t)\|^2_\rho + \frac{1}{\delta}\|\boldsymbol{h}_v(t)\|^2_\rho \leq K_T\left(\|\boldsymbol{h}_u(0)\|^2_\rho + \frac{1}{\delta}\|\boldsymbol{h}_v(0)\|^2_\rho\right), \qquad 0\leq t\leq T,$$

where $K_T = \max\{1, e^{\lambda_2 T}\}$. □

18.1.3 *Existence and uniqueness of solutions in* $\ell_\rho^2 \times \ell_\rho^2$

Fix $T > 0$ in $\mathbb{R}$. Then by Lemma 18.2 there exists a mapping $\mathcal{S}$ from $\ell^2 \times \ell^2$ into $\mathcal{C}\left([t_0, T], \ell_\rho^2 \times \ell_\rho^2\right)$ such that $\mathcal{S}(\boldsymbol{u}_o, \boldsymbol{v}_o)$ for each $(\boldsymbol{u}_o, \boldsymbol{v}_o) \in \ell^2 \times \ell^2$ is the unique solution of the initial value problem for the system (18.3)–(18.4). Further, the mapping $\mathcal{S}$ is continuous from $\ell^2 \times \ell^2 \subset \ell_\rho^2 \times \ell_\rho^2$ into $\mathcal{C}\left([0, T], \ell_\rho^2 \times \ell_\rho^2\right)$.

Since ℓ^2 is dense ℓ_ρ^2, $\mathcal{S}$ can be extended uniquely to a mapping $\hat{\mathcal{S}}$ from $\ell_\rho^2 \times \ell_\rho^2$ into $\mathcal{C}\left([0, T], \ell_\rho^2 \times \ell_\rho^2\right)$. For $t \geq 0$ define the mappings $\varphi^\delta(t) : \ell_\rho^2 \to \ell_\rho^2$ by

$$\varphi^\delta(t, \boldsymbol{u}_o, \boldsymbol{v}_o) = \hat{\mathcal{S}}(\boldsymbol{u}_o, \boldsymbol{v}_o)(t) \quad t \in [0, T], \quad (\boldsymbol{u}_o, \boldsymbol{v}_o) \in \ell_\rho^2 \times \ell_\rho^2. \tag{18.8}$$

Then by Lemmas 18.2 the mapping φ^δ is continuous in $(\boldsymbol{u}_o, \boldsymbol{v}_o)$. It also inherits the semi-group property of the solutions of the system (18.3)–(18.4) in $\ell^2 \times \ell^2$. Hence $\{\varphi^\delta(t)\}_{t\geq 0}$ is a semi-group on $\ell_\rho^2 \times \ell_\rho^2$. The proof of Theorem 18.1 is complete. □
The semi-group defined by (18.8) will be referred to as the semi-group $\{\varphi^\delta(t)\}_{t\geq 0}$ generated by the lattice system (18.3)–(18.4).

18.2 Existence of a global attractor

In this section we show that the semi-group $\{\varphi^\delta(t)\}_{t\geq 0}$ generated by the lattice system (18.3)–(18.4) possesses a global attractor.

18.2.1 *Existence of an absorbing set*

The following assumption is needed for the existence of an absorbing set.

Assumption 18.4. $\alpha > \max\left\{\nu\gamma_0\gamma_1^2, \frac{\nu}{2}(\gamma_0 + 1)\gamma_1\right\}$.

Lemma 18.3. *Let Assumptions 18.1–18.4 hold and let δ satisfy*

$$0 < \delta < \min\left\{1, \frac{2\alpha - \nu\gamma_0\gamma_1^2}{2\mu}\right\}. \tag{18.9}$$

Then the semi-group $\{\varphi^\delta(t)\}_{t\geq 0}$ on $\ell_\rho^2 \times \ell_\rho^2$ generated by the system (18.3)–(18.4) *has a closed and bounded absorbing subset in $\ell_\rho^2 \times \ell_\rho^2$, which is positively invariant.*

Proof. Taking the inner product in ℓ_ρ^2 of the system (18.3) with $\delta\boldsymbol{u}$ and using Lemma 5.5 in Chapter 5 gives

$$\begin{aligned}
\frac{\mathrm{d}}{\mathrm{d}t}\delta\|\boldsymbol{u}\|_\rho^2 &= 2\delta\nu\left\langle \Lambda\boldsymbol{u}, \boldsymbol{u}\right\rangle_\rho + 2\delta\langle F(\boldsymbol{u}), \boldsymbol{u}\rangle_\rho - 2\delta\langle \boldsymbol{v}, \boldsymbol{u}\rangle_\rho \\
&\leq -\delta\nu\left\|D^+\boldsymbol{u}\right\|_\rho + \nu\delta\gamma_0\gamma_1^2\left\|\boldsymbol{u}\right\|_\rho^2 - 2\delta\alpha\left\|\boldsymbol{u}\right\|_\rho^2 + 2\delta\beta\sum_{i\in\mathbb{Z}}\rho_i - 2\delta\langle \boldsymbol{v}, \boldsymbol{u}\rangle_\rho \\
&\leq \delta\nu\gamma_0\gamma_1^2\left\|\boldsymbol{u}\right\|_\rho^2 - 2\delta\alpha\left\|\boldsymbol{u}\right\|_\rho^2 + 2\delta\beta\rho_\Sigma - 2\delta\langle \boldsymbol{v}, \boldsymbol{u}\rangle_\rho,
\end{aligned}$$

while the inner product in ℓ_ρ^2 of the system (18.4) with $\boldsymbol{v}$ gives

$$\frac{\mathrm{d}}{\mathrm{d}t}\|\boldsymbol{v}\|_\rho^2 = 2\delta\left\langle \boldsymbol{v}, \boldsymbol{u}\right\rangle_\rho - 2\delta\mu\left\langle \boldsymbol{v}, \boldsymbol{v}\right\rangle_\rho = -2\delta\mu\|\boldsymbol{v}\|_\rho^2 + 2\delta\left\langle \boldsymbol{v}, \boldsymbol{u}\right\rangle_\rho.$$

Hence

$$
\begin{aligned}
\frac{d}{dt}\left(\delta\|\boldsymbol{u}\|_\rho^2 + \|\boldsymbol{v}\|_\rho^2\right) &\leq \delta\nu\gamma_0\gamma_1^2\,\|\boldsymbol{u}\|_\rho^2 - 2\delta\alpha\,\|\boldsymbol{u}\|_\rho^2 + 2\delta\beta\rho_\Sigma - 2\delta\mu\|\boldsymbol{v}\|_\rho^2 \\
&\leq -2\left(\left(\alpha - \frac{\nu}{2}\gamma_0\gamma_1^2\right)\delta\,\|\boldsymbol{u}\|_\rho^2 + \delta\mu\|\boldsymbol{v}\|_\rho^2\right) + 2\delta\beta\rho_\Sigma \\
&\leq -2\min\left\{\alpha - \frac{\nu}{2}\gamma_0\gamma_1^2, \delta\mu\right\}\left(\delta\|\boldsymbol{u}\|_\rho^2 + \|\boldsymbol{v}\|_\rho^2\right) + 2\delta\beta\rho_\Sigma \\
&\leq -2\delta\mu\left(\delta\|\boldsymbol{u}\|_\rho^2 + \|\boldsymbol{v}\|_\rho^2\right) + 2\delta\beta\rho_\Sigma
\end{aligned}
$$

under the Assumption 18.4 and condition (18.9).

The Gronwall inequality then gives

$$
\delta\|\boldsymbol{u}(t;\boldsymbol{u}_o,\boldsymbol{v}_o)\|_\rho^2 + \|\boldsymbol{v}(t;\boldsymbol{u}_o,\boldsymbol{v}_o)\|_\rho^2 \leq \left(\delta\|\boldsymbol{u}_o\|_\rho^2 + \|\boldsymbol{v}_o\|_\rho^2\right)e^{-2\delta\mu t} + \frac{\beta\rho_\Sigma}{\mu}\left(1 - e^{-2\delta\mu t}\right). \tag{18.10}
$$

Therefore, the closed and bounded subset of $\ell_\rho^2 \times \ell_\rho^2$

$$
Q^\delta := \left\{(\boldsymbol{u},\boldsymbol{v}) \in \ell_\rho^2 \times \ell_\rho^2 : \delta\|\boldsymbol{u}\|_\rho^2 + \|\boldsymbol{v}\|_\rho^2 \leq R^2 := 1 + \frac{\beta\rho_\Sigma}{\mu}\right\} \tag{18.11}
$$

is positively invariant under $\{\varphi^\delta(t)\}_{t\geq 0}$ and is an absorbing set for $\{\varphi^\delta(t)\}_{t\geq 0}$ on $\ell_\rho^2 \times \ell_\rho^2$. □

18.2.2 *Asymptotic tails and asymptotic compactness*

The next step of the proof is to derive an asymptotic tails estimate for the semi-group $\{\varphi^\delta(t)\}_{t\geq 0}$ on $\ell_\rho^2 \times \ell_\rho^2$ in the positive invariant absorbing set Q^δ for each $\delta < 1$.

Lemma 18.4. *For every $\varepsilon > 0$ there exist $T(\varepsilon) > 0$ and $I(\varepsilon) \in \mathbb{N}$ such that*

$$
\sum_{|i|>I(\varepsilon)}\left(\delta\rho_i\,|u_i(t;\boldsymbol{u}_o,\boldsymbol{v}_o)|^2 + \rho_i\,|v_i(t;\boldsymbol{u}_o,\boldsymbol{v}_o)|^2\right) \leq \varepsilon^2
$$

for all $(\boldsymbol{u}_o,\boldsymbol{v}_o) \in Q^\delta$ and $t \geq T(\varepsilon)$.

Proof. Consider the smooth function $\xi : \mathbb{R} \to [0,1]$ defined in (3.8) and recall that there exists a constant C_0 such that $|\xi'(s)| \leq C_0$ for all $s \geq 0$. Then define $\xi_k(s) = \xi(\frac{s}{k})$ for all $s \in \mathbb{R}$ and a fixed $k \in \mathbb{N}$ (its value will be specified later).

Given $(\boldsymbol{u},\boldsymbol{v}) \in \ell_\rho^2 \times \ell_\rho^2$ define $(\tilde{\boldsymbol{u}},\tilde{\boldsymbol{v}}) = (\tilde{u}_i,\tilde{v}_i)_{i\in\mathbb{Z}} \in \ell_\rho^2 \times \ell_\rho^2$ component wise as

$$
\tilde{u}_i := \xi_k(|i|)u_i, \quad \tilde{v}_i := \xi_k(|i|)v_i, \quad i \in \mathbb{Z}.
$$

Taking the inner product in ℓ_ρ^2 of equation (18.3) with $\delta\tilde{\boldsymbol{u}}$ gives

$$
\frac{d}{dt}\delta\langle\boldsymbol{u},\tilde{\boldsymbol{u}}\rangle_\rho = \nu\delta\left\langle\Lambda\boldsymbol{u},\tilde{\boldsymbol{u}}\right\rangle_\rho + \delta\langle F(\boldsymbol{u}),\tilde{\boldsymbol{u}}\rangle + \delta\langle\boldsymbol{v},\tilde{\boldsymbol{u}}\rangle_\rho,
$$

that is,

$$\frac{\mathrm{d}}{\mathrm{d}t}\delta\sum_{i\in\mathbb{Z}}\xi_k(|i|)\rho_i|u_i|^2 = 2\nu\delta\left\langle\Lambda\boldsymbol{u},\tilde{\boldsymbol{u}}\right\rangle_\rho + 2\delta\sum_{i\in\mathbb{Z}}\xi_k(|i|)\rho_i u_i f(u_i) + 2\delta\sum_{i\in\mathbb{Z}}\xi_k(|i|)u_i v_i.$$

Now by Lemma 5.10 in Chapter 5

$$\left\langle\Lambda\boldsymbol{u},\tilde{\boldsymbol{u}}\right\rangle_\rho \le (\gamma_0+1)\gamma_1\sum_{j\in\mathbb{Z}}\xi_k(|j|)\rho_j u_j^2 + \frac{C_0(\gamma_0+1)(\gamma_1+1)}{k}R^2, \tag{18.12}$$

where R is defined in (18.11), since $(\boldsymbol{u}_o,\boldsymbol{v}_o)\in Q^\delta$.

On the other hand, by Assumption 18.1,

$$2\sum_{i\in\mathbb{Z}}\xi_k(|i|)\rho_i u_i f(u_i) \le -2\alpha\sum_{i\in\mathbb{Z}}\xi_k(|i|)\rho_i|u_i|^2 + 2\beta\sum_{i\in\mathbb{Z}}\xi_k(|i|)\rho_i.$$

Adding the above estimates gives

$$\begin{aligned}&\frac{\mathrm{d}}{\mathrm{d}t}\delta\sum_{|i|>k}\rho_i|u_i|^2 + 2\delta\lambda_3\sum_{|i|>k}\rho_i|u_i|^2\\ &\le \frac{2C_0(\gamma_0+1)(\gamma_1+1)\nu\delta}{k}R^2 + 2\delta\beta\sum_{|i|>k}\rho_i + 2\delta\sum_{i\in\mathbb{Z}}\xi_k(|i|)u_i v_i,\end{aligned} \tag{18.13}$$

where $\lambda_3 = \alpha - \nu(\gamma_0+1)\gamma_1 > 0$ due to Assumption 18.4.

Similarly, the inner product in ℓ_ρ^2 of equation (18.4) with $\tilde{\boldsymbol{v}}$ gives

$$\begin{aligned}\frac{\mathrm{d}}{\mathrm{d}t}\sum_{|i|>k}\rho_i|v_i|^2 &= \frac{\mathrm{d}}{\mathrm{d}t}\langle\boldsymbol{v},\tilde{\boldsymbol{v}}\rangle_\rho = 2\delta\left\langle\boldsymbol{u},\tilde{\boldsymbol{v}}\right\rangle_\rho - 2\delta\mu\left\langle\boldsymbol{v},\tilde{\boldsymbol{v}}\right\rangle_\rho\\ &= 2\delta\sum_{i\in\mathbb{Z}}\xi_k(|i|)u_i v_i - 2\delta\mu\sum_{|i|>k}\rho_i|v_i|^2.\end{aligned} \tag{18.14}$$

Combining the estimates (18.12)–(18.14) then gives

$$\begin{aligned}&\frac{\mathrm{d}}{\mathrm{d}t}\left(\delta\sum_{|i|>k}\rho_i|u_i|^2 + \sum_{|i|>k}\rho_i|v_i|^2\right)\\ &\le -2\delta\left(\alpha-\nu(\gamma_0+1)\gamma_1\right)\|\boldsymbol{u}\|_\rho^2 - 2\delta\mu\|\boldsymbol{v}\|_\rho^2 + \frac{\nu C_0(\gamma_0+1)(\gamma_1+1)\delta}{k}R^2 + 2\delta\beta\sum_{|i|>k}\rho_i\\ &\le -2\delta\mu\left(\delta\|\boldsymbol{u}\|_\rho^2 + \|\boldsymbol{v}\|_\rho^2\right) + \frac{\nu C_0(\gamma_0+1)(\gamma_1+1)\delta}{k}R^2 + 2\delta\beta\sum_{|i|>k}\rho_i\\ &\le -2\delta\mu\left(\delta\|\boldsymbol{u}\|_\rho^2 + \|\boldsymbol{v}\|_\rho^2\right) + \frac{\nu C_0(\gamma_0+1)(\gamma_1+1)\delta}{k}R^2 + 2\delta\beta\sum_{|i|>k}\rho_i,\end{aligned}$$

under the condition (18.9).

Since the sum of weights is a convergent series due to Assumption 18.2, then for every $\varepsilon > 0$ there exists $N(\varepsilon)$ such that

$$\frac{\nu C_0(\gamma_0+1)(\gamma_1+1)}{k}R^2 + 2\beta \sum_{|i|>k} \rho_i \leq \varepsilon, \quad k \geq N(\varepsilon),$$

and thus

$$\frac{\mathrm{d}}{\mathrm{d}t}\left(\delta \sum_{|i|>k} \rho_i |u_i|^2 + \sum_{|i|>k} \rho_i |v_i|^2\right) + \delta\mu\left(\delta \sum_{|i|>k} \rho_i |u_i|^2 + \sum_{|i|>k} \rho_i |v_i|^2\right) \leq \delta\varepsilon.$$

It then follows from Gronwall's lemma that

$$\begin{aligned}\delta \sum_{|i|>k} \rho_i |u_i(t)|^2 + \sum_{|i|>k} \rho_i |v_i|(t)^2 &\leq e^{-2\delta\mu t} \sum_{|i|>k} \left(\delta\, |u_{o,i}|_\rho^2 + |v_{o,i}|_\rho^2\right) + \frac{\varepsilon}{2\mu} \\ &\leq e^{-2\delta\mu t}\left(\delta \|\boldsymbol{u}_o\|_\rho^2 + \|\boldsymbol{v}_o\|_\rho^2\right) + \frac{\varepsilon}{2\mu}.\end{aligned}$$

Hence, for $(\boldsymbol{u}_o, \boldsymbol{v}_o) \in Q^\delta$,

$$\delta \sum_{|i|>k} \rho_i |\boldsymbol{u}(t; \boldsymbol{u}_o, \boldsymbol{v}_o)_i|^2 + \sum_{|i|>k} \rho_i |\boldsymbol{v}(t; \boldsymbol{u}_o, \boldsymbol{v}_o)_i|^2 \leq e^{-2\delta\mu t} R^2 + \frac{\varepsilon}{2\mu},$$

so

$$\delta \sum_{|i|>k} \rho_i |\boldsymbol{u}(t; \boldsymbol{u}_o, \boldsymbol{v}_o)_i|^2 + \sum_{|i|>k} \rho_i |\boldsymbol{v}(t; \boldsymbol{u}_o, \boldsymbol{v}_o)_i|^2 \leq \frac{\varepsilon}{\mu}.$$

for $t \geq T(\varepsilon) := \frac{1}{2\delta\mu} \ln \frac{2\mu R^2}{\varepsilon} > 0$. This is the desired pullback asymptotic tails property.

The asymptotic compactness in $\ell_\rho^2 \times \ell_\rho^2$ of φ^δ in the absorbing set Q^δ then follows by Lemma 2.5. This completes the proof of Lemma 18.4. □

The following theorem follows directly from Lemmas 18.3 and 18.4.

Theorem 18.2. *Let Assumptions 18.1–18.4 hold. Then the semi-group* $\{\varphi^\delta(t)\}_{t\geq 0}$ *generated by the lattice system* (18.3)–(18.4) *has a global attractor* $\mathcal{A}_\delta$ *in* $\ell_\rho^2 \times \ell_\rho^2$.

18.3 Limit of the global attractors as $\delta \to 0$

The parameter δ in the equation (18.2) characterises the slow time–fast time nature of the system. The limiting system as $\delta \to 0$ is given by

$$\frac{\mathrm{d}u_i}{\mathrm{d}t} = \nu(u_{i-1} - 2u_i + u_{i+1}) + f(u_i) - v_i, \tag{18.15}$$

$$\frac{\mathrm{d}v_i}{\mathrm{d}t} = 0. \tag{18.16}$$

A particular interest is to see how the behaviour of the global attractors $\mathcal{A}_\delta$ of the lattice system (18.1)–(18.2) changes as $\delta \to 0$, and what this says about the behaviour of the limiting system (18.15)–(18.16).

Let $\mathfrak{S}$ to be the limiting set of $\bigcup_{0<\delta<\delta_0} \mathcal{A}_\delta$ (where δ_0 will be given below) defined by

$$\mathfrak{S} = \left\{ (\boldsymbol{u}, \boldsymbol{v}) \in \ell^2_\rho \times \ell^2_\rho \ : \begin{array}{l} \text{there exists } (\boldsymbol{u}_{\delta_n}, \boldsymbol{v}_{\delta_n}) \in \mathcal{A}_{\delta_n} \text{ with } \delta_n \to 0 \\ \text{such that } (\boldsymbol{u}_{\delta_n}, \boldsymbol{v}_{\delta_n}) \to (\boldsymbol{u}, \boldsymbol{v}) \end{array} \right\},$$

and denote the projection of $\mathfrak{S}$ onto the $\boldsymbol{v}$-space by

$$\mathfrak{V} = \left\{ \boldsymbol{u} \in \ell^2_\rho \ : \text{ there exists } \boldsymbol{v} \in \ell^2_\rho \text{ such that } (\boldsymbol{u}, \boldsymbol{v}) \in \mathfrak{S} \right\}.$$

Finally, let $\mathcal{A}^{\boldsymbol{v}}$ be the global attractor in ℓ^2_ρ of (18.15) for a fixed $\boldsymbol{v} \in \mathfrak{V}$, which exists since (18.15) is dissipative in ℓ^2_ρ under the given assumptions. The proof of the following theorem is similar to that of Theorem 18.1 and is thus omitted here.

Theorem 18.3. *Let Assumptions 18.1–18.4 hold. Then the limiting system* (18.15)–(18.16) *has a local attractor $\mathcal{A}_0$ in $\ell^2_\rho \times \ell^2_\rho$, which is a compact invariant set that attracts all bounded subsets of $\ell^2_\rho \times \mathfrak{V}$ in the norm topology and is characterised by*

$$\mathcal{A}_0 = \left\{ \boldsymbol{u}, \boldsymbol{v}) \in \ell^2_\rho \times \ell^2_\rho \ : \textit{ there exists } \boldsymbol{v} \in \ell^2_\rho \textit{ such that } \boldsymbol{v} \in \mathfrak{S} \textit{ and } \boldsymbol{u} \in \mathcal{A}^{\boldsymbol{v}} \right\}.$$

The main result of this section is the following theorem, which establishes the upper semi-continuous convergence of the global attractors $\mathcal{A}_\delta$ to the local attractor $\mathcal{A}_0$ as $\delta \to 0$.

Theorem 18.4. *Let Assumptions 18.1–18.4 hold. Then the limiting system* (18.15)–(18.16) *has a local attractor $\mathcal{A}_0$ in $\mathcal{A}$ in $\ell^2_\rho \times \ell^2_\rho$ such that*

$$\lim_{\delta \to 0} \operatorname{dist}(\mathcal{A}_\delta, \mathcal{A}_0) = 0. \tag{18.17}$$

The proof of Theorem 18.4 will be given in the final subsection after some preparatory results have been established.

18.3.1 *Uniform bound on the global attractors*

Lemma 18.5. *Let Assumptions 18.1–18.4 hold and suppose that*

$$0 < \delta < \delta_0 := \min\left\{1, \frac{\alpha - \nu\gamma_0\gamma_1^2}{4\mu}\right\}. \tag{18.18}$$

Then the solution $(\boldsymbol{u}(t; \boldsymbol{u}_o, \boldsymbol{v}_o), \boldsymbol{v}(t; \boldsymbol{u}_o, \boldsymbol{v}_o))$ of the lattice system (18.1)–(18.2) *satisfies*

$$\|\boldsymbol{u}(t)\|^2_\rho \le \frac{(2\alpha\mu + 1)\beta\rho_\Sigma}{\alpha\mu\lambda_4} + \left(1 + \frac{2}{\lambda_4}\right)\left(\|\boldsymbol{u}_o\|^2_\rho + \|\boldsymbol{v}_o\|^2_\rho\right) e^{-2\delta\mu t}, \quad t \ge 0,$$

$$\|\boldsymbol{v}(t)\|^2_\rho \le \left(\|\boldsymbol{u}_o\|^2_\rho + \|\boldsymbol{v}_o\|^2_\rho\right) e^{-2\delta\mu t} + \frac{\beta\rho_\Sigma}{\mu}, \quad t \ge 0,$$

where $\lambda_4 := \alpha - \nu\gamma_0\gamma_1^2 > 0$.

Proof. First, it follows from (18.10) that

$$\|\boldsymbol{v}(t;\boldsymbol{u}_o,\boldsymbol{v}_o)\|_\rho^2 \le \left(\delta\|\boldsymbol{u}_o\|_\rho^2 + \|\boldsymbol{v}_o\|_\rho^2\right) e^{-2\delta\mu t} + \frac{\beta\rho_\Sigma}{\mu}.$$

Since $\delta < 1$ this yields

$$|\boldsymbol{v}(t;\boldsymbol{u}_o,\boldsymbol{v}_o)\|_\rho^2 \le \frac{\beta\rho_\Sigma}{\mu} + \left(\|\boldsymbol{u}_o\|_\rho^2 + \|\boldsymbol{v}_o\|_\rho^2\right) e^{-2\delta\mu t},$$

as required.

Then, similar to the proof of Lemma 18.3, divide both sides by δ to get

$$\frac{\mathrm{d}}{\mathrm{d}t}\|\boldsymbol{u}\|_\rho^2 \le -(2\alpha - \nu\gamma_0\gamma_1^2)\,\|\boldsymbol{u}\|_\rho^2 + 2\beta\rho_\Sigma - 2\langle \boldsymbol{v},\boldsymbol{u}\rangle_\rho.$$

Hence

$$\begin{aligned}\frac{\mathrm{d}}{\mathrm{d}t}\|\boldsymbol{u}\|_\rho^2 &\le -(2\alpha - \nu\gamma_0\gamma_1^2)\,\|\boldsymbol{u}\|_\rho^2 + 2\beta\rho_\Sigma + \alpha\,\|\boldsymbol{u}\|_\rho^2 + \frac{1}{\alpha}\,\|\boldsymbol{v}\|_\rho^2 \\ &\le -(\alpha - \nu\gamma_0\gamma_1^2)\,\|\boldsymbol{u}\|_\rho^2 + \frac{(2\alpha\mu+1)\beta\rho_\Sigma}{\alpha\mu} + \frac{1}{\alpha}\left(\|\boldsymbol{u}_o\|_\rho^2 + \|\boldsymbol{v}_o\|_\rho^2\right) e^{-2\delta\mu t},\end{aligned}$$

which can be rewritten as

$$\frac{\mathrm{d}}{\mathrm{d}t}\|\boldsymbol{u}\|_\rho^2 + \lambda_4\,\|\boldsymbol{u}\|_\rho^2 \le \frac{(2+\mu)\beta\rho_\Sigma}{\mu} + \left(\|\boldsymbol{u}_o\|_\rho^2 + \|\boldsymbol{v}_o\|_\rho^2\right) e^{-2\delta\mu t}. \tag{18.19}$$

Notice that under the condition (18.18),

$$\frac{1}{2}\lambda_4 \le \lambda_4 - 2\delta\mu.$$

Thus integrating the differential inequality (18.19) gives

$$\|\boldsymbol{u}(t)\|_\rho^2 \le \|\boldsymbol{u}_o\|_\rho^2 e^{-\lambda_4 t} + \frac{(2\alpha\mu+1)\beta\rho_\Sigma}{\alpha\mu\lambda_4} + \frac{1}{\lambda_4 - 2\delta\mu}\left(\|\boldsymbol{u}_o\|_\rho^2 + \|\boldsymbol{v}_o\|_\rho^2\right) e^{-2\delta\mu t}.$$

In addition, since $2\delta\mu < \lambda_4$, then

$$\|\boldsymbol{u}(t)\|_\rho^2 \le \|\boldsymbol{u}_o\|_\rho^2 e^{-2\delta\mu t} + \frac{(2\alpha\mu+1)\beta\rho_\Sigma}{\alpha\mu\lambda_4} + \frac{2}{\lambda_4}\left(\|\boldsymbol{u}_o\|_\rho^2 + \|\boldsymbol{v}_o\|_\rho^2\right) e^{-2\delta\mu t}.$$

Consequently,

$$\|\boldsymbol{u}(t)\|_\rho^2 \le \frac{(2\alpha\mu+1)\beta\rho_\Sigma}{\alpha\mu\lambda_4} + \left(1 + \frac{2}{\lambda_4}\right)\left(\|\boldsymbol{u}_o\|_\rho^2 + \|\boldsymbol{v}_o\|_\rho^2\right) e^{-2\delta\mu t}.$$

□

The following theorem requires a further restriction on the parameter δ.

Theorem 18.5. *Let Assumptions 18.1–18.4 hold and suppose that*

$$0 < \delta < \delta_1 := \min\left\{1, \frac{\alpha - \nu\gamma_0\gamma_1^2}{4\mu}, \frac{\alpha - \nu(\gamma_0+1)\gamma_1}{4\mu}\right\}. \tag{18.20}$$

Then the global attractors $\mathcal{A}_\delta$ are uniformly bounded in $\ell_\rho^2 \times \ell_\rho^2$ for $0 < \delta < \delta_1$. Specifically, for every $0 < \delta < \delta_1$ and $(\boldsymbol{u},\boldsymbol{v}) \in \mathcal{A}_\delta$,

$$\|\boldsymbol{u}\|_\rho^2 \le \frac{(2\alpha\mu+1)\beta\rho_\Sigma}{\alpha\mu\lambda_4} + 1, \quad \|\boldsymbol{v}\|_\rho^2 \le \frac{\beta\rho_\Sigma}{\mu} + 1,$$

where $\lambda_4 := \alpha - \nu\gamma_0\gamma_1^2 > 0$.

Proof. Fix an arbitrary δ with $0<\delta<\delta_1$ and let $(\boldsymbol{u},\boldsymbol{v})\in\mathcal{A}_\delta$. Consider a sequence $\{t_n\}_{n=1}^\infty$ with $t_n\to\infty$. Then due the φ^δ-invariance of $\mathcal{A}_\delta$ there exists $\left(\boldsymbol{u}_o^{(n)},\boldsymbol{v}_o^{(n)}\right)\in\mathcal{A}_\delta$ such that

$$(\boldsymbol{u},\boldsymbol{v})=\varphi^\delta(t_n,\boldsymbol{u}_o^{(n)},\boldsymbol{v}_o^{(n)})=:(\boldsymbol{u}^{(n)}(t_n),\boldsymbol{v}^{(n)}(t_n))\quad\text{for all}\quad n\geq 1. \tag{18.21}$$

Since $\mathcal{A}_\delta\subset Q^\delta$, where Q^δ is the bounded absorbing set defined in (18.11), it follows that

$$\|\boldsymbol{u}_o^{(n)}\|_\rho^2+\|\boldsymbol{v}_o^{(n)}\|_\rho^2\leq\frac{1}{\delta}R^2\quad\text{for all}\quad n\geq 1, \tag{18.22}$$

where R is independent of n. Applying Lemma 18.5 to $(\boldsymbol{u}^{(n)}(t_n),\boldsymbol{v}^{(n)}(t_n))$ and using the inequality (18.22) gives

$$\begin{aligned}\|\boldsymbol{u}\|_\rho^2=\|\boldsymbol{u}^{(n)}(t_n)\|_\rho^2&\leq\frac{(2\alpha\mu+1)\beta\rho_\Sigma}{\alpha\mu\lambda_4}+\left(1+\frac{2}{\lambda_4}\right)\left(\|\boldsymbol{u}_o^{(n)}\|_\rho^2+\|\boldsymbol{v}_o^{(n)}\|_\rho^2\right)e^{-2\delta\mu t_n}\\&\leq\frac{(2\alpha\mu+1)\beta\rho_\Sigma}{\alpha\mu\lambda_4}+\left(1+\frac{2}{\lambda_4}\right)\frac{1}{\delta}R^2e^{-2\delta\mu t_n},\end{aligned} \tag{18.23}$$

$$\begin{aligned}\|\boldsymbol{v}\|_\rho^2=\|\boldsymbol{v}^{(n)}(t_n)\|_\rho^2&\leq\frac{\beta\rho_\Sigma}{\mu}+\left(\|\boldsymbol{u}_o^{(n)}\|_\rho^2+\|\boldsymbol{v}_o^{(n)}\|_\rho^2\right)e^{-2\delta\mu t_n}\\&\leq\frac{\beta\rho_\Sigma}{\mu}+\frac{1}{\delta}R^2e^{-2\delta\mu t_n}.\end{aligned} \tag{18.24}$$

Taking the limits in (18.23) and (18.24) as $n\to\infty$, it follows by (18.21) that

$$\|\boldsymbol{u}\|_\rho^2\leq\frac{(2\alpha\mu+1)\beta\rho_\Sigma}{\alpha\mu\lambda_4}+1,\qquad\|\boldsymbol{v}\|_\rho^2\leq\frac{\beta\rho_\Sigma}{\mu}+1.$$

Since $(\boldsymbol{u},\boldsymbol{v})$ was arbitrary in $\mathcal{A}_\delta$ and δ was arbitrary in $(0,\delta_1)$, these bounds are uniform in these terms. □

18.3.2 *Pre-compactness of the union of the global attractors*

First an asymptotic tails property is established in the following lemma.

Lemma 18.6. *Let Assumptions 18.1–18.4 and the condition* (18.20) *hold. Then for every* $\varepsilon>0$ *there exist* $I(\varepsilon)>0$ *such that*

$$\sum_{|i|>I(\varepsilon)}\rho_i\left(|u_i|^2+|v_i|^2\right)\leq\varepsilon$$

for all $(\boldsymbol{u},\boldsymbol{v})\in\mathcal{A}_\delta$ *and* $\delta\in(0,\delta_1)$.

Proof. Let $(\boldsymbol{u},\boldsymbol{v})\in\mathcal{A}_\delta$ and take a sequence $\{t_n\}_{n=1}^\infty$ with $t_n\to\infty$. By the φ^δ-invariance of $\mathcal{A}_\delta$ there exists $(\boldsymbol{u}_o^{(n)},\boldsymbol{v}_o^{(n)})\in\mathcal{A}_\delta$ such that

$$(\boldsymbol{u},\boldsymbol{v})=\varphi^\delta(t_n,\boldsymbol{u}_o^{(n)},\boldsymbol{v}_o^{(n)}),\quad\text{for all}\quad n\geq 1. \tag{18.25}$$

Write $(\boldsymbol{u}^{(n)}(t),\boldsymbol{v}^{(n)}(t))=\varphi^\delta(t,\boldsymbol{u}_o^{(n)},\boldsymbol{v}_o^{(n)})$. Then by Theorem 18.5 there exists a constant C which is independent of δ such that

$$\left\|\boldsymbol{u}^{(n)}(t_n)\right\|_\rho^2+\left\|\boldsymbol{v}^{(n)}(t_n)\right\|_\rho^2\leq C,\quad\text{for all}\quad n\geq 1,\,t\geq 0. \tag{18.26}$$

Using (18.26) and proceeding as in Lemma 18.4, for every $\varepsilon > 0$, there exists $K(\varepsilon) > 0$ such that

$$\frac{\mathrm{d}}{\mathrm{d}t}\left(\delta\sum_{|i|>k}\rho_i|u_i(t)|^2+\sum_{|i|>k}\rho_i|v_i|(t)^2\right)+2\delta\mu\left(\delta\sum_{|i|>k}\rho_i|u_i(t)|^2+\sum_{|i|>k}\rho_i|v_i(t)|^2\right)$$
$$\leq\frac{\delta C_2}{k}+2\delta\beta\sum_{|i|>k}\rho_i\leq 2\delta\varepsilon,\quad k\geq 2K(\varepsilon),\qquad t\geq 0.$$

Hence it follows by Gronwall's inequality that

$$\delta\sum_{|i|>k}\rho_i|u_i(t)|^2+\sum_{|i|>k}\rho_i|v_i|(t)^2$$
$$\leq e^{-2\delta\mu t}\left(\delta\sum_{|i|>k}\rho_i|u_{o,i}|^2+\sum_{|i|>k}\rho_i|v_{o,i}|^2\right)+\frac{\varepsilon}{\mu}$$
$$\leq C_1e^{-2\delta\mu t}+\frac{\varepsilon}{\mu}.$$

In particular, this gives

$$\sum_{|i|>k}\rho_i|v_i(t)|^2\leq C_1e^{-2\delta\mu t}+\frac{\varepsilon}{\mu}. \tag{18.27}$$

Now multiply the equation (18.1) by $2\rho_iu_i$ and sum over $|i|>k$. Then, following similar steps to the derivation of inequality (18.13) (without the common multiplier δ), gives for $k\geq 2K(\varepsilon)$

$$\frac{\mathrm{d}}{\mathrm{d}t}\sum_{|i|>k}\rho_i|u_i|^2+\lambda_3\sum_{|i|>k}\rho_i|u_i|^2\leq 2\varepsilon+2\sum_{|i|>k}\rho_iu_iv_i,$$

where $\lambda_3 := \alpha-\nu(\gamma_0+1)\gamma_1$. Hence using the estimate (18.27) for the last term yields

$$\frac{\mathrm{d}}{\mathrm{d}t}\sum_{|i|>k}\rho_i|u_i|^2+\lambda_3\sum_{|i|>k}\rho_i|u_i|^2\leq 2\varepsilon+2C_1e^{-2\delta\mu t}+\frac{2\varepsilon}{\mu}.$$

Then by Gronwall's inequality

$$\sum_{|i|>k}\rho_i|u_i|^2(t)\leq\frac{1}{\lambda_3}\sum_{|i|>k}\rho_i|u_{o,i}|^2e^{-\lambda_3t}+\frac{2(1+\mu)\varepsilon}{\mu\lambda_3}e^{-\lambda_3t}+\frac{2C_1}{\lambda_3-2\delta\mu}e^{-2\delta\mu t}.$$

On the other side, by the definition $\lambda_3 := \alpha-\nu(\gamma_0+1)\gamma_1$, we have $\frac{1}{2}\lambda_3 < \lambda_3-2\delta\mu$, and thus

$$\sum_{|i|>k}\rho_i|u_i(t)|^2\leq\frac{1}{\lambda_3}\sum_{|i|>k}\rho_i|u_{o,i}|^2e^{-\lambda_3t}+\frac{2(1+\mu)\varepsilon}{\mu\lambda_3}+\frac{4C_1}{\lambda_3}e^{-2\delta\mu t}$$
$$\leq C_2e^{-\lambda_3t}+\frac{2(1+\mu)\varepsilon}{\mu\lambda_3}+\frac{4C_1}{\lambda_3}e^{-2\delta\mu t}.$$

Finally $(\boldsymbol{u},\boldsymbol{v}) = \varphi^\delta(t_n, \boldsymbol{u}_o^{(n)}, \boldsymbol{v}_o^{(n)})$ by (18.25), so the above estimates give

$$\sum_{|i|>k} \rho_i |u_i|^2 = \sum_{|i|>k} \rho_i |u_i^{(n)}(t_n)|^2 \leq C_2 e^{-\lambda_3 t_n} + \frac{2(1+\mu)\varepsilon}{\mu\lambda_3} + \frac{4C_1}{\lambda_3} e^{-2\delta\mu t_n},$$

$$\sum_{|i|>k} \rho_i |v_i|^2 = \sum_{|i|>k} \rho_i |v_i^{(n)}(t_n)|^2 \leq C_1 e^{-2\delta\mu t_n} + \frac{\varepsilon}{\mu},$$

so taking the limit $t_n \to \infty$,

$$\sum_{|i|>k} \rho_i |u_i|^2 + \sum_{|i|>k} \rho_i |v_i|^2 \leq \frac{2(1+\mu)\varepsilon}{\mu\lambda_3} + \frac{\varepsilon}{\mu} =: \hat{\varepsilon}.$$

Redefining the parameter ε in the statement of the lemma to be $\hat{\varepsilon}$ gives the desired assertion. □

Theorem 18.6. *Let Assumptions 18.1–18.4 and the condition* (18.20) *hold. Then* $\bigcup_{0<\delta<\delta_0} \mathcal{A}_\delta$ *is pre-compact in* $\ell_\rho^2 \times \ell_\rho^2$.

Proof. Given $\varepsilon > 0$, it suffices to show that $\bigcup_{0<\delta<\delta_0} \mathcal{A}_\delta$ can be covered by a finite number of balls of radius less than ε. From Lemma 18.6 there is an $I(\varepsilon)$ depending on ε such that for all $(\boldsymbol{u},\boldsymbol{v}) \in \bigcup_{0<\delta<\delta_0} \mathcal{A}_\delta$,

$$\sum_{|i|>I(\varepsilon)} \rho_i \left(|u_i|^2 + |v_i|^2\right) \leq \frac{\varepsilon}{4}. \tag{18.28}$$

On the other hand, by Theorem 18.5, the set $\bigcup_{0<\delta<\delta_1} \mathcal{A}_\delta$ is bounded in $\ell_\rho^2 \times \ell_\rho^2$. Therefore the set

$$\mathfrak{S} := \left\{ (\rho_i u_i, \rho_i v_i)_{|i|\leq I(\varepsilon)} : (\boldsymbol{u},\boldsymbol{v}) \in \bigcup_{0<\delta<\delta_1} \mathcal{A}_\delta \right\}$$

is bounded in the finite dimensional space $\mathbb{R}^{2I(\varepsilon)+1}$ and hence pre-compact. In other words, $\mathfrak{S}$ has a finite covering of balls of radius less than $\frac{\varepsilon}{4}$. Together with (18.28) this implies that the set $\bigcup_{0<\delta<\delta_1} \mathcal{A}_\delta$ has a finite covering of balls of radius less than ε in $\ell_\rho^2 \times \ell_\rho^2$. □

18.3.3 *Upper semi-continuity of the global attractors*

The first result relates the solutions of the lattice system (18.1)–(18.2) to those of the limiting system (18.15)–(18.16).

Lemma 18.7. *Let Assumptions 18.1–18.4 hold. Let* $(\boldsymbol{u}^\delta, \boldsymbol{v}^\delta)$ *be the solution of* (18.1)–(18.2) *with the initial data* $(\boldsymbol{u}_o^\delta, \boldsymbol{v}_o^\delta) \in \ell^2 \times \ell^2$ *and let* $(\boldsymbol{u},\boldsymbol{v})$ *be the solution of the limiting system* (18.15)–(18.16) *with the initial value* $(\boldsymbol{u}_o, \boldsymbol{v}_o) \in \ell^2 \times \ell^2$. *Then, there exists* $C > 0$ *independent of* δ *such that*

$$\|\boldsymbol{u}^\delta(t) - \boldsymbol{u}(t)\|_\rho^2 + \|\boldsymbol{v}^\delta(t) - \boldsymbol{v}(t))\|_\rho^2 \leq e^{Ct}\left(\|\boldsymbol{u}_o^\delta - \boldsymbol{u}_o\|_\rho^2 + \|\boldsymbol{v}_o^\delta - \boldsymbol{v}_o\|_\rho^2\right)$$
$$+\delta \int_0^t e^{C(t-s)} \|\boldsymbol{u}(s) - \mu \boldsymbol{v}(s)\|_\rho^2 ds, \quad t \geq 0.$$

Proof. For $t \geq 0$ set $\boldsymbol{h}_u(t) = \boldsymbol{u}^\delta(t) - \boldsymbol{u}(t)$ and $\boldsymbol{h}_v(t) = \boldsymbol{v}^\delta(t) - \boldsymbol{v}(t)$. Then it follows from (18.1)–(18.2) and (18.15)–(18.16) that

$$\frac{\mathrm{d}\boldsymbol{h}_u(t)}{dt} = \nu\Lambda\boldsymbol{h}_u + F(\boldsymbol{u}^\delta) - F(\boldsymbol{u}) - \boldsymbol{h}_v,$$
$$\frac{\mathrm{d}\boldsymbol{h}_v(t)}{dt} = \delta\left(\boldsymbol{h}_u - \mu\boldsymbol{h}_v\right) + \delta\left(\boldsymbol{u} - \mu\boldsymbol{v}\right).$$

Then similar to the proof of Lemma 18.2 it can be shown that $\boldsymbol{h}_u$ and $\boldsymbol{h}_v$ satisfy

$$\frac{\mathrm{d}}{\mathrm{d}t}\left(\|\boldsymbol{h}_u\|_\rho^2 + \|\boldsymbol{h}_v\|_\rho^2\right) \leq C\left(\|\boldsymbol{h}_u\|_\rho^2 + \|\boldsymbol{h}_v\|_\rho^2\right) + \delta\|\boldsymbol{u} - \mu\boldsymbol{v}\|_\rho^2.$$

An application of Gronwall's Lemma then gives the desired assertion. □

Denote by $\{\varphi^\delta(t)\}_{t\geq 0}$ the semi-group associated with the lattice system (18.1)–(18.2) and $\{\varphi^0(t)\}_{t\geq 0}$ the semi-group associated with the limiting lattice system (18.15)–(18.16). Note that these semi-groups map $\ell_\rho^2 \times \ell_\rho^2$ to $\ell_\rho^2 \times \ell_\rho^2$ and are extensions of the corresponding solution mappings on the space $\ell^2 \times \ell^2$, which is dense in $\ell_\rho^2 \times \ell_\rho^2$.

It follows by a continuity argument that Lemma 18.7 holds for any initial data $(\boldsymbol{u}_o^\delta, \boldsymbol{v}_o^\delta)$, $(\boldsymbol{u}_o, \boldsymbol{v}_o) \in \ell^2 \times \ell^2$. This implies that if $\delta_n \to 0$ and $(\boldsymbol{u}_o^{\delta_n}, \boldsymbol{v}_o^{\delta_n}) \to (\boldsymbol{u}_o, \boldsymbol{v}_o)$ in $\ell^2 \times \ell^2$ as $n \to \infty$, then

$$\varphi^\delta(t, \boldsymbol{u}_o^{\delta_n}, \boldsymbol{v}_o^{\delta_n}) \to \varphi^0(t, \boldsymbol{u}_o, \boldsymbol{v}_o), \quad n \to \infty. \tag{18.29}$$

Lemma 18.8. *Let Assumptions 18.1–18.4 hold. If* $(\boldsymbol{u}^{\delta_n}, \boldsymbol{v}^{\delta_n}) \in \mathcal{A}_{\delta_n}$ *as* $\delta_n \to 0$, *then there exists* $(\boldsymbol{u}^*, \boldsymbol{v}^*) \in \mathcal{A}_0$ *such that (up to a subsequence)* $(\boldsymbol{u}^{\delta_n}, \boldsymbol{v}^{\delta_n}) \to (\boldsymbol{u}^*, \boldsymbol{v}^*)$ *in* $\ell_\rho^2 \times \ell_\rho^2$.

Proof. It was shown in Theorem 18.5 that the set $\bigcup_{0<\delta<\delta_1} \mathcal{A}_\delta$ is pre-compact. Hence there exists $(\boldsymbol{u}^*, \boldsymbol{v}^*)$ in $\ell_\rho^2 \times \ell_\rho^2$ and a subsequence of $(\boldsymbol{u}^{\delta_n}, \boldsymbol{v}^{\delta_n})$, which is still denoted by $(\boldsymbol{u}^{\delta_n}, \boldsymbol{v}^{\delta_n})$, such that

$$(\boldsymbol{u}^{\delta_n}, \boldsymbol{v}^{\delta_n}) \to (\boldsymbol{u}^*, \boldsymbol{v}^*) \text{ in } \ell_\rho^2 \times \ell_\rho^2. \tag{18.30}$$

Consider a sequence $\{t_m\}_{m=1}^\infty$ with $t_m \to \infty$. Then by the φ^{δ_n}-invariance of $\mathcal{A}_{\delta_n}$ it follows that

$$\varphi^{\delta_n}(-t_m, \boldsymbol{u}^{\delta_n}, \boldsymbol{v}^{\delta_n}) \in \mathcal{A}_{\delta_n} \text{ for all } n, m \geq 1.$$

Consider $m = 1$. Since $\bigcup_{n=1}^\infty \mathcal{A}_{\delta_n}$ is pre-compact, there exist $(\boldsymbol{u}^{(1)}, \boldsymbol{v}^{(1)})$ in $\ell_\rho^2 \times \ell_\rho^2$ and a subsequence of $(\boldsymbol{u}^{\delta_{n_k}}, \boldsymbol{v}^{\delta_{n_k}})$ of $(\boldsymbol{u}^{\delta_n}, \boldsymbol{v}^{\delta_n})$ such that

$$\varphi^{\delta_{n_k}}(-t_1, \boldsymbol{u}^{\delta_{n_k}}, \boldsymbol{v}^{\delta_{n_k}}) \to (\boldsymbol{u}^{(1)}, \boldsymbol{v}^{(1)}) \text{ in } \ell_\rho^2 \times \ell_\rho^2 \text{ as } n_k \to \infty.$$

Similarly, for $m = 2$ there exist $(\boldsymbol{u}^{(2)}, \boldsymbol{v}^{(2)})$ in $\ell_\rho^2 \times \ell_\rho^2$ and a subsequence of $(\boldsymbol{u}^{\delta_{n_{k_l}}}, \boldsymbol{v}^{\delta_{n_{k_l}}})$ of $(\boldsymbol{u}^{\delta_{n_k}}, \boldsymbol{v}^{\delta_{n_k}})$ such that

$$\varphi^{\delta_{n_{k_l}}}(-t_2, \boldsymbol{u}^{\delta_{n_{k_l}}}, \boldsymbol{v}^{\delta_{n_{k_l}}}) \to (\boldsymbol{u}^{(2)}, \boldsymbol{v}^{(2)}) \text{ in } \ell_\rho^2 \times \ell_\rho^2 \text{ as } n_{k_l} \to \infty.$$

Repeating the procedure above, it follows by a standard diagonal argument that there exist $(\boldsymbol{u}^{(m)}, \boldsymbol{v}^{(m)})$ in $\ell^2_\rho \times \ell^2_\rho$ and a subsequence of $(\boldsymbol{u}^{\delta_{n_r}}, \boldsymbol{v}^{\delta_{n_r}})$ of $(\boldsymbol{u}^{\delta_n}, \boldsymbol{v}^{\delta_n})$ such that for each $m \geq 1$

$$\varphi^{\delta_{n_r}}(-t_m, \boldsymbol{u}^{\delta_{n_r}}, \boldsymbol{v}^{\delta_{n_r}}) \to (\boldsymbol{u}^{(m)}, \boldsymbol{v}^{(m)}) \text{ in } \ell^2_\rho \times \ell^2_\rho \text{ as } r \to \infty. \tag{18.31}$$

Hence, by (18.29) and (18.31), as $r \to \infty$,

$$(\boldsymbol{u}^{\delta_{n_r}}, \boldsymbol{v}^{\delta_{n_r}}) = \varphi^{\delta_{n_r}}(t_m, \varphi^{\delta_{n_r}}(-t_m, \boldsymbol{u}^{\delta_{n_r}}, \boldsymbol{v}^{\delta_{n_r}})) \to \varphi^0(t_m, \boldsymbol{u}^{(m)}, \boldsymbol{v}^{(m)}). \tag{18.32}$$

The two equations (18.30) and (18.32) together imply that

$$(\boldsymbol{u}^*, \boldsymbol{v}^*) = \varphi^0(-t_m, \boldsymbol{u}^{(m)}, \boldsymbol{v}^{(m)}).$$

Since $t_m \to \infty$, this implies that $(\boldsymbol{u}^*, \boldsymbol{v}^*) \in \mathcal{A}_0$, which completes the proof. □

We are now ready to complete the proof of Theorem 18.4 by contradiction. Suppose that the limit (18.17) does not hold. Then there exist $\varepsilon_0 > 0$ and a sequence $(\boldsymbol{u}^{\delta_n}, \boldsymbol{v}^{\delta_n}) \in \mathcal{A}_{\delta_n}$ such that

$$\operatorname{dist}\left((\boldsymbol{u}^{\delta_n}, \boldsymbol{v}^{\delta_n}), \mathcal{A}_0\right) \geq \varepsilon_0. \tag{18.33}$$

On the other side, by Lemma 18.8, there exists a subsequence $(\boldsymbol{u}^{\delta_{n_k}}, \boldsymbol{v}^{\delta_{n_k}})$ of $(\boldsymbol{u}^{\delta_n}, \boldsymbol{v}^{\delta_n})$ such that

$$\lim_{k \to \infty} \operatorname{dist}\left((\boldsymbol{u}^{\delta_{n_k}}, \boldsymbol{v}^{\delta_{n_k}}), \mathcal{A}_0\right) = 0.$$

This contradicts (18.33). Hence $\mathcal{A}_\delta$ converge upper semi-continuously to $\mathcal{A}_0$ as $\delta \to 0$. The proof of Theorem 18.4 is done. □

18.4 End notes

This chapter is based on the paper [Van Vleck and Wang (2005)]. As in Chapter 5, a general class of weights with appropriate properties is used rather than a particular weight in [Van Vleck and Wang (2005)] and [Wang (2006)], which means that the proofs here do not need the auxiliary function introduced there. See [Wang (2007b)] for almost periodic behavior and also [Chow and Shen (1995)]. [Zhang (2022)] considers FitzHugh-Nagumo lattice models with double time-delays.

Stochastic FitzHugh-Nagumo lattice models were investigated in [Huang (2007); Wang and Zhou (2016)]. See also [Yang *et al.* (2023)] where delayed FitzHugh-Nagumo lattice models driven by Wong-Zakai noise are investigated in a weighted norm sequence space.

Applications of FitzHugh-Nagumo lattice in neural information can be found in [Marquié *et al.* (2004)].

18.5 Problems

Problem 18.1. Does the FitzHugh-Nagumo lattice model have travelling wave solutions?

Problem 18.2. For small nonzero δ, rescale the time t in the FitzHugh-Nagumo lattice system (18.1)–(18.2) by t/δ. Write out the rescaled FitzHugh-Nagumo lattice system, and study its attractors as $\delta \to 0$.

Chapter 19

The Amari lattice neural field model

Neural field models are often represented as evolution equations generated as continuum limits of computational models of neural field theory. They are tissue level models that describe the spatio-temporal evolution of coarse grained variables such as synaptic or firing rate activity in populations of neurons. See [Coombes *et al.* (2014)] and the literature therein.

A particularly influential model is that proposed by S. Amari [Amari (1977)] (see also Chapter 3 of [Coombes *et al.* (2014)] by Amari):

$$\partial_t u(t,x) = f(u(t,x)) + \int_\Omega \mathcal{K}(x-y)\mathrm{H}\left(u(t,y)-\varsigma\right)\mathrm{d}y, \qquad x \in \Omega \subset \mathbb{R},$$

where $\varsigma > 0$ is a given threshold and $\mathrm{H} : \mathbb{R} \to \mathbb{R}$ is the Heaviside function defined by

$$\mathrm{H}(x) = \begin{cases} 1, \, x \geq 0, \\ 0, \, x < 0, \end{cases} \qquad x \in \mathbb{R}.$$

Continuum neural models may lose their validity in capturing detailed dynamics at discrete sites when the discrete structures of neural systems become dominant. Lattice field models are also evolution equations, but avoid having to take the continuum limit. In [Han and Kloeden (2019a)] the following Amari neural field lattice model was introduced:

$$\frac{\mathrm{d}}{\mathrm{d}t}u_{\mathrm{i}}(t) = f_{\mathrm{i}}(u_{\mathrm{i}}(t)) + \sum_{\mathrm{j}\in\mathbb{Z}^d} \kappa_{\mathrm{i},\mathrm{j}}\mathrm{H}(u_{\mathrm{j}}(t)-\varsigma) + g_{\mathrm{i}}, \quad \mathrm{i} \in \mathbb{Z}^d, \quad \varsigma > 0, \tag{19.1}$$

with vectorial indices $\mathrm{i} = (i_1, \ldots, i_d) \in \mathbb{Z}^d$.

The neural field lattice system (19.1) describes dynamics of the membrane potential u_{i} of each nerve cell at the ith position of a neural network. The infinite dimensional matrix $(\kappa_{\mathrm{i},\mathrm{j}})_{\mathrm{i},\mathrm{j}\in\mathbb{Z}^d}$ is a discretised kernel function and the term $\kappa_{\mathrm{i},\mathrm{j}}\mathrm{H}(u_{\mathrm{j}}(t)-\varsigma)$ describes the nonlocal interactions between the ith and jth neurons. More precisely, the membrane potential of the ith neuron is affected by all those neurons with membrane potential above a certain threshold ς.

The main difficulty in studying system (19.1) lies in the discontinuity in the vector field caused by the Heaviside function H. This will be handled mathematically by reformulating the lattice system (19.1) as a set-valued lattice system

$$\frac{\mathrm{d}}{\mathrm{d}t}u_{\mathrm{i}}(t) \in f_{\mathrm{i}}(u_{\mathrm{i}}(t)) + \sum_{\mathrm{j}\in\mathbb{Z}^d} \kappa_{\mathrm{i},\mathrm{j}}\chi(u_{\mathrm{j}}(t) - \varsigma) + g_{\mathrm{i}}, \quad \mathrm{i} \in \mathbb{Z}^d, \quad \varsigma > 0, \tag{19.2}$$

with the set-valued mapping χ on $\mathbb{R}$ defined by

$$\chi(s) = \begin{cases} \{0\}, & s < 0, \\ [0,1], & s = 0, \\ \{1\}, & s > 0, \end{cases} \qquad s \in \mathbb{R}. \tag{19.3}$$

To avoid such complications, the Heaviside function is often replaced by a simplifying sigmoidal function (e.g., [Ciuca and Jitaru (2006)]) such as

$$\sigma_\epsilon(x) = \frac{1}{1+e^{-x/\epsilon}}, \quad x \in \mathbb{R}, \quad 0 < \epsilon < 1,$$

which leads to the lattice system

$$\frac{\mathrm{d}}{\mathrm{d}t}u_{\mathrm{i}}(t) = f_{\mathrm{i}}(u_{\mathrm{i}}(t)) + \sum_{\mathrm{j}\in\mathbb{Z}^d} \kappa_{\mathrm{i},\mathrm{j}}\sigma_\epsilon(u_{\mathrm{j}}(t) - \varsigma) + g_{\mathrm{i}}, \quad \mathrm{i} \in \mathbb{Z}^d. \tag{19.4}$$

19.1 Preliminaries

The above lattice systems will be investigated here in the weighted space of bi-infinite real-valued sequences with vectorial indices $\mathrm{i} = (i_1, \ldots, i_d) \in \mathbb{Z}^d$, specifically, ℓ^2_ρ defined as in Section 1.3. The weights ρ_{i} are chosen to satisfy the Assumption 1.1, which is restated here for the reader's convenience.

Assumption 19.1. $\rho_{\mathrm{i}} > 0$ for all $\mathrm{i} \in \mathbb{Z}^d$ and $\rho_\Sigma := \sum_{\mathrm{i}\in\mathbb{Z}^d} \rho_{\mathrm{i}} < \infty$.

Recall that the Assumption 19.1 implies that $\ell^2 \subset \ell^\infty \subset \ell^2_\rho$, where

$$\ell^\infty = \left\{ \boldsymbol{u} = (u_{\mathrm{i}})_{\mathrm{i}\in\mathbb{Z}^d} : \sup_{\mathrm{i}\in\mathbb{Z}^d} |u_{\mathrm{i}}| < \infty \right\}$$

with $\|\boldsymbol{u}\|_\infty := \sup_{\mathrm{i}\in\mathbb{Z}^d} |u_{\mathrm{i}}|$. In addition, recall that ℓ^2_ρ is a Hilbert space with norm and inner product

$$\|\boldsymbol{u}\|^2_\rho = \sum_{\mathrm{i}\in\mathbb{Z}^d} \rho_{\mathrm{i}} \boldsymbol{u}^2_{\mathrm{i}}, \quad \langle \boldsymbol{u}, \boldsymbol{v} \rangle_\rho = \sum_{\mathrm{i}\in\mathbb{Z}^d} u^2_{\mathrm{i}} v^2_{\mathrm{i}}, \quad \boldsymbol{u} = (u_{\mathrm{i}})_{\mathrm{i}\in\mathbb{Z}^d}, \quad \boldsymbol{v} = (v_{\mathrm{i}})_{\mathrm{i}\in\mathbb{Z}^d}.$$

For any sets $B \subset \mathbb{R}$ and $U = (U_{\mathrm{i}})_{\mathrm{i}\in\mathbb{Z}^d} \subset \ell^2_\rho$, define

$$|||B||| := \sup_{b\in B} |b|, \qquad |||U|||_\rho := \left(\sup_{\substack{\boldsymbol{u}=(u_{\mathrm{i}})_{\mathrm{i}\in\mathbb{Z}^d}\in\ell^2_\rho \\ u_{\mathrm{i}}\in U_{\mathrm{i}}}} \|\boldsymbol{u}\|^2_\rho \right)^{1/2}.$$

It then follows immediately that

$$|||U|||^2_\rho = \sup_{\substack{\boldsymbol{u}=(u_{\mathrm{i}})_{\mathrm{i}\in\mathbb{Z}^d}\in\ell^2_\rho \\ u_{\mathrm{i}}\in U_{\mathrm{i}}}} \sum_{\mathrm{i}\in\mathbb{Z}^d} \rho_{\mathrm{i}} u^2_{\mathrm{i}} \le \sum_{\mathrm{i}\in\mathbb{Z}^d} \rho_{\mathrm{i}} \sup_{u_{\mathrm{i}}\in U_{\mathrm{i}}} u^2_{\mathrm{i}} = \sum_{\mathrm{i}\in\mathbb{Z}^d} \rho_{\mathrm{i}} \, ||| \, U_{\mathrm{i}} \, |||^2 . \tag{19.5}$$

19.1.1 *Standing assumptions*

The coefficients of the lattice models (19.1), (19.2), (19.4) are assumed to satisfy the following assumptions.

Assumption 19.2. The aggregate interconnection strength is reciprocal-weighted finite in the sense that

$$\sup_{\mathrm{i}\in\mathbb{Z}} \sum_{\mathrm{j}\in\mathbb{Z}^d} \frac{\kappa_{\mathrm{i},\mathrm{j}}^2}{\rho_{\mathrm{j}}} \le \kappa \quad \text{for some } \kappa > 0.$$

Assumption 19.3. There is no reaction when the membrane potential is zero, i.e., $f_{\mathrm{i}}(0) = 0$ for all $\mathrm{i} \in \mathbb{Z}^d$; $f_{\mathrm{i}} : \mathbb{R} \to \mathbb{R}$ is continuously differentiable with weighted equi-locally bounded derivatives, i.e., there exists a non-decreasing function $L(\cdot) \in C(\mathbb{R}^+, \mathbb{R}^+)$ such that

$$\sup_{\mathrm{i}\in\mathbb{Z}^d} \max_{s\in[-r,r]} |f_{\mathrm{i}}'(s)| \le L(\rho_{\mathrm{i}} r), \quad \text{for all } r \in \mathbb{R}^+, \mathrm{i} \in \mathbb{Z}^d.$$

Assumption 19.4. The constant forcing term satisfies $\boldsymbol{g} := (g_{\mathrm{i}})_{\mathrm{i}\in\mathbb{Z}^d} \in \ell_\rho^2$.

Assumption 19.5. f_{i} is dissipative in the sense that there exist constants $\alpha > 0$ and $\boldsymbol{\beta} := (\beta_{\mathrm{i}})_{\mathrm{i}\in\mathbb{Z}^d} \in \ell_\rho^2$ such that

$$s f_{\mathrm{i}}(s) \le -\alpha s^2 + \beta_{\mathrm{i}}^2, \quad \text{for all } \ s \in \mathbb{R} \quad \mathrm{i} \in \mathbb{Z}^d.$$

19.1.2 *Basic estimates*

Assumption 19.3 implies that f_{i} is locally Lipschitz with

$$|f_{\mathrm{i}}(x) - f_{\mathrm{i}}(y)| \le L\left(\rho_{\mathrm{i}}(|x| + |y|)\right) \cdot |x - y|, \quad \text{for all } \ \mathrm{i} \in \mathbb{Z}^d, \ x, y \in \mathbb{R}.$$

In fact, notice that

$$\rho_{\mathrm{i}}|u_{\mathrm{i}}| \le \sqrt{\rho_\Sigma} \cdot \left(\rho_{\mathrm{i}} u_{\mathrm{i}}^2\right)^{1/2} \le \sqrt{\rho_\Sigma}\Big(\sum_{\mathrm{i}\in\mathbb{Z}^d} \rho_{\mathrm{i}} u_{\mathrm{i}}^2\Big)^{1/2} = \sqrt{\rho_\Sigma}\|\boldsymbol{u}\|_\rho,$$

we have for every $\boldsymbol{u} = (u_{\mathrm{i}})_{\mathrm{i}\in\mathbb{Z}^d}$ and $\boldsymbol{v} = (v_{\mathrm{i}})_{\mathrm{i}\in\mathbb{Z}^d}$,

$$\begin{aligned} |f_{\mathrm{i}}(u_{\mathrm{i}}) - f_{\mathrm{i}}(v_{\mathrm{i}})| &\le L(\rho_{\mathrm{i}}(|u_{\mathrm{i}}| + |v_{\mathrm{i}}|)) \cdot |u_{\mathrm{i}} - v_{\mathrm{i}}| \\ &\le L\left(\sqrt{\rho_\Sigma}(\|\boldsymbol{u}\|_\rho + \|\boldsymbol{v}\|_\rho)\right) \cdot |u_{\mathrm{i}} - v_{\mathrm{i}}|. \end{aligned} \tag{19.6}$$

Define the operator F by

$$F(\boldsymbol{u}) := (f_{\mathrm{i}}(u_{\mathrm{i}}))_{\mathrm{i}\in\mathbb{Z}^d}.$$

The following Lemma shows that F takes values in ℓ_ρ^2 for any $\boldsymbol{u} = (u_{\mathrm{i}})_{\mathrm{i}\in\mathbb{Z}^d} \in \ell_\rho^2$, and inherits the local Lipschitz and dissipative properties of the functions f_{i}.

Lemma 19.1. *Let Assumptions 19.1–19.5 hold. Then $F : \ell_\rho^2 \to \ell_\rho^2$ is locally Lipschitz and satisfies the dissipativity condition*

$$\langle F(\boldsymbol{u}), \boldsymbol{u}\rangle_\rho \le -\alpha\|\boldsymbol{u}\|_\rho^2 + \|\boldsymbol{\beta}\|_\rho^2.$$

Proof. For every $\boldsymbol{u} = (u_{\mathrm{i}})_{\mathrm{i}\in\mathbb{Z}^d} \in \ell^2_\rho$ and $\boldsymbol{v} = (v_{\mathrm{i}})_{\mathrm{i}\in\mathbb{Z}^d} \in \ell^2_\rho$, due to (19.6) we have

$$\begin{aligned}\|F(\boldsymbol{u}) - F(\boldsymbol{v})\|_\rho^2 &= \sum_{\mathrm{i}\in\mathbb{Z}^d} \rho_{\mathrm{i}} |f_{\mathrm{i}}(u_{\mathrm{i}}) - f_i(v_i)|^2 \\ &\le \Big(L\left(\sqrt{\rho_\Sigma}(\|\boldsymbol{u}\|_\rho + \|\boldsymbol{v}\|_\rho)\right)\Big)^2 \|\boldsymbol{u}-\boldsymbol{v}\|_\rho^2,\end{aligned}$$

i.e., F is locally Lipschitz.

Taking $\boldsymbol{v} = 0$, it follows immediately from Assumption 19.3 that

$$\|F(\boldsymbol{u})\|_\rho^2 = \sum_{\mathrm{i}\in\mathbb{Z}^d} \rho_{\mathrm{i}} |f_{\mathrm{i}}(u_{\mathrm{i}})|^2 \le \Big(L\left(\sqrt{\rho_\Sigma}(\|\boldsymbol{u}\|_\rho)\right)\Big)^2 \|\boldsymbol{u}\|_\rho^2 < \infty,$$

i.e., $F(\boldsymbol{u}) \in \ell^2_\rho$ for every $\boldsymbol{u} \in \ell^2_\rho$. In addition, by Assumption 19.5,

$$\langle F(\boldsymbol{u}), \boldsymbol{u}\rangle_\rho = \sum_{\mathrm{i}\in\mathbb{Z}^d} \rho_{\mathrm{i}} u_{\mathrm{i}} f_i(u_{\mathrm{i}}) \le -\alpha\|\boldsymbol{u}\|_\rho^2 + \|\boldsymbol{\beta}\|_\rho^2, \quad \text{for all } \boldsymbol{u} = (u_{\mathrm{i}})_{\mathrm{i}\in\mathbb{Z}^d} \in \ell^2_\rho.$$

□

Lemma 19.2. *For $\epsilon > 0$ define $\mathfrak{S}^\epsilon(\boldsymbol{u}) = (\mathfrak{S}^\epsilon_{\mathrm{i}}(\boldsymbol{u}))_{\mathrm{i}\in\mathbb{Z}^d}$ for $\boldsymbol{u} \in \ell^2_\rho$ componentwise by*

$$\mathfrak{S}^\epsilon_{\mathrm{i}}(\boldsymbol{u}) := \sum_{\mathrm{j}\in\mathbb{Z}^d} \kappa_{\mathrm{i},\mathrm{j}} \sigma_\epsilon(u_{\mathrm{j}} - \varsigma), \quad \mathrm{i} \in \mathbb{Z}^d. \tag{19.7}$$

Then $\mathfrak{S}^\epsilon$ is globally Lipschitz continuous on ℓ^2_ρ.

Proof. Let $\boldsymbol{u} = (u_{\mathrm{i}})_{\mathrm{i}\in\mathbb{Z}^d} \in \ell^2_\rho$, and $\boldsymbol{v} = (v_{\mathrm{i}})_{\mathrm{i}\in\mathbb{Z}^d} \in \ell^2_\rho$. Notice that the sigmoidal function σ_ϵ is globally Lipschitz with the Lipschitz constant $1/\epsilon$. Then for each $\mathrm{i} \in \mathbb{Z}^d$

$$|\mathfrak{S}^\epsilon_{\mathrm{i}}(\boldsymbol{u}) - \mathfrak{S}^\epsilon_{\mathrm{i}}(\boldsymbol{v})| \le \sum_{\mathrm{i}\in\mathbb{Z}^d} \kappa_{\mathrm{i},\mathrm{j}} \left|\sigma_\epsilon(u_{\mathrm{j}} - \varsigma) - \sigma_\epsilon(v_{\mathrm{j}} - \varsigma)\right| \le \frac{1}{\epsilon} \sum_{\mathrm{j}\in\mathbb{Z}^d} \kappa_{\mathrm{i},\mathrm{j}} |u_{\mathrm{j}} - v_{\mathrm{j}}|.$$

Consequently, using Cauchy's inequality and Assumptions 19.1 and 19.2,

$$\begin{aligned}\|\mathfrak{S}^\epsilon(\boldsymbol{u}) - \mathfrak{S}^\epsilon(\boldsymbol{v})\|_\rho^2 &\le \frac{1}{\epsilon^2} \sum_{\mathrm{i}\in\mathbb{Z}^d} \rho_{\mathrm{i}} \Big(\sum_{\mathrm{j}\in\mathbb{Z}^d} \kappa_{\mathrm{i},\mathrm{j}} |u_{\mathrm{j}} - v_{\mathrm{j}}|\Big)^2 \\ &\le \frac{1}{\epsilon^2} \sum_{\mathrm{i}\in\mathbb{Z}^d} \rho_{\mathrm{i}} \Big(\sum_{\mathrm{j}\in\mathbb{Z}^d} \frac{\kappa_{\mathrm{i},\mathrm{j}}^2}{\rho_{\mathrm{j}}}\Big)\Big(\sum_{\mathrm{j}\in\mathbb{Z}^d} \rho_{\mathrm{j}} |u_{\mathrm{j}} - v_{\mathrm{j}}|^2\Big) \\ &\le \frac{1}{\epsilon^2} \rho_\Sigma \kappa \|\boldsymbol{u} - \boldsymbol{v}\|_\rho^2,\end{aligned}$$

i.e., $\mathfrak{S}^\epsilon$ is globally Lipschitz continuous with Lipschitz constant $\sqrt{\rho_\Sigma \kappa}/\epsilon$. □

Lemma 19.3. *The function $\mathfrak{F}^\epsilon(\boldsymbol{u}) := F(\boldsymbol{u}) + \mathfrak{S}^\epsilon(\boldsymbol{u}) + \boldsymbol{g}$ is locally Lipschitz on ℓ^2_ρ and satisfies a dissipativity property*

$$\langle \mathfrak{F}^\epsilon(\boldsymbol{u}), \boldsymbol{u}\rangle_\rho \le -\frac{\alpha}{2}\|\boldsymbol{u}\|_\rho^2 + \|\boldsymbol{\beta}\|_\rho^2 + \frac{1}{\alpha}\left(\kappa\rho_\Sigma^2 + \|\boldsymbol{g}\|_\rho^2\right), \qquad \boldsymbol{u} \in \ell^2_\rho.$$

Proof. The Lipschitz property follows from the corresponding properties of the individual terms. Then, by Cauchy inequality we have

$$\sum_{j\in\mathbb{Z}^d}\kappa_{i,j}\le\sum_{j\in\mathbb{Z}^d}|\kappa_{i,j}|\le\Big(\sum_{j\in\mathbb{Z}^d}\frac{\kappa_{i,j}^2}{\rho_j}\Big)^{1/2}\Big(\sum_{j\in\mathbb{Z}^d}\rho_j\Big)^{1/2}\le\sqrt{\kappa\rho_\Sigma}. \tag{19.8}$$

Thus for $\boldsymbol{u}\in\ell^2_\rho$,

$$\begin{aligned}\langle\mathfrak{F}^\epsilon(\boldsymbol{u}),\boldsymbol{u}\rangle_\rho&\le\langle F(\boldsymbol{u}),\boldsymbol{u}\rangle_\rho+\langle\mathfrak{S}^\epsilon(\boldsymbol{u}),\boldsymbol{u}\rangle_\rho+\langle\boldsymbol{g},\boldsymbol{u}\rangle_\rho\\&\le-\alpha\|\boldsymbol{u}\|^2_\rho+\|\boldsymbol{\beta}\|^2_\rho+\frac{1}{\alpha}\sum_{i\in\mathbb{Z}^d}\rho_i\Big(\sum_{j\in\mathbb{Z}^d}\kappa_{i,j}\Big)^2+\frac{\alpha}{4}\|\boldsymbol{u}\|^2_\rho+\frac{\|\boldsymbol{g}\|^2_\rho}{\alpha}+\frac{\alpha}{4}\|\boldsymbol{u}\|^2_\rho\\&\le-\frac{\alpha}{2}\|\boldsymbol{u}\|^2_\rho+\|\boldsymbol{\beta}\|^2_\rho+\frac{1}{\alpha}\left(\kappa\rho^2_\Sigma+\|\boldsymbol{g}\|^2_\rho\right).\end{aligned}$$

□

19.2 Set-valued lattice systems

The discontinuities caused by the Heaviside function H in the lattice system (19.1) will be handled by formulating the lattice equations in (19.1) as the lattice differential inclusions (19.2).

For $\boldsymbol{u}=(u_i)_{i\in\mathbb{Z}^d}\in\ell^2_\rho$ define the set-valued operator $\mathfrak{K}$ by $\mathfrak{K}(\boldsymbol{u}):=(\mathfrak{K}_i(\boldsymbol{u}))_{i\in\mathbb{Z}^d}$ where

$$\mathfrak{K}_i(\boldsymbol{u})=\sum_{j\in\mathbb{Z}^d}\kappa_{i,j}\chi\left(u_j-\varsigma\right)=\sum_{j\in\mathbb{Z}^d}\kappa_{i,j}\begin{cases}\{0\}, & u_j<\varsigma,\\ [0,1], & u_j=\varsigma,\\ \{1\}, & u_j>\varsigma.\end{cases}$$

Then the system (19.2) can be written as

$$\frac{du_i}{dt}\in f_i(u_i)+\mathfrak{K}_i(\boldsymbol{u})+g_i,\qquad i\in\mathbb{Z}^d. \tag{19.9}$$

Lemma 19.4. *The set-valued operator $\mathfrak{K}$ maps ℓ^2_ρ to ℓ^2_ρ.*

Proof. Given any $\boldsymbol{u}=(u_i)_{i\in\mathbb{Z}^d}$, denote by $\mathbb{Z}^{\boldsymbol{u}}_0$, $\mathbb{Z}^{\boldsymbol{u}}_1$, $\mathbb{Z}^{\boldsymbol{u}}_\varsigma$ the set of vectorial integers such that $u_i<\varsigma$, $u_i>\varsigma$ and $u_i=\varsigma$, respectively, i.e.,

$$\mathbb{Z}^{\boldsymbol{u}}_0:=\left\{i\in\mathbb{Z}^d:u_i<\varsigma\right\},\quad\mathbb{Z}^{\boldsymbol{u}}_1:=\left\{i\in\mathbb{Z}^d:u_i>\varsigma\right\},\quad\mathbb{Z}^{\boldsymbol{u}}_\varsigma:=\left\{i\in\mathbb{Z}^d:u_i=\varsigma\right\}.$$

Then for any $i\in\mathbb{Z}^d$,

$$\mathfrak{K}_i(\boldsymbol{u})=\sum_{j\in\mathbb{Z}^{\boldsymbol{u}}_1}\kappa_{i,j}+\sum_{j\in\mathbb{Z}^{\boldsymbol{u}}_\varsigma}\kappa_{i,j}[0,1],$$

and hence for every $\boldsymbol{u}\in\ell^2_\rho$, by (19.8) it follows that

$$|||\mathfrak{K}_i(\boldsymbol{u})|||\le\sum_{j\in\mathbb{Z}^{\boldsymbol{u}}_1\cup\mathbb{Z}^{\boldsymbol{u}}_\varsigma}\kappa_{i,j}\le\sum_{j\in\mathbb{Z}^d}\kappa_{i,j}\le\sqrt{\kappa\rho_\Sigma},\quad\text{for all } i\in\mathbb{Z}^d.$$

Then by (19.5) and Assumption 19.3

$$|||\mathfrak{K}(\boldsymbol{u})|||^2_\rho\le\sum_{i\in\mathbb{Z}^d}\rho_i\,|||\,\mathfrak{K}_i(\boldsymbol{u})|||^2\le\kappa\rho^2_\Sigma<\infty.$$

□

Lemma 19.4 ensures that the set-valued lattice system (19.9) can be written as a differential inclusion on ℓ_ρ^2:

$$\frac{\mathrm{d}\boldsymbol{u}(t)}{\mathrm{d}t} \in \mathfrak{F}(\boldsymbol{u}(t)) := F(\boldsymbol{u}(t)) + \mathfrak{K}(\boldsymbol{u}(t)) + \boldsymbol{g}. \tag{19.10}$$

A solution to the differential inclusion (19.10) is defined componentwise as follows.

Definition 19.1. A function $\boldsymbol{u}(\cdot) = (u_i(t\cdot))_{i\in\mathbb{Z}^d} : [t_0, t_0+T) \to \ell_\rho^2$ is called a solution to the differential inclusion (19.10) if it is an absolutely continuous function $\boldsymbol{u}(t) : [t_0, t_0+T] \to \ell_\rho^2$ such that

$$\frac{\mathrm{d}u_i(t)}{\mathrm{d}t} \in f_i(u_i(t)) + \mathfrak{K}_i(\boldsymbol{u}(t)) + g_i, \quad \text{for all } i \in \mathbb{Z}^d, \text{ a.e..}$$

In particular, this means [Aubin and Cellina (1984); Smirnov (2002)] that there is a measurable selection $\Upsilon(t) \in \mathfrak{K}(\boldsymbol{u}(t))$ such that

$$\frac{\mathrm{d}\boldsymbol{u}(t)}{\mathrm{d}t} = F(\boldsymbol{u}(t)) + \Upsilon(t) + \boldsymbol{g}, \quad \text{a.e..} \tag{19.11}$$

19.2.1 *Inflated lattice systems*

Inflated systems are used to compare perturbed or approximated systems with the original system, see [Kloeden and Kozyakin (2000)]. The set-valued mapping χ defined in (19.3) is inflated to give a new set-valued function χ_ϵ defined on $\mathbb{R}$ for $\epsilon \in (0, 1]$ by

$$\chi_\epsilon(s) = \begin{cases} [0, \epsilon], & s < -\delta(\epsilon), \\ [0, 1], & -\delta(\epsilon) \le s \le \delta(\epsilon), \\ [1-\epsilon, 1], & s > \delta(\epsilon), \end{cases} \qquad s \in \mathbb{R}, \tag{19.12}$$

where $\delta(\epsilon) > 0$ solves the algebraic equation

$$\sigma_\epsilon(\delta(\epsilon)) = 1 - \epsilon, \quad \text{i.e.,} \quad \frac{1}{1 + e^{-\delta(\epsilon)/\epsilon}} = 1 - \epsilon.$$

It is straightforward to check that

$$\frac{\delta(\epsilon)}{\epsilon} = \ln\frac{1-\epsilon}{\epsilon} \to \infty \quad \text{and} \quad \delta(\epsilon) = \epsilon \ln\frac{1-\epsilon}{\epsilon} \to 0 \text{ as } \epsilon \to 0.$$

The neural lattice inclusion (19.9) is inflated to give the inflated lattice inclusion

$$\frac{\mathrm{d}u_i(t)}{\mathrm{d}t} \in f_i(u_i(t)) + \mathfrak{K}_i^\epsilon(\boldsymbol{u}(t)) + g_i, \qquad i \in \mathbb{Z}^d, \tag{19.13}$$

where

$$\mathfrak{K}_i^\epsilon(\boldsymbol{u}) := \sum_{j\in\mathbb{Z}^d} \kappa_{i,j}\chi_\epsilon(u_j - \varsigma). \tag{19.14}$$

Define the set-valued mapping $\mathfrak{K}^\epsilon$ component wise as $\mathfrak{K}^\epsilon(\boldsymbol{u}) = (\mathfrak{K}_{\rm i}^\epsilon(\boldsymbol{u}))_{{\rm i}\in\mathbb{Z}^d}$ where $\mathfrak{K}_{\rm i}^\epsilon(\boldsymbol{u})$ is defined in (19.14). Since Lemma 19.4 also applies to $\mathfrak{K}^\epsilon$, the inflated lattice inclusion (19.13) can be written as the differential inclusion on ℓ_ρ^2:

$$\frac{{\rm d}\boldsymbol{u}(t)}{{\rm d}t} \in \mathfrak{G}^\epsilon(\boldsymbol{u}(t)) := F(\boldsymbol{u}(t)) + \mathfrak{K}^\epsilon(\boldsymbol{u}(t)) + \boldsymbol{g}. \tag{19.15}$$

19.2.2 *Relations between Heaviside, sigmoid, and inflated*

The range of the Heaviside function H is a subset of the set-valued function χ defined by (19.3), as well as the set-valued function χ_ϵ defined by (19.12). Given any $\epsilon > 0$ the range of the sigmoid function σ_ϵ is a subset of the the set-valued function χ_ϵ defined by (19.12), which approaches the set-valued function χ defined by (19.3) as $\epsilon \to 0$ (see Fig. 19.1 below).

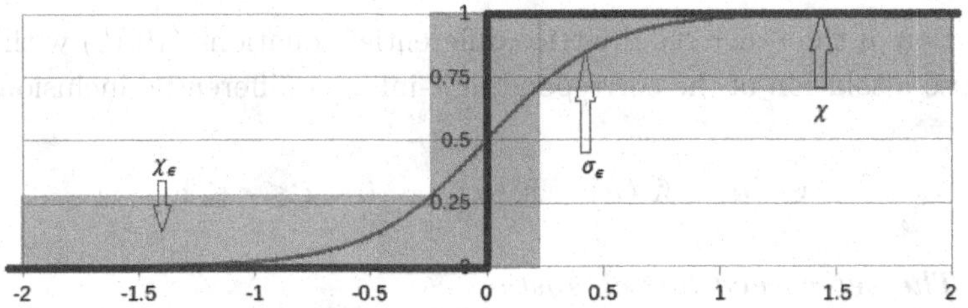

Fig. 19.1 Relations among σ_ϵ, χ and χ_ϵ

19.3 Existence of solutions

This section includes the existence results for the sigmoidal lattice system (19.4), the inflated system (19.15), and the set-valued lattice system (19.10).

19.3.1 *The sigmoidal lattice system*

The sigmoidal lattice model (19.4) can be formulated as an ODE

$$\frac{{\rm d}\boldsymbol{u}(t)}{{\rm d}t} = \mathfrak{F}^\epsilon(\boldsymbol{u}(t)) := F(\boldsymbol{u}(t)) + \mathfrak{S}^\epsilon(\boldsymbol{u}(t)) + \boldsymbol{g} \tag{19.16}$$

on the sequence space ℓ_ρ^2, where $\mathfrak{S}^\epsilon$ is defined in (19.7). By Lemmas 19.1–19.3 for every $\epsilon > 0$ the vector field $\mathfrak{F}^\epsilon(\boldsymbol{u})$ of (19.16) is locally Lipschitz from ℓ_ρ^2 into ℓ_ρ^2 and satisfies a dissipativity property and hence has bounded growth. Standard existence and uniqueness theorems of ordinary differential equations in Banach spaces (see, e.g., [Deimling (1977)]) can then be applied to give the following result.

Theorem 19.1. *Let $T > 0$ be arbitrary and suppose that Assumptions 19.1–19.5 hold. Then given any $\epsilon \in (0, 1]$ and initial data $\boldsymbol{u}_o = (u_{o,{\rm i}})_{{\rm i}\in\mathbb{Z}^d} \in \ell_\rho^2$, the ODE* (19.16) *admits a unique solution $\boldsymbol{u}^\epsilon(t;\boldsymbol{u}_o)$ in ℓ_ρ^2 which exists on $[0, T]$ with*

$\boldsymbol{u}^\epsilon(0; \boldsymbol{u}_o) = \boldsymbol{u}_o$. *Moreover, the solutions of* (19.16) *generate a semi-dynamical system* $\{\varphi^\epsilon(t)\}_{t\geq 0}$ *on* ℓ^2_ρ.

Note that the components of the solution to the equation (19.16) are solutions to the sigmoidal lattice differential equations (19.4).

19.3.2 *The inflated system*

Since $\mathfrak{S}^\epsilon(\boldsymbol{u}) \in \mathfrak{K}^\epsilon(\boldsymbol{u})$ it follows that a solution of the ϵ-sigmoidal ODE (19.16) is a solution of the ϵ-inflated lattice system. Hence the following theorem holds.

Theorem 19.2. *Let* $T > 0$ *and* $\epsilon \in (0, 1]$ *be arbitrary and suppose that Assumptions 19.1–19.5 hold. Then for any initial data* $\boldsymbol{u}_o = (u_{o,\mathrm{i}})_{\mathrm{i}\in\mathbb{Z}^d} \in \ell^2_\rho$, *the* ϵ*-inflated differential inclusion* (19.15) *admits a solution* $\boldsymbol{u}^\epsilon(t; \boldsymbol{u}_o)$ *in* ℓ^2_ρ *which exists on* $[0, T]$ *with* $\boldsymbol{u}^\epsilon(0; \boldsymbol{u}_o) = \boldsymbol{u}_o$.

A solution of the $\tilde{\epsilon}$-sigmoidal lattice differential equations (19.16) with $0 < \tilde{\epsilon} \leq \epsilon \leq 1$ is also a solution of the corresponding ϵ-inflated differential inclusion (19.15), since

$$\mathfrak{S}^{\tilde{\epsilon}}(\boldsymbol{u}) \in \mathfrak{K}^{\tilde{\epsilon}}(\boldsymbol{u}) \subset \mathfrak{K}^\epsilon(\boldsymbol{u}), \qquad 0 < \tilde{\epsilon} \leq \epsilon \leq 1.$$

19.3.3 *The set-valued lattice system*

The existence of solutions of the lattice inclusion (19.10) is stated in Theorem 19.3. Its proof uses a convergent subsequence of solutions of the sigmoidal lattice systems (19.4). The proof, which requires some preparation, is given in the next two sections.

Theorem 19.3. *Let* $T > 0$ *and assume that Assumptions 19.1–19.5 hold. Then for any* $\boldsymbol{u}_o = (u_{o,\mathrm{i}})_{\mathrm{i}\in\mathbb{Z}^d} \in \ell^2_\rho$, *the lattice inclusion* (19.10) *admits a solution* $\boldsymbol{u}(\cdot; \boldsymbol{u}_o)$ *existing on* $[t_0, t_0 + T]$ *with* $\boldsymbol{u}(0; \boldsymbol{u}_o) = \boldsymbol{u}_o$.

19.4 Convergence of sigmoidal solutions

The goal of this section is to show that solutions of the lattice system (19.4) with sigmoidal functions converge to the solution of the lattice system (19.1) with the Heaviside function as $\epsilon \to 0$. To this end first a series of lemmas that are crucial for this goal are established.

19.4.1 *A priori estimates*

This subsection is devoted to some preparatory lemmas on relation of the properties of the sigmoidal lattice system (19.4), the differential inclusion (19.9) and the inflated differential inclusion (19.15).

For each $N \in \mathbb{N}$ let

$$\mathbb{Z}^d_N := \{\mathrm{i} = (i_1, \ldots, i_d) \in \mathbb{Z}^d : |i_1|, \ldots, |i_d| \leq N\},$$

and introduce the truncated set-valued operator

$$\mathfrak{K}_{\mathrm{i}}^{\epsilon,N}(\boldsymbol{u}) = \sum_{\mathrm{j}\in\mathbb{Z}_N^d} \kappa_{\mathrm{i},\mathrm{j}}\chi_\epsilon\left(u_{\mathrm{j}} - \varsigma\right), \qquad \boldsymbol{u} = (u_{\mathrm{i}})_{\mathrm{i}\in\mathbb{Z}^d} \in \ell_\rho^2, \quad \epsilon \in (0,1], \tag{19.17}$$

where χ_ϵ is defined as in (19.12). Note that $\mathfrak{K}_{\mathrm{i}}^{\epsilon_1,N}(\boldsymbol{u}) \subset \mathfrak{K}_{\mathrm{i}}^{\epsilon_2,N}(\boldsymbol{u})$ if $\epsilon_1 \leq \epsilon_2$.

The proofs of the lemmas below require the following inequality [Diamond and Kloeden (1994), Proposition 2.4.1(ii)]: for any nonempty compact subsets A_1, A_2, B_1, B_2 of $\mathbb{R}^d$, it holds

$$\mathrm{dist}_{\mathbb{R}^d}(A_1 + B_1, A_2 + B_2) \leq \mathrm{dist}_{\mathbb{R}^d}(A_1, A_2) + \mathrm{dist}_{\mathbb{R}^d}(B_1, B_2) \tag{19.18}$$

where $A + B := \{a + b : a \in A, b \in B\}$. Note that the inequality (19.18) can be generalised to for any $N > 1$, i.e.,

$$\mathrm{dist}_{\mathbb{R}^d}\left(\frac{1}{N}\sum_{r=1}^{N} A_r, \frac{1}{N}\sum_{r=1}^{N} B_r\right) \leq \frac{1}{N}\sum_{r=1}^{N}\mathrm{dist}_{\mathbb{R}^d}(A_r, B_r) \tag{19.19}$$

for any nonempty compact subsets $A_1, \ldots, A_N$, $B_1, \ldots, B_N$ of $\mathbb{R}^d$.

Lemma 19.5. *For each* $\mathrm{i} \in \mathbb{Z}^d$ *and every* $\mathcal{E} > 0$ *there exists* $N_0(\mathcal{E}, \mathrm{i}) \in \mathbb{N}$ *such that*

$$\mathrm{dist}_{\mathbb{R}^1}\left(\mathfrak{K}_{\mathrm{i}}^{\epsilon}(\boldsymbol{u}), \mathfrak{K}_{\mathrm{i}}^{\epsilon,N}(\boldsymbol{u})\right) \leq \mathcal{E} \quad \textit{for all} \quad N \geq N_0(\mathcal{E}, \mathrm{i}), \quad \boldsymbol{u} \in \ell_\rho^2, \ \epsilon \in (0,1].$$

Proof. For each $\mathrm{i} \in \mathbb{Z}^d$, define the operator $\mathfrak{R}_{\mathrm{i}}^{\epsilon,N}$ by

$$\mathfrak{R}_{\mathrm{i}}^{\epsilon,N}(\boldsymbol{u}) := \sum_{\mathrm{j}\in\mathbb{Z}^d\setminus\mathbb{Z}_N^d} \kappa_{\mathrm{i},\mathrm{j}}\chi_\epsilon\left(u_{\mathrm{j}} - \varsigma\right). \tag{19.20}$$

Then $\mathfrak{K}_{\mathrm{i}}^{\epsilon}(\boldsymbol{u}) = \mathfrak{K}_{\mathrm{i}}^{\epsilon,N}(\boldsymbol{u}) + \mathfrak{R}_{\mathrm{i}}^{\epsilon,N}(\boldsymbol{u})$ and by the inequality (19.18)

$$\begin{aligned}
\mathrm{dist}_{\mathbb{R}^1}\left(\mathfrak{K}_{\mathrm{i}}^{\epsilon}(\boldsymbol{u}), \mathfrak{K}_{\mathrm{i}}^{\epsilon,N}(\boldsymbol{u})\right) &= \mathrm{dist}_{\mathbb{R}^1}\left(\mathfrak{K}_{\mathrm{i}}^{\epsilon,N}(\boldsymbol{u}) + \mathfrak{R}_{\mathrm{i}}^{\epsilon,N}(\boldsymbol{u}), \mathfrak{K}_{\mathrm{i}}^{\epsilon,N}(\boldsymbol{u}) + \{0\}\right) \\
&\leq \mathrm{dist}_{\mathbb{R}^1}\left(\mathfrak{R}_{\mathrm{i}}^{\epsilon,N}(\boldsymbol{u}), \{0\}\right) \\
&\leq |||\mathfrak{R}_{\mathrm{i}}^{\epsilon,N}(\boldsymbol{u})||| \leq \sum_{\mathrm{j}\in\mathbb{Z}^d\setminus\mathbb{Z}_N^d} \kappa_{\mathrm{i},\mathrm{j}} \quad \text{for all} \quad \boldsymbol{u} \in \ell_\rho^2,
\end{aligned}$$

because $|||\chi_\epsilon\left(u_{\mathrm{j}} - \varsigma\right)||| \leq 1$ for all $\boldsymbol{u} \in \ell_\rho^2$.

In addition, since $\sum_{\mathrm{j}\in\mathbb{Z}^d} \kappa_{\mathrm{i},\mathrm{j}} \leq \kappa\rho_\Sigma^2 < \infty$, there exists $N_0(\mathcal{E}, \mathrm{i}) \in \mathbb{N}$ such that

$$\sum_{\mathrm{j}\in\mathbb{Z}^d\setminus\mathbb{Z}_N^d} \kappa_{\mathrm{i},\mathrm{j}} \leq \mathcal{E} \quad \text{for all} \quad N \geq N_0(\mathcal{E}, \mathrm{i}),$$

which implies the assertion of this lemma. □

Lemma 19.6. *For every fixed $N \in \mathbb{N}$, $\mathrm{i} \in \mathbb{Z}^d$ and $\epsilon \in (0,1]$, the set-valued mapping $\boldsymbol{u} \mapsto \mathfrak{K}_{\mathrm{i}}^{\epsilon,N}(\boldsymbol{u})$ is upper semi-continuous from ℓ_ρ^2 into the space of nonempty compact convex subsets of $\mathbb{R}^1$, i.e.,*

$$\operatorname{dist}_{\mathbb{R}^1}\left(\mathfrak{K}_{\mathrm{i}}^{\epsilon,N}(\boldsymbol{u}^{(m)}), \mathfrak{K}_{\mathrm{i}}^{\epsilon,N}(\boldsymbol{u}^*)\right) \to 0 \ \text{for} \ \ \boldsymbol{u}^{(m)} \to \boldsymbol{u}^* \ \text{in} \ \ell_\rho^2, \quad \text{as} \ m \to \infty.$$

Proof. Since $\boldsymbol{u}^{(m)} \to \boldsymbol{u}^*$ in ℓ_ρ^2, for every $\epsilon > 0$ there exists $M(\epsilon)$ such that

$$\left\|\boldsymbol{u}^{(m)} - \boldsymbol{u}^*\right\|_\rho^2 = \sum_{\mathrm{j}\in\mathbb{Z}^d} \rho_{\mathrm{i}} \left|u_{\mathrm{j}}^{(m)} - u_{\mathrm{j}}^*\right|^2 < \epsilon^2 \quad \text{for all } m \geq M(\epsilon).$$

Considering only $\mathrm{j} \in \mathbb{Z}_{\mathbb{N}}^d$ as appearing in $\mathfrak{K}_{\mathrm{i}}^{\epsilon,N}$, then

$$\left|u_{\mathrm{j}}^{(m)} - u_{\mathrm{j}}^*\right| < \epsilon/\sqrt{\rho_N} \quad \text{for all} \ \ m \geq M_\epsilon, \ \mathrm{j} \in \mathbb{Z}_N^d.$$

where $\rho_N := \min_{\mathrm{j}\in\mathbb{Z}_N^d} \rho_{\mathrm{j}}$.

The set-valued mapping $s \mapsto \chi_\epsilon(s - \varsigma)$ is upper semi-continuous for $s \in \mathbb{R}$. Since there are a finite number of terms in the sum in the definition of $\mathfrak{K}_{\mathrm{i}}^{\epsilon,N}$ it follows from the inequality (19.18) that the set-valued mapping $\boldsymbol{u} \mapsto \mathfrak{K}_{\mathrm{i}}^{\epsilon,N}(\boldsymbol{u})$ is also upper semi-continuous. □

The following lemma is crucial to the analysis in the sequel.

Lemma 19.7. *The set-valued mapping $\mathfrak{K}^\epsilon : \ell_\rho^2 \to \ell_\rho^2$ with $\mathfrak{K}^\epsilon(\boldsymbol{u}) = (\mathfrak{K}_{\mathrm{i}}^\epsilon(\boldsymbol{u}))_{\mathrm{i}\in\mathbb{Z}}$ is upper semi-continuous on ℓ_ρ^2.*

Proof. For each $N \in \mathbb{N}$ write

$$\mathfrak{K}^\epsilon(\boldsymbol{u}) = \mathfrak{K}^{\epsilon,N}(\boldsymbol{u}) + \mathfrak{R}^{\epsilon,N}(\boldsymbol{u})$$

where the operator $\mathfrak{K}^{\epsilon,N}$ is defined in (19.17) and $\mathfrak{R}^{\epsilon,N} = (\mathfrak{R}_{\mathrm{i}}^{\epsilon,N})_{\mathrm{i}\in\mathbb{Z}^d}$ is defined component wise by (19.20).

Let $\boldsymbol{u}^{(m)} \to \boldsymbol{u}^*$ as $m \to \infty$ in ℓ_ρ^2. Then for each $\mathrm{i} \in \mathbb{Z}^d$ by repeatedly using the inequality (19.18) gives

$$\begin{aligned}
&\operatorname{dist}_{\mathbb{R}^1}\left(\mathfrak{K}_{\mathrm{i}}^\epsilon(\boldsymbol{u}^{(m)}), \mathfrak{K}_{\mathrm{i}}^\epsilon(\boldsymbol{u}^*)\right) \\
&= \operatorname{dist}_{\mathbb{R}^1}\left(\mathfrak{K}_{\mathrm{i}}^{\epsilon,N}(\boldsymbol{u}^{(m)}) + \mathfrak{R}_{\mathrm{i}}^{\epsilon,N}(\boldsymbol{u}^{(m)}), \mathfrak{K}_{\mathrm{i}}^{\epsilon,N}(\boldsymbol{u}^*) + \mathfrak{R}_{\mathrm{i}}^{\epsilon,N}(\boldsymbol{u}^*)\right) \\
&\leq \operatorname{dist}_{\mathbb{R}^1}\left(\mathfrak{K}_{\mathrm{i}}^{\epsilon,N}(\boldsymbol{u}^{(m)}), \mathfrak{K}_{\mathrm{i}}^{\epsilon,N}(\boldsymbol{u}^*)\right) + \operatorname{dist}_{\mathbb{R}^1}\left(\mathfrak{R}_{\mathrm{i}}^{\epsilon,N}(\boldsymbol{u}^{(m)}), \mathfrak{R}_{\mathrm{i}}^{\epsilon,N}(\boldsymbol{u}^*)\right) \\
&\leq \operatorname{dist}_{\mathbb{R}^1}\left(\mathfrak{K}_{\mathrm{i}}^{\epsilon,N}(\boldsymbol{u}^{(m)}), \mathfrak{K}_{\mathrm{i}}^{\epsilon,N}(\boldsymbol{u}^*)\right) + \operatorname{dist}_{\mathbb{R}^1}\left(\mathfrak{R}_{\mathrm{i}}^{\epsilon,N}(\boldsymbol{u}^{(m)}), \{0\}\right) \\
&\quad + \operatorname{dist}_{\mathbb{R}^1}\left(\{0\}, \mathfrak{R}_{\mathrm{i}}^{\epsilon,N}(\boldsymbol{u}^*)\right).
\end{aligned}$$

Noticing that $|||\chi_\epsilon(u_j - \varsigma)||| := \sup_{x\in\chi_\epsilon(u_j-\varsigma)} |x| \le 1$, we have

$$\mathrm{dist}_{\mathbb{R}^1}\left(\mathfrak{R}_i^{\epsilon,N}(\boldsymbol{u}^{(m)}), \{0\}\right) \le \sum_{j\in\mathbb{Z}^d\setminus\mathbb{Z}_N^d} |\kappa_{i,j}|,$$

$$\mathrm{dist}_{\mathbb{R}^1}\left(\{0\}, \mathfrak{R}_i^{\epsilon,N}(\boldsymbol{u}^*)\right) \le \sum_{j\in\mathbb{Z}^d\setminus\mathbb{Z}_N^d} |\kappa_{i,j}|,$$

and thus

$$\mathrm{dist}_{\mathbb{R}^1}\left(\mathfrak{K}_i^{\epsilon}(\boldsymbol{u}^{(m)}), \mathfrak{K}_i^{\epsilon}(\boldsymbol{u}^*)\right) \le \mathrm{dist}_{\mathbb{R}^1}\left(\mathfrak{K}_i^{\epsilon,N}(\boldsymbol{u}^{(m)}), \mathfrak{K}_i^{\epsilon,N}(\boldsymbol{u}^*)\right) + 2\sum_{j\in\mathbb{Z}^d\setminus\mathbb{Z}_N^d} |\kappa_{i,j}|.$$

First by recalling the inequality (19.8) that $\sum_{j\in\mathbb{Z}^d} |\kappa_{i,j}| \le \sqrt{\kappa\rho_\Sigma}$, for any $\mathcal{E} > 0$ there exists $N_0(\mathcal{E}) \in \mathbb{N}$ such that

$$\sum_{j\in\mathbb{Z}^d\setminus\mathbb{Z}_N^d} |\kappa_{i,j}| \le \frac{\mathcal{E}}{4}, \qquad \text{for all } N \ge N_0(\mathcal{E}),\ i \in \mathbb{Z}^d. \tag{19.21}$$

Hence, for any $N \ge N_0(\mathcal{E})$,

$$\begin{aligned}
\left(\mathrm{dist}_{\ell^2_\rho}\left(\mathfrak{K}^{\epsilon}(\boldsymbol{u}^{(m)}), \mathfrak{K}^{\epsilon}(\boldsymbol{u}^*)\right)\right)^2 &\le \sum_{i\in\mathbb{Z}^d} \rho_i \left(\mathrm{dist}_{\mathbb{R}^1}\left(\mathfrak{K}_i^{\epsilon}(\boldsymbol{u}^{(m)}), \mathfrak{K}_i^{\epsilon}(\boldsymbol{u}^*)\right)\right)^2 \\
&\le \sum_{i\in\mathbb{Z}^d} \rho_i \left(\mathrm{dist}_{\mathbb{R}^1}\left(\mathfrak{K}_i^{\epsilon,N}(\boldsymbol{u}^{(m)}), \mathfrak{K}_i^{\epsilon,N}(\boldsymbol{u}^*)\right) + 2\mathcal{E}\right)^2 \\
&\le 2\sum_{i\in\mathbb{Z}^d} \rho_i \left(\mathrm{dist}_{\mathbb{R}^1}\left(\mathfrak{K}_i^{\epsilon,N}(\boldsymbol{u}^{(m)}), \mathfrak{K}_i^{\epsilon,N}(\boldsymbol{u}^*)\right)\right)^2 + 8\mathcal{E}^2 \sum_{i\in\mathbb{Z}^d} \rho_i \\
&\le 2\sum_{i\in\mathbb{Z}^d} \rho_i \left(\mathrm{dist}_{\mathbb{R}^1}\left(\mathfrak{K}_i^{\epsilon,N}(\boldsymbol{u}^{(m)}), \mathfrak{K}_i^{\epsilon,N}(\boldsymbol{u}^*)\right)\right)^2 + 8\mathcal{E}^2 \rho_\Sigma.
\end{aligned}$$

From Lemma 19.6 the set-valued mapping $\boldsymbol{u} \mapsto \mathfrak{K}_i^{\epsilon,N}(\boldsymbol{u})$ is upper semi-continuous from ℓ^2_ρ into the nonempty compact convex subsets of $\mathbb{R}^1$ for each $i \in \mathbb{Z}^d$, i.e., given $\mathcal{E} > 0$ there exists $M(\mathcal{E}, \mathbb{N}, i) \in \mathbb{N}$ such that

$$\mathrm{dist}_{\mathbb{R}^1}\left(\mathfrak{K}_i^{\epsilon,N}(\boldsymbol{u}^{(m)}), \mathfrak{K}_i^{\epsilon,N}(\boldsymbol{u}^*)\right) \le \mathcal{E} \quad \text{for} \quad m \ge M(\mathcal{E}, N, i).$$

Given $\mathcal{E} > 0$ fix an $N_{\mathcal{E}} \ge N_0(\mathcal{E})$ so the inequality (19.21) holds. Then for this $N_{\mathcal{E}}$ define

$$\widehat{M}(\mathcal{E}) := \max\left\{M(\mathcal{E}, N_{\mathcal{E}}, i) \ :\ i \in \mathbb{Z}^d_{N_{\mathcal{E}}}\right\}.$$

Hence

$$\mathrm{dist}_{\mathbb{R}^1}\left(\mathfrak{K}_i^{\epsilon,N_{\mathcal{E}}}(\boldsymbol{u}^{(m)}), \mathfrak{K}_i^{\epsilon,N_{\mathcal{E}}}(\boldsymbol{u}^*)\right) \le \mathcal{E} \quad \text{for} \quad m \ge \widehat{M}(\mathcal{E}),\ i \in \mathbb{Z}^d_{N_{\mathcal{E}}}.$$

Thus combining the above results, for $m \ge \widehat{M}(\mathcal{E})$,

$$\begin{aligned}
\mathrm{dist}_{\ell^2_\rho}\left(\mathfrak{K}^{\epsilon}(\boldsymbol{u}^{(m)}), \mathfrak{K}^{\epsilon}(\boldsymbol{u}^*)\right)^2 &\le 2\sum_{i\in\mathbb{Z}^d} \rho_i \left[\mathrm{dist}_{\mathbb{R}^1}\left(\mathfrak{K}_i^{\epsilon,N_{\mathcal{E}}}(\boldsymbol{u}^{(m)}), \mathfrak{K}_i^{\epsilon,N_{\mathcal{E}}}(\boldsymbol{u}^*)\right)\right]^2 + 8\mathcal{E}^2\rho_\Sigma \\
&\le 2\sum_{i\in\mathbb{Z}^d} \rho_i \mathcal{E}^2 + 8\mathcal{E}^2\rho_\Sigma \le 10\mathcal{E}^2\rho_\Sigma.
\end{aligned}$$

This completes the proof. □

The upper semi-continuity of the set-valued mapping $\mathfrak{K}_{\mathrm{i}}^{\epsilon}$ follows directly from the above lemma since convergence in the norm of ℓ_ρ^2 implies component wise convergence.

Corollary 19.1. *For every* $\mathrm{i} \in \mathbb{Z}^d$ *and* $\epsilon \in (0,1]$, *the set-valued mapping* $\boldsymbol{u} \mapsto \mathfrak{K}_{\mathrm{i}}^{\epsilon}(\boldsymbol{u})$ *is upper semi-continuous from* ℓ_ρ^2 *into the nonempty compact convex subsets of* $\mathbb{R}^1$, *i.e.,*

$$\mathrm{dist}_{\mathbb{R}^1}\left(\mathfrak{K}_{\mathrm{i}}^{\epsilon}(\boldsymbol{u}^{(m)}), \mathfrak{K}_{\mathrm{i}}^{\epsilon}(\boldsymbol{u}^*)\right) \to 0 \ \textit{for} \ \ \boldsymbol{u}^{(m)} \to \boldsymbol{u}^* \ \textit{in} \ \ell_\rho^2, \quad \textit{as} \ m \to \infty.$$

Lemma 19.8. *For every* $\boldsymbol{u} \in \ell_\rho^2$ *and* $\mathrm{i} \in \mathbb{Z}^d$

$$\lim_{\epsilon \to 0} \mathrm{dist}_{\mathbb{R}^1}\left(\mathfrak{K}_{\mathrm{i}}^{\epsilon}(\boldsymbol{u}), \mathfrak{K}_{\mathrm{i}}(\boldsymbol{u})\right) = 0.$$

Proof. The terminology of Lemma 19.7 will be used. For each $N \in \mathbb{N}$ write

$$\mathfrak{K}^{\epsilon}(\boldsymbol{u}) = \mathfrak{K}^{\epsilon,N}(\boldsymbol{u}) + \mathfrak{R}^{\epsilon,N}(\boldsymbol{u}), \quad \mathfrak{K}(\boldsymbol{u}) = \mathfrak{K}^{N}(\boldsymbol{u}) + \mathfrak{R}^{N}(\boldsymbol{u}).$$

Fix $\boldsymbol{u} \in \ell_\rho^2$. Then for each $\mathrm{i} \in \mathbb{Z}^d$, similarly to the proof of Lemma 19.7, we obtain

$$\begin{aligned}
\mathrm{dist}_{\mathbb{R}^1}\left(\mathfrak{K}_{\mathrm{i}}^{\epsilon}(\boldsymbol{u}), \mathfrak{K}_{\mathrm{i}}(\boldsymbol{u})\right) &= \mathrm{dist}_{\mathbb{R}^1}\left(\mathfrak{K}_{\mathrm{i}}^{\epsilon,N}(\boldsymbol{u}) + \mathfrak{R}_{\mathrm{i}}^{\epsilon,N}(\boldsymbol{u}), \mathfrak{K}_{\mathrm{i}}^{N}(\boldsymbol{u}) + \mathfrak{R}_{\mathrm{i}}^{N}(\boldsymbol{u})\right) \\
&\le \mathrm{dist}_{\mathbb{R}^1}\left(\mathfrak{K}_{\mathrm{i}}^{\epsilon,N}(\boldsymbol{u}), \mathfrak{K}_{\mathrm{i}}^{N}(\boldsymbol{u})\right) + \mathrm{dist}_{\mathbb{R}^1}\left(\mathfrak{R}_{\mathrm{i}}^{\epsilon,N}(\boldsymbol{u}), \mathfrak{R}_{\mathrm{i}}^{N}(\boldsymbol{u})\right) \\
&\le \mathrm{dist}_{\mathbb{R}^1}\left(\mathfrak{K}_{\mathrm{i}}^{\epsilon,N}(\boldsymbol{u}), \mathfrak{K}_{\mathrm{i}}^{N}(\boldsymbol{u})\right) + \mathrm{dist}_{\mathbb{R}^1}\left(\mathfrak{R}_{\mathrm{i}}^{\epsilon,N}(\boldsymbol{u}), \{0\}\right) \\
&\quad + \mathrm{dist}_{\mathbb{R}^1}\left(\{0\}, \mathfrak{R}_{\mathrm{i}}^{N}(\boldsymbol{u})\right) \\
&\le \mathrm{dist}_{\mathbb{R}^1}\left(\mathfrak{K}_{\mathrm{i}}^{\epsilon,N}(\boldsymbol{u}), \mathfrak{K}_{\mathrm{i}}^{N}(\boldsymbol{u})\right) + 2 \sum_{\mathrm{j} \in \mathbb{Z}^d \setminus \mathbb{Z}_N^d} |\kappa_{\mathrm{i},\mathrm{j}}|. \qquad (19.22)
\end{aligned}$$

By construction $\mathrm{dist}_{\mathbb{R}^1}(\chi_\epsilon(\boldsymbol{u}), \chi(\boldsymbol{u})) \le \epsilon$ for every $\boldsymbol{u} \in \ell_\rho^2$. Thus

$$\begin{aligned}
\mathrm{dist}_{\mathbb{R}^1}\left(\mathfrak{K}_{\mathrm{i}}^{\epsilon,N}(\boldsymbol{u}), \mathfrak{K}_{\mathrm{i}}^{N}(\boldsymbol{u})\right) &= \mathrm{dist}_{\mathbb{R}^1}\left(\sum_{\mathrm{j} \in \mathbb{Z}_N^d} \kappa_{\mathrm{i},\mathrm{j}} \chi_\epsilon(\boldsymbol{u}), \sum_{\mathrm{j} \in \mathbb{Z}_N^d} \kappa_{\mathrm{i},\mathrm{j}} \chi(\boldsymbol{u})\right) \\
&\le \sum_{\mathrm{j} \in \mathbb{Z}_N^d} \kappa_{\mathrm{i},\mathrm{j}} \mathrm{dist}_{\mathbb{R}^1}\left(\chi_\epsilon(\boldsymbol{u}), \chi(\boldsymbol{u})\right) \\
&\le \sum_{\mathrm{j} \in \mathbb{Z}_N^d} |\kappa_{\mathrm{i},\mathrm{j}}| \epsilon \le \sqrt{\kappa \rho_\Sigma}\, \epsilon. \qquad (19.23)
\end{aligned}$$

Inserting (19.23) into (19.22) results in

$$\begin{aligned}
\mathrm{dist}_{\mathbb{R}^1}\left(\mathfrak{K}_{\mathrm{i}}^{\epsilon}(\boldsymbol{u}), \mathfrak{K}_{\mathrm{i}}(\boldsymbol{u})\right) &\le \sqrt{\kappa \rho_\Sigma}\, \epsilon + 2 \sum_{\mathrm{j} \in \mathbb{Z}^d \setminus \mathbb{Z}_N^d} |\kappa_{\mathrm{i},\mathrm{j}}| \\
&\le \sqrt{\kappa \rho_\Sigma}\, \epsilon + \epsilon = (\sqrt{\kappa \rho_\Sigma} + 1)\epsilon, \quad \text{for all } N \ge N_0(2\epsilon),\ \mathrm{i} \in \mathbb{Z}^d,
\end{aligned}$$

where $N_0(2\epsilon)$ is from the proof of Lemma 19.7 with $\mathcal{E}$ replaced by 2ϵ. □

19.4.2 The convergence theorem

The above preparation now allows a theorem to be stated and proved on the convergence of solutions of the sigmoidal lattice system (19.4) to solutions of the lattice inclusion system(19.9), i.e., the Heaviside system (19.1).

Given a sequence $\epsilon_m \to 0$ as $m \to \infty$ and any $\boldsymbol{u}_o = (u_{o,\mathrm{i}})_{\mathrm{i}\in\mathbb{Z}^d} \in \ell^2_\rho$ let $\boldsymbol{u}^{\epsilon_m}(t;\boldsymbol{u}_o) = (u^{\epsilon_m}_{\mathrm{i}}(t;\boldsymbol{u}_o))_{\mathrm{i}\in\mathbb{Z}^d}$ be the unique solution to the ϵ_m-sigmoidal lattice system (19.4), i.e., $u^{\epsilon_m}_{\mathrm{i}}(t;\boldsymbol{u}_o)$ satisfies (19.4) with $\epsilon = \epsilon_m$:

$$\frac{\mathrm{d}}{\mathrm{d}t}u^{\epsilon_m}_{\mathrm{i}}(t) = f_{\mathrm{i}}(u^{\epsilon_m}_{\mathrm{i}}(t)) + \mathfrak{S}^{\epsilon_m}_{\mathrm{i}}(\boldsymbol{u}^{\epsilon_m}(t)) + g_i, \quad u^{\epsilon_m}_{\mathrm{i}}(0) = u_{o,\mathrm{i}} \quad \text{for all } \ \mathrm{i}\in\mathbb{Z}^d, \tag{19.24}$$

where $\mathfrak{S}^{\epsilon_m}_{\mathrm{i}}(\boldsymbol{u}^{\epsilon_m}) = \sum_{\mathrm{j}\in\mathbb{Z}^d}\kappa_{\mathrm{i},\mathrm{j}}\sigma_{\epsilon_m}(u^{\epsilon_m}_{\mathrm{j}} - \varsigma)$ is defined according to (19.7).

Theorem 19.4. *Let $T>0$, $\boldsymbol{u}_o \in \ell^2_\rho$, $\epsilon_m \to 0$ as $m\to\infty$ and suppose that Assumptions 19.1–19.5 hold. Then for any sequence $\boldsymbol{u}^{\epsilon_m}(\cdot;\boldsymbol{u}_o)$ of solutions of the sigmoidal lattice systems (19.4) there is a subsequence $\boldsymbol{u}^{\epsilon_{m_r}}(\cdot;\boldsymbol{u}_o) \to \boldsymbol{u}^*(\cdot) \in \mathcal{C}([0,T],\ell^2_\rho)$ with $\epsilon_{m_r} \to 0$ as $r\to\infty$, where $\boldsymbol{u}^*(\cdot)$ is a solution of the lattice differential inclusion (19.9) with the initial value $\boldsymbol{u}^*(0) = \boldsymbol{u}_o$.*

Proof. The proof consists of four parts.

1. Componentwise convergent subsequence

First multiplying the equation (19.24) by $u^{\epsilon_m}_{\mathrm{i}}$ and using (19.8) and the assumption 19.5 to obtain

$$\begin{aligned}\frac{1}{2}\frac{\mathrm{d}}{\mathrm{d}t}|u^{\epsilon_m}_{\mathrm{i}}(t)|^2 &= u^{\epsilon_m}_{\mathrm{i}}(t)f_{\mathrm{i}}(u^{\epsilon_m}_{\mathrm{i}}(t)) + u^{\epsilon_m}_{\mathrm{i}}(t)\sum_{\mathrm{j}\in\mathbb{Z}^d}\kappa_{\mathrm{i},\mathrm{j}}\sigma_{\epsilon_m}(u^{\epsilon_m}_{\mathrm{j}}(t) - \varsigma) + g_i u^{\epsilon_m}_{\mathrm{i}}(t)\\
&\le -\alpha|u^{\epsilon_m}_{\mathrm{i}}(t)|^2 + \beta_{\mathrm{i}}^2 + \frac{\alpha}{4}|u^{\epsilon_m}_{\mathrm{i}}(t)|^2 + \frac{1}{\alpha}\Big(\sum_{\mathrm{j}\in\mathbb{Z}^d}|\kappa_{\mathrm{i},\mathrm{j}}|\Big)^2 + \frac{\alpha}{4}|u^{\epsilon_m}_{\mathrm{i}}(t)|^2 + \frac{1}{\alpha}g_{\mathrm{i}}^2\\
&\le -\frac{\alpha}{2}|u^{\epsilon_m}_{\mathrm{i}}(t)|^2 + \beta_{\mathrm{i}}^2 + \frac{1}{\alpha}(\kappa\rho_\Sigma + g_{\mathrm{i}}^2).\end{aligned}$$

Then integrating the above inequality gives

$$\begin{aligned}|u^{\epsilon_m}_{\mathrm{i}}(t)|^2 &\le |u_{o,\mathrm{i}}|^2 e^{-\alpha t} + \frac{2}{\alpha^2}(\alpha\beta_{\mathrm{i}}^2 + \kappa\rho_\Sigma + g_{\mathrm{i}}^2)(1 - e^{-\alpha t})\\
&\le |u_{o,\mathrm{i}}|^2 + \frac{2}{\alpha^2}(\alpha\beta_{\mathrm{i}}^2 + \kappa\rho_\Sigma + g_{\mathrm{i}}^2) := \mu_{\mathrm{i}}^2 \quad \text{for all } \ t\ge 0.\end{aligned}$$

Consequently by using Assumption 19.3,

$$\begin{aligned}\left|\frac{\mathrm{d}}{\mathrm{d}t}u^{\epsilon_m}_{\mathrm{i}}(t)\right| &\le |f_{\mathrm{i}}(u^{\epsilon_m}_{\mathrm{i}}(t))| + \sum_{\mathrm{j}\in\mathbb{Z}^d}|\kappa_{\mathrm{i},\mathrm{j}}|\sigma_{\epsilon_m}(u^{\epsilon_m}_{\mathrm{j}}(t) - \varsigma) + |g_i|\\
&\le L(\rho_{\mathrm{i}}|u^{\epsilon_m}_{\mathrm{i}}(t)|)|u^{\epsilon_m}_{\mathrm{i}}(t)| + \sum_{\mathrm{j}\in\mathbb{Z}^d}|\kappa_{\mathrm{i},\mathrm{j}}| + |g_i|\\
&\le L(\rho_{\mathrm{i}}\mu_{\mathrm{i}})\mu_{\mathrm{i}} + \kappa\rho_\Sigma|g_i| \quad \text{for all } t\ge 0.\end{aligned}$$

Hence the sequence $\{u_{\mathrm{i}}^{\epsilon_m}(t)\}_{m\in\mathbb{N}}$ is uniformly bounded and equi-Lipschitz continuous on $[0,T]$ for each $\mathrm{i}\in\mathbb{Z}^d$. Then by the Ascoli-Arzelà Theorem for each $\mathrm{i}\in\mathbb{Z}^d$, there is a $u_{\mathrm{i}}^*(\cdot)\in\mathcal{C}([0,T],\mathbb{R})$ and a convergent subsequence $\{u_{\mathrm{i}}^{\epsilon_{m_r}}(\cdot)\}_{r\in\mathbb{N}}$ such that

$$u_{\mathrm{i}}^{\epsilon_{m_r}}(\cdot)\to u_{\mathrm{i}}^*(\cdot)\ \text{ strongly in } \mathcal{C}([0,T],\mathbb{R})$$

and

$$\frac{\mathrm{d}}{\mathrm{d}t}u_{\mathrm{i}}^{\epsilon_{m_r}}(\cdot)\to \frac{\mathrm{d}}{\mathrm{d}t}u_{\mathrm{i}}^*(\cdot) \text{ weakly in } \mathcal{L}^1([0,T],\mathbb{R}).$$

The limit function $u_{\mathrm{i}}^*(\cdot)$ shares the equi-Lipschitz continuity of the subsequence $\{u_{\mathrm{i}}^{\epsilon_{m_r}}((\cdot)\}_{r\in\mathbb{N}}$ and hence is absolutely continuous on $[0,T]$. The argument can be strengthened to obtain a common diagonal subsequence that converges for all $\mathrm{i}\in\mathbb{Z}^d$.

2. Convergent subsequence in ℓ_ρ^2

Similar to the above, multiplying the equation (19.24) by $\rho_{\mathrm{i}}u_{\mathrm{i}}^{\epsilon_m}$ and summing over $\mathrm{i}\in\mathbb{Z}^d$ we get

$$\frac{1}{2}\frac{\mathrm{d}}{\mathrm{d}t}\|\boldsymbol{u}^{\epsilon_m}(t)\|_\rho^2\le -\frac{\alpha}{2}\|\boldsymbol{u}^{\epsilon_m}(t)\|_\rho^2+\|\boldsymbol{\beta}\|_\rho^2+\frac{1}{\alpha}(\kappa\rho_\Sigma^2+\|\boldsymbol{g}\|_\rho^2),$$

which can be integrated to obtain

$$\|\boldsymbol{u}^{\epsilon_m}(t)\|_\rho^2\le\|\boldsymbol{u}_o\|_\rho^2+\frac{2}{\alpha^2}(\alpha\|\boldsymbol{\beta}\|_\rho^2+\kappa\rho_\Sigma^2+\|\boldsymbol{g}\|_\rho^2):=\nu^2,\quad \forall\, t\ge 0.$$

Now squaring the equation (19.24), then multiplying by $\rho_{\mathrm{i}}u_{\mathrm{i}}^{\epsilon_m}$ and summing over $\mathrm{i}\in\mathbb{Z}^d$ we have

$$\left\|\frac{\mathrm{d}}{\mathrm{d}t}\boldsymbol{u}^{\epsilon_m}(t)\right\|_\rho^2\le 3\Big(\sum_{\mathrm{i}\in\mathbb{Z}^d}\rho_{\mathrm{i}}f_{\mathrm{i}}^2(u_{\mathrm{i}}^{\epsilon_m})+\sum_{\mathrm{i}\in\mathbb{Z}^d}\rho_{\mathrm{i}}\Big(\sum_{\mathrm{j}\in\mathbb{Z}^d}|\kappa_{\mathrm{i},\mathrm{j}}|\Big)^2+\|\boldsymbol{g}\|_\rho^2\Big).$$

Noticing that

$$\rho_{\mathrm{i}}u_{\mathrm{i}}^{\epsilon_m}=\sqrt{\rho_{\mathrm{i}}}\sqrt{\rho_{\mathrm{i}}}u_{\mathrm{i}}^{\epsilon_m}\le\rho_\Sigma\|\boldsymbol{u}^{\epsilon_m}\|_\rho\le\sqrt{\rho_\Sigma}\nu,$$

and using Assumption 19.3 and the inequality (19.8) again we obtain

$$\left\|\frac{\mathrm{d}}{\mathrm{d}t}\boldsymbol{u}^{\epsilon_m}(t)\right\|_\rho^2\le 3\Big(L^2(\sqrt{\rho_\Sigma}\nu)\nu^2+\kappa\rho_\Sigma^2+\|\boldsymbol{g}\|_\rho^2\Big).$$

Therefore by the Ascoli-Arzelà Theorem again there exists $\hat{\boldsymbol{u}}(\cdot)\in\mathcal{C}([0,T],\ell_\rho^2)$ and a convergent subseqence $\{\boldsymbol{u}^{\epsilon_{m_r}}(\cdot)\}_{r\in\mathbb{N}}$ such that

$$\sup_{t\in[0,T]}\|\boldsymbol{u}^{\epsilon_{m_r}}(t)-\hat{\boldsymbol{u}}(t)\|_\rho\to 0\quad\text{as } r\to\infty.$$

3. Equivalence of limit points

It can be assumed without loss of generality that the two subsequences $\{\epsilon_{m_r}\}$ for the above convergence are the same. Since $\boldsymbol{u}^{\epsilon_{m_r}} \to \hat{\boldsymbol{u}}(t)$ in ℓ^2_ρ, for any $\mathcal{E} > 0$ there exists $N_1(\mathcal{E}) \in \mathbb{N}$ such that

$$\|\boldsymbol{u}^{\epsilon_{m_r}} - \hat{\boldsymbol{u}}(t)\|_\rho^2 = \sum_{\mathfrak{i}\in\mathbb{Z}^d} \rho_\mathfrak{i} \left|u_\mathfrak{i}^{\epsilon_{m_r}}(t) - \hat{u}_\mathfrak{i}(t)\right|^2 < \mathcal{E}^2, \quad r \geq N_1(\mathcal{E}).$$

It then follows that

$$\left|u_\mathfrak{i}^{\epsilon_{m_r}}(t) - \hat{u}_\mathfrak{i}(t)\right| < \mathcal{E}/\sqrt{\rho_\mathfrak{i}}, \quad r \geq N_1(\mathcal{E}), \mathfrak{i} \in \mathbb{Z}^d.$$

Also, by part 1 for any $\mathcal{E} > 0$ there exists $N_2(\mathcal{E}) \in \mathbb{N}$ such that

$$\left|u_\mathfrak{i}^{\epsilon_{m_r}}(t) - u_\mathfrak{i}^*(t)\right| < \mathcal{E}, \quad r \geq N_2(\mathcal{E}), \mathfrak{i} \in \mathbb{Z}^d.$$

In particular, pick $M = \max\{m_{N_1(\mathcal{E})}, m_{N_2(\mathcal{E})}\}$ Then for every fixed $\mathfrak{i} \in \mathbb{Z}^d$,

$$|\hat{u}_\mathfrak{i}(t) - u_\mathfrak{i}^*(t)| \leq \left|\hat{u}_\mathfrak{i}(t) - u_\mathfrak{i}^{\epsilon_M}(t)\right| + \left|u_\mathfrak{i}^{\epsilon_M}(t) - u_\mathfrak{i}^*(t)\right| \leq \mathcal{E}/\sqrt{\rho_\mathfrak{i}} + \mathcal{E}.$$

Thus $\hat{u}_\mathfrak{i}(t) = u_\mathfrak{i}^*(t)$ for every $\mathfrak{i} \in \mathbb{Z}^d$ and $t \in [0, T]$.

4. The limit as solution of the lattice inclusion

Rearranging the lattice system (19.24) for the convergent subsequence $\{\boldsymbol{u}^{m_r}(\cdot)\}_{r\in\mathbb{N}}$ gives

$$\Upsilon_\mathfrak{i}^{\epsilon_{m_r}}(t) = \frac{\mathrm{d}}{\mathrm{d}t} u_\mathfrak{i}^{\epsilon_{m_r}}(t) - f_\mathfrak{i}(u_\mathfrak{i}^{\epsilon_{m_r}}(t)) - g_\mathfrak{i}, \quad \mathfrak{i} \in \mathbb{Z}^d, \text{ a.e..} \tag{19.25}$$

Then with the identical limits $u_\mathfrak{i}^*(\cdot) = \hat{u}_\mathfrak{i}(\cdot)$ constructed above define

$$\Upsilon_\mathfrak{i}^*(t) := \frac{\mathrm{d}}{\mathrm{d}t} u_\mathfrak{i}^*(t) - f_\mathfrak{i}(u_\mathfrak{i}^*(t)) - g_\mathfrak{i}, \quad \mathfrak{i} \in \mathbb{Z}^d, \text{ a.e..} \tag{19.26}$$

For $N > 1$ sum both sides of the equation (19.25) from $r = 1$ to N to obtain

$$\frac{1}{N}\sum_{r=1}^N \Upsilon_\mathfrak{i}^{\epsilon_{m_r}}(t) = \frac{1}{N}\sum_{r=1}^N \frac{\mathrm{d}}{\mathrm{d}t} u_\mathfrak{i}^{\epsilon_{m_r}}(t) - \frac{1}{N}\sum_{r=1}^N f_\mathfrak{i}(u_\mathfrak{i}^{\epsilon_{m_r}}(t)) - g_\mathfrak{i}, \quad \mathfrak{i} \in \mathbb{Z}^d, \text{ a.e..} \tag{19.27}$$

Then by the Banach-Saks Theorem (see, e.g., [Banach and Saks (1930); Szlenik (1965)]), the terms on the right side of (19.27) have subsequences (still denoted by the same label) that converge strongly to the terms on the right hand side of (19.26) in $\mathcal{L}^1([0, T], \mathbb{R})$ for each $\mathfrak{i} \in \mathbb{Z}^d$ as $N \to \infty$. Moreover, because $\Upsilon_\mathfrak{i}^{\epsilon_{m_r}}(\cdot) \to \Upsilon_\mathfrak{i}^*(\cdot)$ weakly in $\mathcal{L}^1([0, T], \mathbb{R})$, the Banach-Saks Theorem gives

$$\frac{1}{N}\sum_{r=1}^N \Upsilon_\mathfrak{i}^{\epsilon_{m_r}}(\cdot) \to \Upsilon_\mathfrak{i}^*(\cdot) \text{ strongly in } \mathcal{L}^1([0, T], \mathbb{R}) \text{ as } N \to \infty, \quad \forall\, \mathfrak{i} \in \mathbb{Z}^d. \tag{19.28}$$

We next show that $\Upsilon_\mathfrak{i}^*(t) \in \mathfrak{K}_\mathfrak{i}(\boldsymbol{u}^*(t))$. Fixing an arbitrary $\epsilon \in (0, 1]$, we can assume without loss of generality that $\epsilon_{m_r} \leq \epsilon$ for all $r \in \mathbb{N}$. Then for any $T > 0$ we have

$$\int_0^T \mathrm{dist}_{\mathbb{R}^1}\left(\Upsilon_\mathfrak{i}^*(t), \mathfrak{K}_\mathfrak{i}(\boldsymbol{u}^*(t))\right) \mathrm{d}t \leq \int_0^T (I_1 + I_2 + I_3 + I_4)\, \mathrm{d}t, \tag{19.29}$$

where

$$I_1 := \Big|\Upsilon_i^*(t) - \frac{1}{N}\sum_{r=1}^{N}\Upsilon_i^{\epsilon_{m_r}}(t)\Big|;$$

$$I_2 := \mathrm{dist}_{\mathbb{R}^1}\Big(\frac{1}{N}\sum_{r=1}^{N}\Upsilon_i^{\epsilon_{m_r}}(t), \frac{1}{N}\sum_{r=1}^{N}\mathfrak{K}_i^{\epsilon}(\boldsymbol{u}^{\epsilon_{m_r}}(t))\Big);$$

$$I_3 := \mathrm{dist}_{\mathbb{R}^1}\Big(\frac{1}{N}\sum_{r=1}^{N}\mathfrak{K}_i^{\epsilon}(\boldsymbol{u}^{\epsilon_{m_r}}(t)), \frac{1}{N}\sum_{r=1}^{N}\mathfrak{K}_i^{\epsilon}(\boldsymbol{u}^*(t))\Big);$$

$$I_4 := \mathrm{dist}_{\mathbb{R}^1}\Big(\frac{1}{N}\sum_{r=1}^{N}\mathfrak{K}_i^{\epsilon}(\boldsymbol{u}^*(t)), \mathfrak{K}_i(\boldsymbol{u}^*(t))\Big).$$

First, due to (19.28),

$$\int_0^T I_1 \mathrm{d}t \to 0 \quad \text{as } N \to \infty. \tag{19.30}$$

Second, since $\Upsilon_i^{\epsilon_{m_r}}(t) \subset \mathfrak{K}_i^{\epsilon_{m_r}}(\boldsymbol{u}^{\epsilon_{m_r}}(t)) \subset \mathfrak{K}_i^{\epsilon}(\boldsymbol{u}^{\epsilon_{m_r}}(t))$,

$$\mathrm{dist}_{\mathbb{R}^1}(\Upsilon_i^{\epsilon_{m_r}}(t), \mathfrak{K}_i^{\epsilon}(\boldsymbol{u}^{\epsilon_{m_r}}(t))) = 0.$$

Then by (19.19) we have

$$I_2 \le \frac{1}{N}\sum_{r=1}^{N}\mathrm{dist}_{\mathbb{R}^1}\left(\Upsilon_i^{\epsilon_{m_r}}(t), \mathfrak{K}_i^{\epsilon}(\boldsymbol{u}^{\epsilon_{m_r}}(t))\right) = 0. \tag{19.31}$$

Next, by the upper semi-continuity of $\mathfrak{K}_i^{\epsilon}$ in Corollary 19.1,

$$\mathrm{dist}_{\mathbb{R}^1}\left(\mathfrak{K}_i^{\epsilon}(\boldsymbol{u}^{\epsilon_{m_r}}(t)), \mathfrak{K}_i^{\epsilon}(\boldsymbol{u}^*(t))\right) \to 0, \quad \text{as } r \to \infty,$$

thus by using (19.19) we get

$$I_3 \le \frac{1}{N}\sum_{r=1}^{N}\mathrm{dist}_{\mathbb{R}^1}\left(\mathfrak{K}_i^{\epsilon}(\boldsymbol{u}^{\epsilon_{m_r}}(t)), \mathfrak{K}_i^{\epsilon}(\boldsymbol{u}^*(t))\right) \to 0 \text{ as } N \to \infty. \tag{19.32}$$

Finally, since $\mathfrak{K}_i^{\epsilon}(\boldsymbol{u}^*(t))$ is a convex set, so $\frac{1}{N}\sum_{r=1}^{N}\mathfrak{K}_i^{\epsilon}(\boldsymbol{u}^*(t)) \subset \mathfrak{K}_i^{\epsilon}(\boldsymbol{u}^*(t))$ and hence by (19.19) again

$$I_4 \le \mathrm{dist}_{\mathbb{R}^1}\left(\mathfrak{K}_i^{\epsilon}(\boldsymbol{u}^*(t)), \mathfrak{K}_i(\boldsymbol{u}^*(t))\right) \to 0 \quad \text{as } \epsilon \to 0. \tag{19.33}$$

Inserting (19.30)–(19.33) into (19.29) results in

$$\int_0^T \mathrm{dist}_{\mathbb{R}^1}\left(\Upsilon_i^*(t), \mathfrak{K}_i(\boldsymbol{u}^*(t))\right) \mathrm{d}t = 0,$$

and hence that

$$\Upsilon_i^*(t) \in \mathfrak{K}_i(\boldsymbol{u}^*(t)), \quad t \in [0,T], \text{ a.e..}$$

Finally, the equation for Υ_i^* can be rewritten as

$$\frac{\mathrm{d}}{\mathrm{d}t}u_i^*(t) = f_i(u_i^*(t)) + \Upsilon_i^*(t) + g_i(t), \quad i \in \mathbb{Z}^d, \text{ a.e..}$$

Since $\Upsilon_i^*(t) \in \mathfrak{K}_i(\boldsymbol{u}^*(t))$ this implies that $\boldsymbol{u}^*(t) = (u_i^*(t))_{i\in\mathbb{Z}^d}$ is a solution of the lattice differential inclusion with initial value $\boldsymbol{u}_o \in \ell_\rho^2$. The proof of Theorem 19.4 is complete. □

A slight modification of the above proof gives the following compactness theorem.

Theorem 19.5. *Let* $T > 0$, $\boldsymbol{u}_o^{(m)} \to \boldsymbol{u}_o \in \ell_\rho^2$ *as* $m \to \infty$ *and suppose that Assumptions 19.1–19.5 hold. Then for any sequence* $\boldsymbol{u}^\epsilon(\cdot, \boldsymbol{u}_o^{(m)})$ *of solutions of the inflated system* (19.15) *for* $\epsilon \in (0,1]$, *there is a subsequence* $\boldsymbol{u}^\epsilon(\cdot, \boldsymbol{u}_o^{m_j}) \to \boldsymbol{u}^*(\cdot) \in \mathcal{C}([0,T], \ell_\rho^2)$ *with* $m_j \to \infty$ *as* $j \to \infty$, *where* $\boldsymbol{u}^*(\cdot)$ *is a solution of* (19.15) *with the initial value* $\boldsymbol{u}^*(0) = \boldsymbol{u}_o$.

19.5 Set-valued dynamical systems with compact values

There is a large literature for autonomous set-valued dynamical systems, which are often called set-valued semi-groups or *general dynamical systems*, see e.g., [Szegö and Treccani (1969)]. Such systems are often generated by differential inclusions or differential equations without uniqueness [Aubin and Cellina (1984); Smirnov (2002)]. This theory was mainly developed on the locally compact state space $\mathbb{R}^d$, but much of it holds here in the Hilbert space ℓ_ρ^2, when the system takes compact values.

Denote by $\mathcal{P}_c(\ell_\rho^2)$ the family of all nonempty compact subsets of ℓ_ρ^2.

Definition 19.2. A set-valued dynamical system on ℓ_ρ^2 with compact attainability sets is given by a mapping $(t,x) \mapsto \Phi(t,x) \in \mathcal{P}_c(\ell_\rho^2)$ defined on $\mathbb{R}^+ \times \ell_\rho^2$ such that

(i) $\Phi(0, \boldsymbol{u}_o) = \{\boldsymbol{u}_o\}$ for all $\boldsymbol{u}_o \in \ell_\rho^2$;
(ii) $\Phi(s+t, \boldsymbol{u}_o) = \Phi(s, \Phi(t, \boldsymbol{u}_o))$ for all t, $s \in \mathbb{R}^+$ and all $\boldsymbol{u}_o \in \ell_\rho^2$;
(iii) $(t, \boldsymbol{u}) \mapsto \Phi(t, \boldsymbol{u})$ is upper semi-continuous in $(t, \boldsymbol{u}) \in \mathbb{R}^+ \times \ell_\rho^2$ with respect to the Hausdorff semi-distance $\mathrm{dist}_{\ell_\rho^2}$, i.e.

$$\mathrm{dist}_{\ell_\rho^2}\left(\Phi(t,\boldsymbol{u}), \Phi(t_0, \boldsymbol{u}_o)\right) \to 0 \quad \text{as} \quad (t,\boldsymbol{u}) \to (t_0, \boldsymbol{u}_o) \ \text{ in } \ \mathbb{R}^+ \times \ell_\rho^2;$$

(iv) $t \mapsto \Phi(t, \boldsymbol{u}_o)$ is continuous in $t \in \mathbb{R}^+$ with respect to the Hausdorff metric $\mathcal{H}_{\ell_\rho^2}$ uniformly in $\boldsymbol{u}_o$ in compact subsets $B \in \mathcal{C}(\ell_\rho^2)$, i.e.,

$$\sup_{\boldsymbol{u}_o \in B} \mathcal{H}_{\ell_\rho^2}\left(\Phi(t, \boldsymbol{u}_o), \Phi(t_0, \boldsymbol{u}_o)\right) \to 0 \quad \text{as} \quad t \to t_0 \ \text{ in } \ \mathbb{R}^+.$$

For $\epsilon \in [0,1]$, the attainability set of the inflated lattice inclusion (19.15) is defined as

$$\Phi^\epsilon(t, \boldsymbol{u}_o) := \big\{ \boldsymbol{v} \in \ell_\rho^2 : \text{there exists a solution } \boldsymbol{u}^\epsilon(\cdot; \boldsymbol{u}_o) \text{ of (19.15) with } \boldsymbol{u}^\epsilon(0; \boldsymbol{u}_o) = \boldsymbol{u}_o \text{ such that } \boldsymbol{v} = \boldsymbol{u}^\epsilon(t; \boldsymbol{u}_o) \big\}.$$

Lemma 19.9. *The attainability sets* Φ^ϵ *generate a set-valued dynamical system* $\{\Phi^\epsilon(t)\}_{t \geq 0}$ *with values in* $\mathcal{P}_c(\ell_\rho^2)$ *for every* $\epsilon \in [0,1]$.

Proof. First, it follows from the definition that $\Phi^\epsilon(0, \boldsymbol{u}_o) = \{\boldsymbol{u}_o\}$ for every $\boldsymbol{u}_o \in \ell_\rho^2$.

1. Compactness of attainability sets: To show that $\Phi^\epsilon(T,\boldsymbol{u}_o)$ is a compact set for any given $T>0$, consider $\boldsymbol{v}^{(k)}\in\Phi^\epsilon(T,\boldsymbol{u}_o)$ for $k\in\mathbb{N}$. Then there exist solutions $\boldsymbol{u}^{(k)}(t;\boldsymbol{u}_o)$ of (19.15) with $\boldsymbol{u}^{(k)}(T;\boldsymbol{u}_o)=\boldsymbol{v}^{(k)}$ and by Theorem 19.5 there exists a convergent subsequence in $\mathcal{C}([0,T],\ell^2_\rho)$ converging to a solution $\bar{\boldsymbol{u}}(t)$ of (19.15). Clearly $\bar{\boldsymbol{u}}(0)=\boldsymbol{u}_o$, so $\bar{\boldsymbol{u}}(T)\in\Phi^\epsilon(T,\boldsymbol{u}_o)$.

2. Semi-goup property: Recall that $\boldsymbol{u}^\epsilon(t)\in\Phi^\epsilon(t,\boldsymbol{u}_o)$ is a unique solution to the lattice ODE

$$\frac{\mathrm{d}\boldsymbol{u}^\epsilon(t)}{\mathrm{d}t}=F(\boldsymbol{u}(t))+\Upsilon^\epsilon(t)+\boldsymbol{g},\quad \text{a.e.},\tag{19.34}$$

in ℓ^2_ρ for a selection $\Upsilon^\epsilon(t)\in\mathfrak{K}^\epsilon(\boldsymbol{u}(t))$. For every $\boldsymbol{v}\in\Phi^\epsilon(s,\Phi^\epsilon(t,\boldsymbol{u}_o))$, there exists $\boldsymbol{w}\in\Phi^\epsilon(t,\boldsymbol{u}_o)$ and a solution $\boldsymbol{u}^\epsilon_{(1)}(\cdot;\boldsymbol{w})$ of (19.34) such that $\boldsymbol{v}=\boldsymbol{u}^\epsilon_{(1)}(s;\boldsymbol{w})$. At the same time since $\boldsymbol{w}\in\Phi^\epsilon(t,\boldsymbol{u}_o)$, there exists a solution $\boldsymbol{u}^\epsilon_{(2)}(\cdot;\boldsymbol{u}_o)$ of (19.34) such that $\boldsymbol{w}=\boldsymbol{u}^\epsilon_{(2)}(t,\boldsymbol{u}_o)$. Let $\boldsymbol{u}^\epsilon_*(\cdot)$ be the concatenation of $\boldsymbol{u}^\epsilon_{(2)}(\cdot)$ and $\boldsymbol{u}^\epsilon_{(1)}(\cdot)$ and their corresponding selections $\Upsilon^\epsilon_{(2)}(\tau)\in\mathfrak{K}^\epsilon(U^\epsilon_{(2)}(\tau))$ for $\tau\in[0,t]$ and $\Upsilon^\epsilon_{(1)}(\tau)\in\mathfrak{K}(\boldsymbol{u}^\epsilon_{(1)}(\tau))$ for $\tau\in[t,s+t]$. Then $\boldsymbol{u}^\epsilon_*(\cdot;\boldsymbol{u}_o)$ is also a solution to (19.34) with the corresponding selection $\boldsymbol{u}^\epsilon_*$ given by the concatenation of $\Upsilon^\epsilon_{(2)}$ and $\Upsilon^\epsilon_{(1)}$. This implies that $\boldsymbol{v}=\boldsymbol{u}^\epsilon_*(s+t,\boldsymbol{u}_o)\in\Phi^\epsilon(s+t,\boldsymbol{u}_o)$ and hence

$$\Phi^\epsilon(s+t,\Phi^\epsilon(t,\boldsymbol{u}_o))\subset\Phi^\epsilon(t,\boldsymbol{u}_o).$$

On the other hand, for every $\boldsymbol{v}\in\Phi^\epsilon(s+t,\boldsymbol{u}_o)$ there exists a solution $\boldsymbol{u}^\epsilon(\cdot,\boldsymbol{u}_o)$ of (19.34) such that $\boldsymbol{v}=\boldsymbol{u}^\epsilon(s+t;\boldsymbol{u}_o)$. There also exists a selection $\Upsilon^\epsilon(\tau)\in\mathfrak{K}^\epsilon(\boldsymbol{u}^\epsilon(\tau))$ on $[0,s+t]$ such that $\boldsymbol{u}^\epsilon(\cdot)$ is a unique solution of the ODE (19.34). Define $\Upsilon^\epsilon_{(1)}(\tau)=\Upsilon^\epsilon(\tau)$, $\boldsymbol{u}^\epsilon_{(1)}(\tau)=\boldsymbol{u}^\epsilon(\tau)$ on $[0,t]$ and $\Upsilon^\epsilon_{(2)}(\tau)=\Upsilon^\epsilon(\tau)$, $\boldsymbol{u}^\epsilon_{(2)}(\tau)=\boldsymbol{u}^\epsilon(\tau)$ on $[t,s+t]$. Then $\Upsilon^\epsilon_{(1)}(\tau)\in\mathfrak{K}^\epsilon(\boldsymbol{u}^\epsilon_{(1)}(\tau))$ on $[0,t]$ and $\Upsilon^\epsilon_{(2)}(\tau)\in\mathfrak{K}^\epsilon(\boldsymbol{u}^\epsilon_{(2)}(\tau))$ on $[t,s+t]$, and thus $\boldsymbol{u}^\epsilon_{(1)}(\cdot)$ and $\boldsymbol{u}^\epsilon_{(2)}(\cdot)$ are the corresponding solutions to the ODE (19.34). Then, by the uniqueness of solutions to (19.34), $\boldsymbol{u}^\epsilon_{(1)}(0)=\boldsymbol{u}_o$, $\boldsymbol{u}^\epsilon_{(1)}(t)=\boldsymbol{u}^\epsilon_{(2)}(t)=\boldsymbol{w}$ and $\boldsymbol{u}^\epsilon_{(2)}(s+t)=\boldsymbol{v}$. This implies that

$$\boldsymbol{v}\in\Phi^\epsilon(s,\boldsymbol{u}(t,\boldsymbol{u}_o))\subset\Phi^\epsilon(s,\Phi^\epsilon(t,\boldsymbol{u}_o)).$$

Hence $\Phi^\epsilon(s+t,\boldsymbol{u}_o)\subset\Phi^\epsilon(s,\Phi^\epsilon(t,\boldsymbol{u}_o))$. Combining the above results gives $\Phi^\epsilon(s+t,\boldsymbol{u}_o)=\Phi^\epsilon(s,\Phi^\epsilon(t,\boldsymbol{u}_o))$.

3. Upper semi-continuous convergence in $\boldsymbol{u}_o$: Consider $\{\boldsymbol{u}_o^{(m)}\}_{m\in\mathbb{N}}$ such that $\boldsymbol{u}_o^{(m)}\to\boldsymbol{u}_o$ in ℓ^2_ρ as $m\to\infty$. It can be assumed without loss of generality that

$$\left\|\boldsymbol{u}_o^{(m)}\right\|_\rho\le\|\boldsymbol{u}_o\|_\rho+1\quad\text{for all}\quad m\in\mathbb{N}.$$

Given $\tau\in[0,T]$, suppose (for contradiction) that the sequence of attainability sets $\{\Phi^\epsilon(\tau,\boldsymbol{u}_o^{(m)})\}_{m\in\mathbb{N}}$ does not converge upper semi-continuously to $\Phi^\epsilon(\tau,\boldsymbol{u}_o)$ as $m\to\infty$. Then there exists a $\mathcal{E}_0>0$ and a subsequence of $\{\boldsymbol{u}_o^{(m)}\}$, still denoted by $\{\boldsymbol{u}_o^{(m)}\}$, such that

$$\operatorname{dist}_{\ell^2_\rho}\left(\Phi^\epsilon(\tau,\boldsymbol{u}_o^{(m)}),\Phi^\epsilon(\tau,\boldsymbol{u}_o)\right)\ge\mathcal{E}_0\quad\text{for all}\quad m\in\mathbb{N}.\tag{19.35}$$

For each $m \in \mathbb{N}$, since the set $\Phi^\epsilon(\tau, \boldsymbol{u}_o^{(m)})$ is compact, there exists $\boldsymbol{v}^{(m)} \in \Phi^\epsilon(\tau, \boldsymbol{u}_o^{(m)})$ such that

$$\text{dist}_{\ell^2_\rho}\left(\boldsymbol{v}^{(m)}, \Phi^\epsilon(\tau, \boldsymbol{u}_o)\right) = \text{dist}_{\ell^2_\rho}\left(\Phi^\epsilon(\tau, \boldsymbol{u}_o^{(m)}), \Phi^\epsilon(\tau, \boldsymbol{u}_o)\right) \geq \mathcal{E}_0 \quad \text{for all } m \in \mathbb{N}.$$

In addition, there exists a solution $\boldsymbol{u}^\epsilon_{(m)}(\cdot)$ of the differential inclusion (19.34) such that $\boldsymbol{u}^\epsilon_{(m)}(0) = \boldsymbol{u}_o^{(m)}$ and $\boldsymbol{u}^\epsilon_{(m)}(\tau) = \boldsymbol{v}^{(m)}$.

Then by Theorem 19.5 there exists a subsequence $\{\boldsymbol{u}_{(m_k)}(\cdot)\}_{k\in\mathbb{N}}$ and a solution $\boldsymbol{u}^\epsilon_*(\cdot)$ of the inflated inclusion (19.15) such that

$$\left\|\boldsymbol{u}^\epsilon_{(m_k)}(\tau) - \boldsymbol{u}^\epsilon_*(\tau)\right\|_\rho \leq \frac{1}{m_k}, \qquad k \text{ large}.$$

Hence the sequence $\{\boldsymbol{v}^{(m_k)}\}_{k\in\mathbb{N}}$ converges to $\boldsymbol{u}^\epsilon_*(\tau)$ in ℓ^2_ρ. On the other hand, $\boldsymbol{u}^\epsilon_*(0) = \boldsymbol{u}_o$, so $\boldsymbol{u}^\epsilon_*(\tau) \in \Phi^\epsilon(\tau, \boldsymbol{u}_o)$. This contradicts the inequality (19.35). Thus $\Phi^\epsilon(\tau, \boldsymbol{u}_o)$ is upper semi-continuous in $\boldsymbol{u}_o$ for every $\tau \in [0, T]$ as claimed.

4. Continuous convergence in t: Suppose that $\Phi^\epsilon(t, \boldsymbol{u}_o)$ does not converge upper semi-continuously to $\Phi^\epsilon(t_0, \boldsymbol{u}_o)$ as $t \to t_0$. Then there exists $\mathcal{E}_0 > 0$ and $t_n \to t_0$ such that

$$\text{dist}_{\ell^2_\rho}\left(\Phi^\epsilon(t_n, \boldsymbol{u}_o), \Phi^\epsilon(t_0, \boldsymbol{u}_o)\right) \geq \mathcal{E}_0 \quad \text{for all } n \in \mathbb{N}. \tag{19.36}$$

By compactness of the attainability sets there exist $a_n \in \Phi^\epsilon(t_n, \boldsymbol{u}_o)$ such that

$$\text{dist}_{\ell^2_\rho}\left(a_n, \Phi^\epsilon(t_0, \boldsymbol{u}_o)\right) = \text{dist}_{\ell^2_\rho}\left(\Phi^\epsilon(t_n, \boldsymbol{u}_o), \Phi^\epsilon(t_0, \boldsymbol{u}_o)\right) \geq \mathcal{E}_0 \quad \text{for all } n \in \mathbb{N}.$$

Moreover, there exist continuous trajectories ϕ_n of Φ^ϵ with $\phi_n(0) = \boldsymbol{u}_o$ and $\phi_n(t_n) = a_n$. Let t_0, $t_n \in [0, T]$ for some T. Then, by Theorem 19.5 there is a uniformly convergent subsequence $\phi_{n_j} \to \phi^*$ in $C([0,T], \ell^2_\rho)$, where ϕ^* is a trajectory of Φ^ϵ. In particular, $\phi_{n_j}(t_0) = \boldsymbol{u}_o = \phi^*(0)$, so $\phi^*(t_0) \in \Phi^\epsilon(t_0, \boldsymbol{u}_o)$. Then

$$\begin{aligned}
\|\phi_{n_j}(t_{n_j}) - \phi^*(t_0)\|_{\ell^2_\rho} &\leq \|\phi_{n_j}(t_{n_j}) - \phi^*(t_{n_j})\|_{\ell^2_\rho} + \|\phi^*(t_{n_j}) - \phi^*(t_0)\|_{\ell^2_\rho} \\
&\leq \|\phi_{n_j} - \phi^*\|_{C([0,T],\ell^2_\rho)} + \|\phi^*(t_{n_j}) - \phi^*(t_0)\|_{\ell^2_\rho} \\
&\to 0 \quad \text{as } n_j \to \infty,
\end{aligned}$$

since ϕ^* is continuous. Thus

$$\text{dist}_{\ell^2_\rho}\left(a_n, \Phi^\epsilon(t_0, \boldsymbol{u}_o)\right) \leq \|\phi_{n_j}(t_{n_j}) - \phi^*(t_0)\|_{\ell^2_\rho} \to 0 \quad \text{as } n_j \to \infty,$$

which contradicts (19.36), so $\Phi^\epsilon(t, \boldsymbol{u}_o)$ does converge upper semi-continuously to $\Phi^\epsilon(t_0, \boldsymbol{u}_o)$ as $t \to t_0$.

Suppose now that $\Phi^\epsilon(t, \boldsymbol{u}_o)$ does not converge lower semi-continuously to $\Phi^\epsilon(t_0, \boldsymbol{u}_o)$ as $t \to t_0$. Then there exist $\mathcal{E}_0 > 0$ and $t_n \to t_0$ such that

$$\text{dist}_{\ell^2_\rho}\left(\Phi^\epsilon(t_0), \boldsymbol{u}_o), \Phi^\epsilon(t_n, \boldsymbol{u}_o)\right) \geq \mathcal{E}_0, \quad \text{for all } n \in \mathbb{N}. \tag{19.37}$$

By compactness of the attainability set there exist $a_n \in \Phi^\epsilon(t_0, \boldsymbol{u}_o)$ such that

$$\text{dist}_{\ell^2_\rho}\left(a_n, \Phi^\epsilon(t_n, \boldsymbol{u}_o)\right) = \text{dist}_{\ell^2_\rho}\left(\Phi^\epsilon(t_0, \boldsymbol{u}_o), \Phi^\epsilon(t_n, \boldsymbol{u}_o)\right) \geq \mathcal{E}_0, \quad \text{for all } n \in \mathbb{N}.$$

Moreover, there is a convergent subsequence (still denoted by the original sequence) $a_n \to \bar{a} \in \Phi^\epsilon(t_0, \boldsymbol{u}_o)$ and there exist continuous trajectories ϕ_n of Φ^ϵ with $\phi_n(0) = \boldsymbol{u}_o$ and $\phi_n(t_0) = a_n$. Let $t_0, t_n \in [0, T]$ for some T. Then, by Theorem 19.5 there is a uniformly convergent subsequence $\phi_{n_j} \to \bar{\phi}$ in $\mathcal{C}([0,T], \ell_\rho^2)$, where $\bar{\phi}$ is a trajectory of Φ^ϵ. In particular, $\phi_{n_j}(0) = \boldsymbol{u}_o = \bar{\phi}(0)$, so $\bar{\phi}(t_0) = \bar{a} \in \Phi^\epsilon(t_0, \boldsymbol{u}_o)$ and $\bar{\phi}(t_{n_j}) \in \Phi^\epsilon(t_{n_j}, \boldsymbol{u}_o)$. Then

$$
\begin{aligned}
\|a_{n_j} - \bar{\phi}(t_{n_j})\|_{\ell_\rho^2} &= \|\phi_{n_j}(t_0) - \bar{\phi}(t_{n_j})\|_{\ell_\rho^2} \\
&\leq \|\phi_{n_j}(t_0) - \bar{\phi}(t_0)\|_{\ell_\rho^2} + \|\bar{\phi}(t_0) - \bar{\phi}(t_{n_j})\|_{\ell_\rho^2} \\
&\leq \|\phi_{n_j} - \bar{\phi}\|_{\mathcal{C}([0,T],\ell_\rho^2)} + \|\bar{\phi}(t_0) - \bar{\phi}(t_{n_j})\|_{\ell_\rho^2} \\
&\to 0 \quad \text{as } n_j \to \infty,
\end{aligned}
$$

since $\bar{\phi}$ is continuous. Thus

$$
\operatorname{dist}_{\ell_\rho^2}\left(a_{n_j}, \Phi^\epsilon(t_{n_j}, \boldsymbol{u}_o)\right) \leq \|a_{n_j} - \bar{\phi}(t_{n_j})\|_{\ell_\rho^2} \to 0 \quad \text{as } n_j \to \infty,
$$

which contradicts (19.37), so $\Phi^\epsilon(t, \boldsymbol{u}_o)$ does converge lower semi-continuously to $\Phi^\epsilon(t_0, \boldsymbol{u}_o)$ as $t \to t_0$.

Combining both results gives the asserted continuous convergence and completes the proof of the lemma. □

Lemma 19.10. *Let Assumptions 19.1–19.5 hold. Then the set-valued dynamical system $\{\Phi^\epsilon(t)\}_{t\in\mathbb{R}}$ is asymptotically upper semi-compact.*

Proof. This proof has two parts. The first part provides a tail estimate for the solutions and the second part shows the asymptotic compactness.

1. Tails estimate. Let ξ be the smooth cut-off function defined in (3.8), and for a fixed and large number k (to be specified later) set

$$
v_i(t) = \xi_k(|i|)\, u_i(t) \quad \text{with } \xi_k(|i|) = \xi\left(\frac{|i|}{k}\right), \qquad i \in \mathbb{Z}^d,
$$

where $|\cdot|$ denotes the Euclidean norm. Multiply both side of the lattice differential inclusion (19.10) by $\rho_i v_i(t)$ to obtain

$$
\begin{aligned}
\rho_i\xi_k(|i|)\, u_i(t)\frac{\mathrm{d}}{\mathrm{d}t}u_i(t) \in \rho_i\xi_k(|i|)\, f_i(u_i(t))u_i(t) + \rho_i\xi_k(|i|)\, \mathfrak{K}_i^\epsilon(\boldsymbol{u}(t))u_i(t) \\
+\rho_i\xi_k(|i|)\, g_i u_i(t). \qquad (19.38)
\end{aligned}
$$

First by Assumption 19.4,

$$
\rho_i\xi_k(|i|)\, f_i(u_i(t))u_i(t) \leq -\alpha\rho_i\xi_k(|i|)\, u_i^2(t) + \rho_i\xi_k(|i|)\, \beta_i^2. \qquad (19.39)
$$

Then, by the definition of $\mathfrak{K}_i^\epsilon$ and Young's inequality

$$
\begin{aligned}
|||\rho_i\xi_k(|i|)\, \mathfrak{K}_i^\epsilon(\boldsymbol{u}(t))u_i(t)||| &\leq \frac{\alpha}{4}\rho_i\xi_k(|i|)\, u_i^2(t) + \frac{1}{\alpha}\rho_i\xi_k(|i|)\, |||\, \mathfrak{K}_i^\epsilon(\boldsymbol{u}(t))\, |||^2 \\
&\leq \frac{\alpha}{4}\rho_i\xi_k(|i|)\, u_i^2(t) + \frac{\kappa\rho_\Sigma}{\alpha}\rho_i\xi_k(|i|). \qquad (19.40)
\end{aligned}
$$

Inserting the estimations (19.39) and (19.40) into (19.38), then summing over $\mathrm{i} \in \mathbb{Z}^d$ gives

$$\frac{1}{2}\frac{\mathrm{d}}{\mathrm{d}t}\sum_{\mathrm{i}\in\mathbb{Z}^d}\rho_{\mathrm{i}}\xi_k(|\mathrm{i}|)\,u_{\mathrm{i}}^2(t) \le -\frac{3\alpha}{4}\sum_{\mathrm{i}\in\mathbb{Z}^d}\rho_{\mathrm{i}}\xi_k(|\mathrm{i}|)\,u_{\mathrm{i}}^2(t) + \sum_{\mathrm{i}\in\mathbb{Z}^d}\rho_{\mathrm{i}}\xi_k(|\mathrm{i}|)\,\beta_{\mathrm{i}}^2$$
$$+\frac{\kappa\rho_\Sigma}{\alpha}\sum_{\mathrm{i}\in\mathbb{Z}^d}\rho_{\mathrm{i}}\xi(|\mathrm{i}|) + \sum_{\mathrm{i}\in\mathbb{Z}^d}\rho_{\mathrm{i}}\xi_k(|\mathrm{i}|)\,g_{\mathrm{i}}u_{\mathrm{i}}(t). \tag{19.41}$$

Each term on the right hand side of this inequality will now be estimated. Notice that

$$\sum_{\mathrm{i}\in\mathbb{Z}^d}\rho_{\mathrm{i}}\xi_k(|\mathrm{i}|)\,\beta_{\mathrm{i}}^2 = \sum_{|\mathrm{i}|\ge k}\rho_{\mathrm{i}}\xi_k(|\mathrm{i}|)\,\beta_{\mathrm{i}}^2 \le \sum_{|\mathrm{i}|\ge k}\rho_{\mathrm{i}}\beta_{\mathrm{i}}^2.$$

Since $\boldsymbol{\beta} = (\beta_{\mathrm{i}})_{\mathrm{i}\in\mathbb{Z}^d} \in \ell_\rho^2$, then for every $\mathcal{E} > 0$ there exists $I_1(\mathcal{E}) > 0$ such that

$$\sum_{\mathrm{i}\in\mathbb{Z}^d}\rho_{\mathrm{i}}\xi_k(|\mathrm{i}|)\,\beta_{\mathrm{i}}^2 \le \sum_{|\mathrm{i}|\ge k}\rho_{\mathrm{i}}\beta_{\mathrm{i}}^2 < \frac{1}{6}\mathcal{E} \quad \text{when } k \ge I_1(\mathcal{E}). \tag{19.42}$$

Similarly, since $\rho_\Sigma = \sum_{\mathrm{i}\in\mathbb{Z}^d}\rho_{\mathrm{i}} < \infty$, then for every $\mathcal{E} > 0$ there exists $I_2(\mathcal{E}) > 0$ such that

$$\sum_{\mathrm{i}\in\mathbb{Z}^d}\rho_{\mathrm{i}}\xi_k(|\mathrm{i}|) = \sum_{|\mathrm{i}|\ge k}\rho_{\mathrm{i}}\xi_k(|\mathrm{i}|) \le \sum_{|\mathrm{i}|\ge k}\rho_{\mathrm{i}} < \frac{\alpha}{6\kappa\rho_\Sigma}\mathcal{E} \quad \text{when } k \ge I_2(\mathcal{E}). \tag{19.43}$$

In addition, since $\boldsymbol{g} = (g_{\mathrm{i}})_{\mathrm{i}\in\mathbb{Z}^d} \in \ell_\rho^2$, then for every $\mathcal{E} > 0$ there exists $I_3(\mathcal{E}, \boldsymbol{g}) > 0$ such that

$$\sum_{|\mathrm{i}|\ge k}\rho_{\mathrm{i}}g_{\mathrm{i}}^2 \le \frac{\alpha}{6}\mathcal{E}, \quad \text{when } k \ge I_3(\mathcal{E}, \boldsymbol{g}).$$

Consequently, when $k \ge I_3(\mathcal{E}, \boldsymbol{g})$,

$$\begin{aligned}\sum_{\mathrm{i}\in\mathbb{Z}^d}\rho_{\mathrm{i}}\xi_k(|\mathrm{i}|)\,g_{\mathrm{i}}u_{\mathrm{i}} &= \sum_{|\mathrm{i}|\ge k}\rho_{\mathrm{i}}\xi_k(|\mathrm{i}|)\,g_{\mathrm{i}}u_{\mathrm{i}} \\ &\le \frac{\alpha}{4}\sum_{|\mathrm{i}|\ge k}\rho_{\mathrm{i}}\xi_k^2(|\mathrm{i}|)\,u_{\mathrm{i}}^2 + \frac{1}{\alpha}\sum_{|\mathrm{i}|\ge k}\rho_{\mathrm{i}}g_{\mathrm{i}}^2 \\ &\le \frac{\alpha}{4}\sum_{\mathrm{i}\in\mathbb{Z}^d}\rho_{\mathrm{i}}\xi_k(|\mathrm{i}|)\,u_{\mathrm{i}}^2 + \frac{1}{6}\mathcal{E}.\end{aligned} \tag{19.44}$$

Finally, for any $\mathcal{E} > 0$, choosing $I(\mathcal{E}, \boldsymbol{g}) := \max\{I_1(\mathcal{E}), I_2(\mathcal{E}), I_3(\mathcal{E}, \boldsymbol{g})\}$, collecting estimates (19.42), (19.43), (19.44) and putting into (19.41) results in

$$\frac{\mathrm{d}}{\mathrm{d}t}\sum_{\mathrm{i}\in\mathbb{Z}^d}\rho_{\mathrm{i}}\xi_k(|\mathrm{i}|)\,u_{\mathrm{i}}^2(t) \le -\alpha\sum_{\mathrm{i}\in\mathbb{Z}^d}\rho_{\mathrm{i}}\xi_k(|\mathrm{i}|)\,u_{\mathrm{i}}^2(t) + \mathcal{E} \qquad \text{for all } k \ge I(\mathcal{E}, \boldsymbol{g}).$$

It follows immediately from Gronwall's lemma that

$$\begin{aligned}\sum_{\mathrm{i}\in\mathbb{Z}^d}\rho_{\mathrm{i}}\xi_k(|\mathrm{i}|)\,u_{\mathrm{i}}^2(t) &\le e^{-\alpha t}\sum_{\mathrm{i}\in\mathbb{Z}^d}\rho_{\mathrm{i}}\xi_k(|\mathrm{i}|)\,u_{o,\mathrm{i}}^2 + \frac{\mathcal{E}}{\alpha} \\ &\le e^{-\alpha t}\|\boldsymbol{u}_o\|_\rho^2 + \frac{\mathcal{E}}{\alpha},\end{aligned}$$

which implies that there exists $T = T(\mathcal{E})$ such that

$$\sum_{|\mathrm{i}|\geq 2k} \rho_{\mathrm{i}} u_{\mathrm{i}}^2(t) \leq \sum_{\mathrm{i}\in\mathbb{Z}^d} \rho_{\mathrm{i}}\xi_k(|\mathrm{i}|)\, u_{\mathrm{i}}^2(t) \leq \frac{2\mathcal{E}}{\alpha} \quad \text{for all } \ k \geq I(\mathcal{E}, \boldsymbol{g}),\ t \geq T(\mathcal{E}). \qquad (19.45)$$

2. Asymptotic compactness. Using estimates similar to those in the first part of the proof of Theorem 19.4 above, it follows that the closed and bounded set

$$Q := \left\{ \boldsymbol{u} \in \ell_\rho^2 : \|\boldsymbol{u}\|_\rho^2 \leq \frac{2}{\alpha^2}\left(\alpha\|\boldsymbol{\beta}\|_\rho^2 + \kappa\rho_\Sigma^2 + \|\boldsymbol{g}\|_\rho^2\right) + 1 =: R^2 \right\} \qquad (19.46)$$

is a positive invariant absorbing set for Φ^ϵ for every $\epsilon \in [0, 1]$.

It will be shown here that for any fixed $\tau \in [0, T]$, every sequence $\{\boldsymbol{v}^{(m)}\}_{m\in\mathbb{N}}$ in $\Phi^\epsilon(\tau, \boldsymbol{u}_o^{(m)})$ has a convergence subsequence in ℓ_ρ^2 for any $\boldsymbol{u}_o^{(m)} \in Q$. For this it only needs to be shown that $\{\boldsymbol{v}^{(m)}\}_{m\in\mathbb{N}}$ is precompact.

For any $\boldsymbol{v}^{(m)} \in Q$ and for each m there exists a solution $\boldsymbol{u}^{(m)}(\cdot)$ of the inflated inclusion (19.15) on the interval $[0, \tau]$ with $\boldsymbol{u}^{(m)}(0) = \boldsymbol{u}_o^{(m)}$ and $\boldsymbol{u}^{(m)}(\tau) = \boldsymbol{v}^{(m)}$. By similar analysis to Part II in the proof of Theorem 19.3, for m large and any given $t \in [0, T]$, it follows that

$$\left|\frac{\mathrm{d}}{\mathrm{d}t}\boldsymbol{u}_{\mathrm{i}}^{(m)}(t)\right| \leq L\left(\sqrt{\rho_\Sigma}\|\boldsymbol{u}^{(m)}(t)\|_\rho\right)\left|u_{\mathrm{i}}^{(m)}(t)\right| + \kappa + |g_{\mathrm{i}}|.$$

Hence

$$\begin{aligned}\left\|\frac{\mathrm{d}}{\mathrm{d}t}\boldsymbol{u}^{(m)}(t)\right\|_\rho^2 &= \sum_{\mathrm{i}\in\mathbb{Z}^d} \rho_{\mathrm{i}}\left|\frac{\mathrm{d}}{\mathrm{d}t}u_{\mathrm{i}}^{(m)}(t)\right|^2 \\ &\leq 4\left\{\left[L\left(\sqrt{\rho_\Sigma}\|\boldsymbol{u}^{(m)}(t)\|_\rho\right)\right]^2\left\|\boldsymbol{u}^{(m)}(t)\right\|_\rho^2 + \kappa^2 + \|\boldsymbol{g}\|_\rho^2\right\},\end{aligned}$$

and, in particular,

$$\left\|\frac{\mathrm{d}}{\mathrm{d}t}\boldsymbol{u}^{(m)}(\tau)\right\|_\rho^2 \leq 4\left\{\left[L\left(\sqrt{\rho_\Sigma}R\right)\right]^2 R^2 + \kappa^2 + \|\boldsymbol{g}\|_\rho^2\right\}.$$

Therefore by the Ascoli-Arzelà theorem applied on the interval $[0, \tau]$, there exist a subsequence $\{m_l\}_{l\geq 1}$, still denoted by $\{m\}$, and a function $\hat{\boldsymbol{u}}(\cdot) := (\hat{u}_{\mathrm{i}}(\cdot))_{\mathrm{i}\in\mathbb{Z}^d} : \mathbb{R} \to \ell_\rho^2$ such that $\boldsymbol{u}^{(m)}(\tau) \to \hat{\boldsymbol{u}}(\tau)$. This implies that

$$\boldsymbol{u}^{(m)}(t) \rightharpoonup \hat{\boldsymbol{u}}(t) \quad \text{weakly in } \ell_\rho^2 \quad \text{as } m \to \infty. \qquad (19.47)$$

By the tails estimate (19.45) in Part 1, for every $\mathcal{E} > 0$ there exist $M_1(\mathcal{E}) \in \mathbb{N}$ and $N_1(\mathcal{E}) \in \mathbb{N}$ such that

$$\sum_{|\mathrm{i}|\geq M_1(\mathcal{E})} \rho_{\mathrm{i}}(u_{\mathrm{i}}^{(m)}(t))^2 \leq \mathcal{E} \quad \text{when } m \geq N_1(\mathcal{E}). \qquad (19.48)$$

Also, since $\hat{\boldsymbol{u}}(t) \in \ell_\rho^2$, for every $\mathcal{E} > 0$ there exists $M_2(\mathcal{E}) \in \mathbb{N}$ such that

$$\sum_{|\mathrm{i}|\geq M_2(\mathcal{E})} \rho_{\mathrm{i}}\hat{u}_{\mathrm{i}}^2(t) \leq \mathcal{E}. \qquad (19.49)$$

Thus, picking $M(\mathcal{E}) = \max\{M_1(\mathcal{E}), M_2(\mathcal{E})\}$ and using (19.48) and (19.49),

$$\begin{aligned} \sum_{|\mathrm{i}|\geq M(\mathcal{E})} \rho_{\mathrm{i}} \left(u_{\mathrm{i}}^{(m)}(t) - \hat{u}_{\mathrm{i}}(t)\right)^2 &\leq 2\left(\sum_{|\mathrm{i}|\geq M(\mathcal{E})} \rho_{\mathrm{i}} \left((u_{\mathrm{i}}^{(m)}(t))_{\mathrm{i}}^2 + \hat{u}_{\mathrm{i}}^2(t)\right)\right) \\ &\leq 4\mathcal{E} \qquad \text{when } m \geq N_1(\mathcal{E}). \end{aligned} \tag{19.50}$$

On the other hand, by the weak convergence (19.47), for every $\mathcal{E}$ there exists $N_2(\mathcal{E})$ such that

$$\sum_{|\mathrm{i}|\leq M(\mathcal{E})} \rho_{\mathrm{i}} \left(u_{\mathrm{i}}^{(m)}(t) - \hat{u}_{\mathrm{i}}(t)\right)^2 \leq \mathcal{E} \quad \text{when } m \geq N_2(\mathcal{E}). \tag{19.51}$$

Summarising the above and letting $N(\mathcal{E}) = \max\{N_1(\mathcal{E}), N_2(\mathcal{E})\}$, it follows from the estimates (19.50) and (19.51) that

$$\left\|u_{\mathrm{i}}^{(m)}(t) - \hat{u}_{\mathrm{i}}(t)\right\|_\rho^2 \leq \left(\sum_{|\mathrm{i}|\geq M(\mathcal{E})} + \sum_{|\mathrm{i}|\leq M(\mathcal{E})}\right) \rho_{\mathrm{i}} \left(u_{\mathrm{i}}^{(m)}(t) - \hat{u}_{\mathrm{i}}(t)\right)^2 \leq 5\mathcal{E}.$$

This implies that the convergence of $\boldsymbol{u}^{(m)}(\cdot)$ to $\hat{\boldsymbol{u}}(\cdot)$ is strong. As a result the family of set-valued mappings $\{\Phi^\epsilon(t)\}_{t\in\mathbb{R}}$ is asymptotically compact. □

Remark 19.1. The above arguments also work for $\epsilon = 0$ and the un-inflated set-valued semi-group Φ.

19.6 Attractors of the sigmoidal and lattice systems

Denote by $\Phi^\epsilon = \{\Phi^\epsilon(t)\}_{t\geq 0}$, $\Phi = \{\Phi(t)\}_{t\geq 0}$ and $\varphi^\epsilon = \{\varphi^\epsilon(t)\}_{t\geq 0}$ the dynamical systems defined by the solutions of the ϵ-inflated lattice differential inclusions (19.15), the Heaviside lattice differential inclusions (19.10), and the sigmoidal lattice ODEs (19.4), respectively.

As shown above, the closed and bounded set Q defined in (19.46) is a positive invariant absorbing set for Φ^ϵ for every $\epsilon \in [0,1]$. It was shown in Lemma 19.10 that Φ^ϵ is asymptotically compact. Hence by an autonomous version of Proposition 2.2 the set-valued dynamical system Φ^ϵ has unique global attractor $\mathfrak{A}^\epsilon \subset Q$ given by

$$\mathfrak{A}^\epsilon = \bigcap_{t\geq 0} \overline{\bigcup_{\tau\geq t} \Phi^\epsilon(\tau, Q)}, \qquad 0 \leq \epsilon \leq 1.$$

Since $\Phi(t, \boldsymbol{u}_o) \subset \Phi^\epsilon(t, \boldsymbol{u}_o)$ for all $t \geq 0$ and $\boldsymbol{u}_o \in \ell_\rho^2$ it follows that the set Q is also positive invariant and absorbing for Φ and that particularly, the omega-limit points of Φ are omega-limit points of Φ^ϵ. Hence Φ has a unique global attractor $\mathfrak{A} = \mathfrak{A}^0 \subset \mathfrak{A}^\epsilon$.

Moreover, $\varphi^\epsilon(t, \boldsymbol{u}_o) \subset \Phi^\epsilon(t, \boldsymbol{u}_o)$ for all $t \geq 0$ and $\boldsymbol{u}_o \in \ell_\rho^2$, since the solutions of the sigmoidal lattice system are also solutions of the corresponding inflated lattice system. Hence the set Q is also positive invariant and absorbing for φ^ϵ. In particular, the omega-limit points of φ^ϵ are omega-limit points of Φ^ϵ, so φ^ϵ has a unique global attractor $\mathcal{A}^\epsilon \subset \mathfrak{A}^\epsilon$.

19.6.1 *Comparison of the attractors*

Summarising from the above subsection, the Heaviside system (19.10), the inflated system (19.15) and the sigmoidal system (19.16) have global attractors $\mathfrak{A}$, $\mathfrak{A}^\epsilon$ and $\mathcal{A}^\epsilon$, respectively, for which

$$\mathfrak{A} = \mathfrak{A}^0 \subset \mathfrak{A}^\epsilon \quad \text{and } \mathcal{A}^\epsilon \subset \mathfrak{A}^\epsilon \quad \text{for all } \epsilon \in (0,1].$$

In fact, stronger properties hold.

Theorem 19.6. *The attractors $\mathfrak{A}^\epsilon$ for the inflated lattice systems converge to the attractor $\mathfrak{A}$ for the Heaviside lattice system upper semi-continuously in ℓ^2_ρ, i.e.,*

$$\mathrm{dist}_{\ell^2_\rho}(\mathfrak{A}^\epsilon, \mathfrak{A}) \to 0 \quad as \quad \epsilon \to 0. \tag{19.52}$$

Proof. The global attractor $\mathfrak{A}^\epsilon$ of the set-valued dynamical system Φ^ϵ consists of entire trajectories, i.e., continuous functions $\boldsymbol{u}^\epsilon : \mathbb{R} \to \ell^2_\rho$ such that $\boldsymbol{u}^\epsilon(t) \in \Phi^\epsilon(t-s, \boldsymbol{u}^\epsilon(s))$ for all $s \le t$ and $\boldsymbol{u}^\epsilon(t) \in \mathfrak{A}^\epsilon$ for all $t \in \mathbb{R}$.

Suppose (for contradiction) that the convergence (19.52) does not hold. Then there are ε_0, $\epsilon_n \in (0,1]$ with $\epsilon_n \to 0$ such that

$$\mathrm{dist}_{\ell^2_\rho}(\mathfrak{A}^{\epsilon_n}, \mathfrak{A}) \ge 2\varepsilon_0 \quad \text{for all } n \in \mathbb{N}. \tag{19.53}$$

Moreover, since the $\mathfrak{A}^{\epsilon_n}$ is compact there is $a_n \in \mathfrak{A}^{\epsilon_n}$ such that

$$\mathrm{dist}_{\ell^2_\rho}(a_n, \mathfrak{A}) = \mathrm{dist}_{\ell^2_\rho}(\mathfrak{A}^{\epsilon_n}, \mathfrak{A}) \ge 2\varepsilon_0 \quad \text{for all } n \in \mathbb{N}.$$

Choose an arbitrary sequence of entire solutions $\boldsymbol{u}^{\epsilon_n}$ of Φ^{ϵ_n} with $\boldsymbol{u}^{\epsilon_n}(0) = a_n$. These are also entire solutions of Φ^ϵ.

Similar to the proof of the convergence Theorem 19.4, there exists a convergent subsequence $\boldsymbol{u}^{\epsilon_n}$ converging uniformly on any closed and bounded time interval to a continuous function $\boldsymbol{u}^* : \mathbb{R} \to \ell^2_\rho$, which is also an entire trajectory of Φ. In particular, $\boldsymbol{u}^*(0) \in \mathfrak{A}$. Moreover, $\boldsymbol{u}^{\epsilon_{n_r}}(0) = a_{n_r} \to \boldsymbol{u}^*(0)$, so

$$\mathrm{dist}_{\ell^2_\rho}(\boldsymbol{u}^*(0), \mathfrak{A}) \ge \varepsilon_0,$$

which is a contradiction. The proof is complete. □

Remark 19.2. Note that from the argument above, $\mathfrak{A} = \bigcap_{\epsilon>0} \mathfrak{A}^\epsilon$. Hence

$$\mathrm{dist}_{\ell^2_\rho}(\mathfrak{A}, \mathfrak{A}^\epsilon) = 0,$$

so the convergence above is in fact continuous convergence, i.e., the equation (19.52) holds with Hausdorff distance $\mathcal{H}_{\ell^2_\rho}$ instead of just the semi-distance $\mathrm{dist}_{\ell^2_\rho}$.

In addition, the following corollary holds.

Corollary 19.2. *The attractors for the sigmoidal lattice systems $\mathcal{A}^\epsilon$ converge to the attractor $\mathfrak{A}$ for the Heaviside lattice system upper semi-continuously in ℓ^2_ρ, i.e.,*

$$\mathrm{dist}_{\ell^2_\rho}(\mathcal{A}^\epsilon, \mathfrak{A}) \to 0 \quad as \quad \epsilon \to 0.$$

Proof. This follows immediately from the inequality

$$\text{dist}_{\ell^2_\rho}(\mathcal{A}^\epsilon, \mathfrak{A}) \leq \text{dist}_{\ell^2_\rho}(\mathcal{A}^\epsilon, \mathfrak{A}^\epsilon) + \text{dist}_{\ell^2_\rho}(\mathfrak{A}^\epsilon, \mathfrak{A}) \leq 0 + \text{dist}_{\ell^2_\rho}(\mathfrak{A}^\epsilon, \mathfrak{A})$$

and Theorem 19.6 above. □

19.7 End notes

The results in this chapter are based on [Han and Kloeden (2020)] and were first established in [Han and Kloeden (2019a)] for nonautonomous systems and pullback attractors. The existence proof there was similar but involved approximations by solutions of truncated finite dimensional inclusion systems rather than sigmoidal solutions.

Part (iv) of the proof of the convergence theorem, Theorem 19.4, given in [Han and Kloeden (2020)], has been corrected here using the Banach-Saks Theorem, see corrigendum to [Han and Kloeden (2020)].

[Wang *et al.* (2020a,b)] considers the above model with delays, while [Kloeden and Villarragut (2020)] considers a related neural field lattice model with delays.

Traveling wave fronts for the Amari lattice neural field equations were studied by Faye [Faye (2018)].

19.8 Problems

Problem 19.1. How much do the proofs need to be changed when the sigmoidal function σ_ϵ is replaced by the piecewise linear function

$$p_\epsilon(x) := \begin{cases} 0, & x \leq -\frac{\epsilon}{2} \\ \frac{1}{\epsilon}x + \frac{1}{2}, & -\frac{\epsilon}{2} \leq x \leq \frac{\epsilon}{2} \\ 1, & x \geq \frac{\epsilon}{2} \end{cases}.$$

Problem 19.2. Construct an alternative proof for Part 4 of Theorem 19.4, without using the Banach-Saks Theorem.

Problem 19.3. Write out a proof for Theorem 19.5.

Chapter 20

Stochastic neural field models with nonlinear noise

Stochastic lattice systems with additive white noise or finite linear multiplicative noise, such as those considered in Chapters 12 and 13, can be studied using the theory of random dynamical systems and pathwise analysis. Such analysis, however, is not straightforward to apply to stochastic systems with infinite noise or nonlinear noise. Recently, the concepts and theory of mean random dynamical system have been used in studying stochastic lattice systems with infinite-dimensional nonlinear noise [Wang and Wang (2020)].

Consider the following non-autonomous stochastic neural field lattice model with infinite state-dependent nonlinear noise defined on $\mathbb{Z}^d$ with $d \in \mathbb{N}$:

$$\begin{cases} \mathrm{d}u_{\mathrm{i}}(t) = \Big(f_{\mathrm{i}}(u_{\mathrm{i}}) + \sum\limits_{\mathrm{j}\in\mathbb{Z}^d} \kappa_{\mathrm{i},\mathrm{j}}\eta(u_{\mathrm{j}}) + g_{\mathrm{i}}(t)\Big)\mathrm{d}t + (\sigma_{\mathrm{i}}(u_{\mathrm{i}}) + h_{\mathrm{i}}(t))\,\mathrm{d}W_{\mathrm{i}}(t), \ t > t_0, \\ u_{\mathrm{i}}(t_0) = u_{o,\mathrm{i}}, \end{cases} \tag{20.1}$$

where $\mathrm{i} = (i_1, \ldots, i_d) \in \mathbb{Z}^d$ is the vectorial index of integers, $t_0 \in \mathbb{R}$, and $\boldsymbol{u}_o := (u_{o,\mathrm{i}})_{\mathrm{i}\in\mathbb{Z}^d}$ is the corresponding initial data.

Here u_{i} represents the neural activity such as neural synapse of the ith node, $f_{\mathrm{i}} : \mathbb{R} \to \mathbb{R}$ is a continuously differentiable function describing the attenuation of neural activity of the ith node, and $\eta : \mathbb{R} \to \mathbb{R}$ is the activation function that defines the output of a node given an input. The quantity $\kappa_{\mathrm{i},\mathrm{j}}$ describes the connection strength from the jth to the ith node, that can take either positive or negative value for stimulation or inhibition of the jth neuron on the ith neuron, respectively. The external forcing at the ith location for the drift and diffusion are described by the time dependent (possibly random) functions $g_{\mathrm{i}} : \mathbb{R} \to \mathbb{R}$ and $h_{\mathrm{i}} : \mathbb{R} \to \mathbb{R}$, respectively.

20.1 Well-posedness of the LDS in ℓ^2_ρ

The lattice system (20.1) will be considered in the weighted space ℓ^2_ρ of bi-infinite square summable real-valued sequences on vectorial indices $\mathrm{i} = (i_1, \ldots, i_d) \in \mathbb{Z}^d$ with the weights satisfying

Assumption 20.1. $\rho_{\rm i} > 0$ for all $\rm i \in \mathbb{Z}^d$ and $\rho_\Sigma := \sum_{\rm i\in\mathbb{Z}^d} \rho_{\rm i} < \infty$.

One complex character of the model (20.1) is the global interactions among neurons, modeled by the infinite sum $\sum_{\rm j\in\mathbb{Z}^d} \kappa_{\rm i,j}\eta(u_{\rm j})$ that indicates each neuron takes activations from all other neurons in the whole network. A natural assumption is that interactions between any pair of neurons that are extremely far away from each other are extremely small. In particular, we assume that the infinite dimensional interconnection matrix $(\kappa_{\rm i,j})_{\rm i,j\in\mathbb{Z}^d}$ satisfy

Assumption 20.2. there exists a constant $\kappa > 0$ such that

$$\sum_{\rm j\in\mathbb{Z}^d} \frac{\kappa_{\rm i,j}^2}{\rho_{\rm j}} \le \kappa \quad \text{for each } \rm i \in \mathbb{Z}^d.$$

In addition to the assumptions above, the following assumptions are imposed on the nonlinear drift and diffusion terms.

Assumption 20.3. For each $\rm i \in \mathbb{Z}^d$ the attenuation function $f_{\rm i} : \mathbb{R} \to \mathbb{R}$ is continuously differentiable with $f_{\rm i}(0) = 0$ and weighted equi-locally bounded derivatives, i.e., there exists a non-decreasing function $L_f(\cdot) \in \mathcal{C}(\mathbb{R}^+, \mathbb{R}^+)$ such that

$$\max_{\rho_{\rm i} x\in[-r,r]} |f_{\rm i}'(x)| \le L_f(r), \quad \forall\, r \in \mathbb{R}^+, \rm i \in \mathbb{Z}^d.$$

Assumption 20.4. For each $\rm i \in \mathbb{Z}^d$ the state dependent nonlinear diffusion term $\sigma_{\rm i} : \mathbb{R} \to \mathbb{R}$ is continuously differentiable, and there exists a non-decreasing function $L_\sigma(\cdot) \in \mathcal{C}(\mathbb{R}^+, \mathbb{R}^+)$ such that

$$\max_{\rho_{\rm i} x\in[-r,r]} |\sigma_{\rm i}'(x)| \le L_\sigma(r), \quad \forall\, r \in \mathbb{R}^+, \rm i \in \mathbb{Z}^d.$$

In addition, there exist $\boldsymbol{a} = \{a_{\rm i}\}_{\rm i\in\mathbb{Z}^d} \in \ell^\infty$ and $\boldsymbol{b} = \{b_{\rm i}\}_{\rm i\in\mathbb{Z}^d} \in \ell^2_\rho$ such that

$$|\sigma_{\rm i}(x)| \le a_{\rm i}|x| + b_{\rm i}, \quad \forall\, x \in \mathbb{R}.$$

Assumption 20.5. The activation function η is globally Lipschitz continuous with Lipchitz constant L_η, and there exist $a_\eta \in \mathbb{R}$ and $b_\eta > 0$ such that

$$|\eta(x)| \le a_\eta|x| + b_\eta, \quad \forall\, x \in \mathbb{R}.$$

To reformulate system (20.1) as an abstract stochastic differential equation on ℓ^2_ρ, define the operators F, $\mathcal{K}$ and $\mathfrak{S}$ by

$$F(\boldsymbol{u}) = (f_{\rm i}(u_{\rm i}))_{\rm i\in\mathbb{Z}^d}, \qquad \mathfrak{S}(\boldsymbol{u}) = (\sigma_{\rm i}(u_{\rm i}))_{\rm i\in\mathbb{Z}^d}$$
$$\mathcal{K}(\boldsymbol{u}) = (K_{\rm i}(\boldsymbol{u}))_{\rm i\in\mathbb{Z}^d} \text{ with } K_{\rm i}(\boldsymbol{u}) := \sum_{\rm j\in\mathbb{Z}^d} \kappa_{\rm i,j}\eta(u_{\rm j}).$$

Lemma 20.1. *Under the Assumptions 20.1–20.5, the operators $F, \mathcal{K}, \mathfrak{S}$ map ℓ^2_ρ to ℓ^2_ρ. Moreover, the operator $\mathcal{K}$ is globally Lipschitz and the operators F and $\mathfrak{S}$ are locally Lipschitz.*

Proof. First by Assumption 20.3, we have

$$\begin{aligned}\|F(\boldsymbol{u})\|_\rho^2 &= \sum_{\mathrm{i}\in\mathbb{Z}^d} \rho_\mathrm{i}|f_\mathrm{i}(u_\mathrm{i})|^2 = \sum_{\mathrm{i}\in\mathbb{Z}^d} \rho_\mathrm{i}|f_\mathrm{i}(u_\mathrm{i}) - f_\mathrm{i}(0)|^2 \\ &\le \sum_{\mathrm{i}\in\mathbb{Z}^d} \rho_\mathrm{i} L_f^2(\sqrt{\rho_\Sigma}\|\boldsymbol{u}\|_\rho)|u_\mathrm{i}|^2 \le L_f^2(\sqrt{\rho_\Sigma}\|\boldsymbol{u}\|_\rho)\|\boldsymbol{u}\|_\rho^2,\end{aligned}$$

i.e., $F(\boldsymbol{u}) \in \ell_\rho^2$ for all $\boldsymbol{u} \in \ell_\rho^2$. In addition, notice that for any $\boldsymbol{u} \in \ell_\rho^2$ with $\|\boldsymbol{u}\|_\rho \le R$ we have

$$\rho_i|u_i| \le \sum_{\mathrm{i}\in\mathbb{Z}^d} \rho_\mathrm{i}|u_i| \le \sqrt{\sum_{\mathrm{i}\in\mathbb{Z}^d} \rho_\mathrm{i} \sum_{\mathrm{i}\in\mathbb{Z}^d} \rho_\mathrm{i} u_i^2} \le \sqrt{\rho_\Sigma} R. \tag{20.2}$$

Therefore for any $\boldsymbol{u}, \boldsymbol{v} \in \ell_\rho^2$ with $\|\boldsymbol{u}\|_\rho, \|\boldsymbol{v}\|_\rho \le R$ it follows from Assumption 20.3 that

$$\begin{aligned}\|F(\boldsymbol{u}) - F(\boldsymbol{v})\|_\rho^2 &\le \sum_{\mathrm{i}\in\mathbb{Z}^d} \rho_\mathrm{i} L_f^2(\rho_\mathrm{i}(|u_\mathrm{i}| + |v_\mathrm{i}|))|u_\mathrm{i} - v_\mathrm{i}|^2 \\ &\le \sum_{\mathrm{i}\in\mathbb{Z}^d} \rho_\mathrm{i} L_f^2(2\sqrt{\rho_\Sigma}R)|u_\mathrm{i} - v_\mathrm{i}|^2 = L_f^2(2\sqrt{\rho_\Sigma}R)\|\boldsymbol{u} - \boldsymbol{v}\|_\rho^2, \end{aligned}\tag{20.3}$$

which implies that F is locally Lipschitz.

Next, notice it follows from Assumptions 20.1 and 20.2 that

$$\sum_{\mathrm{j}\in\mathbb{Z}^d} |\kappa_{\mathrm{i},\mathrm{j}}| = \sum_{\mathrm{j}\in\mathbb{Z}^d} \frac{|\kappa_{\mathrm{i},\mathrm{j}}|}{\sqrt{\rho_\mathrm{j}}}\sqrt{\rho_\mathrm{j}} \le \left(\sum_{\mathrm{j}\in\mathbb{Z}^d} \frac{\kappa_{\mathrm{i},\mathrm{j}}^2}{\rho_\mathrm{j}}\right)^{1/2} \left(\sum_{\mathrm{j}\in\mathbb{Z}^d} \rho_\mathrm{j}\right)^{1/2} \le \sqrt{\kappa\rho_\Sigma}.$$

Then due to Assumptions 20.1, 20.2 and 20.5,

$$\begin{aligned}\|\mathcal{K}(\boldsymbol{u})\|_\rho^2 &\le \sum_{\mathrm{i}\in\mathbb{Z}^d} \rho_\mathrm{i}\Big(\sum_{\mathrm{j}\in\mathbb{Z}^d} |\kappa_{\mathrm{i},\mathrm{j}}|(a_\eta|u_\mathrm{j}| + b_\eta)\Big)^2 \\ &\le \sum_{\mathrm{i}\in\mathbb{Z}^d} \rho_\mathrm{i} \sum_{\mathrm{j}\in\mathbb{Z}^d} \frac{\kappa_{\mathrm{i},\mathrm{j}}^2}{\rho_\mathrm{j}} \sum_{\mathrm{j}\in\mathbb{Z}^d} \rho_\mathrm{j}(a_\eta|u_\mathrm{j}| + b_\eta)^2 \\ &\le 2\kappa \sum_{\mathrm{i}\in\mathbb{Z}^d} \rho_\mathrm{i} \left(a_\eta^2 \sum_{\mathrm{j}\in\mathbb{Z}^d} \rho_\mathrm{j}|u_\mathrm{j}^2| + b_\eta^2 \sum_{\mathrm{j}\in\mathbb{Z}^d} \rho_\mathrm{j}\right) \\ &\le 2\kappa\rho_\Sigma\left(a_\eta^2\|\boldsymbol{u}\|_\rho^2 + \rho_\Sigma b_\eta^2\right). \end{aligned}\tag{20.4}$$

In addition, for any $\boldsymbol{u}, \boldsymbol{v} \in \ell_\rho^2$ due to Assumption 20.5 we have

$$\begin{aligned}\|\mathcal{K}(\boldsymbol{u}) - \mathcal{K}(\boldsymbol{v})\|_\rho^2 &\le \sum_{\mathrm{i}\in\mathbb{Z}^d} \rho_\mathrm{i}\Big(\sum_{\mathrm{j}\in\mathbb{Z}^d} |\kappa_{\mathrm{i},\mathrm{j}}| L_\eta |u_\mathrm{j} - v_\mathrm{j}|\Big)^2 \\ &\le L_\eta^2 \sum_{\mathrm{i}\in\mathbb{Z}^d} \rho_\mathrm{i} \sum_{\mathrm{j}\in\mathbb{Z}^d} \frac{\kappa_{\mathrm{i},\mathrm{j}}^2}{\rho_\mathrm{j}} \sum_{\mathrm{j}\in\mathbb{Z}^d} \rho_\mathrm{j}|u_\mathrm{j} - v_\mathrm{j}|^2 \\ &\le \kappa\rho_\Sigma L_\eta^2 \|\boldsymbol{u} - \boldsymbol{v}\|_\rho^2, \end{aligned}\tag{20.5}$$

i.e., $\mathcal{K} : \ell_\rho^2 \to \ell_\rho^2$ is globally Lipschitz.

Last, for any $\boldsymbol{u} \in \ell^2_\rho$, by Assumption 20.4, we have

$$\begin{aligned}\|\mathfrak{S}(\boldsymbol{u})\|^2_\rho &= \sum_{\mathfrak{i}\in\mathbb{Z}^d} \rho_i|\sigma_\mathfrak{i}(u_\mathfrak{i})|^2 \le 2\sum_{\mathfrak{i}\in\mathbb{Z}^d} \rho_i(|a_\mathfrak{i}|^2|u_\mathfrak{i}|^2 + |b_\mathfrak{i}|^2)\\ &\le 2(\|\boldsymbol{a}\|^2_\infty\|\boldsymbol{u}\|^2_\rho + \|\boldsymbol{b}\|^2_\rho),\end{aligned}$$

i.e., $\mathfrak{S}$ maps ℓ^2_ρ to ℓ^2_ρ. Then using (20.2) again and Assumption 20.4, similar to (20.3) we have

$$\|\mathfrak{S}(\boldsymbol{u}) - \mathfrak{S}(\boldsymbol{v})\|^2_\rho \le L^2_\sigma(2\sqrt{\rho_\Sigma}R)\|\boldsymbol{u}-\boldsymbol{v}\|^2_\rho, \quad \forall\ \|\boldsymbol{u}\|_\rho, \|\boldsymbol{v}\|_\rho \le R,$$

which implies that $\mathfrak{S}$ is locally Lipschitz. The proof is complete. □

In the end to rewrite the terms $(\sigma_\mathfrak{i}(u_\mathfrak{i}) + h_\mathfrak{i}(t))\,\mathrm{d}W_\mathfrak{i}(t)$ as vectors in ℓ^2_ρ, define two sequence of operators $\mathfrak{S}_\mathfrak{i}$ and $H_\mathfrak{i}$ by

$$\mathfrak{S}_\mathfrak{i}(\boldsymbol{u}) = (\sigma_\mathfrak{i}(u_\mathfrak{i}))\boldsymbol{e}^\mathfrak{i}, \qquad H_\mathfrak{i}(t) = (h_\mathfrak{i}(t))\boldsymbol{e}^\mathfrak{i}, \qquad \mathfrak{i}\in\mathbb{Z}^d,$$

where $\boldsymbol{e}^\mathfrak{i}$ represents the infinite sequence with 1 at position $\mathfrak{i}$ and 0 elsewhere.

Assumption 20.6. The forcing processes $\boldsymbol{g}(t) := (g_\mathfrak{i}(t))_{\mathfrak{i}\in\mathbb{Z}^d}$ and $H(t) := (h_\mathfrak{i}(t))_{\mathfrak{i}\in\mathbb{Z}^d}$ are progressively measurable in ℓ^2_ρ and satisfy

$$\int_{t_0}^{t_0+T} \mathbb{E}\left[\|\boldsymbol{g}(t)\|^2_\rho + \|H(t)\|^2_\rho\right]\mathrm{d}t < \infty, \quad \forall\ t_0\in\mathbb{R},\ T>0.$$

It then follows directly from Lemma 20.1 and Assumption 20.6 that the lattice system (20.1) is equivalent to the following stochastic differential equation that is well-defined on ℓ^2_ρ.

$$\begin{cases}\mathrm{d}\boldsymbol{u}(t) = \big(F(\boldsymbol{u}(t)) + \mathcal{K}(\boldsymbol{u}(t)) + \boldsymbol{g}(t)\big)\mathrm{d}t + \sum_{\mathfrak{i}\in\mathbb{Z}^d}(\mathfrak{S}_\mathfrak{i}(\boldsymbol{u}) + H_\mathfrak{i}(t))\mathrm{d}W_\mathfrak{i}(t),\ t>t_0,\\ \boldsymbol{u}(t_0) = \boldsymbol{u}_o := (u_{o,\mathfrak{i}})_{\mathfrak{i}\in\mathbb{Z}^d}.\end{cases} \tag{20.6}$$

20.2 Existence of mean-square solutions

This section concerns the existence of mean-square solutions to the problem (20.6), defined as follows.

Definition 20.1. For every $t_0\in\mathbb{R}$ and $\mathcal{F}_{t_0}$-measurable initial data $\boldsymbol{u}_o$, a continuous ℓ^2_ρ-valued $\mathcal{F}_t$-adapted stochastic process $\boldsymbol{u}(t)$ is called a mean-square solution of the problem (20.6) if for every $T>0$, $\boldsymbol{u}\in\mathcal{L}^2(\Omega, \mathcal{C}([t_0, t_0+T], \ell^2_\rho))$ and satisfies the following integral equation in ℓ^2_ρ

$$\boldsymbol{u}(t) = \boldsymbol{u}_o + \int_{t_0}^t (F(\boldsymbol{u}(s)) + \mathcal{K}(\boldsymbol{u}(s)) + \boldsymbol{g}(s))\,\mathrm{d}s + \sum_{\mathfrak{i}\in\mathbb{Z}^d}\int_{t_0}^t (\mathfrak{S}_\mathfrak{i}(\boldsymbol{u}(s)) + H_i(s))\,\mathrm{d}W_i(s)$$

for all $t\ge t_0$ and almost every $\omega\in\Omega$.

20.2.1 *Solutions of the truncated system*

Notice that by the definitions of $\mathfrak{S}$ and $\mathfrak{S}_{\mathfrak{i}}$ we have $\mathfrak{S}(\boldsymbol{u}) = \sum_{\mathfrak{i}\in\mathbb{Z}^d} \mathfrak{S}_{\mathfrak{i}}(\boldsymbol{u})$, and moreover

$$\|\mathfrak{S}(\boldsymbol{u})\|_\rho^2 = \sum_{\mathfrak{i}\in\mathbb{Z}^d} \|\mathfrak{S}_{\mathfrak{i}}(\boldsymbol{u})\|_\rho^2, \sum_{\mathfrak{i}\in\mathbb{Z}^d} \|\mathfrak{S}_{\mathfrak{i}}(\boldsymbol{u}) - \mathfrak{S}_{\mathfrak{i}}(\boldsymbol{v})\|_\rho^2 = \|\mathfrak{S}(\boldsymbol{u}) - \mathfrak{S}(\boldsymbol{v})\|_\rho^2, \forall\, \boldsymbol{u}, \boldsymbol{v} \in \ell_\rho^2. \tag{20.7}$$

The operators F and $\mathfrak{S}$ are locally Lipschitz continuous that guarantees local in time solutions along with the relation (20.7). To obtain global-in-time solutions in the sense of Definition 20.1, for each $n \in \mathbb{N}$ define a truncation function $\Xi_n : \mathbb{R} \to \mathbb{R}$ by

$$\Xi_n(x) = \begin{cases} -n, & x \in (-\infty, -n), \\ x, & x \in [-n, n], \\ n, & x \in (n, +\infty), \end{cases} \qquad x \in \mathbb{R}.$$

It is straightforward to check Ξ_n is increasing and has the following properties:

$$\Xi_n(0) = 0, \quad \Xi_n(x) \le n, \quad |\Xi_n(x_1) - \Xi_n(x_2)| \le |x_1 - x_2|, \quad \forall x, x_1, x_2 \in \mathbb{R}. \tag{20.8}$$

Next, for every $n \in \mathbb{N}$ and $\boldsymbol{u} = (u_{\mathfrak{i}})_{\mathfrak{i}\in\mathbb{Z}^d} \in \ell_\rho^2$ set $u_{\mathfrak{i}}^{(n)} = \Xi_n(u_{\mathfrak{i}})$ and define two truncated operators, F^n and $\mathfrak{S}^n$, by

$$F^n(\boldsymbol{u}) = (f_{\mathfrak{i}}(u_{\mathfrak{i}}^{(n)}))_{\mathfrak{i}\in\mathbb{Z}^d}, \qquad \mathfrak{S}^n(\boldsymbol{u}) = (\sigma_{\mathfrak{i}}(u_{\mathfrak{i}}^{(n)}))_{\mathfrak{i}\in\mathbb{Z}^d}.$$

Then for each fixed $n \in \mathbb{N}$ the operators $F^n, \mathfrak{S}^n : \ell_\rho^2 \to \ell_\rho^2$ are globally Lipschitz under Assumptions 20.1–20.4. In fact, for any $\boldsymbol{u}, \boldsymbol{v} \in \ell_\rho^2$, due to Assumptions 20.3, 20.4 and property (20.8) we have

$$\|F^n(\boldsymbol{u}) - F^n(\boldsymbol{v})\|_\rho^2 \le L_f^2(2\sqrt{\rho_\Sigma}n)\|\boldsymbol{u} - \boldsymbol{v}\|_\rho^2, \tag{20.9}$$

$$\|\mathfrak{S}^n(\boldsymbol{u}) - \mathfrak{S}^n(\boldsymbol{v})\|_\rho^2 \le L_\sigma^2(2\sqrt{\rho_\Sigma}n)\|\boldsymbol{u} - \boldsymbol{v}\|_\rho^2. \tag{20.10}$$

Also, it follows directly from (20.9) and $f_{\mathfrak{i}}(0) = \Xi_n(0) = 0$ that $F^n(\boldsymbol{u}) \in \ell_\rho^2$ for all $\boldsymbol{u} \in \ell_\rho^2$. In the end, by Assumption 20.4 and property (20.8), we have

$$\begin{aligned}\|\mathfrak{S}^n(\boldsymbol{u})\|_\rho^2 &\le \sum_{\mathfrak{i}\in\mathbb{Z}^d} \rho_{\mathfrak{i}}(a_{\mathfrak{i}}|u_{\mathfrak{i}}^{(n)}| + b_{\mathfrak{i}})^2 \le 2(\|\boldsymbol{a}\|_\infty^2 \sum_{\mathfrak{i}\in\mathbb{Z}^d} \rho_{\mathfrak{i}}|u_{\mathfrak{i}}|^2 + \sum_{\mathfrak{i}\in\mathbb{Z}^d} \rho_{\mathfrak{i}} b_{\mathfrak{i}}^2) \\ &\le 2\|\boldsymbol{a}\|_\infty^2\|\boldsymbol{u}\|_\rho^2 + 2\|\boldsymbol{b}\|_\rho^2,\end{aligned}$$

which implies that $\mathfrak{S}^n(\boldsymbol{u}) \in \ell_\rho^2$ for all $\boldsymbol{u} \in \ell_\rho^2$.

For $\boldsymbol{u} \in \ell_\rho^2$ and $n \in \mathbb{N}$ define $\mathfrak{S}_{\mathfrak{i}}^n(\boldsymbol{u}) := (\sigma_{\mathfrak{i}}(\Xi_n(u_{\mathfrak{i}})))\boldsymbol{e}^{\mathfrak{i}}$, and consider the following auxiliary stochastic system in ℓ_ρ^2 of $\boldsymbol{z}^{(n)}(t) = (z_{\mathfrak{i}}^{(n)}(t))_{\mathfrak{i}\in\mathbb{Z}^d}$ for $t_0 \in \mathbb{R}$

$$\begin{cases} \mathrm{d}\boldsymbol{z}^{(n)}(t) = (F^n(\boldsymbol{z}^{(n)}) + \mathcal{K}(\boldsymbol{z}^{(n)}) + \boldsymbol{g}(t))\mathrm{d}t + \sum_{\mathfrak{i}\in\mathbb{Z}^d} (\mathfrak{S}_{\mathfrak{i}}^n(\boldsymbol{z}^{(n)}) + H_i(t))\mathrm{d}W_i(t), \ t > t_0, \\ \boldsymbol{z}^{(n)}(t_0) = \boldsymbol{u}_o. \end{cases} \tag{20.11}$$

Noticing that for all $\boldsymbol{u}, \boldsymbol{v} \in \ell^2_\rho$ due to (20.7) there holds

$$\sum_{\mathrm{i}\in\mathbb{Z}^d} \|\mathfrak{S}^n_\mathrm{i}(\boldsymbol{u})\|^2_\rho = \|\mathfrak{S}^n(\boldsymbol{u})\|^2_\rho, \quad \sum_{\mathrm{i}\in\mathbb{Z}^d} \|\mathfrak{S}^n_\mathrm{i}(\boldsymbol{u}) - \mathfrak{S}^n_\mathrm{i}(\boldsymbol{v})\|^2_\rho = \|\mathfrak{S}^n(\boldsymbol{u}) - \mathfrak{S}^n(\boldsymbol{v})\|^2_\rho.$$

The existence and uniqueness of solutions to the stochastic equation (20.11) in the sense of Definition 20.1 then follows directly from the global Lipschitz property of operators $F^n, \mathcal{K}$ and $\mathfrak{S}^n$, Assumption 20.6, and standard theory for stochastic differential equations (see, e.g., [Arnold (1974)]).

20.2.2 *Existence of a global mean-square solution*

We next investigate the existence and uniqueness of solutions of system (20.6) in the sense of Definition 20.1 by considering the limit of the sequence $\{\boldsymbol{z}^{(n)}(t)\}_{n\in\mathbb{N}}$ of solutions of system (20.11) as $n \to \infty$. To that end, we impose the following dissipativity assumption on the attenuation functions.

Assumption 20.7. There exist constants $\alpha > 0$ and $\boldsymbol{\beta} = (\beta_\mathrm{i})_{\mathrm{i}\in\mathbb{Z}^d} \in \ell^2_\rho$ such that

$$x f_\mathrm{i}(x) \le -\alpha x^2 + \beta_\mathrm{i}^2, \quad \forall\, x \in \mathbb{R},\ \mathrm{i} \in \mathbb{Z}^d.$$

Theorem 20.2 below uses the Burkholder-Davis-Gundy inequality [Burkholder (1966)], stated as follows.

Theorem 20.1. *(Burkholder-Davis-Gundy inequalities) Let $T > 0$ and $\{M_t\}_{0\le t\le T}$ be a continuous local martingale such that $M_0 = 0$. Then for every $0 < p < \infty$ there exist universal constants c_p and C_p independent of T and $\{M_t\}_{0\le t\le T}$ such that*

$$c_p \mathbb{E}\left[\langle M\rangle_t\right] \le \mathbb{E}\left[\left(\sup_{s\in[0,t]} |M_s|\right)^p\right] \le C_p \mathbb{E}\left[\langle M\rangle_t\right],$$

where the process $\{\langle M\rangle_t\}_{0\le t\le T}$ is the quadratic variation process of $\{M_t\}_{0\le t\le T}$.

Theorem 20.2. *Let Assumptions 20.1–20.7 hold. Then for any $t_0 \in \mathbb{R}$ and $\mathcal{F}_{t_0}$-measurable initial data $\boldsymbol{u}_o \in \mathcal{L}^2(\Omega, \ell^2_\rho)$, the stochastic system (20.6) possesses a unique solution $\boldsymbol{u}$ in the sense of Definition 20.1. Moreover, there exist $\mu_1 = \mu_1(\boldsymbol{u}_o, t_0, T, \boldsymbol{\beta}, \alpha, \kappa, b_\eta, \rho_\Sigma, \boldsymbol{b}, C_p, G, H)$ and $\mu_2 = \mu_2(\boldsymbol{a}, C_p, a_\eta, \kappa, \rho_\Sigma)$ in which C_p is a generic constant due to the Burkholder-Davis-Gundy inequality, such that*

$$\mathbb{E}\left[\|\boldsymbol{u}\|^2_{C([t_0,t_0+T],\ell^2_\rho)}\right] \le \mu_1 e^{\mu_2 T}, \quad \textit{for every } T > 0. \tag{20.12}$$

The proof of Theorem 20.2 consists of four parts. For clarity of exposition, the first two parts are presented in Lemma 20.2 and Lemma 20.3 below, respectively. For $n \in \mathbb{N}$, define a stopping time τ_n for the solution to the auxiliary system (20.11) by

$$\tau_n = \begin{cases} \inf\{t \ge t_0 : \|\boldsymbol{z}^{(n)}(t)\|_\rho > n\}, & \{t \ge t_0 : \|\boldsymbol{z}^{(n)}(t)\|_\rho > n\} \ne \emptyset \\ +\infty, & \{t \ge t_0 : \|\boldsymbol{z}^{(n)}(t)\|_\rho > n\} = \emptyset \end{cases}. \tag{20.13}$$

Lemma 20.2. *Let Assumptions 20.1–20.7 hold. Then for any $t_0 \in \mathbb{R}$ and $\mathcal{F}_{t_0}$-measurable initial data $\boldsymbol{u}_o \in \mathcal{L}^2(\Omega, \ell^2_\rho)$, the solution to the auxiliary system* (20.11) *satisfies*

$$\boldsymbol{z}^{(n+1)}(t \wedge \tau_n) = \boldsymbol{z}^{(n)}(t \wedge \tau_n) \quad \textit{for all} \quad t \geq t_0 \quad \textit{almost surely.}$$

Proof. First by (20.11) we have

$$\begin{aligned} \mathrm{d}(\boldsymbol{z}^{(n+1)} - \boldsymbol{z}^{(n)}) &= \big(F^{n+1}(\boldsymbol{z}^{(n+1)}) - F^n(\boldsymbol{z}^n)\big)\mathrm{d}t + \big(\mathcal{K}(\boldsymbol{z}^{(n+1)}) - \mathcal{K}(\boldsymbol{z}^{(n)})\big)\mathrm{d}t \\ &\quad + \sum_{\mathfrak{i} \in \mathbb{Z}^d} \big(\mathfrak{S}_{\mathfrak{i}}^{n+1}(\boldsymbol{z}^{(n+1)}) - \mathfrak{S}_{\mathfrak{i}}^n(\boldsymbol{z}^{(n)})\big)\mathrm{d}W_i(t). \end{aligned} \tag{20.14}$$

Taking the inner product $\langle \cdot, \cdot \rangle_\rho$ of (20.14) with $\boldsymbol{z}^{(n+1)} - \boldsymbol{z}^{(n)}$, applying Itô's formula, and integrating the resultant stochastic differential equation from t_0 to $t \wedge \tau_n$ we obtain

$$\begin{aligned} &\|\boldsymbol{z}^{(n+1)}(t \wedge \tau_n) - \boldsymbol{z}^{(n)}(t \wedge \tau_n)\|_\rho^2 \\ &\quad = \underbrace{2\int_{t_0}^{t \wedge \tau_n} \langle \boldsymbol{z}^{(n+1)}(s) - \boldsymbol{z}^{(n)}(s), F^{n+1}(\boldsymbol{z}^{(n+1)}(s)) - F^n(\boldsymbol{z}^{(n)}(s))\rangle_\rho \mathrm{d}s}_{\textbf{(i)}} \\ &\qquad + \underbrace{2\int_{t_0}^{t \wedge \tau_n} \langle \boldsymbol{z}^{(n+1)}(s) - \boldsymbol{z}^{(n)}(s), \mathcal{K}(\boldsymbol{z}^{(n+1)}(s)) - \mathcal{K}(\boldsymbol{z}^{(n)}(s))\rangle_\rho \mathrm{d}s}_{\textbf{(ii)}} \\ &\qquad + \underbrace{\int_{t_0}^{t \wedge \tau_n} \|\mathfrak{S}^{n+1}(\boldsymbol{z}^{(n+1)}(s)) - \mathfrak{S}^n(\boldsymbol{z}^{(n)}(s))\|_\rho^2 \mathrm{d}s}_{\textbf{(iii)}} \\ &\qquad + 2\sum_{\mathfrak{i} \in \mathbb{Z}^d} \rho_{\mathfrak{i}} \int_{t_0}^{t \wedge \tau_n} (z_{\mathfrak{i}}^{(n+1)}(s) - z_{\mathfrak{i}}^{(n)}(s))(\sigma_{\mathfrak{i}}^{n+1}(z_{\mathfrak{i}}^{(n+1)}) - \sigma_{\mathfrak{i}}^n(z_{\mathfrak{i}}^{(n)}))\mathrm{d}W_i(s). \end{aligned} \tag{20.15}$$

Note that $\|\boldsymbol{z}^{(n)}(s)\|_\rho \leq n$ for $s \in [t_0, \tau_n)$, we have for all $s \in [t_0, \tau_n)$,

$$F^{n+1}(\boldsymbol{z}^{(n)}(s)) = F^n(\boldsymbol{z}^{(n)}(s)) = F(\boldsymbol{z}^{(n)}(s)), \tag{20.16}$$

$$\mathfrak{S}^{n+1}(\boldsymbol{z}^{(n)}(s)) = \mathfrak{S}^n(\boldsymbol{z}^{(n)}(s)) = \mathfrak{S}(\boldsymbol{z}^{(n)}(s)). \tag{20.17}$$

Then by (20.9) and (20.16) the term **(i)** on the right-hand side of (20.15) satisfies

$$\textbf{(i)} \leq L_f(2\sqrt{\rho_\Sigma}(n+1)) \int_{t_0}^{t \wedge \tau_n} \|\boldsymbol{z}^{(n+1)}(s) - \boldsymbol{z}^{(n)}(s)\|_\rho^2 \mathrm{d}s. \tag{20.18}$$

Similarly, it follows from (20.10) and (20.17) that the term **(iii)** on the right-hand side of (20.15) satisfies

$$\textbf{(iii)} \leq L_\sigma(2\sqrt{\rho_\Sigma}(n+1)) \int_{t_0}^{t \wedge \tau_n} \|\boldsymbol{z}^{(n+1)}(s) - \boldsymbol{z}^{(n)}(s)\|_\rho^2 \mathrm{d}s. \tag{20.19}$$

Next, it follows immediately from (20.5) that

$$(\mathbf{ii}) \le \sqrt{\rho_\Sigma \kappa} L_\eta \|\boldsymbol{z}^{(n+1)} - \boldsymbol{z}^{(n)}\|_\rho^2. \tag{20.20}$$

Inserting (20.18)–(20.20) into (20.15) and taking expectation of the supremum over t_0 and t of the resultant inequality gives

$$\begin{aligned}
&\mathbb{E}\Big[\sup_{t_0 \le s \le t} \|\boldsymbol{z}^{(n+1)}(s \wedge \tau_n) - \boldsymbol{z}^{(n)}(s \wedge \tau_n)\|_\rho^2\Big] \\
&\le C_1 \int_{t_0}^t \mathbb{E}\Big[\sup_{t_0 \le s \le r} \|\boldsymbol{z}^{(n+1)}(s \wedge \tau_n) - \boldsymbol{z}^{(n)}(s \wedge \tau_n)\|_\rho^2\Big] \mathrm{d}r \\
&\quad + 2\mathbb{E}\left[\sup_{t_0 \le s \le t \wedge \tau_n} \left| \sum_{\mathrm{i} \in \mathbb{Z}^d} \rho_\mathrm{i} \int_{t_0}^s \mathcal{I}_\mathrm{i}(r) \mathrm{d}W_i(r) \right| \right],
\end{aligned} \tag{20.21}$$

where $C_1 = 2L_f(2\sqrt{\rho_\Sigma}(n+1)) + L_\sigma(2\sqrt{\rho_\Sigma}(n+1)) + 2\sqrt{\rho_\Sigma \kappa} L_\eta$ and

$$\mathcal{I}_\mathrm{i}(r) = (z_\mathrm{i}^{(n+1)}(r) - z_\mathrm{i}^{(n)}(r))(\sigma_\mathrm{i}^{n+1}(z_\mathrm{i}^{(n+1)}(r)) - \sigma_\mathrm{i}^n(z_\mathrm{i}^{(n)}(r))).$$

We next estimate the last term on the right hand side of the inequality (20.21). In fact, due to the Burkholder-Davis-Gundy inequality, there exists $C_p > 0$ such that

$$\mathbb{E}\left[\sup_{t_0 \le s \le t \wedge \tau_n} \left| \sum_{\mathrm{i} \in \mathbb{Z}^d} \rho_\mathrm{i} \int_{t_0}^s \mathcal{I}_\mathrm{i}(r) \mathrm{d}W_i(r) \right| \right] \le C_p \sum_{\mathrm{i} \in \mathbb{Z}^d} \rho_\mathrm{i} \mathbb{E}\left[\left(\int_{t_0}^{t \wedge \tau_n} \mathcal{I}_\mathrm{i}^2(r) \mathrm{d}r\right)^{1/2}\right], \tag{20.22}$$

in which

$$\begin{aligned}
\left(\int_{t_0}^{t \wedge \tau_n} \mathcal{I}_\mathrm{i}^2(r) \mathrm{d}r\right)^{1/2} &\le L_\sigma(2\rho_\Sigma(n+1)) \sup_{t_0 \le s \le t} \left| z_\mathrm{i}^{(n+1)}(s \wedge \tau_n) - z_\mathrm{i}^{(n)}(s \wedge \tau_n) \right| \\
&\quad \cdot \left(\int_{t_0}^{t \wedge \tau_n} \left| z_\mathrm{i}^{(n+1)}(r) - z_\mathrm{i}^{(n)}(r) \right|^2 \mathrm{d}r \right)^{\frac{1}{2}} \\
&\le \frac{1}{4C_p} \sup_{t_0 \le s \le t} \left| z_\mathrm{i}^{(n+1)}(s \wedge \tau_n) - z_\mathrm{i}^{(n)}(s \wedge \tau_n) \right|^2 \\
&\quad + C_p L_\sigma^2(2\rho_\Sigma(n+1)) \int_{t_0}^{t \wedge \tau_n} \left| z_\mathrm{i}^{(n+1)}(r) - z_\mathrm{i}^{(n)}(r) \right|^2 \mathrm{d}r.
\end{aligned}$$

Applying the above inequality to (20.22) then plugging into (20.21) results in

$$\begin{aligned}
&\mathbb{E}\Big[\sup_{t_0 \le s \le t} \|\boldsymbol{z}^{(n+1)}(s \wedge \tau_n) - \boldsymbol{z}^{(n)}(s \wedge \tau_n)\|_\rho^2\Big] \\
&\le 2\left(C_1 + 2C_p^2 L_\sigma^2(2\rho_\Sigma(n+1))\right) \int_{t_0}^t \mathbb{E}\Big[\sup_{t_0 \le s \le r} \|\boldsymbol{z}^{(n+1)}(s \wedge \tau_n) - \boldsymbol{z}^{(n)}(s \wedge \tau_n)\|_\rho^2\Big] \mathrm{d}r.
\end{aligned} \tag{20.23}$$

It then follows from Gronwall's lemma that

$$\mathbb{E}\Big[\sup_{t_0\le s\le t}\|\boldsymbol{z}^{(n+1)}(s\wedge\tau_n)-\boldsymbol{z}^{(n)}(s\wedge\tau_n)\|_\rho^2\Big]=0,\quad \forall\, t\ge t_0,$$

which implies the desired assertion. □

Lemma 20.3. *Let Assumptions 20.1–20.7 hold. Then the stopping time defined by* (20.13) *satisfies*

$$\lim_{n\to\infty}\tau_n=\infty,\quad \textit{almost surely.}$$

Proof. Notice that due to Lemma 20.2 and (20.13) we can infer that $\tau_{n+1}\ge\tau_n$ almost surely. We next derive a uniform estimates for the solution $\boldsymbol{z}^{(n)}$ of the auxiliary equation (20.11). To that end, taking the inner product $\langle\cdot,\cdot\rangle_\rho$ of (20.11) with $\boldsymbol{z}^{(n)}$, applying Itô's formula, and integrating the resultant stochastic differential equation from t_0 to $t\wedge\tau_n$ to obtain

$$\begin{aligned}\|\boldsymbol{z}^{(n)}(t\wedge\tau_n)\|_\rho^2&=\|\boldsymbol{u}_o\|_\rho^2+2\underbrace{\int_{t_0}^{t\wedge\tau_n}\langle\boldsymbol{z}^{(n)}(s),F^n(\boldsymbol{z}^{(n)}(s))\rangle_\rho\mathrm{d}s}_{\textbf{(iv)}}\\&\quad+\underbrace{2\int_{t_0}^{t\wedge\tau_n}\langle\boldsymbol{z}^{(n)}(s),\mathcal{K}(\boldsymbol{z}^{(n)}(s))\rangle_\rho\mathrm{d}s}_{\textbf{(v)}}+\underbrace{2\int_{t_0}^{t\wedge\tau_n}\langle\boldsymbol{z}^{(n)}(s),\boldsymbol{g}(s)\rangle_\rho\mathrm{d}s}_{\textbf{(vi)}}\\&\quad+\underbrace{\int_{t_0}^{t\wedge\tau_n}\big\|\mathfrak{S}^n(\boldsymbol{z}^{(n)}(s))+H(s)\big\|_\rho^2\mathrm{d}s}_{\textbf{(vii)}}\\&\quad+\underbrace{2\sum_{\mathrm{i}\in\mathbb{Z}^d}\rho_\mathrm{i}\int_{t_0}^{t\wedge\tau_n}z_\mathrm{i}^{(n)}(s)(\sigma_\mathrm{i}^n(z_\mathrm{i}^{(n)}(s))+h_\mathrm{i}(s))\mathrm{d}W_\mathrm{i}(s)}_{\textbf{(viii)}}.\qquad(20.24)\end{aligned}$$

We next estimate terms **(iv)**–**(vii)** in the above equation.

First by (20.16) and Assumption 20.7 we have

$$\begin{aligned}\textbf{(iv)}&=\int_{t_0}^{t\wedge\tau_n}\sum_{\mathrm{i}\in\mathbb{Z}^d}\rho_\mathrm{i}z_\mathrm{i}^{(n)}(s)f_\mathrm{i}(z_\mathrm{i}^{(n)}(s))\mathrm{d}s\\&\le\int_{t_0}^{t\wedge\tau_n}\sum_{\mathrm{i}\in\mathbb{Z}^d}\rho_\mathrm{i}(-\alpha|z_\mathrm{i}^{(n)}(s)|^2+\beta_\mathrm{i}^2)\mathrm{d}s\\&\le-\alpha\int_{t_0}^{t\wedge\tau_n}\|\boldsymbol{z}^{(n)}(s)\|_\rho^2\mathrm{d}s+\|\boldsymbol{\beta}\|_\rho^2T,\quad\forall\, t\in[t_0,t_0+T).\qquad(20.25)\end{aligned}$$

Next, by Assumptions 20.1, 20.2, and 20.5 we have

$$
\begin{aligned}
(\mathbf{v}) &= \int_{t_0}^{t\wedge\tau_n} \sum_{\mathrm{i}\in\mathbb{Z}^d} \rho_{\mathrm{i}} z_{\mathrm{i}}^{(n)}(s) \sum_{\mathrm{j}\in\mathbb{Z}^d} \kappa_{\mathrm{i},\mathrm{j}} \eta(z_{\mathrm{j}}^{(n)}(s))\mathrm{d}s \\
&\le \int_{t_0}^{t\wedge\tau_n} \sum_{\mathrm{i}\in\mathbb{Z}^d} \rho_{\mathrm{i}} z_{\mathrm{i}}^{(n)}(s) \sum_{\mathrm{j}\in\mathbb{Z}^d} |\kappa_{\mathrm{i},\mathrm{j}}|(a_\eta |z_{\mathrm{j}}^{(n)}(s)| + b_\eta)\mathrm{d}s \\
&\le \int_{t_0}^{t\wedge\tau_n} \big\|\boldsymbol{z}^{(n)}(s)\big\|_\rho \Big(\sum_{\mathrm{i}\in\mathbb{Z}^d} \rho_{\mathrm{i}} \big(\sum_{\mathrm{j}\in\mathbb{Z}^d} \frac{\kappa_{\mathrm{i},\mathrm{j}}^2}{\rho_{\mathrm{j}}} \sum_{\mathrm{j}\in\mathbb{Z}^d} \rho_{\mathrm{j}} (a_\eta |z_{\mathrm{j}}^{(n)}(s)| + b_\eta)^2 \big) \Big)^{\frac{1}{2}} \mathrm{d}s \\
&\le \sqrt{\kappa\rho_\Sigma} \int_{t_0}^{t\wedge\tau_n} \big\|\boldsymbol{z}^{(n)}(s)\big\|_\rho (2a_\eta \|\boldsymbol{z}^{(n)}(s)\|_\rho + 2b_\eta\sqrt{\rho_\Sigma})\mathrm{d}s,
\end{aligned}
$$

which can be further estimated to give

$$
(\mathbf{v}) \le \Big(2a_\eta\sqrt{\kappa\rho_\Sigma} + \frac{\alpha}{2}\Big) \int_{t_0}^{t\wedge\tau_n} \|\boldsymbol{z}^{(n)}(s)\|_\rho^2 \mathrm{d}s + \frac{2}{\alpha} b_\eta^2 \rho_\Sigma^2 \kappa T, \quad \forall\, t \in [t_0, t_0 + T). \tag{20.26}
$$

The term (**vi**) satisfies

$$
(\mathbf{vi}) \le \frac{\alpha}{2} \int_{t_0}^{t\wedge\tau_n} \big\|\boldsymbol{z}^{(n)}(s)\big\|_\rho^2 \mathrm{d}s + \frac{2}{\alpha} \int_{t_0}^{t\wedge\tau_n} \big\|\boldsymbol{g}(s)\big\|_\rho^2 \mathrm{d}s. \tag{20.27}
$$

In the end by Assumption 20.4 and (20.8), for all $t \in [t_0, t_0 + T)$ term (**vii**) satisfies

$$
\begin{aligned}
(\mathbf{vii}) &\le 2\int_{t_0}^{t\wedge\tau_n} \|\mathfrak{S}(\boldsymbol{z}^{(n)}(s))\|_\rho^2 \mathrm{d}s + 2\int_{t_0}^{t\wedge\tau_n} \|H(s)\|_\rho^2 \mathrm{d}s \\
&\le 2\int_{t_0}^{t\wedge\tau_n} \sum_{\mathrm{i}\in\mathbb{Z}^d} \rho_{\mathrm{i}} (a_{\mathrm{i}} |z_{\mathrm{i}}^{(n)}| + b_{\mathrm{i}})^2 \mathrm{d}s + 2\int_{t_0}^{t\wedge\tau_n} \|H(s)\|_\rho^2 \mathrm{d}s \\
&\le 4\|\boldsymbol{a}\|_\infty^2 \int_{t_0}^{t\wedge\tau_n} \|\boldsymbol{z}^{(n)}(s)\|_\rho^2 \mathrm{d}s + 4\|\boldsymbol{b}\|_\rho^2 T + 2\int_{t_0}^{t\wedge\tau_n} \|H(s)\|_\rho^2 \mathrm{d}s.
\end{aligned} \tag{20.28}
$$

Collecting the estimates (20.25) through (20.28) into (20.24) results in

$$
\begin{aligned}
\|\boldsymbol{z}^{(n)}(t\wedge\tau_n)\|_\rho^2 \le \|\boldsymbol{u}_o\|_\rho^2 &+ (4a_\eta\sqrt{\kappa\rho_\Sigma} + 4\|\boldsymbol{a}\|_\infty^2) \int_{t_0}^{t\wedge\tau_n} \|\boldsymbol{z}^{(n)}(s)\|_\rho^2 \mathrm{d}s \\
&+ \frac{1}{\alpha} \int_{t_0}^{t\wedge\tau_n} \|\boldsymbol{g}(s)\|_\rho^2 \mathrm{d}s + 2\int_{t_0}^{t\wedge\tau_n} \|H(s)\|_\rho^2 \mathrm{d}s + 2\cdot(\mathbf{viii}) \\
&+ \left(2\|\boldsymbol{\beta}\|_\rho^2 + \frac{4b_\eta^2 \kappa\rho_\Sigma^2}{\alpha} + 4\|\boldsymbol{b}\|_\rho^2 \right) T, \quad \forall\, t \in [t_0, t_0 + T).
\end{aligned} \tag{20.29}
$$

Taking supremum of (20.29) over (t_0, t) and taking expectation of the resultant inequality we obtain

$$\mathbb{E}\Big[\sup_{t_0\le r\le t}\|\boldsymbol{z}(r\wedge\tau_n)\|_\rho^2\Big] \le \mathbb{E}[\|\boldsymbol{u}_o\|_\rho^2] + \left(2\|\boldsymbol{\beta}\|_\rho^2 + \frac{4b_\eta^2\kappa\rho_\Sigma^2}{\alpha} + 4\|\boldsymbol{b}\|_\rho^2\right)T$$
$$+\left(4a_\eta\sqrt{\kappa\rho_\Sigma} + 4\|\boldsymbol{a}\|_\infty^2\right)\int_{t_0}^{t}\mathbb{E}\Big[\sup_{t_0\le r\le s}\|\boldsymbol{z}^{(n)}(r\wedge\tau_n)\|_\rho^2\Big]\mathrm{d}s$$
$$+\frac{1}{\alpha}\int_{t_0}^{t_0+T}\mathbb{E}[\|\boldsymbol{g}(s)\|_\rho^2]\mathrm{d}s + 2\int_{t_0}^{T}\mathbb{E}[\|H(s)\|_\rho^2]\mathrm{d}s$$
$$+\underbrace{2\,\mathbb{E}\Big[\sup_{t_0\le r\le t\wedge\tau_n}\Big|\sum_{\mathfrak{i}\in\mathbb{Z}^d}\rho_{\mathfrak{i}}\int_{t_0}^{t\wedge\tau_n} z_{\mathfrak{i}}^{(n)}(s)(\sigma_{\mathfrak{i}}^n(z_{\mathfrak{i}}^{(n)}(s)) + h_{\mathfrak{i}}(s))\mathrm{d}W_{\mathfrak{i}}(s)\Big|\Big]}_{(\mathbf{ix})}. \tag{20.30}$$

Now using the Burkholder-Davis-Gundy inequality again, there exists $C_p > 0$ such that

$$(\mathbf{ix}) \le C_p\sum_{\mathfrak{i}\in\mathbb{Z}^d}\rho_{\mathfrak{i}}\mathbb{E}\left[\left(\int_{t_0}^{t\wedge\tau_n}\left(z_{\mathfrak{i}}^{(n)}(s)(\sigma_{\mathfrak{i}}^n(z_{\mathfrak{i}}^{(n)}(s)) + h_{\mathfrak{i}}(s))\right)^2\mathrm{d}s\right)^{1/2}\right],$$

after which it follows from Hölder's inequality and Assumption 20.4 that

$$(\mathbf{ix}) \le C_p\mathbb{E}\left[\sum_{\mathfrak{i}\in\mathbb{Z}^d}\rho_{\mathfrak{i}}\left(\int_{t_0}^{t\wedge\tau_n}\left(z_{\mathfrak{i}}^{(n)}(s)(\sigma_{\mathfrak{i}}^n(z_{\mathfrak{i}}^{(n)}(s)) + h_{\mathfrak{i}}(s))\right)^2\mathrm{d}s\right)^{1/2}\right]$$
$$\le \frac{1}{4}\mathbb{E}\left[\sup_{t_0\le s\le t}\|\boldsymbol{z}^{(n)}(s\wedge\tau_n)\|_\rho^2\right] + C_p^2\int_{t_0}^{t\wedge\tau_n}\mathbb{E}[\|\mathfrak{S}^n(\boldsymbol{z}^{(n)}(s)) + H(s)\|_\rho^2]\mathrm{d}s$$
$$\le \frac{1}{4}\mathbb{E}\left[\sup_{t_0\le s\le t}\|\boldsymbol{z}^{(n)}(s\wedge\tau_n)\|_\rho^2\right] + 2C_p^2\int_{t_0}^{t\wedge\tau_n}\mathbb{E}[\|H(s)\|_\rho^2]\mathrm{d}s$$
$$+4C_p^2\left(\|\boldsymbol{a}\|_\infty^2\int_{t_0}^{t\wedge\tau_n}\mathbb{E}[\|\boldsymbol{z}^{(n)}(s)\|_\rho^2]\mathrm{d}s + \|\boldsymbol{b}\|_\rho^2T\right). \tag{20.31}$$

Now inserting (20.31) into (20.30) and using Assumption 20.6 we obtain

$$\mathbb{E}\Big[\sup_{t_0\le r\le t}\|\boldsymbol{z}^{(n)}(r\wedge\tau_n)\|_\rho^2\Big] \le \mu_2\int_{t_0}^{t}\mathbb{E}[\sup_{t_0\le r\le s}\|\boldsymbol{z}^{(n)}(r\wedge\tau_n)\|_\rho^2]\mathrm{d}s + \mu_1,$$

where

$$\mu_1 = 2\mathbb{E}[\|\boldsymbol{u}_o\|_\rho^2] + 4\left(\|\boldsymbol{\beta}\|_\rho^2 + \frac{2\kappa b_\eta^2\rho_\Sigma^2}{\alpha} + 2\|\boldsymbol{b}\|_\rho^2 + 4C_p^2\|\boldsymbol{b}\|_\rho^2\right)T$$
$$+\frac{2}{\alpha}\int_{t_0}^{t_0+T}\mathbb{E}\left[\|\boldsymbol{g}(s)\|_\rho^2\right]\mathrm{d}s + 4(2C_p^2+1)\int_{t_0}^{t_0+T}\mathbb{E}[\|H(s)\|_\rho^2]\mathrm{d}s$$
$$\mu_2 = 8\|\boldsymbol{a}\|_\infty^2(1+2C_p^2) + 8a_\eta\sqrt{\kappa\rho_\Sigma}.$$

It then follows immediately from Gronwall's Lemma that

$$\mathbb{E}\left[\sup_{t_0\le r\le t}(\|z^{(n)}(r\wedge\tau_n)\|_\rho^2)\right]\le \mu_1 e^{\mu_2 T},\quad \forall\, t\in[t_0,t_0+T]. \tag{20.32}$$

Finally, consider an arbitrary number $T\in\mathbb{N}$. Due to (20.13) we have

$$\{\omega:\tau_n(\omega)<T\}\subseteq\left\{\omega:\sup_{t_0\le r\le t_0+T}\|z(r\wedge\tau_n(\omega))\|_\rho\ge n\right\}.$$

The relation above, together with Chebychev's inequality and (20.32), imply

$$\begin{aligned}\mathbb{P}[\tau_n<T]&\le\mathbb{P}\left[\sup_{t_0\le r\le t_0+T}\|z^{(n)}(r\wedge\tau_n)\|_\rho\ge n\right]\\&\le\frac{1}{n^2}\mathbb{E}\left[\sup_{t_0\le r\le t_0+T}\|z^{(n)}(r\wedge\tau_n)\|_\rho^2\right]\le\frac{\mu_1 e^{\mu_2 T}}{n^2}.\end{aligned}$$

Note that both μ_2 and μ_1 are independent of n. Then it further follows

$$\sum_{n=1}^{\infty}\mathbb{P}[\tau_n<T]\le\mu_1 e^{\mu_2 T}\sum_{n=1}^{\infty}\frac{1}{n^2}<\infty. \tag{20.33}$$

Let $\Omega_T=\cap_{m=1}^{\infty}\cup_{n=m}^{\infty}\{\omega:\tau_n(\omega)<T\}$. Then it follows from the Borel-Cantelli Lemma and (20.33) that

$$\mathbb{P}[\Omega_T]=\mathbb{P}\left[\cap_{m=1}^{\infty}\cup_{n=m}^{\infty}\{\tau_n<T\}\right]=0.$$

Thus for every $\omega\in\Omega\backslash\Omega_T$, there exists $N_0=N_0(\omega)>0$ such that $\tau_n(\omega)\ge T$ for all $n\ge N_0$. Noticing that (20.13) implies τ_n is increasing in n, we have $\tau(\omega)\ge T$ for all $\Omega\backslash\Omega_T$. Let $\Omega_0=\cup_{T=1}^{\infty}\Omega_T$, then

$$\mathbb{P}(\Omega_0)=0\quad\text{and}\quad\tau(\omega)\ge T\quad\forall\,\omega\in\Omega\backslash\Omega_0,\ T\in\mathbb{N}.$$

Consequently, $\tau(\omega)=\infty$ for all $\omega\in\Omega\backslash\Omega_0$, and hence

$$\tau:=\lim_{n\to\infty}\tau_n=\sup_{n\in\mathbb{N}}\tau_n=\infty\quad a.s.$$

□

We are now ready to complete the proof of Theorem 20.2.

Proof of Theorem 20.2. The remaining proof consists of two parts, on the existence and uniqueness of solutions of system (20.6), respectively.

1. Existence. Due to Lemmas 20.2 and 20.3, there exists $\mathfrak{D}\subset\Omega$ with $\mathbb{P}(\Omega\backslash\mathfrak{D})=0$ such that

$$\tau(\omega)=\lim_{n\to\infty}\tau_n(\omega)=\infty,\ \forall\,\omega\in\mathfrak{D}, \tag{20.34}$$

$$z^{(n+1)}(t\wedge\tau_n,\omega)=\quad z^{(n)}(t\wedge\tau_n,\omega),\quad\forall\,\omega\in\mathfrak{D},\ n\in\mathbb{N},\ t\ge t_0. \tag{20.35}$$

It follows from (20.34)–(20.35) that for every $\omega\in\mathfrak{D}$ and $t\ge t_0$, there exists $N_1=N_1(t,\omega)\ge 1$ such that

$$\tau_n(\omega)>t,\quad z^{(n)}(t,\omega)=z^{(N_1)}(t,\omega)\quad\text{for all } n\ge N_1.$$

Now define the mapping $\boldsymbol{u} : [t_0, \infty) \times \Omega \to \ell^2_\rho$ by

$$\boldsymbol{u}(t,\omega) = \begin{cases} \boldsymbol{z}^{(n)}(t,\omega), \omega \in \mathfrak{D} \quad \text{and} \quad t \in [t_0, \tau_n(\omega)], \\ \boldsymbol{u}_o(\omega), \quad \omega \in \Omega\backslash\mathfrak{D} \quad \text{and} \quad t \in [t_0, \infty). \end{cases} \tag{20.36}$$

Note that $\boldsymbol{z}^{(n)}$ is a continuous ℓ^2_ρ-valued process, it can be inferred from (20.36) that $\boldsymbol{u}$ is also almost surely continuous with respect to t in ℓ^2_ρ. Moreover, (20.36) implies that

$$\lim_{n\to\infty} \boldsymbol{z}^{(n)}(t,\omega) = \boldsymbol{u}(t,\omega), \quad \omega \in \mathfrak{D}, \quad t \geq t_0. \tag{20.37}$$

Since $\boldsymbol{z}^{(n)}$ is $\mathcal{F}_t$-adapted for every $n \in \mathbb{N}$, it follows from (20.37) that $\boldsymbol{u}$ is also $\mathcal{F}_t$-adapted. Furthermore, using (20.37), (20.32) and Fatou's lemma, we conclude that for every $T > 0$,

$$\mathbb{E}\left[\|\boldsymbol{u}\|^2_{C([t_0,t_0+T],\ell^2_\rho)}\right] \leq \mu_1 e^{\mu_2 T}.$$

Note that (20.36) implies $\boldsymbol{z}^{(n)}(t\wedge\tau_n) = \boldsymbol{u}(t\wedge\tau_n)$ almost surely, and hence

$$F^n(\boldsymbol{z}^{(n)}(t\wedge\tau_n)) = F(\boldsymbol{z}^{(n)}(t\wedge\tau_n)) = F(\boldsymbol{u}(t\wedge\tau_n)) \quad a.s.$$

$$\mathcal{K}(\boldsymbol{z}^{(n)}(t\wedge\tau_n)) = \mathcal{K}(\boldsymbol{u}(t\wedge\tau_n)) \quad a.s.$$

$$\mathfrak{S}^n_{\mathfrak{i}}(\boldsymbol{z}^{(n)}(t\wedge\tau_n)) = \mathfrak{S}_{\mathfrak{i}}(\boldsymbol{z}^{(n)}(t\wedge\tau_n)) = \mathfrak{S}_{\mathfrak{i}}(\boldsymbol{u}(t\wedge\tau_n)) \quad a.s. \ \forall\, \mathfrak{i} \in \mathbb{Z}^d.$$

Then due to (20.11), we obtain that $\boldsymbol{u}(t)$ satisfies

$$\begin{aligned} \boldsymbol{u}(t\wedge\tau_n) = \boldsymbol{u}_o &+ \int_{t_0}^{t\wedge\tau_n} (F(\boldsymbol{u}(s)) + \mathcal{K}(\boldsymbol{u}(s)) + \boldsymbol{g}(s))\mathrm{d}s \\ &+ \int_{t_0}^{t\wedge\tau_n} \sum_{\mathfrak{i}\in\mathbb{Z}^d} (\mathfrak{S}_{\mathfrak{i}}(\boldsymbol{u}(s)) + H_{\mathfrak{i}}(s))\mathrm{d}W_{\mathfrak{i}}(s) \quad a.s. \end{aligned} \tag{20.38}$$

Equation (20.38) along with $\lim_{n\to\infty}\tau_n = \infty$ proved in Lemma 20.3 immediately implies that $\boldsymbol{u}$ is a solution of (20.6) in the sense of Definition 20.1.

2. Uniqueness. Let $\boldsymbol{u}(t; t_0, \boldsymbol{u}_o)$, $\boldsymbol{v}(t; t_0, \boldsymbol{v}_o)$ be two solutions of the system (20.6) with initial conditions $\boldsymbol{u}(t_0) = \boldsymbol{u}_o$ and $\boldsymbol{v}(t_0) = \boldsymbol{v}_o$, respectively. For every $n \in \mathbb{N}$ and $T > 0$, define a stopping time T_n by

$$T_n = (t_0 + T) \wedge \inf\{t \geq t_0 : \|\boldsymbol{u}(t)\| \geq n \text{ or } \|\boldsymbol{v}(t)\| \geq n\}.$$

Then by the system (20.6), we have

$$\begin{aligned} \boldsymbol{u}(t\wedge T_n) - \boldsymbol{v}(t\wedge T_n) = \boldsymbol{u}_o - \boldsymbol{v}_o &+ \int_{t_0}^{t\wedge T_n} (F(\boldsymbol{u}(s)) - F(\boldsymbol{v}(s)))\mathrm{d}s \\ &+ \int_{t_0}^{t\wedge T_n} (\mathcal{K}(\boldsymbol{u}(s)) - \mathcal{K}(\boldsymbol{v}(s)))\mathrm{d}s \\ &+ \int_{t_0}^{t\wedge T_n} \sum_{\mathfrak{i}\in\mathbb{Z}^d} (\mathfrak{S}_{\mathfrak{i}}(\boldsymbol{u}(s)) - \mathfrak{S}_{\mathfrak{i}}(\boldsymbol{v}(s)))\mathrm{d}W_i(s). \end{aligned}$$

The equation above along with Itô's formula give that

$$\begin{aligned}\|\boldsymbol{u}(t\wedge T_n)-\boldsymbol{v}(t\wedge T_n)\|_\rho^2 \le \|\boldsymbol{u}_o-\boldsymbol{v}_{t_0}\|_\rho^2 &+ \int_{t_0}^{t\wedge T_n}\|\mathfrak{S}(\boldsymbol{u}(s))-\mathfrak{S}(\boldsymbol{v}(s))\|_\rho^2 \mathrm{d}s\\ &+2\int_{t_0}^{t\wedge T_n}\langle \boldsymbol{u}(s)-\boldsymbol{v}(s), F(\boldsymbol{u}(s))-F(\boldsymbol{v}(s))\rangle_\rho \mathrm{d}s\\ &+2\int_{t_0}^{t\wedge T_n}\langle \boldsymbol{u}(s)-\boldsymbol{v}(s), \mathcal{K}(\boldsymbol{u}(s))-\mathcal{K}(\boldsymbol{v}(s))\rangle_\rho \mathrm{d}s\\ &+2\int_{t_0}^{t\wedge T_n}\sum_{i\in\mathbb{Z}^d}\rho_i(u_i(s)-v_i(s))(\sigma_i(u_i(s))\\ &-\sigma_i(v_i(s)))\mathrm{d}W_i(s), \ a.s.\end{aligned}$$

Following similar analysis to reach (20.23), we obtain

$$\begin{aligned}\mathbb{E}\Big[\sup_{t_0\le s\le t}\|\boldsymbol{u}(s\wedge T_n)-\boldsymbol{v}(s\wedge T_n)\|_\rho^2\Big] &\le C_2\mathbb{E}\left[\|\boldsymbol{u}_o-\boldsymbol{v}_{t_0}\|_\rho^2\right]\\ &+C_2\int_{t_0}^{t}\mathbb{E}\Big[\sup_{t_0\le s\le r}\|\boldsymbol{u}(s\wedge T_n)-\boldsymbol{v}(s\wedge T_n)\|_\rho^2\Big]\mathrm{d}r, \qquad (20.39)\end{aligned}$$

where $C_2 = 2\big(2L_f(2\sqrt{\rho_\Sigma}n)+L_\sigma(2\sqrt{\rho_\Sigma}n)+2\sqrt{\rho_\Sigma}\kappa L_\eta+2C_p^2L_\sigma^2(2\rho_\Sigma n)\big)+1$ and C_p is a generic constant from the Burkholder-Davis-Gundy inequality. Applying the Gronwall's Lemma to (20.39) results in

$$\mathbb{E}\left[\sup_{t_0\le s\le t_0+T}\|\boldsymbol{u}_1(s\wedge T_n)-\boldsymbol{u}_2(s\wedge T_n)\|_\rho^2\right]\le C_2e^{C_2T}\mathbb{E}\left[\|\boldsymbol{u}_o-\boldsymbol{v}_o\|_\rho^2\right],$$

which implies that $\|\boldsymbol{u}(t\wedge T_n)-\boldsymbol{v}(t\wedge T_n)\|_\rho=0$ for all $t\in[t_0,t_0+T]$ almost surely when $\boldsymbol{u}_o=\boldsymbol{v}_o$.

Note that due to the continuity of $\boldsymbol{u}$, $\boldsymbol{v}$ in t, we have $T_n=t_0+T$ for n large enough. Then we have $\|\boldsymbol{u}(t)-\boldsymbol{v}(t)\|_\rho=0$ for all $t\in[t_0,t_0+T]$ almost surely and consequently, for every $T>0$,

$$\mathbb{P}\Big[\|\boldsymbol{u}_1(t)-\boldsymbol{u}_2(t)\|_\rho=0 \text{ for all } \ t\in[t_0,t_0+T]\Big]=1.$$

Since T is an arbitrary number, the above relation implies

$$\mathbb{P}\Big[\|\boldsymbol{u}_1(t)-\boldsymbol{u}_2(t)\|_\rho=0 \ \text{ for all } \ t\ge t_0\Big]=1,$$

that ensures the uniqueness of the solutions. The proof of Theorem 20.2 is complete.

□

20.3 Weak pullback mean random attractors

20.3.1 *Preliminaries on mean random dynamical systems*

The following definitions of mean random dynamical system and weak pullback mean random attractor, along with a proposition on the existence of weak pullback mean attractors are taken from [Wang (2018)].

Definition 20.2. A family $\psi = \{\psi(t, t_0)\}_{(t,t_0)\in\mathbb{R}^+\times\mathbb{R}}$ of mappings is called a mean random dynamical system (MRDS) on $\mathcal{L}^2(\Omega, \mathcal{F}; \ell^2_\rho)$ over $(\Omega, \mathcal{F}, \{\mathcal{F}_t\}_{t\in\mathbb{R}}, \mathbb{P})$ if for all $t_0 \in \mathbb{R}$ and $t, s, \in \mathbb{R}^+$ it satisfies

(i) $\psi(t, t_0)$ maps $\mathcal{L}^2(\Omega, \mathcal{F}_{t_0}; \ell^2_\rho)$ to $\mathcal{L}^2(\Omega, \mathcal{F}_{t+t_0}; \ell^2_\rho)$;
(ii) $\psi(0, t_0)$ is the identity operator on $\mathcal{L}^2(\Omega, \mathcal{F}_{t_0}; \ell^2_\rho)$;
(iii) $\psi(t + s, t_0) = \Psi(t, t_0 + s, \psi(s, t_0))$.

Remark 20.1. The mean random dynamical system $\{\psi(t, t_0)\}_{(t,t_0)\in\mathbb{R}^+\times\mathbb{R}}$ here is different from the non-autonomous dynamical system $\{\psi(t, t_0)\}_{(t,t_0)\in\mathbb{R}^2_\geq}$ defined in Chapter 2. In particular, $\psi(t, t_0)$ here corresponds to $\psi(t+t_0, t_0)$ in the terminology in Chapter 2.

Let $\mathfrak{D}$ be a collection of families of nonempty bounded subsets defined by

$$\mathfrak{D} = \Big\{\mathcal{D} = \{D_{t_0} \subset \mathcal{L}^2(\Omega, \mathcal{F}_{t_0}; \ell^2_\rho) : D_{t_0} \neq \emptyset \text{ for } t_0 \in \mathbb{R}\} \\ : \lim_{t\to+\infty} e^{-ct}\|D_{t_0-t}\|_{\mathcal{L}^2(\Omega,\mathcal{F};\ell^2_\rho)} = 0, \quad \forall c > 0\Big\}. \tag{20.40}$$

Definition 20.3. A family $\mathcal{A} = (A_{t_0})_{t_0\in\mathbb{R}} \in \mathfrak{D}$ is called a $\mathfrak{D}$-pullback weakly attracting set of an MRDS ψ on $\mathcal{L}^2(\Omega, \mathcal{F}; \ell^2_\rho)$ over $(\Omega, \mathcal{F}, \{\mathcal{F}_t\}_{t\in\mathbb{R}}, \mathbb{P})$ if for every $t_0 \in \mathbb{R}$, $D \in \mathfrak{D}$ and every weak neighborhood $\mathcal{N}(A_{t_0})$ of A_{t_0}, there exists $T = T(t_0, D, \mathcal{N}(A_{t_0})) > 0$ such that

$$\psi(t, t_0 - t, D_{t_0-t}) \subset \mathcal{N}(A_{t_0}), \quad \forall\, t \geq T.$$

Definition 20.4. A family $\mathcal{A} = (A_{t_0})_{t_0\in\mathbb{R}} \in \mathfrak{D}$ is called a weak $\mathfrak{D}$-pullback mean random attractor for an MRDS ψ on $\mathcal{L}^2(\Omega, \mathcal{F}, \ell^2_\rho)$ over $(\Omega, \mathcal{F}, \{\mathcal{F}_t\}_{t\in\mathbb{R}}, \mathbb{P})$ if the following three conditions are satisfied:

(i) for each $t_0 \in \mathbb{R}$, A_{t_0} is a weakly compact subset of $\mathcal{L}^2(\Omega, \mathcal{F}_{t_0}; \ell^2_\rho)$;
(ii) $\mathcal{A}$ is a $\mathfrak{D}$-pullback weakly attracting set of ψ;
(iii) $\mathcal{A}$ is the minimal element of $\mathfrak{D}$ with properties (i) and (ii).

The existence of a weak $\mathfrak{D}$-pullback mean random attractor relies on the existence of a $\mathfrak{D}$-pullback absorbing set defined as follows.

Definition 20.5. A family $\mathcal{Q} = (Q_{t_0})_{t_0\in\mathbb{R}} \in \mathfrak{D}$ is called a $\mathfrak{D}$-pullback absorbing set for an MRDS ψ on $\mathcal{L}^2(\Omega, \mathcal{F}; \ell^2_\rho)$ if for every $t_0 \in \mathbb{R}$ and $D \in \mathfrak{D}$ such that there exists $T = T(t_0, D) > 0$ such that $\psi(t, t_0 - t, D_{t_0-t}) \subset Q_{t_0}$ for all $t \geq T$.

The following result on existence of a weak $\mathfrak{D}$-pullback mean random attractor follows from Theorem 2.13 in [Wang (2018)] with $\mathfrak{X} = \ell^2_\rho$, $p = 2$ and $\mathfrak{D}$ defined in (20.40).

Proposition 20.1. *A mean random dynamical system $\{\psi(t,t_0)\}_{(t,t_0)\in\mathbb{R}^+\times\mathbb{R}}$ in $\mathcal{L}^2(\Omega,\mathcal{F};\ell^2_\rho)$ possesses a unique weak $\mathfrak{D}$-pullback mean random attractor $\mathcal{A} \in \mathfrak{D}$ provided ψ has a weakly compact $\mathfrak{D}$-pullback absorbing set $\mathcal{Q} = (Q_t)_{t\in\mathbb{R}} \in \mathfrak{D}$. In particular, the component set of the attractor $\mathcal{A} = (A_{t_0})_{t_0\in\mathbb{R}}$ is given by*

$$A_{t_0} = \bigcap_{r\geq 0}\overline{\bigcup_{t\geq r}\psi(t,t_0-t,Q_{t_0-t})} \quad \text{for each } t_0\in\mathbb{R},$$

where the closure is taken with respect to the weak topology of $\mathcal{L}^2(\Omega,\mathcal{F};\ell^2_\rho)$.

20.3.2 *Existence of absorbing sets*

First by Theorem 20.2, we see that for each $t_0 \in \mathbb{R}$ and $\boldsymbol{u}_o \in \mathcal{L}^2(\Omega,\ell^2_\rho)$, the system (20.6) possesses a unique solution $\boldsymbol{u} = \boldsymbol{u}(t;t_0,\boldsymbol{u}_o) \in \mathcal{L}^2(\mathcal{C}(t_0,\infty),\ell^2_\rho)$ $\mathbb{P}$-a.s. Then due to the uniform estimates (20.12) and the Lebesgue dominated convergence theorem it can be shown that $\boldsymbol{u} \in \mathcal{C}([t_0,\infty),\mathcal{L}^2(\Omega,\ell^2_\rho))$. For each $t_0 \in \mathbb{R}$ and $t \in \mathbb{R}^+$, we define a mapping $\psi(t,t_0) : \mathcal{L}^2(\Omega,\mathcal{F}_{t_0};\ell^2_\rho) \to \mathcal{L}^2(\Omega,\mathcal{F}_{t_0+t};\ell^2_\rho)$ by

$$\psi(t,t_0,\boldsymbol{u}_o) = \boldsymbol{u}(t+t_0;t_0,\boldsymbol{u}_o) \quad \text{for} \quad \boldsymbol{u}_o \in \mathcal{L}^2(\Omega,\mathcal{F}_{t_0};\ell^2_\rho). \tag{20.41}$$

Then $\{\psi(t,t_0)\}_{(t,t_0)\in\mathbb{R}^+\times\mathbb{R}}$ defines an MRDS, referred to as the MRDS generated by the system (20.6) on $\mathcal{L}^2(\Omega,\mathcal{F},\ell^2_\rho)$ over $(\Omega,\mathcal{F},\{\mathcal{F}_t\}_{t\in\mathbb{R}},\mathbb{P})$.

We next show that the MRDS $\{\psi(t,t_0)\}_{(t,t_0)\in\mathbb{R}^+\times\mathbb{R}}$ possesses a pullback absorbing set. To that end, the following additional assumptions are needed.

Assumption 20.8. $2a_\eta\sqrt{2\kappa\rho_\Sigma} + 4\|\boldsymbol{a}\|^2_\infty < \alpha$.

Assumption 20.9. $\int_{-\infty}^{t_0} e^{\frac{\alpha}{2}s}\mathbb{E}[\|\boldsymbol{g}(s)\|^2_\rho + \|H(s)\|^2_\rho]\mathrm{d}s < \infty$ for all $t_0 \in \mathbb{R}$.

The following Lemma provides a $\mathfrak{D}$-uniform estimate of solutions of system (20.6).

Lemma 20.4. *Let Assumptions 20.1–20.9 hold. Then for every $t_0 \in \mathbb{R}$ and $\mathcal{D} = (D_{t_0})_{t_0\in\mathbb{R}} \in \mathfrak{D}$, there exist $T = T(t_0,\mathcal{D})$ and $R > 0$ depending on $\mathfrak{a},\boldsymbol{\beta},\rho_\Sigma,a_\eta,b_\eta,\kappa,\boldsymbol{b}$ such that for all $t \geq T$, the solution $\boldsymbol{u}(t)$ of system (20.6) satisfies*

$$\mathbb{E}\left[\|\boldsymbol{u}(t_0;t_0-t,\boldsymbol{u}_o)\|^2_\rho\right] \leq R + R\int_{-\infty}^{t_0} e^{\frac{\alpha}{2}(s-t_0)}\mathbb{E}[\|\boldsymbol{g}(s)\|^2_\rho + \|H(s)\|^2_\rho]\mathrm{d}s$$

for every $\boldsymbol{u}_o = \boldsymbol{u}(t_0-t) \in D_{t_0-t}$.

Proof. First, take inner product $\langle\cdot,\cdot\rangle_\rho$ of (20.6) with $\boldsymbol{u}$ and use Itô's formula to obtain

$$\begin{aligned}\mathrm{d}\|\boldsymbol{u}(t)\|^2_\rho &= 2\langle\boldsymbol{u}(t),F(\boldsymbol{u}(t))+\mathcal{K}(\boldsymbol{u}(t))+\boldsymbol{g}(t)\rangle_\rho\mathrm{d}t + \|\mathfrak{S}(\boldsymbol{u}(t))+H(t)\|^2_\rho\mathrm{d}t\\ &\quad +2\sum_{\mathfrak{i}\in\mathbb{Z}^d}\langle\boldsymbol{u}(t),(\mathfrak{S}_\mathfrak{i}(\boldsymbol{u}(t))+H_\mathfrak{i}(t))\rangle_\rho\mathrm{d}W_\mathfrak{i} \quad a.s.\end{aligned}$$

Taking expectation of the above equality gives

$$\begin{aligned}\frac{\mathrm{d}}{\mathrm{d}t}\mathbb{E}\left[\|\boldsymbol{u}(t)\|_\rho^2\right] =& 2\mathbb{E}\left[\langle \boldsymbol{u}(t), F(\boldsymbol{u}(t))\rangle_\rho\right] + 2\mathbb{E}\left[\langle \boldsymbol{u}(t), \mathcal{K}(\boldsymbol{u}(t))\rangle_\rho\right] \\ &+ 2\mathbb{E}\left[\langle \boldsymbol{u}(t), \boldsymbol{g}(t)\rangle_\rho\right] + \mathbb{E}\left[\|\mathfrak{S}(\boldsymbol{u}(t)) + H(t)\|_\rho^2\right].\end{aligned} \tag{20.42}$$

First due to Assumption 20.7, we have

$$\mathbb{E}\left[\langle \boldsymbol{u}(t), F(\boldsymbol{u}(t))\rangle_\rho\right] \le -\alpha\mathbb{E}[\|\boldsymbol{u}(t)\|_\rho^2] + \|\boldsymbol{\beta}\|_\rho^2. \tag{20.43}$$

Then using Assumptions 20.1, 20.2 and 20.5, we get

$$\begin{aligned}\mathbb{E}\left[\langle \boldsymbol{u}(t), \mathcal{K}(\boldsymbol{u}(t))\rangle_\rho\right] &\le \mathbb{E}\Big[\sum_{\mathrm{i}\in\mathbb{Z}^d} \rho_{\mathrm{i}}|u_{\mathrm{i}}(t)| \sum_{\mathrm{j}\in\mathbb{Z}^d} |\kappa_{\mathrm{i},\mathrm{j}}|(a_\eta|u_{\mathrm{j}}(t)| + b_\eta)\Big] \\ &\le \frac{a_\eta}{2}\sqrt{2\kappa\rho_\Sigma}\mathbb{E}[\|\boldsymbol{u}(t)\|_\rho^2] + \frac{1}{2a_\eta\sqrt{2\kappa\rho_\Sigma}}\mathbb{E}\Big[\sum_{\mathrm{i}\in\mathbb{Z}^d} \rho_{\mathrm{i}}\Big(\sum_{\mathrm{j}\in\mathbb{Z}^d} |\kappa_{\mathrm{i},\mathrm{j}}|(a_\eta|u_{\mathrm{j}}(t)| + b_\eta)\Big)^2\Big] \\ &\le \frac{a_\eta}{2}\sqrt{2\kappa\rho_\Sigma}\mathbb{E}[\|\boldsymbol{u}(t)\|_\rho^2] + \frac{1}{2a_\eta\sqrt{2\kappa\rho_\Sigma}}\rho_\Sigma\kappa\Big(2a_\eta^2\mathbb{E}[\|\boldsymbol{u}(t)\|_\rho^2] + 2b_\eta^2\rho_\Sigma\Big) \\ &\le a_\eta\sqrt{2\kappa\rho_\Sigma}\mathbb{E}[\|\boldsymbol{u}(t)\|_\rho^2] + \frac{b_\eta^2\rho_\Sigma}{a_\eta}\sqrt{2\kappa\rho_\Sigma}.\end{aligned} \tag{20.44}$$

The third term on the right-hand side of (20.42) satisfies

$$\mathbb{E}\left[\langle \boldsymbol{u}(t), \boldsymbol{g}(t)\rangle_\rho\right] \le \frac{\alpha}{4}\mathbb{E}[\|\boldsymbol{u}(t)\|_\rho^2] + \frac{1}{\alpha}\mathbb{E}[\|\boldsymbol{g}(t)\|_\rho^2], \tag{20.45}$$

and by Assumptions 20.4 we have

$$\begin{aligned}\mathbb{E}[\|\mathfrak{S}(\boldsymbol{u}(t)) + H(t)\|_\rho^2] &\le 2\mathbb{E}\Big[\sum_{\mathrm{i}\in\mathbb{Z}^d} \rho_{\mathrm{i}}\sigma_{\mathrm{i}}^2(u_{\mathrm{i}}(t))\Big] + 2\mathbb{E}[\|H(t)\|_\rho^2] \\ &\le 2\mathbb{E}\Big[\sum_{\mathrm{i}\in\mathbb{Z}^d} \rho_{\mathrm{i}}(a_{\mathrm{i}}|u_{\mathrm{i}}(t)| + b_{\mathrm{i}})^2\Big] + 2\mathbb{E}[\|H(t)\|_\rho^2] \\ &\le 2\mathbb{E}\Big[\sum_{\mathrm{i}\in\mathbb{Z}^d} \rho_{\mathrm{i}}(2a_{\mathrm{i}}^2u_{\mathrm{i}}^2(t) + 2b_{\mathrm{i}}^2)\Big] + 2\mathbb{E}[\|H(t)\|_\rho^2] \\ &\le 4\|\boldsymbol{a}\|_\infty^2\mathbb{E}[\|\boldsymbol{u}(t)\|_\rho^2] + 4\|\boldsymbol{b}\|_\rho^2 + 2\mathbb{E}[\|H(t)\|_\rho^2].\end{aligned} \tag{20.46}$$

Substituting (20.43)–(20.46) into (20.42) gives

$$\begin{aligned}\frac{\mathrm{d}}{\mathrm{d}t}\mathbb{E}[\|\boldsymbol{u}(t)\|_\rho^2] \le& \left(-\frac{3}{2}\alpha + 2a_\eta\sqrt{2\kappa\rho_\Sigma} + 4\|\boldsymbol{a}\|_\infty^2\right)\mathbb{E}[\|\boldsymbol{u}(t)\|_\rho^2] + 2\|\boldsymbol{\beta}\|_\rho^2 \\ &+ \frac{2\rho_\Sigma b_\eta^2}{a_\eta}\sqrt{2\kappa\rho_\Sigma} + \frac{2}{\alpha}\mathbb{E}[\|\boldsymbol{g}(t)\|_\rho^2] + 4\|\boldsymbol{b}\|_\rho^2 + 2\mathbb{E}[\|H(t)\|_\rho^2].\end{aligned}$$

Further, by using Assumption 20.8 it follows from the above inequality that

$$\begin{aligned}\frac{\mathrm{d}}{\mathrm{d}t}\mathbb{E}[\|\boldsymbol{u}(t)\|_\rho^2] + \frac{\alpha}{2}\mathbb{E}[\|\boldsymbol{u}(t)\|_\rho^2] \le& 2\|\boldsymbol{\beta}\|_\rho^2 + \frac{2\rho_\Sigma b_\eta^2}{a_\eta}\sqrt{2\kappa\rho_\Sigma} + \frac{2}{\alpha}\mathbb{E}[\|\boldsymbol{g}(t)\|_\rho^2] \\ &+ 4\|\boldsymbol{b}\|_\rho^2 + 2\mathbb{E}[\|H(t)\|_\rho^2].\end{aligned} \tag{20.47}$$

Multiplying (20.47) by $e^{\frac{\alpha}{2}t}$ and integrating over $(t_0 - t, t_0)$ results by numbers

$$\begin{aligned}\mathbb{E}[\|\boldsymbol{u}(t_0)\|_\rho^2] \le & e^{-\frac{\alpha}{2}t}\mathbb{E}[\|\boldsymbol{u}(t_0-t)\|_\rho^2] + \frac{2}{\alpha}\left(2\|\boldsymbol{\beta}\|_\rho^2 + \frac{2\rho_\Sigma b_\eta^2}{a_\eta}\sqrt{2\kappa\rho_\Sigma} + 4\|\boldsymbol{b}\|_\rho^2\right) \\ & + \left(\frac{2}{\alpha}+2\right)\int_{t_0-t}^{t_0} e^{\frac{\alpha}{2}(s-t_0)}\mathbb{E}[\|\boldsymbol{g}(s)\|_\rho^2 + \|H(s)\|_\rho^2]\mathrm{d}s. \qquad (20.48)\end{aligned}$$

Since $\boldsymbol{u}(t_0 - t) \in D_{t_0-t} \in \mathcal{D}$ with $\mathcal{D} \in \mathfrak{D}$, we know that

$$\lim_{t\to+\infty} e^{-\frac{\alpha}{2}t}\mathbb{E}[\|\boldsymbol{u}(t_0-t)\|_\rho^2] = 0.$$

Then there exists $T = T(\mathcal{D}, t_0) > 0$ such that

$$e^{-\frac{\alpha}{2}t}\mathbb{E}[\|\boldsymbol{u}(t_0-t)\|_\rho^2] \le 1 \quad \forall\, t \ge T. \qquad (20.49)$$

By (20.48), (20.49) and Assumption 20.9, we obtain

$$\mathbb{E}[\|\boldsymbol{u}(t_0)\|_\rho^2] \le R + R\int_{-\infty}^{t_0} e^{\frac{\alpha}{2}(s-t_0)}\mathbb{E}[\|\boldsymbol{g}(s)\|_\rho^2 + \|H(s)\|_\rho^2]\mathrm{d}s,$$

where

$$R = 1 + \frac{2}{\alpha}\left(2\|\boldsymbol{\beta}\|_\rho^2 + \frac{2\rho_\Sigma b_\eta^2\sqrt{2\kappa\rho_\Sigma}}{a_\eta} + 4\|\boldsymbol{b}\|_\rho^2\right) + \frac{2}{\alpha} + 2. \qquad (20.50)$$

The proof is completed. □

20.3.3 *Existence of a mean random attractor*

Theorem 20.3. *Let Assumptions 20.1–20.9 hold. Then the mean random dynamical system* $\{\psi(t,t_0)\}_{(t,t_0)\in\mathbb{R}^+\times\mathbb{R}}$ *generated by the system* (20.6) *defined by* (20.41) *possesses a unique tempered weak* $\mathfrak{D}$*-pullback mean random attractor* $\mathcal{A} = (A_{t_0})_{t_0\in\mathbb{R}} \in \mathfrak{D}$ *in* $\mathcal{L}^2(\Omega, \mathcal{F}; \ell_\rho^2)$ *over* $(\Omega, \mathcal{F}, \{\mathcal{F}_t\}_{t\in\mathbb{R}}, \mathbb{P})$ *in the sense of Definition 20.4 provided*

$$\lim_{t\to+\infty} e^{-ct}\int_{-\infty}^{0} e^{\frac{\alpha}{2}s}\mathbb{E}[\|\boldsymbol{g}(s-t)\|_\rho^2 + \|H(s-t)\|_\rho^2]\mathrm{d}s = 0 \quad \forall\, c > 0. \qquad (20.51)$$

Proof. For any $t_0 \in \mathbb{R}$ define

$$\hat{R}(t_0) = R + R\int_{-\infty}^{t_0} e^{\frac{\alpha}{2}(s-t_0)}\mathbb{E}[\|\boldsymbol{g}(s)\|_\rho^2 + \|H(s)\|_\rho^2]\mathrm{d}s,$$

where R is given by (20.50), and set $\mathcal{Q} := (Q_{t_0})_{t_0\in\mathbb{R}}$ with

$$Q_{t_0} = \{\boldsymbol{u} \in \mathcal{L}^2(\Omega, \mathcal{F}_{t_0}, \ell_\rho^2) : \mathbb{E}[\|\boldsymbol{u}\|_\rho^2] \le \hat{R}(t_0)\}, \quad \forall t_0 \in \mathbb{R}.$$

Note that Q_{t_0} is weakly compact in $\mathcal{L}^2(\Omega, \mathcal{F}_{t_0}, \ell_\rho^2)$. In addition, for every $c > 0$, by the condition (20.51),

$$\lim_{t\to+\infty} e^{-ct}\|Q_{t_0-t}\|_\rho^2 \le \lim_{t\to+\infty} e^{-ct}\hat{R}(t_0-t)$$

$$= Re^{-ct_0}\lim_{t\to+\infty} e^{-ct}\int_{-\infty}^{0} e^{\frac{\alpha}{2}s}\mathbb{E}[\|\boldsymbol{g}(s-t)\|_\rho^2 + \|H(s-t)\|_\rho^2]\mathrm{d}s$$

$$= 0,$$

which implies that $\mathcal{Q} \in \mathcal{D}$. Then we infer from Lemma 20.4 that $\mathcal{Q}$ is a weakly compact $\mathcal{D}$-pullback absorbing set for the MRDS ψ in $\mathcal{L}^2(\Omega, \mathcal{F}, \ell_\rho^2)$. The existence of weak $\mathcal{D}$-pullback mean random attractor the follows from Proposition 20.1. The proof is complete. □

20.3.4 *End notes*

This chapter is based on Wang, Kloeden & Han [Wang *et al.* (2021)], which uses methods from [Wang and Wang (2020)].

Mean-square random dynamical systems and their attractors were introduced in [Kloeden and Lorenz (2012)]. Later Bixiang Wang [Wang (2018)] considered weak mean random attractors, and applied them to LDS with nonlinear noise in [Wang and Wang (2020)].

20.4 Problems

Problem 20.1. Give a characterisation of compactness in the space of mean-square real-valued random variables $\mathcal{L}_2(\Omega, \mathbb{R})$.

Problem 20.2. Show that the term $\sum_{\mathrm{i}\in\mathbb{Z}^d}(\mathfrak{S}_{\mathrm{i}}(\boldsymbol{u}) + H_i(t))\mathrm{d}W_i(t)$ in (20.6) takes value in ℓ_ρ^2 for every $\boldsymbol{u} \in \ell_\rho^2$.

Chapter 21

Lattice systems with switching effects and delayed recovery

This chapter is motivated by the appearance of switching effects and recovery delays in systems of excitable cells [Joyner *et al.* (1991); Shipston (2001)]. It is based on the paper [Han and Kloeden (2016)], that studied the following lattice system with a reaction term which is switched off when a certain threshold is exceeded and restored after a suitable recovery time:

$$\begin{aligned} \frac{du_i}{dt} &= \nu(u_{i-1} - 2u_i + u_{i+1}) \\ &\qquad + f_i(t, u_i) \cdot \chi[\varsigma_i - \max_{-\theta \le s \le 0} u_i(t+s)], \quad i \in \mathbb{Z}, \quad t > t_0 \qquad (21.1) \\ u_i(\tau) &= \phi_i(\tau - t_0) \quad \text{for all } \tau \in [t_0 - \theta, t_0], \quad i \in \mathbb{Z}, \quad t_0 \in \mathbb{R}. \end{aligned}$$

Here $\nu = 1/\varkappa > 0$ is the coupling coefficient where $\varkappa$ is the intercellular resistance [Keener (1987)], χ is an indicator function defined by

$$\chi[x] = \begin{cases} 1, & x \ge 0, \\ 0, & x < 0. \end{cases}$$

For each $i \in \mathbb{Z}$, $u_i \in \mathbb{R}$ represents the membrane potential of the cell at the ith active site. In addition, $\varsigma_i \in \mathbb{R}$ is the threshold triggering the switch-off at the ith site, while $u_i(t + \cdot) \in \mathcal{C}([-\theta, 0], \mathbb{R})$ is the segment of u_i on time interval $[t - \theta, t]$ where θ is a positive constant, and $f_i \in \mathcal{C}(\mathbb{R}, \mathbb{R})$ is a continuous function satisfying appropriate conditions.

Equation (21.1) describes a reaction-diffusion system with or without delay depending on the maximum value achieved at each location during the past θ period of time. More specifically, each u_i evolves with respect to a reaction-diffusion equation with a delay in the reaction term as long as u_i stays below the threshold ς_i starting from t_0. Once the value of u_i reaches or exceeds ς_i at some time, the reaction will be switched off, and stay off for at least θ period of time. The delay in the reaction term can be recovered later, when u_i evolves back to smaller values than ς_i and remains smaller than ς_i for more than θ period of time (see Fig. 21.1).

This type of delay makes biological sense. For example, consider an inhomogeneous cell tissue under the influence of a chemical substance. Obviously, neighbouring cells interact with each other. Whenever the concentration $u(t, x)$ of that

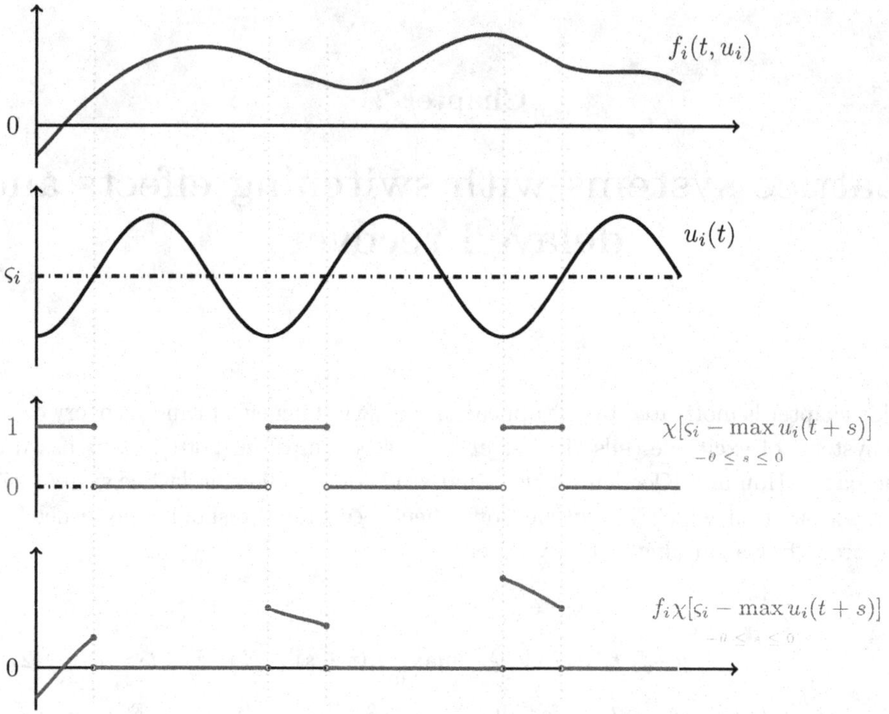

Fig. 21.1 Illustration of the switching reaction function.

substance at a position exceeds a given threshold, the consequences for the cell there are significant, maybe even fatal. Hence it is not just the present concentration, which influences the cells, but also its history, at least for a suitable recovery period. Alternatively, one can think of algae in a pond that asphyxiates when the population density becomes too high and needs a suitable recovery time.

This chapter investigates the long term behaviour of equation (21.1) and proves the existence of a pullback attractor. The switching off and on of the reaction term leads to a relaxation effect, essentially multiplication by a Heaviside function, and thus to the mathematical formulation of the system as a set-valued system, i.e., differential inclusion equation. In view of the recovery time prescribed between switching off and back on again introduces a delay term, which appears only upper semi-continuously.

21.1 Set-valued delay differential inclusions

In this section we introduce a general framework on non-autonomous set-valued delay differential inclusions on Banach spaces. Let $\mathfrak{E}$ be a real separable and reflexive Banach space with norm $\|\cdot\|_{\mathfrak{E}}$ and let $\mathfrak{L}(\mathfrak{E})$ be the set of linear bounded operators from $\mathfrak{E}$ to $\mathfrak{E}$. Denote by $\mathcal{P}_c(\mathfrak{E})$ the family of nonempty closed subsets of $\mathfrak{E}$.

In addition, define $\mathfrak{E}_C := C([-\theta, 0], \mathfrak{E})$ to be the Banach space of all continuous functions from $[-\theta, 0]$ to $\mathfrak{E}$ with the norm

$$\|h\|_{\mathfrak{E}_C} = \sup_{s\in[-\theta,0]} \|h(s)\| \quad \text{for all } h \in \mathfrak{E}_C.$$

Given any continuous function $u(\cdot) : \mathbb{R} \to \mathfrak{E}$, denote by u_t the element in $\mathfrak{E}_C$ given by

$$u_t(s) = u(t+s), \ s \in [-\theta, 0] \quad \text{for all } t \in \mathbb{R}.$$

Recall that $\mathbb{R}^2_{\geq} := \{(t, t_0) \in \mathbb{R}^2 : t \geq t_0\}$, and consider the following general set-valued delay differential inclusion on a Banach space:

$$\frac{\mathrm{d}u}{\mathrm{d}t} \in \mathrm{A}u(t) + \mathcal{G}(t, u_t), \quad (t, t_0) \in \mathbb{R}^2_{\geq} \tag{21.2}$$
$$u_{t_0} = \phi(t) \in \mathfrak{E}_C,$$

where A is the infinitesimal generator of a C_0 semi-group $\{\mathcal{S}(t)\}_{t\geq 0}$ and $\mathcal{G} : [t_0, T] \times \mathfrak{E}_C \to \mathcal{P}_c(\mathfrak{E})$ is a closed, bounded and convex set-valued map satisfying the following standing assumptions.

Assumption 21.1. For each $h \in \mathfrak{E}_C$, $\mathcal{G}(\cdot, h) : [t_0, T] \to \mathcal{P}_c(\mathfrak{E})$ has a Lebesgue measurable selection, i.e., there exists a Lebesgue measurable function $g \in \mathcal{L}^1(t_0, T; \mathfrak{E})$ with $g(t) \in \mathcal{G}(t, h)$ for all $t \in [t_0, T]$.

Assumption 21.2. For almost all $t \in [t_0, T]$, $\mathcal{G}(t, \cdot) : \mathfrak{E}_C \to \mathcal{P}_c(\mathfrak{E})$ is upper semi-continuous, i.e., for each $h \in \mathfrak{E}_C$ the set $\mathcal{G}(t, h)$ is a nonempty and closed subset of $\mathfrak{E}$, and for each open set $\mathcal{D} \subset \mathfrak{E}$ containing $\mathcal{G}(t, h)$, there exists an open neighbourhood $\mathcal{N}_h \subset \mathfrak{E}_C$ of h such that $\mathcal{G}(t, \mathcal{N}_h) \subset \mathcal{D}$.

Assumption 21.3. The growth of $\mathcal{G}$ is bounded, in the sense that there exists a Lebesgue measurable function $M_g(t) \in \mathcal{L}^1(t_0, T; \mathbb{R})$ such that

$$\sup\{\|g\| : g \in \mathcal{G}(t, h)\} \leq M_g(t)\|h\|_{\mathfrak{E}_C},$$

for a.e. $t \in [t_0, T]$ and each $h \in \mathfrak{E}_C$.

For simplicity from now on the collection of all Lebesgue measurable selections of $\mathcal{G}$ will be denoted by

$$\mathcal{M}_{\mathcal{G}} := \left\{g \in \mathcal{L}^1(t_0, T; \mathfrak{E}) : g(t) \in \mathcal{G}(t, u_t), \text{ for } a.e.\ t \in [t_0, T]\right\}.$$

Solutions to (21.2) can be defined via the following auxiliary (non-autonomous) problem:

$$\frac{\mathrm{d}u}{\mathrm{d}t} = \mathrm{A}u + g(t), \quad u(t_0) = u_0 \in \mathfrak{E}, \quad g(t) \in \mathcal{M}_{\mathcal{G}}. \tag{21.3}$$

Definition 21.1. Let A be the infinitesimal generator of a C_0 semi-group $\mathcal{S}(t)$. Let $u_0 \in \mathfrak{E}$ and $g \in \mathcal{L}^1(t_0, T; \mathfrak{E})$. A function $u \in C([t_0, T]; \mathfrak{E})$ is called a mild solution to (21.3) if and only if it satisfies the integral equation

$$u(t) = \mathcal{S}(t)u_0 + \int_{t_0}^{t} \mathcal{S}(t-s)g(s)\mathrm{d}s, \quad t_0 \leq t \leq T. \tag{21.4}$$

Definition 21.2. Let $\mathcal{G} : [t_0, T] \times \mathfrak{E}_C \to \mathcal{P}_c(\mathfrak{E})$ be a set-valued map. A continuous function $u : [t_0, T] \to \mathfrak{E}$ is called a mild solution of problem (21.2) if there exists a selection function $g \in \mathfrak{M}_{\mathcal{G}}$ such that u is a mild solution to the corresponding auxiliary problem (21.3).

Definition 21.3. Given $T > t_0$, $u_0 \in \mathfrak{E}$ and $g \in \mathcal{L}^1(t_0, T; \mathfrak{E})$, a continuous function $u : [t_0, T] \to \mathfrak{E}$ is called a (strong) solution of problem (21.3) if u is absolutely continuous on any compact subinterval of $[t_0, T]$ with $u(t_0) = u_0$ and $\frac{\mathrm{d}u(t)}{\mathrm{d}t} = \mathrm{A}u(t) + g(t)$ in $\mathfrak{E}$ for Lebesgue a.e. $t \in (t_0, T)$.

Definition 21.4. Let $\mathcal{G} : [t_0, T] \times \mathfrak{E}_C \to \mathcal{P}(\mathfrak{E})$ be a set-valued map. A continuous function $u : [t_0, T] \to \mathfrak{E}$ is called a (strong) solution of problem (21.2) if there exists a selection function $g \in \mathfrak{M}_{\mathcal{G}}$ such that u is a (strong) solution to the corresponding auxiliary problem (21.3).

Remark 21.1. In the special case when the solutions to $\frac{\mathrm{d}u}{\mathrm{d}t} = \mathrm{A}u$ form a strongly continuous semi-group of bounded linear operators, a continuous function $u : [t_0, T] \to \mathfrak{E}$ is a mild solution if and only if u is a strong solution, for problem (21.3) with any function $g \in \mathcal{L}^2(t_0, T; \mathfrak{E}) \subset \mathcal{L}^1(t_0, T; \mathfrak{E})$ and $u_0 \in \mathfrak{E}$. (See [Kloeden *et al.* (2014)]).

Definition 21.5. An upper semi-continuous map Γ is said to be condensing, if $\mu(\Gamma(B)) < \mu(B)$ for any nonempty bounded subset B of $\mathfrak{E}$ with $\mu(B) \neq 0$, where μ denotes the Kuratowski measure of non-compactness.

It can be shown that system (21.2) has at least one mild solution, by using the following fixed point theorem for set-valued condensing mappings due to [Martelli (1975)].

Proposition 21.1. *Let $\mathfrak{E}$ be a Banach space and let $\Gamma : \mathfrak{E} \to \mathcal{P}_c(\mathfrak{E})$ be a compact, convex, upper semi continuous and condensing map. Then Γ has a fixed point if the set*

$$\mathcal{D}_\Gamma := \{u \in \mathfrak{E} : \lambda u \in \Gamma(u) \text{ for some } \lambda > 1\}$$

is bounded.

Theorem 21.1. *Assume that Assumptions 21.1–21.3 hold. In addition, assume that the semi-group $\mathcal{S}(t)$ generated by* A *is linear bounded for* $t \geq 0$, *and there exists $\kappa > 1$ such that*

$$\|\mathcal{S}(t)\|_{\mathfrak{L}(\mathfrak{E})} = \sup\{|\mathcal{S}(t)(x)| : |x| = 1\} \leq \kappa.$$

Then equation (21.2) *has at least one mild solution.*

Proof. Recall that $\mathfrak{E}_C = \mathcal{C}([-\theta, 0], \mathfrak{E})$, and denote $\mathfrak{Z} := \mathcal{C}([t_0 - \theta, T], \mathfrak{E})$. Define a set-valued operator $\Gamma : \mathfrak{Z} \to \mathcal{P}_c(\mathfrak{Z})$ by

$$\Gamma(u) = \left\{ z \in \mathfrak{Z} : z(t) = \begin{cases} \phi(t), & t \in [t_0 - \theta, t_0] \\ \mathcal{S}(t)\phi(t_0) + \int_{t_0}^{t} \mathcal{S}(t-s)g(s)\mathrm{d}s, & t \geq t_0 \end{cases} \right\},$$

where $g(t) \in \mathcal{M}_{\mathcal{G}}(u)$ is a Lebesgue measurable selection of $\mathcal{G}(t, u_t)$, and define

$$\|\Gamma(u)\| := \sup\{\|z(t)\| : z(t) \in \Gamma(u)\}.$$

In what follows, we will show that the mapping Γ is (i) compact, (ii) convex, (iii) upper semi-continuous and condensing, and (iv) bounded.

(i) By $\|\mathcal{S}(t)\|_{\mathfrak{L}(\mathfrak{E})} \leq \kappa$, and Assumption 21.3,

$$|z(t)| \leq \kappa\|\phi\|_{\mathfrak{E}_C} + \kappa\|M_g\|_{\mathcal{L}^1} \cdot \|u_t\|_{\mathfrak{E}_C}.$$

For any $u \in \mathfrak{Z}$ with $\|u\| \leq q$, $\|u_t\|_{\mathfrak{E}_C} \leq q$ for any $t \in [t_0, T]$, and hence

$$\|\Gamma(u)\| \leq \kappa\|\phi\|_{\mathfrak{E}_C} + \kappa q\|M_g\|_{\mathcal{L}^1},$$

which implies that Γ maps bounded sets to bounded sets of $\mathfrak{Z}$.

Next for any $t_2 > t_1 \in [t_0, T]$ and $u \in \mathfrak{Z}$ with $\|u\| \leq q$

$$\begin{aligned}\|z(t_2) - z(t_1)\| &\leq \|(\mathcal{S}(t_2) - \mathcal{S}(t_1))\phi(t_0)\| \\ &\quad + \int_{t_1}^{t_2} \|\mathcal{S}(t_2 - s) - \mathcal{S}(t_1 - s)\|_{\mathfrak{L}(\mathfrak{E})} g(s)\mathrm{d}s \\ &\leq \|\mathcal{S}(t_2) - \mathcal{S}(t_1)\|_{\mathfrak{L}(\mathfrak{E})}\|\phi\|_{\mathfrak{E}_C} \\ &\quad + q\|M_g\|_{\mathcal{L}^1} \cdot \int_{t_1}^{t_2} \|\mathcal{S}(t_2 - s) - \mathcal{S}(t_1 - s)\|_{\mathfrak{L}(\mathfrak{E})}\mathrm{d}s,\end{aligned}$$

which implies that Γ maps bounded sets of $\mathfrak{Z}$ into equi-continuous sets of $\mathfrak{Z}$, due to the fact that $\mathcal{S}(t)$ is a compact and strongly continuous operator. Together with the uniform continuity of ϕ on the interval $[t_0 - \theta, t_0]$ and the Ascoli -Arzelà theorem it follows that Γ maps bounded sets into a pre-compact set in $\mathfrak{E}$, and hence is compact.

(ii) For any $z_1, z_2 \in \Gamma(u)$, there exist $g_1, g_2 \in \mathcal{M}_{\mathcal{G}}$ such that for each $t \in [t_0, T]$

$$z_j(t) = \mathcal{S}(t)\phi(t_0) + \int_{t_0}^{t} \mathcal{S}(t - s)g_j(s)\mathrm{d}s, \quad j = 1, 2$$

Then for any $p \in [0, 1]$

$$pz_1 + (1 - p)z_2 = \mathcal{S}(t)\phi(t_0) + \int_{t_0}^{t} \mathcal{S}(t - s)\big(pg_1(s) + (1 - p)g_2(s)\big)\mathrm{d}s.$$

Since $\mathcal{G}$ is convex, $pg_1 + (1 - p)g_2 \in \mathcal{G}$. On the other hand, $g_1, g_2 \in \mathcal{L}^1(t_0, T; \mathfrak{E})$ implies that $pg_1 + (1 - p)g_2 \in \mathcal{L}^1(t_0, T; \mathfrak{E})$. Therefore $pg_1 + (1 - p)g_2 \in \mathcal{M}_{\mathcal{G}}$, and this implies that Γ is convex.

(iii) Let $t \in [t_0, T]$ be fixed and let $\tau \in (t_0, t)$. For any $u \in \mathfrak{Z}$ with $\|u\| \leq q$ define

$$z_\tau(t) = \mathcal{S}(t)\phi(t_0) + \mathcal{S}(\tau)\int_{t_0}^{t-\tau} \mathcal{S}(t - s - \tau)g(s)\mathrm{d}s.$$

Then

$$\|z(t) - z_\tau(t)\| \le \int_{t-\tau}^{t} \|\mathcal{S}(t-s)\|_{\mathfrak{L}(\mathfrak{E})} M_g(s) \mathrm{d}s \le \kappa\tau q \|M_g\|_{\mathcal{L}^2},$$

and thus the set $\{z(t) : z \in \Gamma(u),\ u \text{ is bounded}\}$ is precompact in $\mathfrak{E}$ because there are precompact sets arbitrarily close to it. This shows that Γ is completely continuous which implies that it is condensing.

(iv) It remains to show that the set

$$\mathcal{D}_\Gamma := \{u \in \mathcal{C}([t_0 - \theta, T], \mathfrak{E}) : \lambda u \in \Gamma(u) \text{ for some } \lambda > 1\}$$

is bounded. Since for any $u \in \mathcal{D}_\Gamma$ there is a $\lambda > 1$ such that $\lambda u \in \Gamma(u)$, for each $t \in [t_0, T]$

$$u(t) = \frac{1}{\lambda}\left(\mathcal{S}(t)\phi(t_0) + \int_{t_0}^{t} \mathcal{S}(t-s)g(s)\mathrm{d}s\right),$$

for some $g \in \mathcal{L}^1(t_0, T; \mathfrak{E})$ and $g(t) \in \mathcal{G}(t, u_t)$. Thus,

$$\|u(t)\| \le \kappa\|\phi\|_{\mathfrak{E}_C} + \kappa\int_{t_0}^{t} M_g(s)\mathrm{d}s.$$

Together (i)–(iv) prove that Γ has a fixed point, which corresponds to a mild solution to (21.2). □

It will be shown in the next sections that the lattice system (21.1) has at least one mild solution by using Theorem 21.1, and that the solutions generate a set-valued process that has a pullback attractor.

21.2 Existence of solutions

The focus of this section is to show that solutions for lattice system (21.1) exist. Lattice dynamical systems with delays had been studied in Chapter 6 and also in [Caraballo *et al.* (2014)], [Wang and Bai (2015)] and [Wang *et al.* (2020a,b)], where the continuity of delay terms was assumed. In contrast, here there is a special type of discontinuous delay term of Heaviside type, so the techniques needed for estimations here are different from those previously seen in the literature.

The system (21.1) will be studied in the separable Hilbert space $(\ell^2, \|\cdot\|)$. Define the spaces

$$\mathfrak{H}_1 = \mathcal{C}([-\theta, 0], \mathbb{R}), \quad \mathfrak{H}_C = \mathcal{C}([-\theta, 0], \ell^2), \quad \mathfrak{H}_\infty = \mathcal{C}([-\theta, 0], \ell^\infty),$$

with norms

$$\|\boldsymbol{u}\|_{\mathfrak{H}_1} = \max_{s\in[-\theta,0]} |\boldsymbol{u}(s)|, \quad \|\boldsymbol{u}\|_{\mathfrak{H}_C} = \max_{s\in[-\theta,0]} \|\boldsymbol{u}(s)\|, \quad \|\boldsymbol{u}\|_{\mathfrak{H}_\infty} = \max_{s\in[-\theta,0]} \|\boldsymbol{u}(s)\|_\infty.$$

Notice that $\mathfrak{H}_C \subset \mathfrak{H}_\infty$. In fact,

$$\|\boldsymbol{u}\|_{\mathfrak{H}_\infty} = \max_{s\in[-\theta,0]} \sup_{i\in\mathbb{Z}} |u_i| \le \max_{s\in[-\theta,0]} \left(\sum_{i\in\mathbb{Z}} u_i^2\right)^{\frac{1}{2}} = \|\boldsymbol{u}\|_{\mathfrak{H}_C}$$

and for any $t, s \in [-\theta, 0]$

$$\|\boldsymbol{u}(t) - \boldsymbol{u}(s)\|_\infty = \sup_{i\in\mathbb{Z}} |u_i(t) - u_i(s)| \le \left(\sum_{i\in\mathbb{Z}} |u_i(t) - u_i(s)|^2\right)^{\frac{1}{2}} = \|\boldsymbol{u}(t) - \boldsymbol{u}(s)\|.$$

Let the operators $\Lambda, \mathrm{D}^+, \mathrm{D}^- : \ell^2 \to \ell^2$ be the same as defined in Chapter 3, i.e.,

$$(\Lambda \boldsymbol{u})_i = u_{i-1} - 2u_i + u_{i+1}, \quad (\mathrm{D}^+\boldsymbol{u})_i = u_{i+1} - u_i, \quad (\mathrm{D}^-\boldsymbol{u})_i = \nu^{\frac{1}{2}}(u_{i-1} - u_i).$$

Recall that $-\Lambda = \mathrm{D}^+\mathrm{D}^- = \mathrm{D}^-\mathrm{D}^+$ and that $\langle \mathrm{D}^-\boldsymbol{u}, \boldsymbol{v}\rangle = \langle \boldsymbol{u}, \mathrm{D}^+\boldsymbol{v}\rangle$ for any $\boldsymbol{u}, \boldsymbol{v} \in \mathfrak{H}$, and hence $\langle \Lambda\boldsymbol{u}, \boldsymbol{u}\rangle \le 0$ for any $\boldsymbol{u} \in \mathfrak{H}$. Moreover, Λ is a bounded linear operator which generates a uniformly continuous semi-group.

Denote by $\boldsymbol{u}_t = (u_{it})_{i\in\mathbb{Z}}$ where $u_{it}(s) = u_i(t+s)$, $s \in [-\theta, 0]$, and define the set-valued operator $F(\boldsymbol{u}_t) = (F_i(u_{it}))_{i\in\mathbb{Z}}$, where

$$F_i(u_{it}) = \begin{cases} \{f_i(u_i)\}, & \max_{s\in[-\theta,0]} u_{it}(s) < \varsigma_i \\ f_i(u_i)\cdot[0,1], & \max_{s\in[-\theta,0]} u_{it}(s) = \varsigma_i, \\ \{0\}, & \max_{s\in[-\theta,0]} u_{it}(s) > \varsigma_i \end{cases} \tag{21.5}$$

Then the second term on the right hand side of (21.1) satisfies

$$\left(f_i(u_i)\chi[\varsigma_i - \max_{s\in[-\theta,0]} u_{it}(s)]\right)_{i\in\mathbb{Z}} \in F(\boldsymbol{u}_t).$$

Define the set-valued mapping $\mathfrak{F} : \mathfrak{H}_\mathcal{C} \to \mathcal{P}_c(\mathfrak{H})$ by

$$\mathfrak{F}(\mathcal{U}) = \{F(\mathcal{U}) : \mathcal{U} \in \mathfrak{H}_\mathcal{C}\}, \quad \mathcal{U} \in \mathfrak{H}_\mathcal{C}.$$

Then system (21.1) can be written as the functional differential inclusion

$$\begin{cases} \dfrac{\mathrm{d}\boldsymbol{u}}{\mathrm{d}t} \in \nu\Lambda\boldsymbol{u} + \mathfrak{F}(\boldsymbol{u}_t), & (t, t_0) \in \mathbb{R}^2_{\ge}; \\ \boldsymbol{u}(\tau) = \phi(\tau - t_0), & \text{for all } \tau \in [t_0 - \theta, t_0]. \end{cases} \tag{21.6}$$

The following standing assumptions on function f_i, $i \in \mathbb{Z}$ will be used.

Assumption 21.4. $f_i : \mathbb{R} \to \mathbb{R}$ is continuous for each $i \in \mathbb{Z}$.

Assumption 21.5. There exists a constant $\mathfrak{a} \ge 0$, and $\mathfrak{b} = (\mathfrak{b}_i)_{i\in\mathbb{Z}} \in \ell^2$ such that

$$f_i^2(s) \le \mathfrak{a}s^2 + \mathfrak{b}_i^2 \quad \text{for all } s \in \mathbb{R}.$$

Assumption 21.6. For any $x \in \mathbb{R}$, there exist $\delta > 0$ and $L > 0$ such that

$$|f_i(y) - f_i(x)| \le L|y - x|, \quad i \in \mathbb{Z}, \quad \forall\ |y - x| < \delta.$$

Given any $\mathcal{U} = (\mathcal{U}_i)_{i\in\mathbb{Z}} \in \mathfrak{H}_C$ define the following sets of integers:

$$J_0^{\mathcal{U}} := \left\{ i \in \mathbb{Z} : \max_{s\in[-\theta,0]} \mathcal{U}_i(s) > \varsigma_i \right\},$$

$$J_1^{\mathcal{U}} := \left\{ i \in \mathbb{Z} : \max_{s\in[-\theta,0]} \mathcal{U}_i(s) < \varsigma_i \right\},$$

$$J_\varsigma^{\mathcal{U}} := \left\{ i \in \mathbb{Z} : \max_{s\in[-\theta,0]} \mathcal{U}_i(s) = \varsigma_i \right\}.$$

Theorem 21.2. *Let Assumptions 21.4–21.6 hold. Then for any $t_0 \in \mathbb{R}$, problem* (21.6) *has at least one solution, $\boldsymbol{u}(\cdot) = \boldsymbol{u}(\cdot; t_0, \phi)$.*

Proof. First for any $\mathcal{U} = (\mathcal{U}_i)_{i\in\mathbb{Z}} \in \mathfrak{H}_C$ and each $i \in \mathbb{Z}$, $F_i(\mathcal{U}_i)$ is a closed interval on $\mathbb{R}$ and hence convex. Write $F(\mathcal{U}) = \sum_{i\in\mathbb{Z}} F_i(\mathcal{U}_i)\boldsymbol{e}^i$, where $\boldsymbol{e}^i$ is the element in ℓ^2 with value 1 at the ith position and value 0 elsewhere. Then it follows immediately that $F(\mathcal{U})$ is convex.

Moreover for any $\boldsymbol{g} \in F(\mathcal{U})$, by using Assumptions 21.4 and 21.5,

$$\|\boldsymbol{g}\|^2 \le \sum_{i\in\mathbb{Z}} f_i^2(\mathcal{U}_i(0)) \le \mathfrak{a}^2 \sum_{i\in\mathbb{Z}} \mathcal{U}_i^2(0) + \|\mathfrak{b}\|^2 \le \mathfrak{a}^2\|\mathcal{U}\|_{\mathfrak{H}_C} + \|\mathfrak{b}\|^2.$$

Hence the map $\mathfrak{F}(\cdot) : \mathfrak{H}_C \to \mathcal{P}(\mathfrak{H})$ is bounded, closed and convex.

Next it will be shown that $\mathfrak{F}$ is upper semi-continuous. Clearly for each $\mathcal{U} \in \mathfrak{H}_C$, $\mathfrak{F}(\mathcal{U})$ is a nonempty and closed subset of $\mathfrak{H}_C$. In addition, the following estimates hold for i belonging to different index sets, respectively.

(i) For $i \in J_0^{\mathcal{U}}$ let

$$\epsilon_1 = \min_{i\in J_0^{\mathcal{U}}} \left\{ \max_{s\in[-\theta,0]} \mathcal{U}_i(s) - \varsigma_i \right\} > 0.$$

Then for any $\mathcal{V} = (\mathcal{V}_i)_{i\in\mathbb{Z}} \in \mathfrak{H}_C$ with $\max_{s\in[-\theta,0]} \sum_{i\in J_0^{\mathcal{U}}} (\mathcal{V}_i(s) - \mathcal{U}_i(s))^2 \le \epsilon_1^2$, we have $|\mathcal{V}_i(s) - \mathcal{U}_i(s)| < \epsilon_1$ for any $i \in J_0^{\mathcal{U}}$ and $s \in \mathbb{R}$. Let $s_i \in [-\theta, 0]$ be the point at which $\mathcal{U}_i$ achieves its maximum, i.e., $\max_{s\in[-\theta,0]} \mathcal{U}_i(s) = \mathcal{U}_i(s_i)$. Then

$$\max_{s\in[-\theta,0]} \mathcal{V}_i(s) \ge \mathcal{V}_i(s_i) = \mathcal{V}_i(s_i) - \mathcal{U}_i(s_i) + \mathcal{U}_i(s_i) > \varsigma_i, \quad i \in J_0^{\mathcal{U}}$$

and $F_i(\mathcal{V}_i) = 0$ for any $i \in J_0^{\mathcal{U}}$.

(ii) For $i \in J_1^{\mathcal{U}}$, let

$$\epsilon_2 = \min_{i\in J_1^{\mathcal{U}}} \left\{ \max_{s\in[-\theta,0]} \varsigma_i - \mathcal{U}_i(s) \right\} > 0.$$

Then following a similar procedure as above gives $\sum_{i\in J_1^{\mathcal{U}}} (\mathcal{V}_i(s) - \mathcal{U}_i(s))^2 \le \epsilon_2^2$ for any $\mathcal{V} = (\mathcal{V}_i)_{i\in\mathbb{Z}} \in \mathfrak{H}_C$ with $\max_{s\in[-\theta,0]}$, so

$$\max_{s\in[-\theta,0]} \mathcal{V}_i(s) < \varsigma_i,$$

and $F_i(\mathcal{V}_i) = f_i(\mathcal{V}_i(0))$ for any $i \in J_1^{\mathcal{U}}$.

(iii) For $i \in J_\varsigma^{\mathcal{U}}$, $F_i(\mathcal{U}_i) = f_i(\mathcal{U}_i(0))[0,1]$, consider separately the cases $i \in J_\varsigma^{\mathcal{U}} \cap J_0^{\mathcal{V}}$, $i \in J_\varsigma^{\mathcal{U}} \cap J_1^{\mathcal{V}}$ and $i \in J_\varsigma^{\mathcal{U}} \cap J_\varsigma^{\mathcal{V}}$.

(a) For $i \in J_\varsigma^{\mathcal{U}} \cap J_0^{\mathcal{V}}$, $F_i(\mathcal{V}_i) = 0$ and thus $\mathrm{dist}_{i \in J_\varsigma^{\mathcal{U}} \cap J_0^{\mathcal{V}}}(F(\mathcal{U}), F(\mathcal{V})) = 0$.

(b) For $i \in J_\varsigma^{\mathcal{U}} \cap J_1^{\mathcal{V}}$, $F_i(\mathcal{V}_i) = f_i(\mathcal{V}_i(0))$ and thus

$$\mathrm{dist}_{i \in J_\varsigma^{\mathcal{U}} \cap J_1^{\mathcal{V}}}(F(\mathcal{U}), F(\mathcal{V})) \leq \sum_{i \in J_\varsigma^{\mathcal{U}} \cap J_1^{\mathcal{V}}} (f_i(\mathcal{V}_i(0)) - f_i(\mathcal{U}_i(0)))^2.$$

(c) For $i \in J_\varsigma^{\mathcal{U}} \cap J_\varsigma^{\mathcal{V}}$, $F_i(\mathcal{V}_i) = f_i(\mathcal{V}_i(0))[0,1]$ and thus

$$\mathrm{dist}_{i \in J_\varsigma^{\mathcal{U}} \cap J_\varsigma^{\mathcal{V}}}(F(\mathcal{U}), F(\mathcal{V})) \leq \sum_{i \in J_\varsigma^{\mathcal{U}} \cap J_\varsigma^{\mathcal{V}}} (f_i(\mathcal{V}_i(0)) - f_i(\mathcal{U}_i(0)))^2.$$

Collecting items (i) through (iii), and using Assumption 21.6, define for any $\epsilon > 0$ and $\mathcal{U} \in \mathfrak{H}_C$ the open neighbourhood of $\mathcal{U}$ to be

$$\mathcal{N}_{\mathcal{U}} = \{\mathcal{V} \in \mathfrak{H}_C : \|\mathcal{V} - \mathcal{U}\|_{\mathfrak{H}_C} < \min\{\epsilon, \epsilon_1, \epsilon_2, \delta\}\}.$$

Then for every $\mathcal{V} \in \mathcal{N}_{\mathcal{U}}$

$$\max_{s \in [-\theta, 0]} \sum_{i \in \mathbb{Z}} (\mathcal{V}_i(s) - \mathcal{V}_i(s))^2 < \delta^2,$$

which implies that $\sum_{i \in \mathbb{Z}} (\mathcal{V}_i(0) - \mathcal{U}_i(0))^2 < \delta^2$ and as a result

$$\begin{aligned} \mathrm{dist}_{\ell^2}(F(\mathcal{V}), F(\mathcal{U})) &\leq \left[\sum_{i \in \mathbb{Z}} (f_i(\mathcal{V}_i(0)) - f_i(\mathcal{U}_i(0)))^2\right]^{1/2} \\ &\leq L \left[\sum_{i \in \mathbb{Z}} (\mathcal{V}_i(0) - \mathcal{U}_i(0))^2\right]^{1/2} < L\epsilon. \end{aligned}$$

Therefore the mapping $\mathfrak{F}$ is upper-semicontinuous, and by Theorem 21.1 and properties of the operator Λ, system (21.6) has at least one strong solution. □

21.3 Long term behavior of lattice system

In this section we show that the solutions to the system (21.6) generate a set-valued process $\{\Psi(t, t_0)\}_{(t,t_0) \in \mathbb{R}^2_{\geq}}$, which has a pullback attractor. To that end, the following dissapativity assumptions will be needed.

Assumption 21.7. There exists $\alpha > 0$ and $\boldsymbol{\beta} = (\beta_i)_{i \in \mathbb{Z}} \in \ell^2$ such that

$$f_i(s)s \leq -\alpha s^2 + \beta_i^2, \quad i \in \mathbb{Z}.$$

Assumption 21.8. $\alpha^2 > \mathfrak{a} > 0$.

21.3.1 *Generation of set-valued process*

The following estimates of solutions is crucial to the analysis in the sequel.

Lemma 21.1. *Let Assumptions 21.4–21.8 hold. Then every solution $\boldsymbol{u}(\cdot)$ to (21.6) satisfies*

$$\|\boldsymbol{u}_t\|^2_{\mathfrak{H}_C} \le \|\phi\|^2_{\mathfrak{H}_C} e^{\lambda(\theta+t_0-t)} + \frac{2\alpha\|\boldsymbol{\beta}\|^2 + \|\mathfrak{b}\|^2}{\lambda}\left(1 - e^{\lambda(t_0-t)}\right),$$

where $\lambda = \alpha - \mathfrak{a}/\alpha$.

Proof. First taking inner product of (21.6) with $\boldsymbol{u} = (u_i)_{i\in\mathbb{Z}} \in \ell^2$ to obtain

$$\frac{1}{2}\frac{\mathrm{d}}{\mathrm{d}t}\|\boldsymbol{u}(t)\|^2 \in \nu\langle\Lambda\boldsymbol{u}, \boldsymbol{u}\rangle + \sum_{i\in\mathbb{Z}} F_i(u_{it}).$$

Recalling that that $\langle\Lambda\boldsymbol{u}, \boldsymbol{u}\rangle \le 0$, and using definition of $F_i(u_{it})$ as in (21.5) we have

$$\begin{aligned}
\frac{1}{2}\frac{\mathrm{d}}{\mathrm{d}t}\|\boldsymbol{u}(t)\|^2 &\le \sum_{i\in J_1^{\boldsymbol{u}_t}} f_i(u_i)u_i + \sum_{i\in J_\varsigma^{\boldsymbol{u}_t}} f_i(u_i)u_i[0,1] \\
&= \sum_{i\in\mathbb{Z}} f_i(u_i)u_i + \sum_{i\in J_\varsigma^{\boldsymbol{u}_t}} (f_i(u_i)u_i[0,1] - f_i(u_i)u_i) - \sum_{i\in J_0^{\boldsymbol{u}_t}} f_i(u_i)u_i \\
&\le \sum_{i\in\mathbb{Z}} f_i(u_i)u_i + \sum_{i\in J_\varsigma^{\boldsymbol{u}_t}} |f_i(u_i)u_i| + \sum_{i\in J_0^{\boldsymbol{u}_t}} |f_i(u_i)u_i| \\
&\le \sum_{i\in\mathbb{Z}} f_i(u_i)u_i + \sum_{i\in\mathbb{Z}} |f_i(u_i)u_i|.
\end{aligned}$$

Then, using Assumptions 21.6 and 21.7,

$$\begin{aligned}
\frac{1}{2}\frac{\mathrm{d}}{\mathrm{d}t}\|\boldsymbol{u}(t)\|^2 &\le -\alpha\|\boldsymbol{u}(t)\|^2 + \|\boldsymbol{\beta}\|^2 + \frac{1}{2\alpha}\sum_{i\in\mathbb{Z}} f_i^2(u_i) + \frac{\alpha}{2}\|\boldsymbol{u}(t)\|^2 \\
&\le \left(-\frac{\alpha}{2} + \frac{\mathfrak{a}}{2\alpha}\right)\|\boldsymbol{u}(t)\|^2 + \|\boldsymbol{\beta}\|^2 + \frac{1}{2\alpha}\|\mathfrak{b}\|^2. \qquad (21.7)
\end{aligned}$$

For simplicity, denote by $\lambda := \alpha - \mathfrak{a}/\alpha$. Then by Assumption 21.8, $\lambda > 0$. Multiplying (21.7) by $2e^{\lambda t}$ gives

$$\frac{\mathrm{d}}{\mathrm{d}t}\left(e^{\lambda t}\|\boldsymbol{u}(t)\|^2\right) \le \left(2\|\boldsymbol{\beta}\|^2 + \frac{1}{\alpha}\|\mathfrak{b}\|^2\right)e^{\lambda t}. \qquad (21.8)$$

Integrating (21.8) from t_0 to t gives

$$\|\boldsymbol{u}(t)\|^2 \le \|\boldsymbol{u}(t_0)\|^2 e^{\lambda(t_0-t)} + \frac{2\alpha\|\boldsymbol{\beta}\|^2 + \|\mathfrak{b}\|^2}{\lambda}\left(1 - e^{\lambda(t_0-t)}\right). \qquad (21.9)$$

Let $s \in [-\theta, 0]$. Replacing t by $t+s$ in (21.9) and using $e^{-\lambda s} \in [1, e^{\lambda\theta}]$ gives

$$\begin{aligned}
\|\boldsymbol{u}(t+s)\|^2 &\le \|\boldsymbol{u}(t_0)\|^2 e^{\lambda t_0} e^{-\lambda(t+s)} + \frac{2\alpha\|\boldsymbol{\beta}\|^2 + \|\mathfrak{b}\|^2}{\lambda}\left(1 - e^{\lambda t_0} e^{-\lambda(t+s)}\right) \\
&\le \left(\|\boldsymbol{u}(t_0)\|^2 e^{\lambda\theta} - \frac{2\alpha\|\boldsymbol{\beta}\|^2 + \|\mathfrak{b}\|^2}{\lambda}\right)e^{\lambda(t_0-t)} + \frac{2\alpha\|\boldsymbol{\beta}\|^2 + \|\mathfrak{b}\|^2}{\lambda},
\end{aligned}$$

which implies that

$$\|\boldsymbol{u}_t\|^2_{\mathfrak{H}_C} \le \|\phi\|^2_{\mathfrak{H}_C} e^{\lambda(\theta+t_0-t)} + \frac{2\alpha\|\boldsymbol{\beta}\|^2 + \|\mathfrak{b}\|^2}{\lambda}\left(1 - e^{\lambda(t_0-t)}\right).$$

This completes the proof. □

Lemma 21.1 ensures that every local solution to (21.6) can be defined globally. Thus a set-valued process $\{\Psi(t,t_0)\}_{(t,t_0)\in\mathbb{R}^2_+}$ can be defined on $\mathfrak{H}_C$ via

$$\Psi(t,t_0)(\phi) = \{\boldsymbol{u}_t(\cdot;t_0,\phi) \in \mathfrak{H}_C : \boldsymbol{u}(\cdot) \text{ solves (21.6) with } \phi \in \mathfrak{H}_C\}. \tag{21.10}$$

In fact, it is straight forward to check that

(i) $\Psi(t_0,t_0,\cdot) = \mathrm{Id}_{\mathfrak{H}_C}$;
(ii) $\Psi(t_2,t_0,\cdot) = \Psi(t_2,t_1,\Psi(t_1,t_0,\cdot))$ for any $t_2 \ge t_1 \ge t_0$.

and moreover the map Ψ defined by (21.10) is a set-valued process.

21.3.2 *Existence of an absorbing set*

In this section we show that the set-valued process $\{\Psi(t,t_0)\}_{\mathbb{R}^2_\ge}$ defined by (21.10) has an absorbing set in a collection of nonautonomous sets in $\mathfrak{H}_C$, denoted by $\mathfrak{D}_{\mathfrak{H}_C}$.

Lemma 21.2. *Let Assumptions 21.4–21.8 hold. Then the set-valued process $\{\Psi(t,t_0)\}_{\mathbb{R}^2_\ge}$ generated by system (21.6) has a closed $\mathfrak{D}_{\mathfrak{H}_C}$-pullback absorbing set in $\mathfrak{D}_{\mathfrak{H}_C}$.*

Proof. Denote by $\mathcal{R}$ the set of all functions $R:\mathbb{R}\to(0,\infty)$ satisfying

$$\lim_{t\to-\infty} e^{\lambda t}R^2(t) = 0.$$

Let $\mathfrak{D}_{\mathfrak{H}_C}$ be the collection of all nonautonomous sets $\mathcal{D} = (D_t)_{t\in\mathbb{R}} \subset \mathcal{P}_c(\mathfrak{H}_C)$ such that $D_t \subset \overline{\mathbb{B}(0,R_{\mathcal{D}}(t))}$ for some $R_{\mathcal{D}}(t) \in \mathcal{R}$, where $\overline{\mathbb{B}(0,R)}$ denotes the closed ball in $\mathfrak{H}_C$ centered at 0 with radius R. Then $\mathfrak{D}_{\mathfrak{H}_C}$ is a universe as defined in Definition 2.23.

Now define $R(t)$ by

$$R^2(t) = \|\phi\|^2_{\mathfrak{H}_C} e^{\lambda(\theta-t)} + \frac{2\alpha\|\beta\|^2 + \|\mathfrak{b}\|^2}{\lambda}\left(1 - e^{-\lambda t}\right)$$

and define a closed ball in $\mathfrak{H}_C$ by

$$Q_t = \{\mathcal{U} \in \mathfrak{H}_C : \|\mathcal{U}\|_{\mathfrak{H}_C} \le R(t)\}.$$

It then follows directly from Lemma 21.1 that the nonautonomous set $\mathcal{Q} := (Q_t)_{t\in\mathbb{R}} \in \mathfrak{D}_{\mathfrak{H}_C}$is $\mathfrak{D}_{\mathfrak{H}_C}$-pullback absorbing for the set-valued process $\{\Psi(t,t_0)\}_{\mathbb{R}^2_\ge}$ on $\mathfrak{H}_C$. □

21.3.3 *Tail estimations*

The asymptotic compactness of the set-valued process $\{\Psi(t,t_0)\}_{\mathbb{R}^2_{\geq}}$ requires an estimate of the tails of solutions, which is established in the following lemma.

Lemma 21.3. *Let Assumptions 21.4–21.8 hold. Let B be a bounded set of $\mathfrak{H}_C$. Then for any $\varepsilon > 0$ there exist $T(\varepsilon, B)$ and $I(\varepsilon, B)$ such that*

$$\max_{s\in[-\theta,0]} \Bigg(\sum_{|i|>2I(\varepsilon,B)} u_i^2(t+s) \Bigg)^{\frac{1}{2}} < \varepsilon, \quad t \geq T,$$

for any initial condition $\phi \in B$ and any solution $\boldsymbol{u}(\cdot)$ with $\boldsymbol{u}_{t_0} = \phi$.

Proof. The proof uses again the smooth cut-off function $\xi : \mathbb{R}^+ \to [0,1]$ defined by (3.8). For a large fixed k (to be determined later) define

$$v_i(t) = \xi_k(|i|)u_i(t) \quad \text{with} \quad \xi_k(|i|) = \xi\left(\frac{|i|}{k}\right), \qquad i \in \mathbb{Z}.$$

Taking inner product of (21.1) with $\boldsymbol{v}(t) = (v_i(t))_{i\in\mathbb{Z}}$ gives

$$\sum_{i\in\mathbb{Z}} \frac{\mathrm{d}u_i}{\mathrm{d}t} v_i(t) \in \nu\langle \Lambda\boldsymbol{u}(t), \boldsymbol{v}(t)\rangle + \sum_{i\in\mathbb{Z}} F_i(u_{it})v_i(t).$$

First notice that

$$\sum_{i\in\mathbb{Z}} \frac{\mathrm{d}u_i(t)}{\mathrm{d}t} v_i(t) = \frac{1}{2}\frac{\mathrm{d}}{\mathrm{d}t} \sum_{i\in\mathbb{Z}} \xi_k(|i|)u_i^2 \quad \text{for all } t \geq t_0. \tag{21.11}$$

Second, by Lemma 21.1, there exists a constant C_2 (depending on the bounded subset B and parameters of the problem) such that

$$\nu\langle \Lambda\boldsymbol{u}(t), \boldsymbol{v}(t)\rangle \leq \frac{C_2}{k} \quad \text{for all } t \geq t_0. \tag{21.12}$$

In addition,

$$\begin{aligned}
\sum_{i\in\mathbb{Z}} F_i(u_{it})v_i(t) &= \sum_{i\in J_1^{\boldsymbol{u}_t}} f_i(u_i)v_i(t) + \sum_{i\in J_\varsigma^{\boldsymbol{u}_t}} f_i(u_i)v_i(t)[0,1] \\
&= \sum_{i\in\mathbb{Z}} f_i(u_i)v_i + \sum_{i\in J_\varsigma^{\boldsymbol{u}_t}} \left(f_i(u_i)v_i[0,1] - f_i(u_i)v_i\right) - \sum_{i\in J_0^{\boldsymbol{u}_t}} f_i(u_i)v_i \\
&\leq \sum_{i\in\mathbb{Z}} f_i(u_i)v_i + \sum_{i\in J_\varsigma^{\boldsymbol{u}_t}} |f_i(u_i)v_i| + \sum_{i\in J_0^{\boldsymbol{u}_t}} |f_i(u_i)v_i| \\
&\leq \sum_{i\in\mathbb{Z}} \xi_k(|i|)f_i(u_i)u_i + \sum_{i\in\mathbb{Z}} |\xi_k(|i|)f_i(u_i)u_i|.
\end{aligned} \tag{21.13}$$

Collecting (21.11)–(21.13), using Assumptions 21.7 and 21.8, and following a similar procedure to the proof of Lemma 21.1 yields

$$\begin{aligned}\frac{1}{2}\frac{\mathrm{d}}{\mathrm{d}t}\sum_{i\in\mathbb{Z}}\xi_k(|i|)u_i^2(t) &\leq \sum_{i\in\mathbb{Z}}\left(-\alpha u_i^2+\beta_i^2\right)\xi_k(|i|)+\frac{C_2}{k}+\sum_{i\in\mathbb{Z}}\left(\frac{f_i^2(u_i)}{2\alpha}+\frac{\alpha u_i^2}{2}\right)\xi_k(|i|)\\ &\leq \left(-\frac{\alpha}{2}+\frac{\mathfrak{a}}{2\alpha}\right)\sum_{i\in\mathbb{Z}}\xi_k(|i|)u_i^2+\frac{C_2}{k}\\ &\quad+\sum_{i\in\mathbb{Z}}\xi_k(|i|)\left(\beta_i^2+\frac{\mathfrak{b}_i^2}{2\alpha}\right). \end{aligned} \tag{21.14}$$

Denote again $\lambda=\alpha-\mathfrak{a}/\alpha>0$. Multiplying both sides of (21.14) by $2e^{\lambda t}$ then integrating from t_0 to t yields

$$\begin{aligned}\sum_{i\in\mathbb{Z}}\xi_k(|i|)u_i^2(t) &\leq e^{\lambda(t_0-t)}\sum_{i\in\mathbb{Z}}\xi_k(|i|)u_i^2(t_0)+\frac{2C_2}{k\lambda}\left(1-e^{\lambda(t_0-t)}\right)\\ &\quad+\frac{1}{\lambda}\sum_{i\in\mathbb{Z}}\xi_k(|i|)\left(\beta_i^2+\frac{\mathfrak{b}_i^2}{2\alpha}\right)\left(1-e^{\lambda(t_0-t)}\right). \end{aligned} \tag{21.15}$$

For $s\in[-\theta,0]$, replacing t by $t+s$ in (21.15) gives

$$\begin{aligned}\sum_{i\in\mathbb{Z}}\xi_k(|i|)u_i^2(t+s) &\leq e^{\lambda(t_0-t-s)}\sum_{i\in\mathbb{Z}}\xi_k(|i|)u_i^2(t_0)+\frac{2C_2}{k\lambda}\left(1-e^{\lambda(t_0-t-s)}\right)\\ &\quad+\frac{1}{\lambda}\sum_{i\in\mathbb{Z}}\xi_k(|i|)\left(\beta_i^2+\frac{\mathfrak{b}_i^2}{2\alpha}\right)\left(1-e^{\lambda(t_0-t-s)}\right)\\ &\leq e^{\lambda(\theta+t_0)}e^{-\lambda t}\sum_{i\in\mathbb{Z}}\xi_k(|i|)u_i^2(t_0)+\frac{2C_2}{k\lambda}\left(1-e^{\lambda(t_0-t)}\right)\\ &\quad+\frac{1}{\lambda}\sum_{i\in\mathbb{Z}}\xi_k(|i|)\left(\beta_i^2+\frac{\mathfrak{b}_i^2}{2\alpha}\right)\left(1-e^{\lambda(t_0-t)}\right). \end{aligned}$$

Notice that

$$\sum_{|i|\geq 2k}u_i^2(t+s)\leq\sum_{i\in\mathbb{Z}}\xi_k(|i|)u_i^2(t+s),\qquad \sum_{i\in\mathbb{Z}}\xi_k(|i|)u_i^2(0)\leq\|\phi\|^2_{\mathfrak{H}_C}.$$

Thus

$$\begin{aligned}\sum_{|i|\geq 2k}u_i^2(t+s) &\leq \|\phi\|^2_{\mathfrak{H}_C}e^{\lambda(\theta+t_0)}e^{-\lambda t}+\frac{2C_2}{k\lambda}\left(1-e^{\lambda(t_0-t)}\right)\\ &\quad+\frac{1}{\lambda}\sum_{i\in\mathbb{Z}}\xi_k(|i|)\left(\beta_i^2+\frac{\mathfrak{b}_i^2}{2\alpha}\right)\left(1-e^{\lambda(t_0-t)}\right). \end{aligned} \tag{21.16}$$

For any $\varepsilon>0$ there exists $T(\varepsilon,B)>t_0$ such that

$$\|\phi\|^2_{\mathfrak{H}_C}e^{\lambda(\theta+t_0)}e^{-\lambda t}<\varepsilon^2/3\quad\text{for all } t\geq T(\varepsilon,B), \tag{21.17}$$

and there exists $K_1(\varepsilon, B) > 0$ such that

$$\frac{2C_2}{k\lambda}\left(1 - e^{\lambda(t_0-t)}\right) < \varepsilon^2/3 \quad \text{for all } k \geq K_1(\varepsilon, B),\ t \geq T(\varepsilon, B). \tag{21.18}$$

In addition, since $\boldsymbol{\beta} = (\beta_i)_{i\in Z} \in \ell^2$ and $\mathfrak{b} = (\mathfrak{b}_i)_{i\in\mathbb{Z}} \in \ell^2$, there exists $K_2(\varepsilon, B) > 0$ such that

$$\frac{1}{\lambda}\sum_{i\in\mathbb{Z}} \xi_k(|i|)\left(\beta_i^2 + \frac{\mathfrak{b}_i^2}{2\alpha}\right)\left(1 - e^{\lambda(t_0-t)}\right) < \varepsilon^2/3 \quad \text{for all } k \geq K_2(\varepsilon, B),\ t \geq T(\varepsilon, B). \tag{21.19}$$

Setting $I = \max\{K_1, K_2\}$, and collecting (21.16)–(21.19) gives immediately

$$\max_{s\in[-\theta,0]} \sum_{|i|\geq 2I(\epsilon,B)} u_i^2(t+s) < \varepsilon^2 \quad \text{for all } t \geq T(\varepsilon, B),$$

which completes the proof. □

21.3.4 *Existence of a nonautonomous attractor*

To prove that the set-valued process $\{\Psi(t,t_0)\}_{\mathbb{R}^2_{\geq}}$ has a pullback attractor, it still needs to be proved that Ψ is pullback asymptotically compact, and the map $\phi \mapsto \Psi(t, t_0, \phi)$ is upper semi-continuous with closed values.

For any initial value $\phi \in \mathfrak{H}_C$, define the set

$$\mathcal{D}(\phi) = \{\boldsymbol{u}(\cdot) : \boldsymbol{u}(\cdot) \text{ is a global solution to (21.6) with initial data } \phi\}.$$

Lemma 21.4. *Let Assumptions 21.4–21.8 hold. Then the set-valued mapping $\phi \mapsto \Psi(t, t_0, \phi)$ is upper semi-continuous and has compact values for any $(t, t_0) \in \mathbb{R}^2_{\geq}$.*

Proof. For $\phi \in \mathfrak{H}_C$, suppose (for contradiction) that the map $\phi \mapsto \Psi(t, t_0, \phi)$ is not upper semi-continuous. Then there exist a neighbourhood $\mathcal{N}$ of $\Psi(t, t_0, \phi)$ and a sequence $\mathcal{U}^{(n)} \in \Psi(t, t_0, \phi^{(n)})$ with $\phi^{(n)} \to \phi$ in $\mathfrak{H}_C$, but $\mathcal{U}^{(n)} \notin \mathcal{N}$.

Take $\boldsymbol{u}^{(n)}(\cdot) \in \mathcal{D}(\phi^{(n)})$ such that $\boldsymbol{u}_t^{(n)} = \mathcal{U}^{(n)}$. By using $\phi^{(n)} \to \phi$ in $\mathfrak{H}_C$ and Lemma 21.1, there exists $R > 0$ such that

$$\begin{aligned}\|\boldsymbol{u}^{(n)}(t)\|^2 &\leq \|\boldsymbol{u}_t^{(n)}\|^2_{\mathfrak{H}_C} \leq \|\phi^{(n)}\|^2_{\mathfrak{H}_C} e^{\lambda(\theta+t_0-t)} + \frac{1}{\lambda}\left(2\alpha\|\beta\|^2 + \|\mathfrak{b}\|^2\right)\left(1 - e^{\lambda(t_0-t)}\right) \\ &\leq R \quad \text{for all } t \in [t_0 - \theta, T] \quad \text{and } n \in \mathbb{N}.\end{aligned}$$

Then for any fixed $t \in [t_0, T]$, there is a $\boldsymbol{u}^* \in \ell^2$ and a subsequence of $\{\boldsymbol{u}^{(n)}\}$ (still denoted as $\{\boldsymbol{u}^{(n)}\}$) such that

$$\boldsymbol{u}^{(n)}(\cdot) \to \boldsymbol{u}^*(\cdot) = (u_i^*(\cdot))_{i\in\mathbb{Z}} \text{ in } \ell^2_w,$$

where ℓ^2_w is the space ℓ^2 endowed with the weak topology. It will now be shown that this convergence is in fact strong.

First for any $\varepsilon > 0$, there exists $N_1(\varepsilon) > 0$ such that

$$\left\|\phi^{(n)} - \phi\right\|_{\mathfrak{H}_C} \leq \max_{s\in[-\theta,0]} \sum_{i\in\mathbb{Z}} \left\|\phi_i^{(n)}(s) - \phi_i(s)\right\|^2 < \frac{\varepsilon}{4} \quad \text{for all } n \geq N_1,$$

and there exists $K_1(\varepsilon) > 0$ such that

$$\sum_{i\in\mathbb{Z}} \xi_k(|i|)\phi_i^2(s) < \frac{\varepsilon}{4} \quad \text{for all } s \in [-\theta, 0] \quad \text{for all } k \geq K_1.$$

Then for any $s \in [-\theta, 0]$ we have

$$\begin{aligned}\sum_{i\in\mathbb{Z}} \xi_k(|i|)\left(\phi_i^{(n)}(s)\right)^2 &\leq \sum_{i\in\mathbb{Z}} \xi_k(|i|)\left(\phi_i^{(n)}(s) - \phi_i(s)\right)^2 + \sum_{i\in\mathbb{Z}} \xi_k(|i|)\left(\phi_i(s)\right)^2 \\ &< \frac{\varepsilon}{2} \quad \text{for all } k \geq K_1,\ n \geq N_1.\end{aligned}$$

Following similar arguments to those in Lemma 21.3, for any $\varepsilon > 0$ and every $s \in [-\theta, 0]$, there exists $K_2(\varepsilon, T) \geq K_1$ such that

$$\begin{aligned}\sum_{|i|\geq 2K_2} \left(u_i^{(n)}(t+s)\right)^2 &\leq \sum_{i\in\mathbb{Z}} \xi_k(|i|)\left(u_i^{(n)}(t+s)\right)^2 \\ &\leq e^{\lambda(\theta+t_0)}e^{-\lambda t}\sum_{i\in\mathbb{Z}} \xi_k(|i|)\left(\phi_i(t_0)\right)^2 + \frac{2C_2}{k\lambda}\left(1 - e^{\lambda(t_0-t)}\right) \\ &\quad + \frac{1}{\lambda}\sum_{i\in\mathbb{Z}} \xi_k(|i|)\left(\beta_i^2 + \frac{\mathfrak{b}_i^2}{2\alpha}\right)\left(1 - e^{\lambda(t_0-t)}\right) \\ &< \varepsilon \quad \text{for all } n \geq N_1.\end{aligned}$$

Now for any $\varepsilon > 0$, there exist $K(\varepsilon) > 0$ and $N(\varepsilon) > 0$ such that

$$\sum_{|i|\geq K} \left(u_i^{(n)}(t)\right)^2 < \varepsilon, \quad \sum_{|i|\geq K} (u_i^*)^2 < \varepsilon$$

and

$$\sum_{|i|<K} \left(u_i^{(n)}(t) - u_i^*\right)^2 < \varepsilon \quad \text{for all } n \geq N.$$

It follows immediately that

$$\begin{aligned}\left\|\boldsymbol{u}^{(n)}(t) - \boldsymbol{u}^*\right\|^2 &= \sum_{|i|<K} \left(u_i^{(n)}(t) - u^*\right)^2 + \sum_{|i|\geq K} \left(u_i^{(n)}(t) - u^*\right)^2 \\ &\leq \sum_{|i|<K} \left(u_i^{(n)}(t) - u^*\right)^2 + 2\sum_{|i|\geq K} \left(u_i^{(n)}(t)\right)^2 + 2\sum_{|i|\geq K} (u_i^*)^2 < \varepsilon,\end{aligned}$$

which implies that $\boldsymbol{u}^{(n)}(t) \to \boldsymbol{u}^*$ strongly in ℓ^2. Hence the sequence $\boldsymbol{u}^{(n)}(t)$ is precompact in ℓ^2 for any $t \in [t_0, T]$.

Next for any $t \in [t_0, T]$, it is straight forward to check

$$\left\|\Lambda \boldsymbol{u}^{(n)}(t)\right\|^2 = \nu^2 \sum_{i\in\mathbb{Z}} \left(u_{i-1}^{(n)} - 2u_i^{(n)} + u_{i+1}^{(n)}\right)^2 \leq 16\nu^2 \left\|\boldsymbol{u}^{(n)}\right\|^2 \leq 16\nu^2 R.$$

In addition, for any $t \in [t_0, T]$, and any $\boldsymbol{g}^{(n)} \in \mathfrak{F}(\boldsymbol{u}_t^{(n)})$

$$\|\boldsymbol{g}^{(n)}\|^2 \leq \sum_{i\in\mathbb{Z}} f_i^2\left(u_i^{(n)}\right) \leq \mathfrak{a}\left\|u^{(n)}\right\|^2 + \|\mathfrak{b}\|^2 \leq \mathfrak{a}R + \|\mathfrak{b}\|^2.$$

Therefore for any $t_1, t_2 \in [t_0, T]$, there exists $\tau \in [t_0, T]$ such that

$$\begin{aligned}\left\|\boldsymbol{u}^{(n)}(t_1) - \boldsymbol{u}^{(n)}(t_2)\right\|^2 &= \left\|\frac{\mathrm{d}}{\mathrm{d}t}\boldsymbol{u}^{(n)}(\tau)\right\|^2 \cdot (t_1 - t_2)^2 \\ &\le \left(\left\|\Lambda \boldsymbol{u}^{(n)}(\tau)\right\|^2 + \left\|F(\boldsymbol{u}_\tau^{(n)})\right\|^2\right) \cdot (t_1 - t_2)^2 \\ &\le \left((16\nu^2 + \mathfrak{a})R + \|\mathfrak{b}\|^2\right) \cdot (t_1 - t_2)^2.\end{aligned}$$

Due to the Ascoli-Arzelà theorem, there exists a subsequence $\{\boldsymbol{u}^{(n_j)}(\cdot)\}$ of $\{\boldsymbol{u}^{(n)}(\cdot)\}$ that converges to $\boldsymbol{u}(\cdot)$ in $\mathcal{C}([t_0, T]; \ell^2)$. As a direct consequence, the map $\phi \mapsto \Psi(t, t_0, \phi)$ has closed graph and compact values.

By a standard limiting process it can be checked that $\boldsymbol{u}(\cdot) \in \mathcal{D}(\phi)$. Since $\phi^{(n)} \to \phi$ in $\mathfrak{H}_{\mathcal{C}}$, $\boldsymbol{u}^{n_j}(\cdot) \to \boldsymbol{u}(\cdot)$ in $\mathcal{C}([t_0 - \theta, T]; \ell^2)$ with $\boldsymbol{u}(t+s) = \phi(s)$ for $s \in [-\theta, 0]$. It follows immediately that $\boldsymbol{u}_t^{n_j} \to \boldsymbol{u}_t$ in $\mathfrak{H}_{\mathcal{C}}$ and hence $\mathcal{U}^{n_j} = \boldsymbol{u}_t^{n_j} \to \boldsymbol{u}_t = \mathcal{U}$ in ℓ^2, i.e., $\mathcal{U}^{n_j} \to \mathcal{U} \in \Psi(t, t_0, \phi)$. This contradicts the statement of $\mathcal{U}^{(n)} \notin \mathcal{N}$ and thus completes the proof. □

It remains to verify the pullback asymptotic compactness of Ψ.

Lemma 21.5. *Let Assumptions 21.4–21.8 hold. Then the set-valued mapping* $\{\Psi(t, t_0\}_{(t,t_0) \in \mathbb{R}^2_{\geq}}$ *is pullback asymptotically compact.*

Proof. Let $\{\tau_n\}$ be a sequence with $\tau_n \ge 2\theta$ and $\tau_n \to \infty$. Consider $\mathcal{U}^{(n)} = \boldsymbol{u}_{t_n}^{(n)} \in \Psi(t, t - \tau_n, \phi^{(n)})$ with $\boldsymbol{u}^{(n)}(\cdot) \in \mathcal{D}(\phi^{(n)})$ and $\phi^{(n)}$ belongs to a bounded set in $\mathfrak{H}_{\mathcal{C}}$. According to Lemma 21.1, there exists a constant $C_1 > 0$ such that

$$\left\|\boldsymbol{u}_{t_n}^{(n)}\right\| \le C_1 \quad \text{for all } s \in [-\theta, 0] \quad \text{for all } n \in \mathcal{N}.$$

For each fixed $s \in [-\theta, 0]$, there exists a subsequence $\{\boldsymbol{u}^{n_j}\}$ of $\{\boldsymbol{u}^{(n)}\}$, still denoted by $\{\boldsymbol{u}^{(n)}\}$, such that

$$\boldsymbol{u}^{(n)}(t_n + s) \to \boldsymbol{u}_s^* \quad \text{in} \quad \ell_w^2.$$

It can be shown by similar arguments to those in the proof of Lemma 21.4, that the convergence is strong, i.e., $\boldsymbol{u}^{(n)}(t_n + s) \to u_s^*$ in ℓ^2, and that the sequence $\{\boldsymbol{u}_{t_n}^{(n)}(s)\}$ is precompact for any $s \in [-\theta, 0]$.

To obtain equi-continuity, we consider $s_1, s_2 \in [-\theta, 0]$ and follow again similar arguments to those in the proof of Lemma 21.4 to obtain that there exists $C_2 > 0$ such that

$$\left\|\boldsymbol{u}^{(n)}(t_n + s_1) - \boldsymbol{u}^{(n)}(t_n + s_2)\right\| \le C_2 |s_1 - s_2|.$$

By the Ascoli-Arzelà theorem, $\{\mathcal{U}^{(n)}\}_{n \in \mathbb{N}}$ is precompact in $\mathfrak{H}_{\mathcal{C}}$. □

Finally, combining Proposition 2.2 and Lemmas 21.2, 21.4 and 21.5 leads to

Theorem 21.3. *Let Assumptions 21.4–21.8 hold. Then the set-valued nonautonomous dynamical system generated by* (21.6) *possesses a global pullback attractor.*

21.4 End notes

This chapter is based on the paper [Han and Kloeden (2016)], which is a lattice version of a more complicated PDE model used by Kloeden, Lorenz & Yang [Kloeden *et al.* (2014)].

21.5 Problems

Problem 21.1. What changes are needed in the proofs when a weighted norm space ℓ^2_ρ is used instead of ℓ^2?

Problem 21.2. What changes are needed in the proofs when state-dependent delays are considered, i.e., the constant θ is replaced by θ_i in (21.1).

Bibliography

Afraimovich, V. S. and Nekorkin, V. I. (1994). Chaos of traveling waves in a discrete chain of diffusively coupled maps, *Internat. J. Bifur. Chaos Appl. Sci. Engrg.* **4**, 631–637.

Abdallah, A. Y. (2005). Upper semicontinuity of the attractor for a second order lattice dynamical system, *Disc. Cont. Dyn. Syst. Ser. B* **5**, 899–916.

Abdallah, A. Y. (2006). Asymptotic bahaviour of the Klein–Gordon–Schrödinger lattice dynamical systems, *Commun. Pure Appl. Anal.* **5**, 55–69.

Abdallah, A. Y. (2008). Exponential attractors for first-order lattice dynamical systems, *J. Math. Anal. Appl.* **339**, 217–224.

Abdallah, A. Y. (2009). Exponential attractors for second order lattice dynamical systems, *Commun. Pure Appl. Anal.* **8**, 803–813.

Abdallah, A. Y. (2010). Uniform global attractors for first order non-autonomous lattice dynamical systems, *Proc. Amer. Math. Soc.* **138**, 3219–3228.

Amari, S. I. (1977). Dynamics of pattern formation in lateral-inhibition type neural fields, *Biol. Cybernet.* **27**, 77–87.

Adams, R. A. and Fournier, J. J. (2003). *Sobolev Spaces*, Second Edition, Elsevier.

Ambrosio, L. and Tilli, P. (2004). *Topics on Analysis in Metric Spaces*, Oxford University Press, Oxford.

Amigo, J. M., Gimenes, A., Morillas, F., and Valero, J. (2010). Attractors for a lattice dynamical system generated by non-Newtonian fluids modeling suspensions, *Internat. J. Bifur. Chaos Appl. Sci. Engrg.* **20**, 2681–2700.

Arnold, L. (1974). *Stochastic Differential Equations*, Wiley & Sons, New York.

Arnold, L. (1998). *Random Dynamical Systems*, Springer-Verlag, Berlin.

Aubin, J. P. and Cellina, A. (1984). *Differential Inclusions, Set-Valued Maps and Viability Theory*, Springer-Verlag, Berlin.

Ban, J-C., Hsu, C-H., Lin Y-H. and Yang, T-S. (2009). Pullback and forward attractors for dissipative lattice dynamical systems with additive noise, *Dynamical Systems* **24**, 139–155.

Banach, S. and Saks, S. (1930). Sur la convergence forte dans les champs L^p, *Studia Math.* **2**, 51–57.

Bates, P. W., Chen, X., and Chmaj, A. (2003). Traveling waves of bistable dynamics on a lattice, *SIAM J. Math. Anal.* **35**, 520–546.

Bates, P. W. and Chmaj (1999). A discrete convolution model for phase transitions, *Arch. Ration. Mech. Anal.* **150**, 281–305.

Bates, P. W., Lisei, H., and Lu Kening (2006). Attractors for stochastic lattice dynamical systems, *Stochastics and Dynamics* **6**, 1–21.

Bates, P. W., Lu Kening and Wang Bixiang (2001). Attractors for lattice dynamical systems, *Int. J. Bifur. Chaos App. Sci. Eng.* **11**, 143–153.

Bauer, H. (1996). *Probability Theory,* Walter de Gruyter & Co., Berlin.

Bessaih, H, Garrido-Atienza, M.-J., Han Xiaoying and Schmalfuss, B. (2017). Stochastic lattice dynamical systems with fractional noise, *SIAM J. Math. Anal.* **49**, 1495–1518.

Beyn, W.-J., and Pilyugin, S. Yu. (2003). Attractors of Reaction Diffusion Systems on Infinite Lattices, *Journal of Dynamics and Differential Equations* **5**, 485–515.

Brezis, H. (2011). *Functional Analysis, Sobolev Spaces and Partical Differential Equations,* Springer, Heidelberg.

Bunke, H. (1972). *Gewöhnliche Differentialgleichungen mit zufälligen Parametern,* Akademie-Verlag, Berlin.

Burkholder, D. L. (1966). Martingale transforms, *Annals of Math Stat.* **37**, 1494–1504.

Caraballo, T. and Han Xiaoying (2016). *Applied Nonautonomous and Random Dynamical Systems,* BCAM SpringerBrief, Springer Cham.

Caraballo, T., Han Xiaoying, Schmalfuss, B., and Valero, J. (2016). Random attractors for stochastic lattice dynamical systems with infinite multiplicative white noise, *Nonlinear Anal.* **130**, 255–278.

Caraballo, T. and Kloeden, P. E. (2009). Nonautonomous attractors for integro-differential evolution equations, *Disc. Cont. Dyn. Syst.* **2**, 17–36.

Caraballo, T., Kloeden, P. E. and Schmalfuß, B. (2004). Exponentially stable stationary solutions for stochastic evolution equations and their perturbation, *Applied Mathematics and Optimization* **50**, 183–207.

Caraballo, T. and Lu Kening (2008). Attractors for stochastic lattice dynamical systems with a multiplicative noise, *Front. Math. China* **3**, 317–335.

Caraballo, T., Morillas, F, and Valero, J. (2011). Random attractors for stochastic lattice systems with non-Lipschitz nonlinearity, *J. Diff. Equat. App.* **17**, 161–184.

Caraballo, T., Morillas, F, and Valero, J. (2012). Attractors of stochastic lattice dynamical systems with a multiplicative noise and non-lipschitz nonlinearities, *J. Differential Equations* **253**, 667–693.

Caraballo, T., Morillas, F., and Valero, J. (2014). On differential equations with delay in Banach spaces and attractors for retarded lattice dynamical systems, *Disc. Cont. Dyn. Sys.* **34**, 51–77.

Castaing, C. and Valadier, M. (1977). *Convex Analysis and Measurable Multifunction,* Lecture Notes in Mathematics, Vol. 580. Springer-Verlag, Berlin.

Chen, Y., Gao, H., Garrido-Atiennza, M. J. and Schmalfuß, B. (2014). Pathwise solutions of SPDEs driven by Hölder-continuous integrators with exponent larger than 1/2 and random dynamical systems, *Disc. Cont. Dyn. Sys.* **34**, 79–98.

Chen Yiju and Wang Xiaohu (2021). Asymptotic behavior of non-autonomous fractional stochastic lattice systems with multiplicative noise, *Disc. Cont. Dyn. Sys. - B* **26**, 1–20, doi:10.3934/dcdsb.20212.

Cholewa, J. W. and Czaja, R. (2020). Lattice dynamical systems: dissipative mechanism and fractal dimension of global and exponential attractors, *J. Evol. Equ.* **20**, 485–515.

Chow, S. N. (2003). Lattice dynamical systems, in *Lecture Notes in Math.* **1822** Springer-Verlag, Berlin, pp. 1–102.

Chow, S. N. and Mallet-Paret, J. (1995). Pattern formation and spatial chaos in lattice dynamical systems, I, II, *IEEE Trans. Circuits Systems* **42**, 746–751.

Chow, S. N., Mallet-Paret, J. and Shen, W. (1998). Traveling waves in lattice dynamical systems, *J. Diff. Equations* **149**, 248–291.

Chow, S. N. and W. Shen, W. (1995). Dynamics in a discrete Nagumo equation: Spatial topological chaos, *SIAM J. Appl. Math.* **55**, 1764–1781.

Chueshov, I. (2002). *Monotone random systems theory and applications*, Lecture Notes in Mathematics vol. 1779, Springer-Verlag.

Chueshov, I. (2015). *Dynamics of Quasi-Stable Dissipative Systems*, Springer-Verlag, Berlin.

Ciuca, I. and Jitaru, E. (2006). On the three layer neural networks using sigmoidal functions, *IWANN 1999: Foundations and Tools for Neural Modeling* 321–329.

Coombes, S., beim Graben, P., Potthast, R. and Wright J. (eds.) (2014). *Neural Fields: Theory and Applications* Springer-Verlag, Heidelberg.

Crauel, H. and Flandoli, F. (1994). Attractors for random dynamical systems, *Probability Theory and Related Fields* **100**, 365–393.

Crauel, H. and Kloeden, P. E. (2015). Nonautonomous and random Attractors, *Jahresbericht der Deutschen Mathematiker-Vereinigung* **117**, 173–206.

Cui, Hongyong, and Kloeden, P. E., Invariant forward attractors of non-autonomous random dynamical systems. *J. Differential Equations*, **265** (2018), 6166–6186.

Czaja, R. (2022). Pullback atttractors via quasi-stability for non-autonomous lattice dynamical systems, *Disc. Cont. Dyn. Syst. Series B* **27**, 1–26.

Da Prato, G. and Zabczyk, J. (2014). *Stochastic equations in infinite dimensions, second edition*, Encyclopedia of Mathematics and Its Applications, Cambridge University Press, New York.

Deimling, K. (1977). *Ordinary differential equations in Banach spaces*, Springer Lecture Notes in Mathematics, Vol. 596, Springer-Verlag, Berlin.

Diamond, P. and Kloeden, P. E. (1994). *Metric spaces of fuzzy sets: theory and applications*, World Scientific, Singapore.

Diestel, J. (1977). Remarks on weak compactness in $L_1(\mu, X)$, *Glasg. Math. J.* **18**, 87–91.

Diestel, J., Ruess, W. M. and Schachermayer, W. (1993). On weak compactness in $L^1(\mu, X)$, *Proc. Amer. Math. Soc.* **118**, 447–453.

Van den Driessche, P. and Xingfu Zou (1998). Global attractivity in delayed Hopfield neural network models, *SIAM Applied Math.* **58**, 1878–1890.

Doss, H. (1977). Liens entre équations différentielles stochastiques et ordinaires, *Ann. Inst. Henri Poincaré, Nouv. Sér., Sect. B* **13**, 99–125.

Duan, J., Lv, K., and Schmalfuss, B. (2003). Invariant manifold for stochastic partial differential equations, *The Annals of Probability* **31**, 2109–2135.

Dung, L. and Nicolaenko, B. (2001). Exponential attractors in Banach spaces, *J. Dyn. Diff. Eqs.* **13**, 791–806.

Eden, A., Foias, C. and Kalantarov, V. (1998). A remark on two constructions of exponential attractors for α-contractions, *J. Dyn. Diff. Eqs.* **10**, 37–45.

Eden, A., Foias, C., Nicolaenko, B. and Temam, R. (1994). Exponential Attractors for Dissipative Evolution Equation, *Research in Applied Mathematics* **37**, Masson/John Wiley co-publication, Paris.

Erneux, T. and G. Nicolis, G. (1993). Propagating waves in discrete bistable reaction diffusion systems, *Physica D* **67**, 237–244.

Engler, H. (1989). Global Regular Solutions for the Dynamic Antiplane Shear Problem in Nonlinear Viscoelasticity, *Mathematische Zeitschrift* **202**, 251–259.

Ermentrout, G. B. and McLeod, J. B. (1993). Existence and uniqueness of travelling waves for a neural network, *Proc. R. Soc. Edinburgh* **123A**, 461–478.

Evans, L. C. (2013). *An Introduction to Stochastic Differential Equations*, Amer. Math. Soc., Providence, RI.

Fan, X. and Wang, Y., Attractors for a second order nonautonomous lattice dynamical system with nonlinear damping, *Physics Letters A* **365** (2007), 17–27.

Fan, X., Exponential attractors for a first-order dissipative lattice dynamical systems. *J. Appl. Math.* pp. 1–8 (2008).

Fan, X. and Yang, H., Exponential attractor and its fractal dimension for a second order lattice dynamical system. *J. Math. Anal. Appl.* **367** (2010), 350–359.

Faye, G. (2018). Traveling fronts for lattice neural field equations, *Physica D* **378379**, 2032.

Garrido-Atienza, M., Neuenkirch, A. and Schmalfuss, B. (2018). Asymptotical stability of differential equations driven by Hölder-continuous paths, *J. Dyn. Diff. Eqns* **30**, DOI:10/1007/s10884-017-9574-6.

Green, J. W. and Valentine, F. A. (1960/1961). On the Arzelscoli theorem, *Math. Mag.* **34**, 199–202.

Gu Anhui and Kloeden, P. E. (2016). Asymptotic behavior of a non-autonomous p-Laplacian lattice system, *Inter. J. Bifurcation & Chaos* **26** 1650174 (9 pages).

Han Xiaoying (2011a). Random attractors for stochastic sine-Gordon lattice systems with multiplicative white noise, *J. Math. Anal. Applns.* **376**, 481–493.

Han Xiaoying (2011b). Exponential attractors for lattice dynamical systems in weighted spaces, *Discrete & Cont. Dyn. Syst.* **31**, 445–467.

Han Xiaoying (2011c). Asymptotic behavior of stochastic partly dissipative lattice systems in weighted spaces, *Inter. J. Diff. Eqns*, Article ID 628459, 23 pages.

Han Xiaoying (2012). Random attractors for second order stochastic lattice dynamical systems with multiplicative noise in weighted spaces, *Stochastics & Dynamics* **12**, 1150024.

Han Xiaoying (2013). Asymptotic behaviors for second order stochastic lattice dynamical systems on $\mathbb{Z}^k$ in weighted spaces, *J. Math. Anal. Applns.* **397**, 242–254.

Han Xiaoying (2015). Asymptotic dynamics of stochastic lattice differential equations: a review, *Continuous and distributed systems. II, Stud. Syst. Decision Control* **30**, 121–136.

Han Xiaoying and Kloeden, P. E. (2016). Lattice systems with switching effects and delayed recovery, *J. Differential Eqns.* **261**, 2986–3009.

Han Xiaoying and Kloeden, P. E. (2017a). *Random Ordinary Differential Equations and their Numerical Solution*, Springer Nature, Singapore.

Han Xiaoying and Kloeden, P. E. (2017b) *Attractors under Discretisation*, BCAM SpringerBriefs, Springer, Cham.

Han Xiaoying and Kloeden, P. E. (2019b). Asymptotic behaviour of a neural field lattice model with a Heaviside operator, *Physica D: Nonlinear Phenomena* **389**, 1–12.

Han Xiaoying and Kloeden, P. E. (2019b). Lattice dynamical systems in the biological sciences, in *Modeling, Stochastic Control, Optimization, and Applications*, IMA Volumes in Mathematics and its Applications vol. 164, G. Yin and Q. Zhang (eds.), Springer Nature Switzerland; pp. 201–233.

Han Xiaoying and Kloeden, P. E. (2020). Sigmoidal approximations of Heaviside functions in neural lattice models, *J. Diff. Eq.* **268**, pp. 5283–5300. Corrigendum to "Sigmoidal approximations of Heaviside functions in neural lattice models" *J. Diff. Eq.* (2021) **274**, 1214–1220.

Han Xiaoying, and Kloeden, P. E. (2022). Pullback and forward dynamics of nonautonomous Laplacian lattice systems on weighted spaces, *DCDS Series S* **15** (2022), 2909–2927.

Han Xiaoying, Kloeden, P. E. and Simsen, J. (2019). Sequence spaces with variable exponents for lattice models with nonlinear diffusion, in V. A. Sadovnichiy & M. Zgurovsky (eds.), *Modern Mathematics and Mechanics – Fundamentals, Problems, Challenges* Springer-Verlag, Cham, 195–214.

Han Xiaoying, Kloeden, P. E. and Sonner, S. (2020). Discretisation of the global attractor of a lattice system, *Journal of Dynamics and Differential Equations* **32** 1457–1474.

Han Xiaoying, Kloeden, P. E and Usman, B. (2020). Upper semi-continuous convergence of attractors for a Hopfield-type lattice model, *Nonlinearity* **33**, 1881–1906.

Han Xiaoying, Shen Wenxian and Zhou Shengfan (2011). Random attractors for stochastic lattice dynamical systems in weighted spaces, *J. Differential Equations* **250**, 1235–1266.

Han Xiaoying, Usman, B., and Kloeden, P. E. (2019). Long term behavior of a random Hopfield neural lattice model, *Comm. Pure Appl. Math* **18**, 809–824.

HanZongfei and Zhou Shengfan (2020). Random uniform exponential attractors for non-autonomous stochastic LDS and FitzHugh-Nagumo lattice systems with quasi-periodic forces and multiplicative noise, *Stochastics and Dynamics* **20**, 2050036.

Hopfield, J. J. (1984). Neurons with graded response have collective computational properties like those of two-stage neurons, *Proc. Nat. Acad. Sci. U.S.A* **81**, 3088–3092.

Hu, S. and Papageorgiou, N. S. (2000). *Handbook of multivalued analysis. Vol. II,* Kluwer Academic Publishers, Dordrecht.

Huang, J. (2007). The random attractor of stochastic FitzHugh-Nagumo equations in an infinite lattice with white noise, *Physica D* **233**, 83–94.

Huang, J., Han Xiaoying and Zhou, S. (2009). Uniform attractors for non-autonomous Klein-Gordon-Schrödinger lattice systems, *Applied Mathematics &Mechanics* (English Ed.) **30**, 1597.

Huynh, H, Kloeden, P. E., and Pötzsche, C., Forward and pullback dynamics of nonautonomous integrodifference equations: basic constructions. *Journal of Dynamics and Differential Equations* (2020), doi 10.1007/s10884-020-09887-8T.

Imkeller, P, and Schmalfuß, B. (2001). The conjugacy of stochastic and random differential equations and the existence of global attractors *J. Dyn. Diff. Eqns.* **13**, 215–249.

Joyner, R. W., Ramza, B. M., Osaka, T. and Tan, R. C. (1991). Cellular mechanisms of delayed recovery of excitability in ventricular tissue, *American Journal of Physiology* **260**, 225–233.

Kapral, R. (1991). Discrete models for chemically reacting systems, *J. Math. Chem.* **6**, 113–163.

Karachalios, N. I. and Yannacopoulos, A. N. (2005). Global existence and compact attractors for the discrete nonlinear Schrödinger equation, *J. Differential Equations* **217**, 88–123.

Keener, J. P. (1987). Propagation and its failure in coupled systems of discrete excitable cells, *SIAM J. Appl. Math.* **47**, 556–572.

Kisielewicz, M. (1992). Weak compactness in spaces $\mathcal{C}(S,X)$, in *Information theory, statistical decision functions, random processes*, Kluwer Academic Publishers, Dordrecht.

Kloeden, P. E. (2006). Upper semi continuity of attractors of retarded delay differential equations in the delay, *Bulletin Aus. Math. Soc.* **73**, 299–306.

Kloeden, P. E. (2022). Attractors of deterministic and random lattice difference equations, *Stochastics and Dynamics* **22**, 2240006 16pp.

Kloeden, P. E. (2023). Pullback attractors of nonautonomous lattice difference equations, *Proc. ICDEA 2021*, Sarajevo. to appear.

Kloeden, P. E. and Kozyakin, V. S. (2000). The inflation of attractors and discretization: the autonomous case, *Nonlinear Anal. TMA* **40**, 333–343.

Kloeden, P. E. and Lorenz, J. (1986). Stable attracting sets in dynamical systems and in their one-step discretizations, *SIAM J. Numer. Analysis* **23** 986–995.

Kloeden, P. E. and Lorenz, T. (2012). Mean-square random dynamical systems, *J. Differential Equations* **253**, 1422–1438.

Kloeden, P. E. and Lorenz, T. (2017). Pullback attractors of reaction-diffusion inclusions with space-dependent delay, *Disc. Cont. Dyn. Sys. - Series B* **22**, 1909–1964.

Kloeden, P. E. and Lorenz, T. (2016). Construction of nonautonomous forward attractors, *Proc. Amer. Mat. Soc.* **144**, 259–268.

Kloeden, P. E., Lorenz, T. and Yang, Meihua (2014). Reaction-diffusion equations with a non-reaction zone, *Commun. Pure & Appl. Anal.* **13**, 1907–1933.

Kloeden, P. E. and Platen, E. (1992). *Numerical Solutions of Stochastic Differential Equations*, Springer-Verlag, Berlin.

Kloeden, P. E. and Rodrigues, H. M. (2011). Dynamics of a class of ODEs more general than almost periodic, *Nonlinear Analysis TMA* **74**, 2695–2719.

Kloeden, P. E. and Rasmussen, M. (2011). *Nonautonomous Dynamical Systems*, Amer. Math. Soc., Providence.

Kloeden, P. E. and Simsen, J. (2014) Pullback attractors for non-autonomous evolution equations with spatially variable exponents, *Commun. Pure & Appl. Anal.* **13**, 2543–2557.

Kloeden, P. E. and Simsen, J. (2015), Attractors of asymptotically autonomous quasi-linear parabolic equation with spatially variable exponents, *J. Math. Anal. Appl.* **425**, 911–918.

Kloeden, P. E. and Villarragut, V. M. (2020). Sigmoidal approximations of a nonautonomous neural network with infinite delay and Heaviside function, *Journal of Dynamics and Differential Equations*, 1–25. DOI 10.1007/s10884-020-09899-4.

Kloeden P. E. and Yang Meihua (2021). *Introduction to Nonautonomous Dynamical Systems and their Attractors*, World Scientific Publishing Co. Inc, Singapore.

Kunita, H. (1990). *Stochastic Flows and stochastic Differential Equations*, Cambridge University Press.

Lebl, J. (2016). *Basic Analysis: Introduction to Real Analysis*, CreateSpace Independent Publishing Platform.

Li, X. and Lv, H. (2009). Uniform attractor for the partly dissipative nonautonomous lattice systems.

Li, X., K. Wei, K., and Zhang, H. (2011). Exponential attractors for lattice dynamical systems in weighted spaces. *Acta Appl. Math.* **114**, 157–172.

Li, X. and Wang, D. (2007). Attractors for partly dissipative lattice dynamic systems in weighted spaces. *J. Math. Anal. Appl.* **325**, 141–156.

Li, X. and Zhong, C. (2005). Attractors for partly dissipative lattice dynamic systems in $\ell^2 \times \ell^2$, *J. Comp. Appl. Math.* **177**, 159–174.

Lütscher, F. (2019). *Integrodifference Equations in Spatial Ecology*, Springer, Cham.

Lv, Y. and Sun, J. (2006). Dynamical behavior for stochastic lattice systems, *Chaos, Solitons & Fractals* **27**, 1080–1090.

Mallet-Paret, J. (1999). The global structure of traveling waves in spatially discrete dynamical systems, *J. Dynam. Differential Equations* **11**, 49–127.

Mallet-Paret, J., Wu, J, Yi, Y. and Zhu, H. (2012). *Infinite Dimensional Dynamical Systems*, Fields Institute Communications **64**, Springer, New York.

Marquié, P., Comte, J. C. and Morfu, S. (2004). Analog simulation of neural information propagation using an electrical FitzHugh-Nagumo lattice, *Chaos, Solitons & Fractals* **19**, 27–30.

Mao, X. (1994). *Exponential Stability of Stochastic Differential Equations*, Monogr. Textb. Pure Appl. Math. 182, Marcel Dekker, New York.

Martelli, M. (1975). A Rothe's type theorem for non-compact acyclic-valued maps, *Bollettino della Unione Matematica Italiana* **11**, 70–76.

McBride, A. C., Smith, A. L., and Lamb, W. (2010). Strongly differentiable solutions of the discrete coagulation-fragmentation equation, *Physica D* **239**, 1436–1445.

Mishura, Y. S. (2008). *Stochastic Calculus for Fractional Brownian Motion and Related Processes*, Springer-Verlag, Berlin.

Morillas, F. and Valero, J. (2009). Peano's theorem and attractors for lattice dynamical systems, *Internat. J. Bifur. Chaos* **19**, 557–578.

Morillas, F. and Valero, J. (2012). On the connectedness of the attainability set for lattice dynamical systems, *J. Diff. Eqns. App.* **18**, 675–692.

Neckel, T. and Rupp, F. (2014). *Random Differential Equations in Scientific Computing*, Versita de Gruyter-Verlag.

Nualart, D. and Rascanu, A. (2022). Differential equations driven by fractional Brownian motion, *Universitat de Barcelona. Collectanea Mathematica* **53**, 55–81.

Okada, N. (1984). On the Banach-Saks property, *J. Proc. Japan Acad. Ser. A Math. Sci.* **60**, 246–248.

Oliveira, J. C. and Pereira, J. M. (2010). Global attractor for a class of nonlinear lattices, *J. Math. Anal. Appl.* **370**, 726–739.

Partington, J. R. (1977). On the Banach-Saks property, *Math. Proc. Cambridge Phil. Soc.* **82**, 369–374.

Pazy, A. (2007). *Semigroups of Linear Operators and Applications to Partial Differential Equations*, Springer-Verlag, New York.

Prévôt, C. and Röckner, M. (2007). *A Concise Course on Stochastic Partical Differential Equations*, Springer-Verlag, Berlin Heidelberg.

Robinson, J. C. (2001 Infinite-Dimensional Dynamical Systems, Cambridge University Press, Cambridge.

Samko, S. G., Kilbas, A. A. and Marichev, O. I. (1993). *Fractional Integrals and Derivatives: Theory and Applicatios.* Gordon and Breach Science Publishers, Switzerland and Philadelphia, USA.

Sell, G. R. (1971). *Topological Dynamics and Ordinary Differential Equations.* Van Nostrand Reinhold Mathematical Studies, London.

Shipston, M. J. (2001). Alternative splicing of potassium channels: a dynamic switch of cellular excitability, *Trends in Cell Biology* **11**, 353–358.

Shu, J., Li, P., Zhang, J. and Liao, O. (2015). Random attractors for the stochastic coupled fractional Ginzburg-Landau equation with additive noise, *J. Math. Phys.* **56**, 102702.

Smirnov, G. V. (2002). *Introduction to the Theory of Differential Inclusions*, American Mathematical Society, Providence.

Song Li, Li Yangrong and Wang Fengling (2022). Controller and asymptotic autonomy of random attractors for stochastic p-Laplacian lattice equations. Evolution Equations and Control Theory *******, doi:10.3934/eect.2022010.

Stuart, A. M. and Humphries, A. R. (1996). *Dynamical Systems and Numerical Analysis*, Cambridge University Press, Cambridge.

Su Haijuan, Zhou Shengfan and Wu Luyao (2019). Random exponential attractor for second order non-autonomous stochastic lattice dynamical systems with multiplicative white noise in weighted spaces, *Advances in Difference Equations* **45**, 821.

Sui Meiyu, Wang Yejuan, Han Xiaoying, Kloeden, Peter E. (2020), Random recurrent neural networks with delays, *J. Differential Eqns.* **269**, 8597–8639.

Sui Meiyu, Yejuan Wang and Kloeden, Peter E. (2021a), Pullback attractors for stochastic recurrent neural networks with discrete and distributed delays, *Elec. Res. Arch.* **28**, 2187–2221.

Sui Meiyu, Wang Yejuan, Kloeden, Peter E., and Han Xiaoying (2021b). Dynamics of continuous-time recurrent neural networks with random connection weights and unbounded distributed delays, *European Physical Journal Plus* (2021) 136:811, doi.org/10.1140/epjp/s13360-021-01744-x.

Sussmann, H. J. (1977). On the gap between deterministic and stochastic differential equations, *Ann. Probab.* **6**, 590–603.

Szegö, G. P. and Treccani, G. (1969). *Semigruppi di Trasformazioni Multivoche*, Lecture Notes in Mathematics, Springer-Verlag, Heidelberg.

Szlenik, W. (1965). Sur les suites faiblement convergentes dans l'espace L, *Studia Mathematica* **25**, 337–341.

Teschl, G. (2012). *Ordinary Differential Equations and Dynamical Systems*, American Mathematical Society.

Tolstonogov, A. A. (2000). *Differential inclusions in a Banach space*, Kluwer Academic Publishers, Dordrecht.

Ülger, A. (1991). Weak compactness in $L^1(\mu, X)$, *Proc. Amer. Math. Soc.* **113**, 143–149.

Van Vleck, E. and Wang, B. (2005). Attractors for lattice FitzHugh-Nagumo systems, *Physica D* **212**, 317–336.

Wang Bixiang (2006). Dynamics of systems on infinite lattices, *J. Differential Equations* **221**, 224–245.

Wang Bixiang (2007a). Asymptotic behavior of non-autonomous lattice systems, *J. Math. Anal. Appl.* **331**, 121–136.

Wang Bixiang (2007b). Dynamical behavior of the almost-periodic discrete FitzHugh-Nagumo systems, *Internat. J. Bifurcation Chaos* **17**, 1673–1685.

Wang Bixiang (2018). Weak pullback attractors for mean random dynamical systems in Bochner spaces, *J. Dyn. Diff. Eq.*, https://doi.org/10.1007/s10884-018-9696-5.

Wang Xiaohu, Lu Kening and Wang Bixiang (2016). Exponential stability of non-autonomous stochastic delay lattice systems with multiplicative noise, *J. Dyn. Diff. Eq.* **38**, 1309–1335.

Wang Y. and Bai K. (2015). Pullback attractors for a class of nonlinear lattices with delays, *Discrete Cont. Dyn. Syst. A* **20**, 1213–1230.

Wang, X., Li, S., Xu, D. (2010). Random attractors for second-order stochastic lattice dynamical systems, *Nonlinear Anal.* **7**, 483–494.

Wang Xiaoli, Kloeden, P. E. and Han Xiaoying (2020). Attractors of Hopfield-type lattice models with increasing neuronal input, *Dist. Cont. Dyn. Syst. - B* **25**, 799–813.

Wang Xiaoli, Kloeden, P. E., and Han Xiaoying (2021). Stochastic dynamics of a neural field lattice model with state dependent nonlinear noise, *Nonlinear Differ. Equi. Appl.* **28:43**, 31pp.

Wang Xiaoli, Kloeden, P. E. and Yang Meihua (2020). Asymptotic behaviour of a neural field lattice model with delays, *Disc. Cont. Dyn. Sys. - B* **28**, 1037–1048.

Wang Fengling and Li Yangrong (2021). Random attractors for multi-valued multi-stochastic delayed p-Laplace lattice equations, *J. Diff. Equ. Appl.* **27**, 1232–1258.

Wang Xiaoli, Yang Meihua, and Kloeden, P. E. (2020). Sigmoidal approximations of a delay neural lattice model with of Heaviside functions, *Comm. Pure Appl. Math.* **19**, 2385–2402.

Wang Renhai and Wang Bixiang (2020). Random dynamics of lattice wave equations driven by infinite-dimensional nonlinear noise, *Disc. Cont. Dyn. Sys. - Series B* **25**, 2461–2493.

Wang, Z. and Zhou, S. (2016). Random attractors for non-autonomous stochastic lattice FitzHugh-Nagumo systems with coupled coefficients, *Taiwanese J. Math.* **20**, 589–616.

Yan, W., Li, Y., and Ji, S. (2010). Random attractors for first order stochastic retarded lattice dynamical systems, *J. Math. Phys.* **51**, 032702, https://doi.org/10.1063/1.3319566.

Yang Lin, Wang Yejuan, and Kloeden, P. E. (2022a). Exponential attractors for two-dimensional nonlocal diffusion systems with delay, *Comm. Pure Appl. Anal.* **21**, 1181–1831.

Yang Lu, Wang Yejuan and Kloeden, P. E. (2022b). Robustness of exponential attractors for infinite dimensional dynamical systems with small delay and application to 2D nonlocal diffusion delay lattice systems (to appear).

Yang Shuang and Li Yangrong (2021). Dynamics and invariant measures of multi-stochastic sine-Gordon lattices with random viscosity and nonlinear noise *J. Math. Phys.* **62**, 051510.

Yang Shuang, Li Yangrong and T. Caraballo (2021). Dynamical stability of random delay FitzHugh-Nagumo lattice systems driven by nonlinear Wong-Zakai noise, *J. Math. Phys.* **63**, 111512.

Yin, F., Zhou, S., Yin, S. and Xiao, C. (2007). Global attractor for Klein-Gordon-Schrödinger lattice systems, *Appl. Math. Mech.* (English Ed.) **28**, 695–706.

Zähle, M. (1998). Integration with respect to fractal functions and stochastic calculus, *Probab. Theory Related Fields* **111**, 333–374.

Zhang Qiangheng (2022). Well-posedness and dynamics of double time-delayed lattice FitzHugh-Nagumo systems, *J. Difference Eqns. Appl.* *******, 157–182.

Zhang Sijia and Zhou Shengfan (2022). Random uniform exponential attractors for Schrödinger lattice systems with quasi-periodic forces and multiplicative white noise, *Discrete Cont. Dyn. Syst., Series S* doi:10.3934/dcdss.2022056

Zhao, C. and Duan, J. (2010). Random attractor for the Ladyzhenskaya model with additive noise, *J. Math. Anal. Appl.* **362**, 241–251.

Zhao, C. and Zhou, D. S. (2007). Attractors of retarded first order lattice systems, *Nonlinearity* **20**, 1987–2006.

Zhao, X and Zhou, S. (2008). Kernel sections for processes and nonautonomous lattice systems, *Discrete Cont. Dyn. Syst. Series B* **9**, 763–785.

Zhao, C. and Zhou, S. (2009). Sufficient conditions for the existence of global random attractors for stochastic lattice dynamical systems and applications, *J. Math. Anal. Appl.* **354**, 78–95.

Zhao, C., Zhou, S. and Wang, W. (2009). Compact kernel sections for lattice systems with delays, *Nonlinear Anal. TMA* **70**, 1330–1348.

Zhou, Q., Li, W., Fu, H. and Zhang, Q. (2021). Pullback attractor of Hopfield neural networks with multiple time-varying delays, *AIMS Mathematics,* **6**, pp. 7441–7455.

Zhou, S. (2002a). Attractors for second order lattice dynamical systems, *J. Diff. Eq.* **179**, 605–624

Zhou, S. F. (2002b). Attractors for second-order lattice dynamical systems with damping, *J. Math. Phys.* **43**, 452–465.

Zhou, S. (2003). Attractors for first order dissipative lattice dynamical systems, *Physica D* **178**, 51–61.

Zhou, S. (2004). Attractors and approximations for lattice dynamical systems, *J. Differential Equations* **200**, 342–368.

Zho, S. (2017). Random exponential attractor for cocycle and application to nonautonomous stochastic lattice systems with multiplicative white noise, *J. Differential Equations* **263**, 2247–2279.

Zhou, S. and Han Xiaoying (2012). Pullback exponential attractors for non-autonomous lattice systems, *J. Dyn. Diff. Eqns* **24**, p 601–631.

Zhou, S. and Han Xiaoying (2013). Uniform exponential attractors for non-autonomous KGS and Zakharov lattice systems with quasi-periodic external forces, *Nonlinear Analysis: Theory, Methods & Applications* **78**, 1410–1455.

Zhou, S. and Shi, W. (2006). Attractors and dimension of dissipative lattice systems, *J. Differential Equations* **224**, 172–204.

Zhou Shengfan and Wang Zhaojuan (2013). Random attractors for stochastic retarded lattice systems, *J. Diff. Equ. Appl.* **19**, 1523–1543.

Zhou, S., Zhao, C. and Liao, X. (2007). Compact uniform attractors for dissipative non-autonomous lattice dynamical systems, *Commun. Pure. Appl. Anal.* **6**, 1087–1111.

Zhou, S., Zhao, C. and Wang, Y. (2008). Finite dimensionality and upper semicontinuity of compact kernel sections of non-autonomous lattice systems. *Disc. Conti. Dyn. Syst. Series B* **21**, 1259–1277.

Zhu, Z., Sang, Y., and Zhao, C. (2022). Pullback attractors and invariant measures for the discrete Zakharov equations, *Journal of Applied Analysis and Computation* to appear.

Index

Printed in the United States
by Baker & Taylor Publisher Services